PRACTICAL BACTERIOLOGY, MICROBIOLOGY AND SERUM THERAPY

(MEDICAL AND VETERINARY)

PRACTICAL
BACTERIOLOGY, MICROBIOLOGY
AND SERUM THERAPY

(MEDICAL AND VETERINARY)

A TEXT BOOK FOR LABORATORY USE

BY

DR. A. BESSON

FORMERLY DIRECTOR OF BACTERIOLOGICAL LABORATORIES OF THE MILITARY
HOSPITALS OF FRANCE AND OF THE PÉAN HOSPITAL.
LAUREATE OF THE FRENCH INSTITUTE

TRANSLATED
AND ADAPTED FROM THE FIFTH FRENCH EDITION

BY

H. J. HUTCHENS, D.S.O.

M.A., M.R.C.S., L.R.C.P., D.P.H. (OXFORD)

HEATH PROFESSOR OF COMPARATIVE PATHOLOGY AND BACTERIOLOGY OF THE
UNIVERSITY OF DURHAM; FORMERLY AN ASSISTANT SCIENTIFIC
INVESTIGATOR, ROYAL COMMISSION ON TUBERCULOSIS

WITH 416 *ILLUSTRATIONS,* 149 *OF WHICH ARE COLOURED*

LONGMANS, GREEN, AND CO.
39 PATERNOSTER ROW, LONDON
NEW YORK, BOMBAY, AND CALCUTTA
1913

PREFACE TO THE ENGLISH EDITION.

So far as the aim and scope of the book are concerned they are sufficiently described in the Author's prefaces. It remains to be said that the lack of a similar text book in English, at once sufficiently simple to put into the hands of the beginner and at the same time sufficiently advanced to be of service to the more experienced observer, together with the evident popularity of Dr. Besson's work in French speaking countries, seems to be sufficient justification for preparing the present translation.

A mere translation however of a book dealing with so rapidly advancing a science as Microbiology would have been hardly satisfactory, hence an attempt has been made to bring it up to date by incorporating matter which has appeared since the French edition went to press.

As regards the translation itself the original text has been followed as closely as possible, but the aim throughout has been to reproduce the sense rather than the actual words.

The alterations which have been made may be included under two heads, namely: alterations in the text and alterations in the arrangement of the text. With two exceptions the whole of the French text has been followed. In Chap. VII. the theory of the Microscope has been rewritten as, generally speaking, the Student seems to possess but a very limited knowledge of the instrument and it was thought that an amplification of the French text would be useful. I wish to thank Mr. A. S. Percival, Senior Surgeon to the Eye Infirmary, Newcastle upon Tyne, for the help he has given me in this part of the subject. Chaps. XXV., XXVI., and XXVII. dealing with the Paratyphoid bacilli have also been rewritten in view of the work of the Royal Army Medical Corps in India and of Dr. F. A. Bainbridge in England.

There are many notes and additions. Thus for instance it was found necessary to incorporate the important results obtained by the Royal Commission on Tuberculosis. All such notes and additions are clearly indicated either by a footnote or by being enclosed within square brackets. It should be said also that in a few cases where the authorities were in favour of a different nomenclature from that until recently in use the new names have been substituted thus *Discomyces* appears as the generic name in place of *Streptothrix*.

519

The arrangement of the text in the translation varies from that of the original in some respects. In the first place the subject matter has been divided into seven Parts instead of three. Secondly, the arrangement of the Parts has been subjected to some modification. To take the Bacteria for example it seemed that in a book intended for use in the laboratory it might be an advantage if these organisms were arranged morphologically and then subdivided according to their staining reactions and cultural characteristics. The plan adopted can be readily seen by a reference to the Table of Contents. Of course no classification is perfect and therefore free from criticism but after a good deal of consideration it was felt that practical usefulness merited the attempt.

Reference to any particular point will present, I hope, no difficulty. In addition to an Index and a very full Table of Contents, a summary of the subject matter heads the various Chapters. These of course are quite independent of the French edition.

The illustrations have been carefully revised. Many of them are new though illustrating familiar subjects and were drawn by my former laboratory attendant, Mr. H. Boot, under my supervision from preparations in my laboratory. Some were drawn by Mr. Richard Muir. For others I am indebted to the courtesy of the Controller of His Majesty's Stationery Office, of Professor G. H. F. Nuttall, F.R.S., of Dr. H. G. Adamson, and of the Publishers of Mense's *Handbuch der Tropenkrankheiten*. Miss M. V. Lebour, M.Sc. of the Zoological Department of the University of Leeds kindly undertook to redraw the whole of the line drawings.

In the preparation of this translation I wish to acknowledge my very particular indebtedness to Professor G. A. Lebour, M.A., D.Sc. who has given me at all times most invaluable assistance. To my former colleague Mr. C. F. Fox who had charge of the records of the Royal Commission on Tuberculosis I owe many thanks for the considerable care with which he undertook the thankless task of reading over the whole of the MS. before it went to press and for revising the proofs.

H. J. HUTCHENS.

Newcastle upon Tyne,
  March 31st, 1913.

PREFACE TO THE FIFTH FRENCH EDITION.

RECENT advances in Microbiology have necessitated an entire revision of the text. While still retaining its original form most of the chapters have been recast and much new matter has been incorporated.

The plan adopted when the book was first written of omitting all discussion upon matters of theory has been adhered to but it has nevertheless been thought desirable to include a Chapter on Immunity and the Properties of Immune Serums. The object of this has been to explain as clearly and simply as possible the principles underlying the phenomena of agglutination, of the fixation of the complement and the opsonic index and to describe the practical details in such a manner as to enable the Student to become familiar with the technique employed in these delicate investigations and so be in a position to appreciate the more detailed monographs.

In view of its importance in clinical diagnosis a description of the Ultra-microscope has been included.

Numerous additions and alterations have been made in the second Part. Most of the chapters have been supplemented. The serum treatment of Dysentery and of meningococcal Meningitis has been described as fully as was consistent with the scope of the work. The anaërobic micro-organisms, the paratyphoid bacilli, Sporotrichosis, Syphilis, etc. are all subjects of additions while many modifications have been introduced into the description and classification of the parasitic Protozoa, especially the *Piroplasmata*, *Leishmania* and *Trypanosomata*.

As in former editions the sole object has been to write a clear and concise account of each subject and one which will be abreast of recent knowledge retaining at the same time those characteristic features of the book which have been the subject of favourable comment both here and abroad.

A. BESSON.

15th May, 1911.

PREFACE TO THE FIRST FRENCH EDITION.

So important a place does Microbiology now occupy in the medical curriculum that not only are laboratories fully equipped for research and teaching to be found in all medical Schools, but the Student on leaving his School should have at least sufficient knowledge of the subject to carry out for himself the more simple investigations, such, for instance, as the recognition of the tubercle bacillus and the detection of the diphtheria bacillus.

The present work has been designed purely as a laboratory guide, the one object constantly in view in its preparation having been to make it a true *vade mecum*—a book which would both direct the beginner step by step and, at the same time, afford to the more skilled worker such assistance as would enable him to pursue his researches in a profitable direction.

My experience as a Teacher of Microbiology and as a Director of laboratories has I venture to think given me the qualifications necessary for the task in hand.

All matters of theory and all references to original sources have been studiously avoided since adequate information upon these matters is forthcoming in the many excellent Text books of Bacteriology.

In the first Part of the book the methods applicable to micro-organisms in general are detailed and while in each chapter a number of methods, all of which have been recommended by various authorities, are described, emphasis is laid upon those with which I have obtained the most satisfactory results and which I feel may confidently be recommended to the beginner.

The second Part is concerned with a description of the methods most suitable to the various different micro-organisms. The Bacteria are described first and then the parasitic Fungi and Protozoa the importance of which, however considerable it now may be, threatens to occupy an even greater place in the Pathology of the future.

The third Part which completes the book is devoted to a short account of the methods available for the bacteriological examination of water and air.

Much care has been bestowed upon the illustrations, and in order that the figures may be of as much use as possible to the Student in interpreting his own results they were drawn and coloured by myself from my own preparations and faithfully represent the appearances which should be obtained if the directions in the text are carefully followed.

I wish to take this opportunity of expressing my thanks to those of my Teachers to whom I am indebted for my instruction in the subject; I have drawn largely upon them and should this book be received with some favour I shall be not unmindful of those to whom the credit is due.

A. BESSON.

15th October, 1897.

CONTENTS.

PART I. GENERAL TECHNIQUE.

CONTENTS xiii

XVI. Bacillus pyocyaneus.

XVII. The bacillus of swine erysipelas.

The bacillus of mouse septicæmia, p. 288.

SUB-DIVISION II. THE NON-SPORE-BEARING, GRAM-POSITIVE,
ACID-FAST BACILLI.

XVIII. Bacillus tuberculosis.

XXXI. Bacillus mallei.

SUB-DIVISION IV. NON-SPORE-BEARING, GRAM-NEGATIVE BACILLI
THAT LIQUEFY GELATIN.

XXXII. Vibrio choleræ asiaticæ.

SUB-DIVISION V. NON-SPORE-BEARING, GRAM-NEGATIVE BACILLI
THAT DO NOT GROW ON GELATIN.

XXXIII. Pfeiffer's influenza bacillus.

XXXIV. The bacillus of soft sore.

---

[1] Though morphologically not belonging to the group it will be convenient to include the micrococcus in the same chapter as the streptococcus causing mammitis in cows.

SUB-DIVISION X. THE GRAM-NEGATIVE MICROCOCCI.

PART III. THE PARASITIC FUNGI.

XLIX. The parasitic Phycomycetes. Parasites of the family Mucoracidæ.

L. The parasitic Ascomycetes. Parasites of the family Gymnoascidæ.

IV. THE INFUSORIA.

LXIII. The Infusoria.

PART VI.

LXIV. The filtrable viruses.

PART VII. THE APPLICATION OF BACTERIOLOGICAL METHODS TO THE EXAMINATION OF WATER, SEWAGE AND AIR.

LXV. The bacteriological examination of water.

LXVI. The bacteriological examination of air.

PART I.

GENERAL TECHNIQUE.

CHAPTER I.

STERILIZATION.

Introduction.

For the study of any given micro-organism it is necessary to have a pure culture of the organism, that is to say a culture from which all other organisms have been excluded. Since micro-organisms are universally present in air and water and in the ambient media generally, it is essential that all vessels, culture media, instruments, etc., to be used in the preparation and investigation of pure cultures should themselves be free from living organisms, or in other words be sterile. **Sterilization** therefore means the destruction of living micro-organisms in, [or their removal from,] materials and apparatus used in bacteriological investigations.

It would however be useless to sterilize vessels, instruments and culture media unless steps were also taken to prevent them from again becoming soiled (using the word in its bacteriological sense) before being put to their proper use ; they must therefore be dealt with in such a manner that when sterilized they are completely protected from contact with extraneous organisms.

To accomplish this, vessels with a narrow mouth such as flasks, bottles and tubes are plugged with wool after being washed and before sterilization, such articles as watch-glasses, dishes, etc., are wrapped in paper, [while metal instruments, pipettes, etc., may be placed in a metal cylinder or box, or in a piece of glass tubing of large diameter plugged with wool at the two ends].

1. Plugging with wool.—To plug a narrow-mouthed vessel take a small piece of non-absorbent cotton-wool, fold it by twisting it round and round, insert one end into the mouth of the vessel and then force it gently to a depth of 2 to 3 cm. leaving the other end projecting from the orifice. It is better that the plug should be too large than too small.

2. Paper covers.—For wrapping up vessels and other articles ordinary filter paper may be used, but any common paper of decent texture is equally serviceable and has the merit of being more economical.

(*a*) Watch-glasses, Petri dishes, etc., should be wrapped in several folds of paper.

FIG. 1.—Copper cylinder with deep overlap in which to sterilize Petri dishes.

(*b*) Wide-mouthed cylindrical or conical vessels only need to have the opening covered with a double layer of paper, though this should be large enough to allow of it being turned down and twisted [or tied] round the vessel, so that the greater part of the latter is enveloped. In doing this, be careful not to tear the paper, which is apt to split on the edges of the opening.

[**3. Other methods.**—Petri dishes, pipettes, watch-glasses, metal instruments, etc., may be conveniently enclosed in copper boxes of suitable shape, which should have tightly fitting lids with a deep overlap. For ordinary Petri dishes a circular copper cylinder 25 × 12 cm. (fig. 1) containing a moveable tray may be used ; for pipettes a similar but longer and narrower cylindrical metal vessel or rectangular box is useful. Pipettes may also be enclosed in a piece of large glass tubing, which is then plugged at both ends with wool. The pipettes must of course be themselves plugged at the upper end with wool.]

Sterilization may be effected in one of several ways, the most generally employed being heat and filtration ; chemical antiseptics are seldom used in bacteriology. The methods of sterilization commonly employed will now be considered in detail.

SECTION I.—STERILIZATION BY DRY HEAT.

1. Sterilization in a naked flame.

1. The simplest means of sterilizing a metal instrument is to **heat it to redness** in a spirit flame or Bunsen burner. This method is always adopted for sterilizing platinum wires and iron and nickel spatulas.

Knives and similar instruments can also of course be sterilized by heating them in a flame, but on account of the injury done to the instrument the method is very rarely adopted.

An instrument which has been sterilized by heating to redness must be cooled before it is allowed to touch any material which is to be used for sowing cultures.

2. An instrument may be sterilized by **flaming** it, *i.e.* by passing it rapidly through a hot flame.

Only pipettes, glass rods, and other instruments with polished surfaces devoid of crevices in which organisms might escape destruction can be sterilized in this way, so that the method is of limited application.

2. Sterilization by hot air.

Exposure to hot air is the usual method of sterilizing all glass and porcelain apparatus, instruments with metal handles, etc., but it is not suitable for organic substances, with the exception of wool and paper.

Some form of apparatus in which sterilization can be effected by means of hot air is to be found in all laboratories.

To ensure efficient sterilization, the temperature must be maintained at approximately 180° C. for 30 minutes. Cotton-wool and paper are slightly scorched and browned at this temperature.

Hot air sterilizers.

The various forms of hot air sterilizers differ from one another only in details and in external appearance, the principles of construction and methods of use being the same in all.

1. Pasteur's sterilizer (fig. 2) is a double-walled cylinder of sheet iron with a chimney outlet, and is fitted internally with a wire basket in which the articles to be sterilized are packed. The top is closed by a lid, through

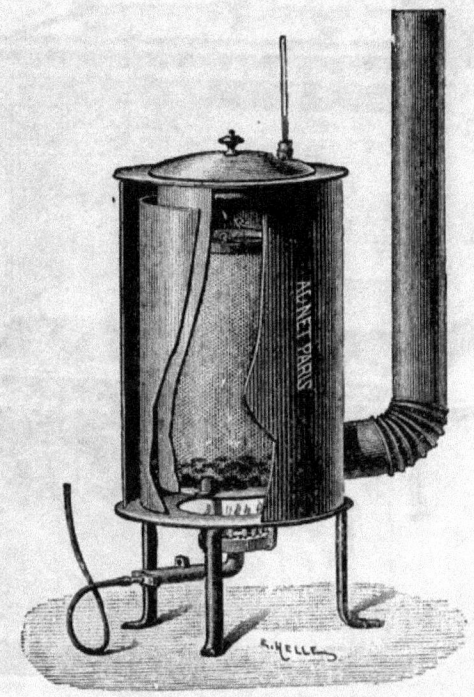

FIG. 2.—Pasteur's hot air sterilizer.

a hole in which a cork carrying a thermometer registering to 200° C. is passed. The heat is derived from a large gas burner below, and when this is lighted the heated air rising from the bottom of the stove circulates between the inner and outer walls and escapes up the chimney.

2. Chantemesse's and **Poupinel's hot air sterilizers** are rectangular and cupboard-shaped. They are fitted internally with moveable shelves on which the glass and other apparatus is arranged.

[**3. Hearson's hot air sterilizer** (fig. 3) is similar in shape to Chantemesse's, but is provided with an arrangement by which the gas is automatically regulated when the temperature has reached the point for which the regulator is set.]

Technique of sterilization by hot air.

(*a*) Carefully wash and rinse in a large volume of water all apparatus, whatever its nature, until all traces of organic matter have been removed. Unless the cleansing of glass for example be very thorough, black stains, due to the charring of organic matter during the heating process, will be found on the surface after sterilization. After washing allow the apparatus to dry, being specially careful in the case of glass, to avoid subsequent breakage during heating. When dry, treat each article in the manner already described, either plugging with wool, wrapping in paper, or packing in a metal box, according to its nature and use.

FIG. 3.—Hearson's hot air sterilizer.

(*b*) Place the articles in the sterilizer, taking care that neither wool nor paper touch the floor or sides, for these substances will char if they come in contact with the heated metal, and a tarry product rich in antiseptic substances will be deposited on the sterilized vessels, which will interfere with the subsequent growth of organisms. If by accident charring should take place, the articles which have been soiled must be washed, first in alcohol, then in water, dried and re-sterilized.

To avoid charring and breakage, it is advisable to place one or two fire bricks on the bottom of the sterilizer to keep the contents from touching the heated metal surface.

(*c*) Close the sterilizer and place the thermometer in position, pushing the latter well down into the interior.

(*d*) Light the gas. It is well to hold a lighted taper to the burner *before* turning on the tap, since if gas escape it will mix with the air between the inner and outer walls of the sterilizer and so tend to cause an explosion.

(*e*) Regulate the flame so that the temperature rises slowly; this is particularly important if the sterilizer contain vessels of thick glass, *e.g.* test-tubes on feet, glass dishes, etc.

(*f*) When the thermometer records a temperature of 175°–180° C. in the interior of the sterilizer, lower the gas gently, leaving sufficient flame to maintain the temperature at 180° C. or thereabouts for half an hour or so.

With a little practice this is easily done. Rather than use the fingers it is better to manipulate the tap by tapping it with some heavy instrument such as the spanner used for tightening the bolts of the autoclave, which will give very delicate control over the supply of gas, and will obviate the annoyance caused by accidentally turning out the gas altogether.

When experience has been gained, a thermometer can be dispensed with; at a temperature of 180° C. wool and paper become slightly scorched, and when this effect is noted the gas is turned down.

(*g*) When sterilization is completed turn out the gas, but allow the temperature to fall considerably before removing the contents, because glass, and especially thick glass, is liable to crack if exposed to a sudden change of temperature.

[With **Hearson's hot air sterilizer** the procedure is the same, except that stage (*f*) is omitted; when the temperature for which the capsule is set is reached, the gas is automatically lowered. It is only necessary therefore to note when the thermometer reaches the point at which sterilization is to be effected, and half an hour later to turn out the gas and proceed as in (*g*).]

SECTION II.—STERILIZATION BY MOIST HEAT.

Sterilization by moist heat may be effected in one of three ways.

1. By heating in water or steam at 100° C.
2. By heating in steam under pressure.
3. By discontinuous heating at low temperatures.

1. Sterilization in steam at 100° C.

Simple boiling or exposure to steam at 100° C., even though the exposure be prolonged, is not a reliable method of sterilization.

When micro-organisms have been dried, their resistance to the effects of heat is much enhanced, and especially is this the case when they are mixed with substances of an albuminoid nature. Further there are certain resistant forms of bacterial protoplasm known as spores, which in the majority of cases at least are not destroyed by heating to 100° C., even when the temperature is maintained for several minutes.

In France sterilization by moist heat at 100° C. is very seldom employed, except for sterilizing syringes for inoculation. In this case a sufficient degree of asepsis is obtained by boiling in water for 15 to 20 minutes at ordinary atmospheric pressure.

[In England on the other hand, and] in Germany, sterilization by moist heat at 100° C. is in general use. The operation is carried out in a Koch's sterilizer or steamer, and must be repeated at intervals of 24 hours on at least two, but ordinarily on three, successive days.

This method is the outcome of an observation by Tyndall to the effect that while it is impossible to sterilize an infusion of hay by boiling it continuously even for a prolonged period, yet by boiling it for a short time on three successive days all living organisms are destroyed.

This process embodies the principle of sterilization by discontinuous heating. The explanation put forward by Tyndall was that the hay infusion contains both bacilli and spores (*B. subtilis*). By heating to 100° C. the bacilli, but not the spores, are killed. The latter germinate as the fluid cools, and are killed during the second heating. A few spores however escape destruction on the second heating : these will have germinated by the time the third heating is due. After the third heating then sterilization is completed. The explanation now given however is that the resistance of micro-organisms is gradually lowered under the influence of repeated heating.

Steamers.

1. Koch's steamer.—Koch's steamer (fig. 4) consists of a cylindrical copper boiler, provided with a water gauge below and closed above by a lid through a hole in which a thermometer can be passed. It is fitted with perforated and moveable metal trays on which to rest the apparatus.

A metal cylinder open at both ends is often supplied with the sterilizer, so that the latter can be lengthened when necessary by fitting the metal cylinder on top.

Technique.— When sterilizing culture media by steam at 100° C., it is advisable to use vessels already sterilized in the hot air sterilizer.

(*a*) Pour sufficient water into the steamer to reach the level marked on the water gauge. Stand the vessels on the trays, and if extra space be needed adjust the lengthening cylinder. Put on the lid, and insert the thermometer.

(*b*) Light the gas under the boiler, note when steam begins to escape from under the lid—the thermometer will then register 98°–100° C.—and maintain the apparatus at this temperature for 30 minutes.

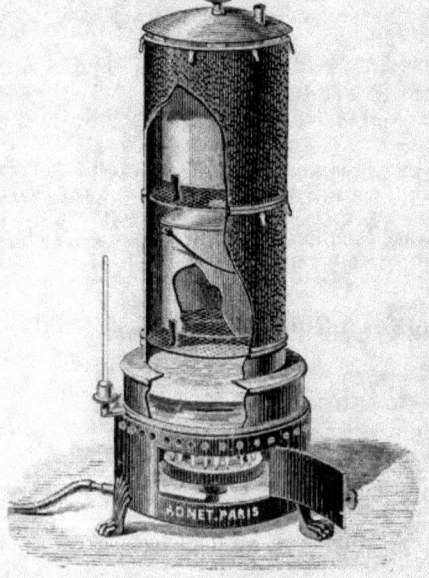

FIG. 4.—Koch's steamer.

(*c*) Heat again in a similar manner on the two following days.

When the flasks, tubes, etc., are taken out of the steamer, the wool plugs are generally wet with water of condensation; and since wool is only efficient as a filter for micro-organisms so long as it is absolutely free from moisture, the vessels may be put in the incubator for an hour or two to dry the plugs.

In many of the newer patterns of steamers the steam circulates between double walls before escaping, thus maintaining an absolutely constant temperature in the steamer. Some forms are further provided with a constant-level adjustment. [One of the most useful of these newer patterns is that made by Hearson.]

[**2. Hearson's steamer.**—Hearson's steamer (fig. 5) consists of two copper cylinders, one suspended within the other, thus conserving the heat. By means of a special regulator the gas is automatically lowered when the inner chamber is full of steam, and this, instead of escaping into the sterilizing room, is condensed and returned to the boiler. A further advantage is that the water is added from the outside.]

2. Sterilization in steam under pressure.

Water, syringes, india-rubber apparatus, filters, etc., are generally sterilized by heating in steam under pressure. This method is also in general use for the sterilization of certain culture media, but is not particularly suitable for steel cutting instruments, as it destroys the edge.

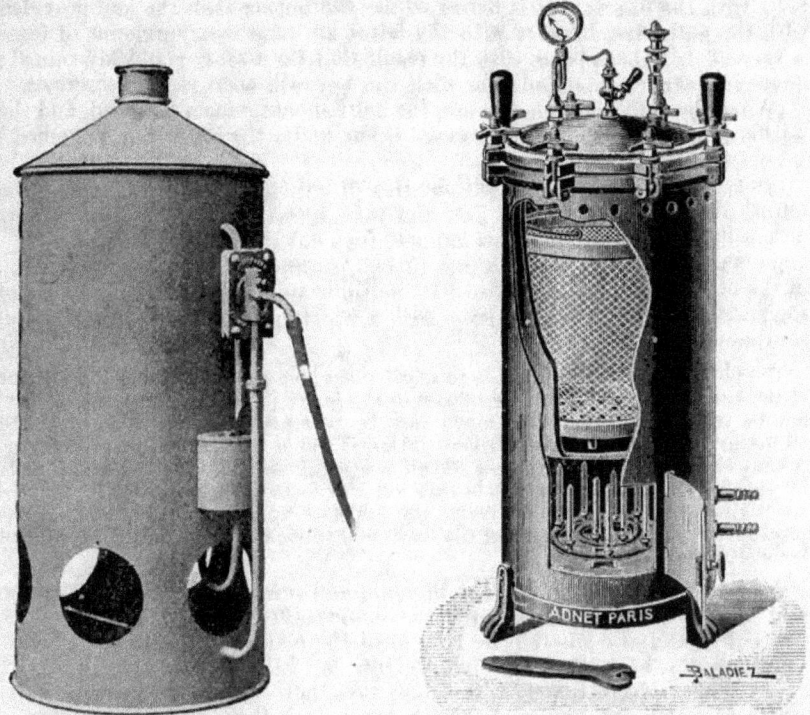

FIG. 5.—Hearson's steamer.   FIG. 6.—Chamberland's autoclave.

Exposure to steam at a temperature of 115° C. for 20 minutes is in most cases sufficient to ensure sterilization, but some media, potato for example, require a temperature of 120° C.

Some of the commoner forms of autoclave may be shortly described here.

Autoclaves.

1. Chamberland's autoclave (fig. 6).—This autoclave consists of a cylindrical copper boiler, the free edge of which is turned out flangewise. A flanged bronze cover is secured to this edge by screw bolts, and the whole is made air-tight by the insertion of an india-rubber washer between the two metal flanges.

The cover is provided with a safety valve, a steam tap, and a manometer which records the pressure in atmospheres and the temperature in degrees centigrade. The boiler contains a removeable copper-wire basket, which rests on short feet (5–6 cm.) on the bottom of the boiler. The boiler itself is supported within a cylindrical sheet-iron or copper furnace provided with one or two rings of Bunsen burners.

Technique.—(*a*) Pour sufficient water into the boiler to reach to just below the bottom of the wire basket; distilled water is preferable, as by its use furring is avoided.

Place the apparatus to be sterilized in the basket, and lay two or three thicknesses of cloth or paper over the wool plugs to prevent condensation water dropping on to them from the cover.

(*b*) Adjust the india-rubber washer, put on the cover, and screw up the bolts with the fingers. It is better to use the fingers than the key provided with the autoclave, because with the latter an unnecessary amount of force is very likely to be applied, with the result that the washer is quickly ruined; moreover, careless manipulation with the key will soon strip the screws.

(When the autoclave is not in use, the bolts should remain loosened, and the washer removed and hung up because if left under the cover it gets crushed.)

(*c*) Open the steam tap.

(*d*) It will be sufficient to light one ring of burners. Hold the taper to the burner before turning on the gas, and take special notice that the burners do not light below: should this happen, turn out the gas and re-light it.

(*e*) As soon as the water begins to boil, steam will escape from the tap in the cover, and must be allowed to continue to do so until the pressure of the steam within causes it to issue with a whistling sound in a powerful and continuous jet.

The object of this manœuvre is to expel the whole of the air from the interior of the autoclave, since if any air remain in the boiler the manometer readings will not be reliable. Still however much care be taken it is impossible to drive out all the air, and the larger the autoclave the larger will be the volume of air remaining. A more effectual means of expelling the air is to compress and decompress repeatedly by opening and shutting the steam tap, but this method should never be adopted when sterilizing fluids because under the influence of a sudden lowering of the pressure the plugs and contents of the flasks and tubes are driven out by the violent boiling of the liquids.

Now close the steam tap. The pressure and temperature will rise rapidly, and when the manometer records the temperature required (115°–120° C.), lower the gas and regulate it by trial until the manometer reading is steady. Continue the heating at this temperature for 20 minutes.

(*f*) When sterilization is completed, turn out the gas; the manometer needle soon falls to zero, and then, but not until then, open the steam tap. When all the steam has escaped unscrew the bolts, raise the cover, and remove the contents. If the plugs be damp it is well to put the flasks, etc., in the incubator until the wool dries.

The following minor practical details in the working of an autoclave may be mentioned. It is important never to open the steam tap until the pressure within the apparatus has fallen to the zero mark on the manometer, for the reason already given, namely that under the influence of sudden decompression the fluid contents of the flasks, etc., are liable to be discharged into the autoclave. Again, to avoid accidents by scalding from an escape of steam beneath the cover, the steam tap must always be opened before the bolts are loosened. Lastly, to obviate any difficulty in lifting the cover owing to the rubber washer sticking to the metal, always open the autoclave before the latter gets quite cold.

Note.—The autoclave is also available for sterilization at 100° C. The procedure will be the same as regards the first four steps *a*, *b*, *c*, *d*, but the steam tap must be left open the whole time (30 minutes), and the gas burners regulated so that the pressure as recorded by the manometer needle does not rise above the zero point.

It is obvious of course that a sufficient quantity of water must be put into the boiler before commencing the sterilization. Heating should never be continued for longer than 30 to 40 minutes in case the boiler should boil dry.

2. Ducretet and Lejeune's autoclave.—The principle and working of the instrument are the same as in the case of Chamberland's autoclave. The tall form of boiler however makes it especially useful for the sterilization of long pieces of apparatus and of porcelain filter bougies ; as many as thirty of the latter can be accommodated at one and the same time by means of a special pattern of support. The autoclave will withstand a pressure of 3 or 4 atmospheres, and is strong enough to be used for sterilization by means of compressed carbonic acid (d'Arsonval).

To facilitate manipulation, some minor alterations have been introduced in the construction of the newest models of autoclaves. For instance, in one made by Adnet the cover is secured by a gearing controlled by a single screw instead of by bolts. In another, made by Rongier, the cover is fitted with an hinge, and in yet another, by Radias, with a lever.

3. Vaillard and Besson's autoclave.—In large laboratories where, for instance, toxins for immunizing horses in the preparation of therapeutic serums or for other purposes are required in large bulk, and the consumption of media is considerable, it is necessary or at least convenient to have some more commodious form of autoclave than Chamberland's. In such cases Vaillard and Besson's pattern is available (fig. 7).

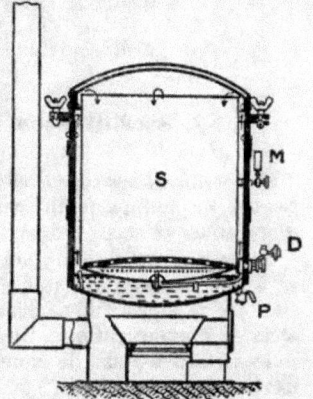

FIG. 7.—Vaillard and Besson's autoclave.

This autoclave[1] consists of a large cylindrical boiler with double walls. The apparatus to be sterilized is arranged on shelves in a central space. The steam rising from the water in the double bottom ascends between the inner and outer walls, passes through the sterilizing chamber from above downwards, and escapes through a safety valve, the escape being regulated in such a manner that the pressure and therefore the temperature rise gradually. When the temperature reaches 115° C., the safety valve automatically allows the steam to escape sufficiently to prevent any further increase of pressure. The boiler is also fitted with a lateral funnel through which the water may be poured in, a tap by which the level of the water is regulated, a manometer and a safety valve. The construction of this apparatus is such that sterilization is effected in a current of steam, and a further advantage is that all the air is expelled without resort to decompression, the disadvantages of which have already been noted.

Technique.—(a) Place the apparatus in the chamber S, and secure the cover firmly by means of the screw bolts. (b) Open the tap of the lateral supply funnel and pour in water until it runs out at P, which must also have been previously opened ; then close both taps and raise the valve D. (c) Light the stove. (In France charcoal is generally used as the source of heat, but the autoclave is also constructed to work with gas.) (d) As soon as the water boils, steam will rise between the inner and outer walls, enter the sterilizing chamber, and escape by way of the tube leading to D. When the pressure is sufficient to cause the steam to issue in a powerful jet, lower the valve D. The temperature and pressure within the autoclave will now rise, and will be registered on the manometer M. The steam escapes more and more violently as the pressure increases, until the temperature for which the valve has been regulated (usually 115° C.) is reached, when the volume of escaping steam is such as to prevent any further rise of temperature. The

[1] *Annales de l'Institut Pasteur*, 1894.

temperature and pressure must be maintained for 20 minutes, reckoning from the moment when it reaches 115° C. (ε) When the necessary time has elapsed, remove the source of heat and allow the autoclave to cool until the pressure reaches zero on the manometer, then open the steam-tap (not shown in the illustration), and raise the cover.

Note.—The autoclave may also be used for sterilization at 100° C. The technique is the same as in the preceding case, except that the valve D is never raised. Under these conditions steam will issue in a powerful jet during the whole operation. The temperature must be maintained for 30 minutes after reaching 100° C.

Sterilization can also be effected at any temperature between 100° and 115° C. by suitably altering the position of the knobbed handle of the valve; the further the handle is from the vertical the less will the temperature rise above 100° C.

Method of controlling the temperatures at which sterilization was effected.

In laboratories where the sterilization of apparatus, etc., is entrusted to laboratory assistants, it is convenient to have a method of controlling the temperature at which sterilization was effected. This may be done by placing a maximum thermometer or, more conveniently, a fragment of fusible alloy or some chemical compound of suitable melting point (110°–120° C.) alongside the apparatus in the autoclave. If a powder be used it may be mixed, as suggested by Demandre, with a trace of some dye, and sealed up in a small glass ampoule. The small amount of dye used is not visible in the powder, but when the latter melts the dye diffuses through it, and on cooling forms a coloured bead.

The following substances are suitable for the purpose. the temperature in brackets indicating the melting point: benzonaphthol (110° C.); antipyrine and sulphur (113° C.); resorcin (119° C.); benzoic acid (121° C.).

For coloured beads the following formulæ may be employed:

Melting at 110° C.	Benzonaphthol,	-	100 grams.
	Safranin,	- - -	0·01 gram.
Melting at 121° C.	Benzoic acid,	- -	100 grams.
	Brilliant green, -	- -	0·01 gram.

3. Sterilization by discontinuous heating at low temperatures.

Some substances used as culture media, being rich in albumin, cannot be heated to boiling point without marked alteration and to some extent destruction of their properties. Serum is a case in point.

Pasteur showed that such media can be better sterilized by heating them at a low temperature (55°–60° C.) for a long time than at a high temperature for a short time. This prolonged heating at a low temperature constitutes **Pasteurization.** In practice however it is found that to be effectual, pasteurization must be combined with the method of discontinuous heating devised by Tyndall (p. 7).

Technique.—Distribute the medium into a series of sterile flasks with long necks (fig. 35, p. 46), each flask being about three-fourths filled, and seal the mouths in a blow-pipe. [Flasks and test-tubes covered with india-rubber caps (p. 29) over the wool plugs can be used equally well.]

Place the flasks in a water bath fitted with a thermometer, slowly raise the temperature and regulate the gas flame so that it remains constant at 56°–57° C. for an hour, then turn out the gas, but leave the flasks in the bath until they are quite cool.

The flasks must be heated in the same way daily for a week before the contents can be regarded as sterile; and even then they ought to be incubated at 37° C. for two or three days, and any flask in which a growth appears must, of course, be rejected.

Water baths.

Conducted in the manner described, this method of sterilization is tedious, and it is difficult to avoid exceeding a temperature of 58° C., with the result that the albumin coagulates, rendering the medium useless for the purpose for which it is required. Hence it is better to have some form of water bath, in which the temperature is automatically controlled by a regulator on the gas supply.

[1. **Hearson's water bath.**—This is a very convenient and reliable form of water bath. It consists of a cylindrical copper vessel (fig. 8) heated below by an ordinary fish-tail gas burner, the temperature being controlled

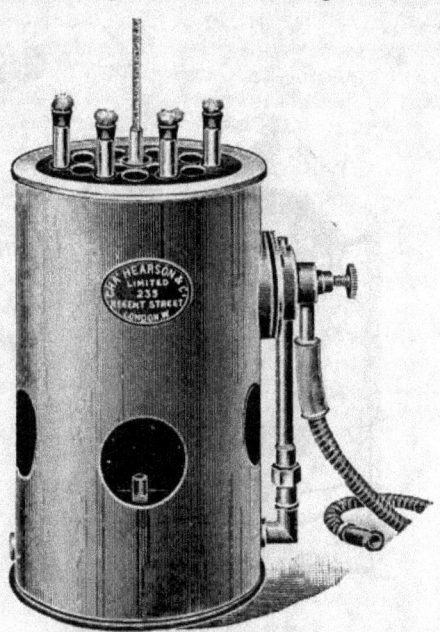

FIG. 8.—Hearson's water bath.

by a capsule attached to the outside of the bath, and through which the gas passes. The capsule has a range of about 10° C., and within these limits the temperature is regulated by means of a milled screw.

[*Technique.*—Pour sufficient water into the bath to reach above the level of the regulator outside. Put on the lid, and pass a thermometer through the hole in it, being careful to see that the bulb is in the water. Light the gas. The temperature gradually rises until it reaches the point for which the regulator is set: the gas is then automatically lowered and the temperature remains stationary. To raise the temperature, turn the screw clockwise, to lower it, contra-clockwise.

[*Note.*—It must be remembered that the capsule has a working limit of about 10° only, the exact limits being indicated when the instrument is supplied. Consequently, if a bath is required to work sometimes at 55°–65° C., and at other times at 75°—85° C., it is necessary either to have two baths, or a single bath in which capsules are interchangeable.

[Be careful always to see that the water is above the level of the top of the capsule, and when filling the bath never add water of a temperature higher than that for which the capsule is regulated.]

2. **Weissnegg's water bath.**—Another form of bath, which is shown in fig. 9, consists of a metal vessel fitted with a Roux's regulator placed in a side chamber, the heat being supplied by a gas burner below.

Technique.—Fill the vessel about three-fourths full of water and immerse the flasks by means of the wire tray, put on the lid, pass a thermometer through the opening provided for the purpose, and light the gas. Watch the thermometer carefully until the desired temperature is reached, then set the regulator in the manner to be described later (Chap. IV.), and no further supervision is required.

The regulator being once set for a given temperature will always work automatically at that temperature until it is again altered, so that beyond lighting the gas and when necessary pouring water into the bath no further manipulation is required.

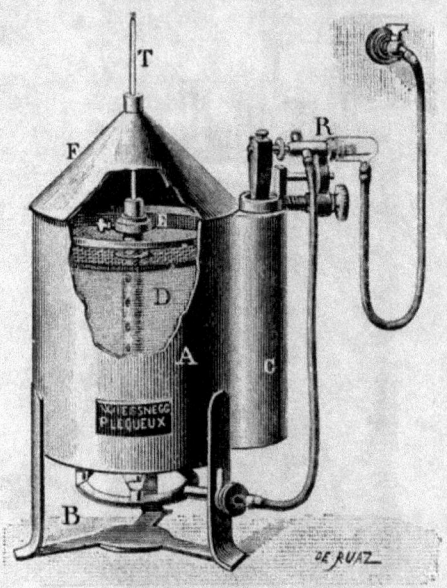

FIG. 9.—Weissnegg's water bath.

Note.—When using the bath for the first time, it is advisable to set the regulator beforehand by means of a blank experiment, thus avoiding accidental overheating of the medium. Sterilization is then carried out as already described, the medium being heated on six or eight consecutive days for an hour each time.

SECTION III. STERILIZATION BY FILTRATION.

The application of heat in some form is the usual method of sterilization used in bacteriological work, but it sometimes happens that fluids have to be dealt with which cannot be subjected to even a moderate degree of heat without profoundly altering their nature. In order to sterilize such a fluid, it is passed through a solid bougie, the pores of which are so fine that while liquids and solids in solution pass through, micro-organisms are retained. Pasteur in his early work utilized plaster plates as the filtering medium, but as a result of Chamberland's researches porous porcelain superseded plaster.

Filters.

The Pasteur-Chamberland filter consists of a porous porcelain tube or bougie closed at one end but open at the other, and finished at the latter with a nozzle of glazed porcelain. The unfiltered liquid traverses the pores of the bougie from without inwards, and issues from the nozzle filtered and sterilized. The Pasteur-Chamberland bougies are made in two grades of porosity.

That known as the **Chamberland "F"** is the more permeable, and the one generally used both for domestic purposes and for ordinary filtration by aspiration. The less porous and harder bougie, the **Chamberland "B,"** is only used for filtration under pressure (*vide infra*), and when manipulating fluids containing exceedingly minute organisms, *e.g.* the organisms of foot and mouth disease, pleuro-pneumonia, horse-sickness, etc. (*vide* "Filtrable Viruses" Chap.LXIV.), which pass through the more porous "F" bougies.

In addition to the Chamberland bougies there are other filters of a similar nature. [The **Doulton white-porcelain filter** (fig. 10) has been found to be "at least as efficient in the retention of micro-organisms as the best material on the market, viz. the Pasteur-Chamberland filter," and to "excel the latter in its rate of filtration."[1]]

Another filter, **Garros'**, is made of infusorial earth. This, [like Doulton's] filter, has all the essential properties of a Chamberland filter, and both are used in exactly the same way. The **Berkefeld bougie** is also made of infusorial earth; it is inferior to the Chamberland "B" in that it wears out more rapidly and does not hold back the smallest organisms; on the other hand it filters more quickly than, and does not retain dissolved organic matter to the same extent as, the Chamberland filters. For the latter reason it is especially useful for the filtration of albuminous fluids. [But it must be pointed out that recent experiments have shown that the Berkefeld is not a trustworthy filter.[2]]

There are several ways in which these unglazed porcelain and similar filters may be used.

FIG. 10.— Doulton's porous porcelain filter with nozzle.

1. Filtration of water.

Every laboratory has a filter attached to a water tap, for the purpose of readily obtaining a supply of sterile water (fig. 11).

The filter (Chamberland F [or Doulton white]) is contained within a metal cylinder, through the lower end of which it is inserted, and then securely fixed by means of a metal screw-cap, an india-rubber washer intervening; both the washer and metal cap are perforated to allow the passage of the glazed nozzle. The upper end of the metal cylinder is screwed on to a tap connected with the water main. When the tap is turned on, water runs into the space between the cylinder and the bougie, traverses the bougie, on the surface of which the solid matter in suspension is deposited, enters the central cavity, and escapes from the mouth of the glazed nozzle.

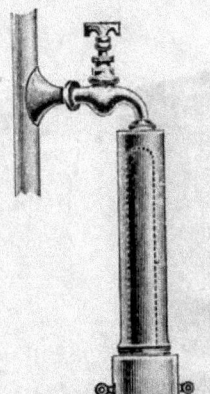

FIG. 11.—A porous porcelain filter attached to the water main : the outline of the filter is shown by the dotted lines and the glazed nozzle is seen projecting below.

Preparation of the filter.—1. Before putting a filter into the metal cylinder it is absolutely necessary to ascertain that it has no fissure or flaw in its substance, because unless it be perfect micro-organisms will quickly find their way through it. To determine whether or no the bougie is sound, attach an india-rubber syringe to the nozzle and immerse all but the nozzle in a cylinder filled with water (fig. 12). By squeezing the syringe air will be driven into the bougie, and if a fissure be present, even

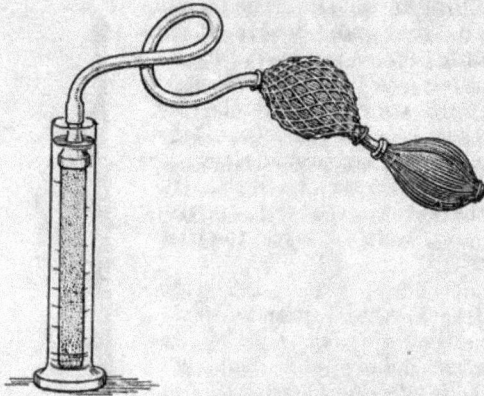

one so small as to be invisible to the naked eye, bubbles of air will stream out through it into the water and will at once render it apparent. All defective bougies must necessarily be rejected.

2. The filter must then be sterilized. After testing the bougie and while it is still wet, plug the nozzle with dry wool and sterilize in the autoclave at 115°–120° C. for 20 minutes. Fix the bougie in the metal cylinder, withdraw the wool plug, and the filter is ready for use.

3. Before drawing off sterile water, flame the nozzle well with a spirit flame.

FIG. 12.—Method of testing a porcelain filter.

Cleansing and renovation of bougies.—1. When in use the external surface of a filter soon gets soiled, and organisms are then likely to find their way through the pores. It is necessary therefore that filters be frequently taken out and cleaned by scrubbing with a stiff brush in a stream of running water, and re-sterilized.

FIG. 13.—A muffle furnace.

2. But this surface cleansing does not prevent the pores of the filter becoming choked after a time, filtration being impeded in consequence; when this occurs a porcelain filter can be renovated by one or other of the following methods.

(a) After scrubbing the filter autoclave it at 120° C., but before taking it out of the autoclave compress and decompress several times. As far as it goes this is an excellent method for unchoking a filter because there is no risk of damaging its structure, but the regeneration is only partial.

(*b*) Clean the filter as above, and dry it thoroughly : then heat it to redness in a Bunsen flame. This involves considerable risk of fissuring the filter.

(*c*) The best method is to heat the bougie to redness in an incinerator [or " muffle " furnace] (fig. 13).

Note.—After regenerating a filter by either of the two last methods it must be re-tested to make certain that it has suffered no damage.

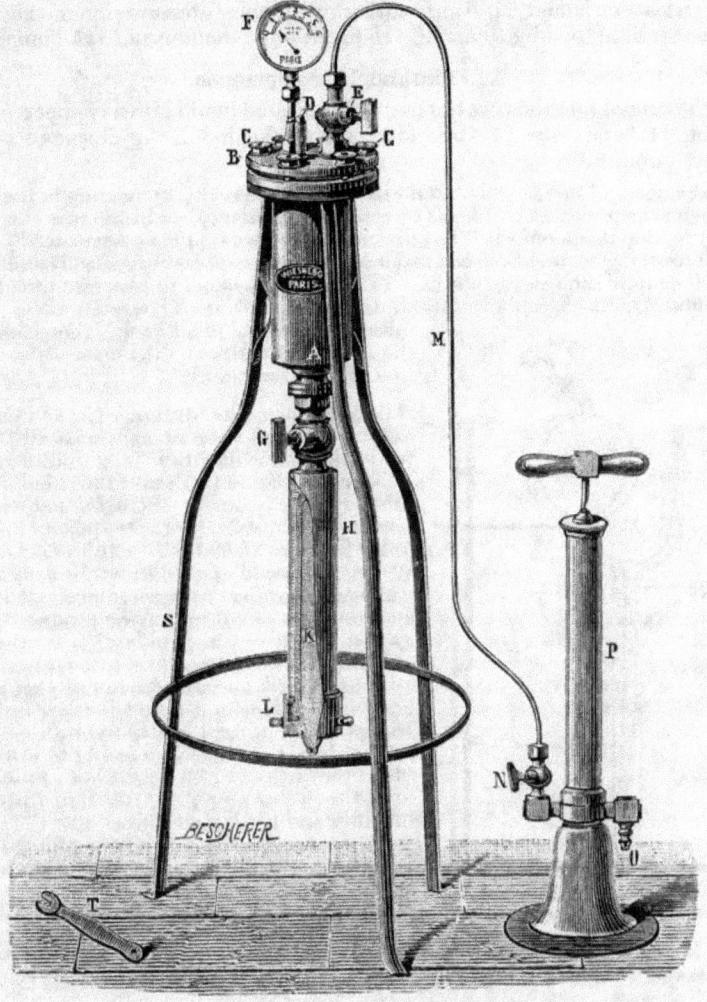

FIG. 14.—Filtration by compression (Gay-Lussac pump).

(*d*) Lastly, a filter may be regenerated by passing through it a 0·5 per cent. solution of potassium permanganate, followed by a 5 per cent. solution of sodium bisulphite (Guinochet). The method is less satisfactory than the foregoing.

3. Whenever culture media containing micro-organisms have been passed through a filter, the latter must be autoclaved immediately.

B

Berkefeld bougie.—The Berkefeld bougie does not lend itself to heating in a flame or incinerator. To clean it, it must be brushed with a stiff brush in a solution of sodium carbonate, washed in running water, and then autoclaved.[1]

2. Filtration of culture media.

Bougies are also used for rendering sterile media which are to be used for growing cultures, and for freeing a culture medium of the organisms which have been grown in it.

This may be effected in one of many ways; but filtration should always be carried out under pressure either by putting pressure upon the liquid to be sterilized, or by aspirating the filtrate at the mouth of the bougie.

A. Filtration by compression.

The original method was to pour the unfiltered liquid into a copper reservoir A (fig. 14), and then to force it through the filter K by means of a Gay-Lussac pump P.

Technique.—Close the tap G, and half fill the reservoir A by pouring in the liquid through the opening D. Close D by screwing on the cap, and compress the air in A by working the pump P. The pressure can be read on the manometer F. When the necessary pressure has been attained (2 or 3 atmospheres is generally sufficient), close the tap E and slowly open G. This allows the liquid to pass into the filtering chamber H, which contains a sterile Chamberland filter K (size B) fitted up as described above. The liquid is forced through the filter and issues at the nozzle where it can be collected aseptically.

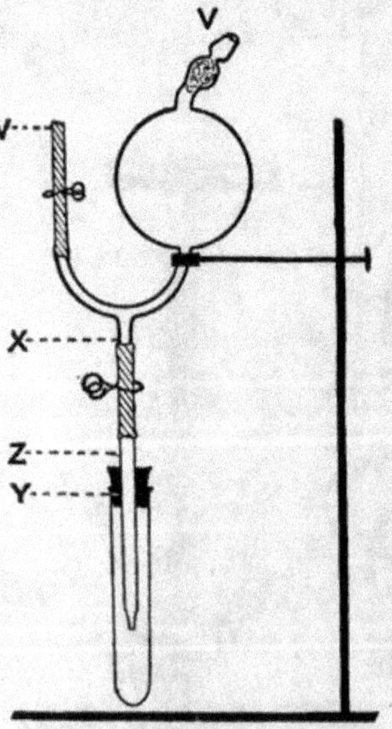

Collection of the filtrate.—1. [A Cobbett's bulb is a useful piece of apparatus with which to collect and distribute the filtrate. The illustration (fig. 15) shows the bulb, which usually has a capacity of about 200 c.c. To render it available for the present purpose, plug the small bulb V with wool, attach W to the nozzle of the filter K by means of stout red rubber pressure-tubing, and with another piece of rubber tubing connect X with a short length of glass tubing Z, the other end of which has been drawn out to a fairly narrow opening. Select a small india-rubber plug Y with one perforation, slip it over the lower end of Z and push it up until it fits tightly round the tube, then enclose the lower end of Z in a test-tube the mouth of which must be of suitable size to fit the rubber plug Y. After thus fitting up the filter and bulb, autoclave at 120° C.

[When required for use, fit the filter K in its metal case H (fig. 14), and screw on the cap L firmly. Support the bulb by clamping it to a retort stand. Clip the tubing between X and Z.

[The filtered liquid will be forced into the bulb, the rate being regulated by the tap G, and when nearly filled turn off the tap G; take off the test-tube, and by releasing the clip between X and Z the fluid can

FIG. 15.—Cobbett's bulb with attachments for filtration by compression.

[1] According to Dr. Andrew Wilson, however, it would appear that Berkefeld bougies must not be autoclaved. "It is a well-known fact that in consequence of the composition and the mounting of the Berkefeld filtering cylinders, they do not stand sterilization in an autoclave at 120° C. The only way effectually to sterilize the cylinder without injuring it is to place it in a vessel with cold or tepid water, and to boil it for about an hour" (*Journal of Hygiene*, 1909, p. 33). It has already been stated that simple boiling at 100° C., though prolonged, cannot be relied upon to destroy all micro-organisms.]

be run off into any suitable sterile vessel. When the bulb is emptied, replace the test-tube, tighten the clip, open G, and repeat the operation.]

2. The filtrate may also be collected through a piece of glass tubing connected by a piece of india-rubber tubing a few centimetres long to the nozzle of the filter (fig. 16). The bougie, with rubber and glass tubing attached, is wrapped in paper

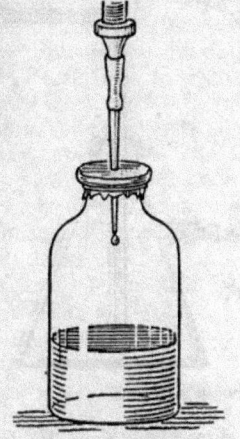

FIG. 16.—Alternative method of collecting the filtrate.

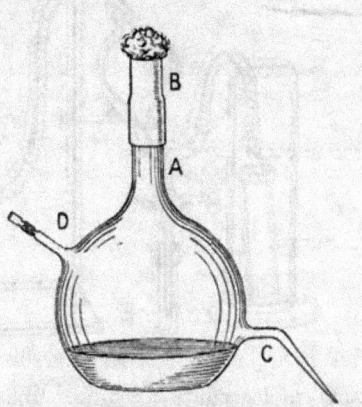

FIG. 17.—Flask with three tubulures for the collection and distribution of the filtrate.

and sterilized. When required for use, the bougie is fixed in its metal cylinder, the paper removed from the rubber and glass collecting tubes, and the latter pushed through the paper cap covering the mouth of the sterile vessel in which the filtered liquid is to be collected. If the collecting vessel be plugged with wool, the tube is inserted between the neck and the plug, the tube being surrounded as completely as possible with wool and pushed downwards until the orifice projects below the wool.

3. Another arrangement is to use a flask with three tubulures, such for instance as that shown in fig. 17. The flask must of course be sterile; the wool in the mouth of the india-rubber tubing B is removed, and the tube itself attached to the nozzle of the filter. When filtration is completed the tubing is removed from A, which is then plugged with a sterile plug, all necessary precautions being taken to prevent contamination. To manipulate the filtrate the tubulure C is broken and the liquid run out by simply inclining the flask. The third tubulure D is plugged with wool.

Filtration by compression involves the use of a costly piece of apparatus, and is limited in practice to the filtration of viscous fluids.

B. Filtration by aspiration.

Aspiration is the means usually employed for the filtration of fluids. The methods by which the principle of filtering by aspiration is applied vary in detail, and the technique of a few of the simplest and easiest devices will be described.

[1. Fit up and sterilize a Cobbett's bulb exactly as described for filtration under pressure (p. 18), and clamp it to a suitable stand. Connect the bulb to a wash-bottle with a piece of red rubber pressure-tubing, but between the bulb and the wash-bottle insert either a three-way tap or a T-piece of glass tubing the vertical limb of which is closed by india-rubber tubing and a clip, then connect the wash-bottle to the pump.

[*Technique.*—Stand the filter F in a tall glass cylinder C which must be rather larger than the filter, and fill up the cylinder with the unfiltered liquid. Tighten the clip K on the vertical limb of the T-piece and also the clip H of the delivery tube, and turn on the water. The liquid is thus aspirated

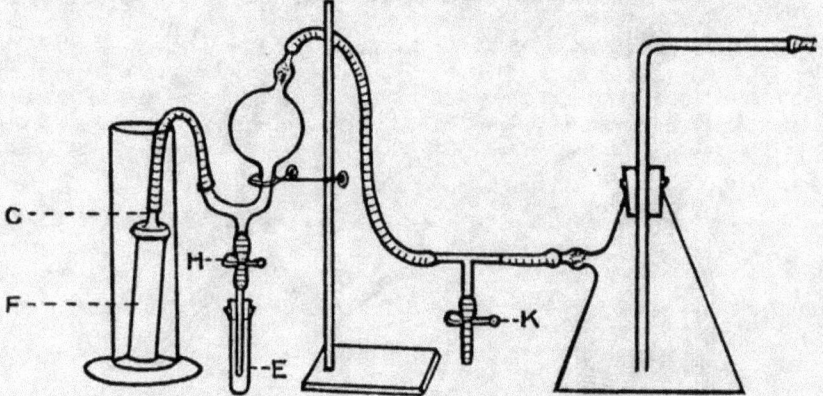

FIG. 18.—Cobbett's bulb fitted up for filtration by aspiration.

through the filter into the bulb. When the bulb is nearly full, gently release the clip K on the vertical limb of the T-piece, and then turn off the water. Then, as before (p. 18), remove the test-tube E and draw off the filtrate into suitable and previously sterilized vessels. Having emptied the bulb replace the test-tube E, tighten the clips K and H, turn on the water and exhaust again. In case the filtrate or a part of it has to be stored for future use, the vessels in which it has been collected may be sealed off in the flame of the blow-pipe, or the wool plugs can be made air-tight to prevent evaporation by sealing them with paraffin or sealing-wax (p. 30).]

[2. Instead of a Cobbett's bulb, an Erlenmeyer flask may be used, but the procedure is a little more complicated (fig. 19). The filtrate is aspirated into an Erlenmeyer flask, and then blown out with a bicycle pump.

[*Technique.*—(*a*) Take an Erlenmeyer flask of sufficient size to contain the filtrate. Plug the lateral tubulure with wool between the constrictions.

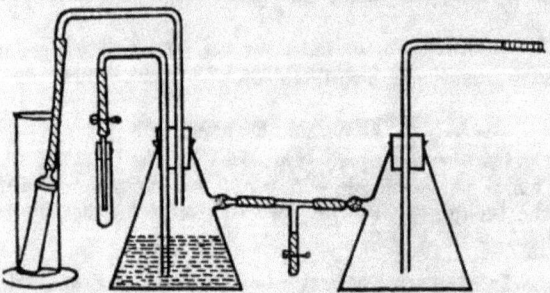

FIG. 19.—A convenient arrangement for filtration by aspiration.

Fit the mouth with an india-rubber bung with two holes; through one hole pass a piece of glass tubing bent at a right angle, the vertical limb of which is long enough to reach to the bottom of the flask, and through the other another piece of glass tubing also bent at a right angle, the vertical limb of which reaches only just below the level of the lateral tubulure. To the horizontal arm of the latter attach a piece of red rubber pressure-tubing, which at its other end is connected to the nozzle of the filter. To the other piece of glass tubing attach another piece of pressure-tubing, into the distal end of which is inserted a short piece of glass tubing drawn out to a narrow orifice; over this glass tube an india-rubber plug is slipped to fit a test-tube which protects the end of the delivery tube.

[(b) Sterilize in the autoclave at 120° C.

[(c) Attach the lateral tubulure of the flask to the lateral tubulure of another Erlenmeyer flask with pressure-tubing, and insert a three-way piece of glass tubing between the two flasks, the vertical limb being closed with rubber tubing and a clip. Pass a right-angled piece of glass tubing through an india-rubber bung in the mouth of the flask, and attach this tubing to the pump by means of pressure-tubing.

[(d) Place the filter in a glass cylinder larger than the filter, and fill up the cylinder with the fluid to be filtered. Secure the two clips.

[(e) Turn on the pump, and the fluid is aspirated from the cylinder to the Erlenmeyer flask. When the fluid reaches nearly up to the level of the lateral tubulure, release the clip on the vertical limb of the three-way piece of glass tubing. Disconnect the second flask, and attach a bicycle pump to the first flask. Clip the tubing attached to the filter.

[By working the pump the flask can be filled with air—filtered by passing through the wool plug in the lateral tubulure—and the contents of the flask thus put under sufficient pressure to allow them to be drawn off through the tube contained in the test-tube, as in the former case.]

3. A third method is to attach one end of a piece of red rubber pressure-tubing B (fig. 20) to the nozzle of a filter, and the other end to a piece of

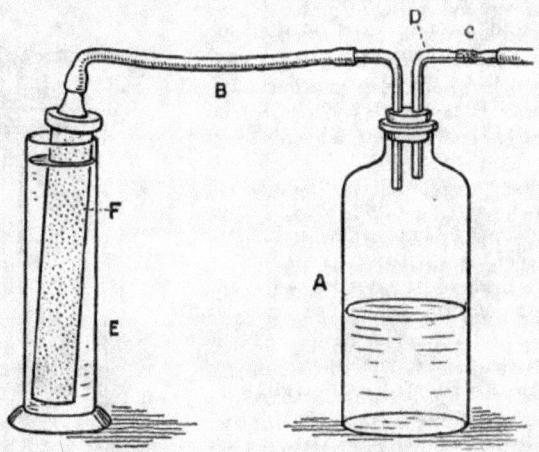

FIG. 20.—Filtration by aspiration.

glass tubing bent at a right angle. Pass the latter through one of the holes in an india-rubber bung. Fit the bung tightly into the neck of a stout white-glass bottle A the capacity of which is equal to the amount of fluid to be filtered (thin-walled flasks will not withstand the pressure, and should therefore never be used for this purpose). Through the other hole in the bung pass another piece of glass tubing also bent at a right angle ; the vertical branch should reach a few centimetres below the bung, while the horizontal arm has two constrictions with a fairly tight plug of wool C between them.

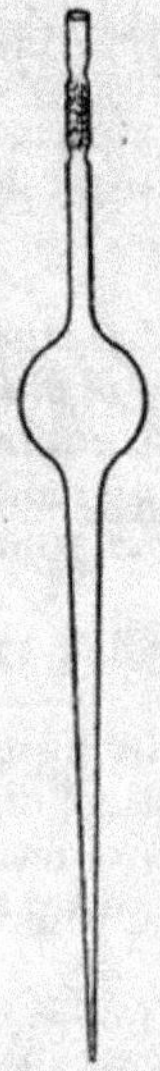

The apparatus thus fitted up is heated in the autoclave at 120° C. for 20 minutes. When cool, it is examined to see that the bung still fits tightly, and the apparatus is then ready for use.

Technique.—Stand the filter F in a glass cylinder E which must be rather larger than the filter, and fill up the cylinder with the unfiltered liquid. Connect the horizontal limb of the tube D by means of pressure-tubing with a water pump (Chap. VI.) and exhaust. The liquid is thus aspirated through the filter into the bottle A.

When the liquid has all passed through, turn off the water and disconnect the tubing connecting the bottle and the pump (the air which will then enter the bottle is filtered through the wool plug C between the constrictions). Flame the neck of the bottle, and replace the bung either by a previously sterilized wool plug or by another bung so arranged that the fluid can be manipulated as described later. The liquid thus sterilized by filtration can be kept sterile indefinitely in the bottle.

There is always a little liquid left in the filter, and if necessary this can be collected by disconnecting the tube B from the nozzle and aspirating the fluid into a long sterile bulb pipette (fig. 21).

Distribution of the filtrate.—Having obtained a sterile filtrate, it follows that the subsequent manipulations must be so devised as not to contaminate it. The methods to be employed will now be described.

FIG. 21.— Bulb pipette.

The bung used during filtration (p. 21) must be replaced by another fitted in the following manner. Take an india-rubber bung perforated with two holes of the same size as the one to be replaced. Through one hole pass a piece of glass tubing A (fig. 22) bent at a right angle and having a cotton-wool plug between constrictions in the horizontal arm. Through the other hole pass another piece of glass tubing B bent in the form of an inverted open U, one limb of which should reach nearly to the bottom of the bottle while the other, which will be outside the bottle, is drawn out to a fine capillary point and sealed. Wrap the bung with its glass tubes *in situ* in paper, and autoclave at 120° C.

Flame the neck of the bottle, remove the paper covering from the sterilized bung, and hold the latter by its upper part in the left hand. Take out of the bottle with the right hand the bung used for filtration, and replace it by the other as quickly as possible in order to prevent dust falling into the bottle. Care must of course be taken that during these manipulations the new bung with its tubes comes in contact with nothing likely to soil it.

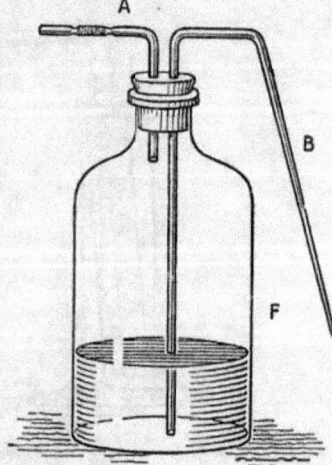

FIG. 22.—Distribution of the filtrate.

To withdraw the fluid, it is only necessary to connect an india-rubber syringe to the tube A, and after flaming the capillary end of B to break off the point with a pair of sterile forceps. By squeezing the syringe a few times the liquid will be forced out through B. When the quantity required has been withdrawn the end B is sealed in a Bunsen flame thus effectually excluding air from the bottle.

It may however sometimes happen that on ceasing to work the syringe some air will enter the bottle through B, and since this may carry organisms with it there is a risk of the liquid in the bottle becoming contaminated. The difficulty is easily overcome by the following simple device.

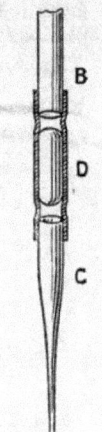

Before being sterilized the external limb of the tube B is cut in the middle, and the two ends B and C (fig. 23) connected by means of a piece of red rubber tubing, into which a short length (1–2 cm.) of glass rod D has already been introduced. When not in use the glass rod completely obliterates the lumen of the rubber tubing, and cuts off all communication between the outside air and the contents of the bottle. But if the rubber tubing be pinched between the thumb and index finger, a small channel is formed through which the liquid can be forced by squeezing the syringe attached to A (fig. 22). The apparatus would then be worked as follows:

The end of the glass tubing C being flamed and the point broken off, the syringe is squeezed a few times, and then the india-rubber tubing between B and C pinched up. The liquid will then run out from the open end. The flow of liquid is stopped by first releasing the finger and thumb from the india-rubber tube, thus cutting off all communication with the outside air, and then but not till then relaxing the pressure on the syringe. Finally, the broken end is sealed and the syringe disconnected.

FIG. 23.— Stopper for use with distributing plug.

4. L. Martin's filtering apparatus (fig. 24).—This consists of a porcelain

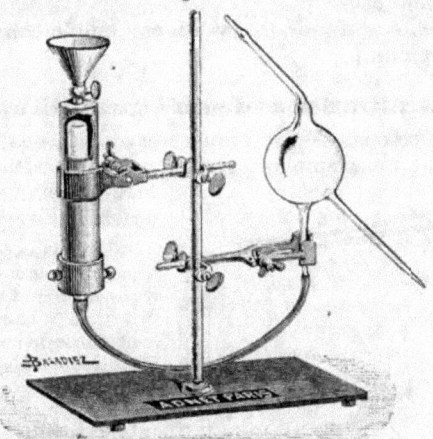

FIG. 24.—L. Martin's filter.
(The upper tubulure of the bulb should have a wool plug between constrictions: this has been accidentally omitted from the figure.)

filter contained in a metal cylinder similar to that described before (p. 15). The cylinder has a tap funnel screwed into the top to facilitate manipulation.

Technique.—Connect the nozzle of a porcelain filter with a bulb of the shape shown in the figure by a length of pressure-tubing. Sterilize in the autoclave. Then fix the filter in its metal case, and connect the upper tubulure of the bulb to a water pump with pressure-tubing.

Fill the cylinder through the funnel with the fluid to be filtered, close the tap and turn on the water. The fluid in the cylinder is aspirated through the filter, along the tubing, and so into the bulb. When all the liquid has been aspirated, turn off the water, open the tap, and disconnect the bulb from the water pump. The lower tubulure of the bulb is sealed during filtration, but is flamed and the point broken off with sterile forceps before distributing the filtrate.

This is a useful piece of apparatus, but its cost is a disadvantage.

5. Chamberland's method.—If water be not available, a small hand pump, *e.g.* Potain's, may be used for aspiration. Place the filter B (fig. 25) in a

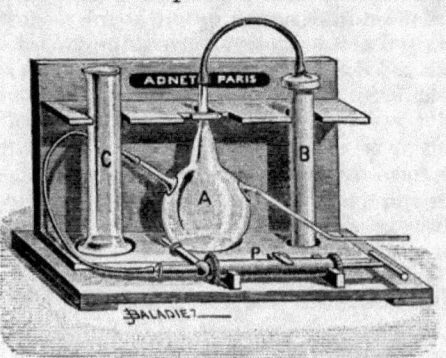

FIG. 25.—Chamberland's filter.

tall glass cylinder C, and fill up the latter with the liquid to be filtered. On working the aspirator the fluid is drawn through the filter into the flask A, which has three tubulures. The filter and flask must both be sterilized in the autoclave before use.

[It will be obvious, of course, that Cobbett's bulb can be used equally with a hand or water pump.]

3. The filtration of small quantities of liquid.

The laboratory bougie.—When only very small quantities of liquid have to be filtered, as for example in testing a toxin, a small thin-walled bougie 12–15 cm. long and without a nozzle is very useful.

Technique.—[A. Slip one end of a piece of stout pressure-tubing over the open end of the bougie and secure it with a rubber ligature, then connect the other end to the free limb of the U-tube of a Cobbett's bulb. Sterilize in the autoclave. Place the filter in a small glass cylinder or test-tube and fill the latter with the fluid to be filtered. Connect the bulb to a water pump and exhaust. Any fluid remaining in the filter can be recovered by holding the filter upside down, and allowing it to run into the bulb.]

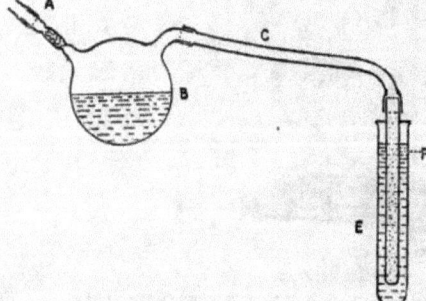

FIG. 26.—Laboratory bougie for filtering small quantities of liquid.

B. The filter may also be arranged as shown in fig. 26. As in **A**, a piece of pressure-tubing is firmly fixed by one end to the upper end of the filter, but the other end is attached to one of the two tubulures of the flask B. The other tubulure A is plugged with cotton-wool and connected to a water pump or a small aspirator, *e.g.* Potain's. Sterilize the apparatus before use.

C. Duclaux's filter.—The filter can also be fitted to a flask with three tubulures (fig. 27). In this case the open end of the filter is wrapped round with cotton-wool, which serves to hold the bougie in position in the neck of the upper tubulure E. The tubulure D is sealed, and B is plugged with wool. After autoclaving, the wool packing around the neck of the filter F is made air-tight by running a little melted

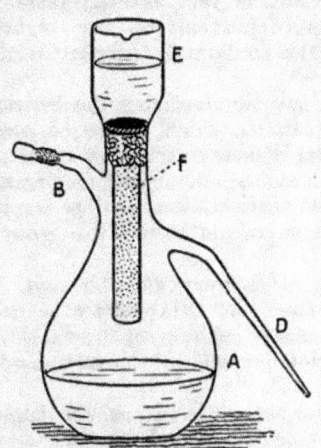

FIG. 27.—Laboratory bougie—Duclaux's arrangement.

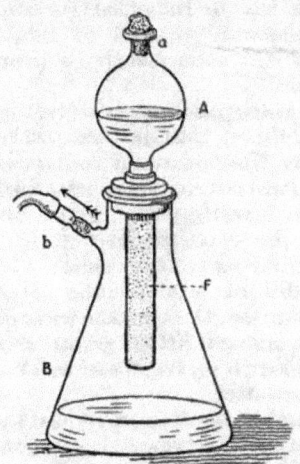

FIG. 28.—Laboratory bougie—Kitasato's arrangement.

wax ([paraffin or] Golaz's) over it. The tubulure B is connected to a water pump, and the liquid to be filtered poured into E. On turning on the water the liquid is drawn through the filter and collects in the flask. To distribute the filtrate, break off the sealed end D and blow air into the flask through the wool-plugged orifice B.

D. Kitasato's filter (fig. 28) consists of a conical flask of thick glass furnished with a lateral tube *b*, which when in use is plugged with wool and connected to a water pump.

The wide neck of the flask is fitted with a perforated india-rubber bung through which the filter F is passed. The mouth of the filter is attached by means of another india-rubber bung to a glass bulb A. The technique is very simple: pour the liquid into A, turn on the water pump, and the filtrate collects in the flask B. The apparatus must of course be autoclaved before use.

E. Martin's filter (fig. 29), as arranged for dealing with small quantities of fluid, consists of a tube R, which can be connected to a water pump through the tubulure A. Within it is a moderately large test-tube T resting upon a pad of cotton-wool. A small filter F is passed into the test-tube, and firmly fixed in the mouth with the open end upwards. The tube R is closed above with an india-rubber bung, through which passes a glass funnel E the lower end of which is connected with the upper (open) end of the filter. The liquid to be filtered is poured into E, and being drawn through the filter collects in the tube T.

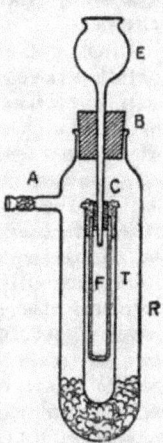

FIG. 29.—Laboratory bougie—Martin's arrangement.

To sum up, *there are three important considerations to be kept in mind when using a filter for purposes of sterilization.*

(1) *In every case the filter must be tested to make sure it is sound and free from fissures;* (2) *filters must always be sterilized immediately before use;* and (3) *subsequent contamination of the filtrate must be carefully guarded against.*

SECTION IV.—STERILIZATION BY ANTISEPTICS.

Sterilization by antiseptics has but limited use in bacteriological work. The addition of antiseptics will it is true destroy micro-organisms in a medium designed for the growth of cultures, but the amount of antiseptic which has to be added to effect this result is very much greater than the amount required to inhibit the multiplication of any organisms which may subsequently be sown in it; the medium is therefore rendered useless.

1. Antiseptics are, however, in general use for sterilizing the interior of glass dishes, bell jars, and other similar articles which are to be used to protect from dust and contamination Petri dishes, culture tubes, etc., and which will not come in contact with any culture medium, or with the organisms under investigation. Fixed non-volatile antiseptics must be employed since the vapours given off by volatile compounds hinder the growth of organisms on culture media.

A 0·1 per cent. solution of perchloride of mercury may be used. The solution should be made with distilled water, but if tap water be used a small amount (0·5–1 gram) of tartaric, acetic or hydrochloric acid must be added to prevent precipitation of the mercury salt by the salts dissolved in the water.

Perchloride of mercury has however now been almost entirely discarded in favour of oxycyanide of mercury in 0·1 per cent. solution. This solution though powerfully antiseptic has no caustic action, it does not precipitate albuminoid substances, neither does it attack instruments and other metal articles.

2. Antiseptics are also in general use for sterilizing the hands, and for washing out vessels and sterilizing instruments during inoculation and other experiments. Solutions of 0·1 per cent. of perchloride or oxycyanide of mercury or 1·5 per cent. of formalin are often used for these purposes.

[Lysol, a solution of the three cresols in soap and water, is a particularly useful antiseptic. In 2 per cent. solution it does not hurt the skin, and the soap in solution makes a lather if the hands be washed in it or if the surface of the skin be rubbed with a sponge soaked in the solution; the presence of the soap makes a solution of lysol a more efficient antiseptic for these purposes than mercury solutions. Lysol does not damage metal instruments, and does not precipitate albuminoid solutions. If made up in large quantities with hard water the soap is liable to be precipitated to some extent, but the antiseptic constituents still remain in solution.]

Solutions of perchloride or oxycyanide of mercury may also be used for sterilizing the surface of the skin before collecting pus, blood, etc., from the living subject (Chap. XII.). Care must of course be taken that, after sterilization, all traces of the antiseptic are removed by washing the part well with alcohol before collecting the material, otherwise the presence of the antiseptic would materially interfere with the subsequent growth of organisms in culture. [At the present time, however, it is more usual to paint the surface of the skin with tincture of iodine (British Pharmacopœia) before penetrating it for the purpose of collecting material for bacteriological investigation.]

3. Antiseptics are also added to sterile filtrates which are no longer required as culture media. For this purpose a small quantity of some antiseptic (such as thymol or camphor) which is without chemical action on the constituents of the fluid is selected.

[Wright adds a trace (0·5 per cent.) of carbolic acid to his vaccines.]

4. Antiseptics are sometimes used to sterilize a culture when the products of micro-organisms are under investigation. Volatile antiseptics such as chloroform, ether, toluol, essence of garlic or mustard, etc., which can be readily driven off afterwards by evaporation, are the most useful in this connexion.

CHAPTER II.

CULTURE MEDIA.

Introduction.

THE substances requisite for the growth of micro-organisms may be obtained by macerating, infusing or boiling tissues of animal or vegetable origin. Saline solutions in which some carbo-hydrate is dissolved also supply all the ingredients essential for a culture medium.

Culture media are either solid or liquid.

Chemically, like all other living cells, micro-organisms consist of organic and inorganic nitrogen and mineral salts; it is therefore necessary in order to grow a micro-organism that these three classes of substances be made available, together with oxygen which is an essential to the life of all living structures. [Finally, a certain amount of moisture is absolutely necessary.]

Micro-organisms are divided into two large groups, the members of one of which derive their oxygen, like more highly organized structures, from the free oxygen of the atmosphere, while the members of the other group cannot multiply in presence of free oxygen, but obtain the oxygen they require by the decomposition of substances containing it (Pasteur). The former are known as the aërobic, the latter as the anaërobic organisms.

These two groups of micro-organisms call for very different methods of artificial cultivation. Aërobic micro-organisms should be grown in vessels in which there is an ample supply of air; anaërobic micro-organisms on the other hand only grow if air be excluded. The latter therefore are cultivated either *in vacuo*, or in presence of some inert gas.

The constituents of culture media are however the same for both aërobic and anaërobic organisms and ought to include nitrogen compounds and salts of the ternary bases. Many organisms can convert inorganic nitrogen (nitrates, etc.) into organic nitrogen, while in some cases organisms will grow in purely inorganic solutions provided these contain a small quantity of some carbohydrate such as sugar.

General Rules.—Every culture medium therefore must

(1) *contain the substances necessary for the growth of the organism to be sown* [(2) *be of suitable reaction*]; (3) *have been previously sterilized*; (4) *be contained in vessels which afford protection from contamination from without.*

Culture vessels.

Vessels of various patterns are used for culture media, and these will be described as occasion for their use arises. In this chapter a description of those most commonly used for the growth of aërobes only will be given.

1. Ordinary test-tubes, but without lips, are in constant use (fig. 30); they must be plugged with wool as already described.

2. Erlenmeyer flasks—conical glass vessels with a flat bottom (fig. 31) and of different sizes—[and **Jena flasks,**] are also in frequent use. Ordinary small

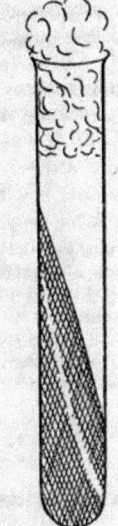

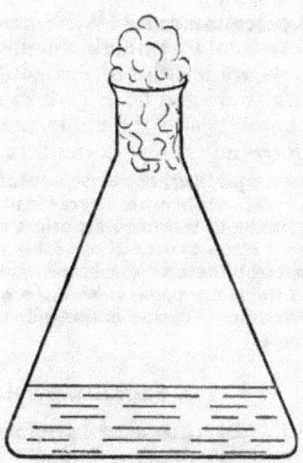

FIG. 30.—Culture tube   FIG. 31.—Erlenmeyer flask.
plugged with wool.

medicine bottles of 30 to 50 c.c. capacity can be used in many cases. Whatever be the shape of the vessel it must be plugged with wool, and as a further protection a paper cap is useful (*vide* p. 4).

3. Small vessels capable of holding 30 to 50 grams and known as Pasteur's flasks, are also in frequent use [in France]. The mouths of these flasks are generally closed by means of an hollow ground-glass stopper fitting a similarly ground-glass surface on the neck of the flask, and having a small orifice above, which must be plugged with wool. This method of plugging effectually preserves the contents of the flasks from contamination, but has the disadvantage of being very fragile, and the glass cap is often broken in the flaming process preliminary to opening the flask.

It is better to cover the mouth of the flask with a small paper hood (this can be done by enveloping the neck in a small strip of filter paper and tightly screwing the projecting part into a point). It is even simpler to plug the flasks with wool in the same way as test-tubes are plugged.

Miquel's flask is merely a conical form of Pasteur's pattern.

Prevention of evaporation.

Evaporation readily takes place through a wool plug, and if a medium—especially a solid medium such as potato, serum, agar, etc.—be stored or incubated for a long time, the amount of evaporation is likely to be excessive. To avoid this, the mouth of the vessel may be closed with an **india-rubber cap.** This must be sterilized before use, because when the cap is slipped over the mouth of the tube or flask the air within being saturated with aqueous vapour will soon make the wool plug moist, and then any organism on the inner surface of the cap will ultimately grow through the moist plug and contaminate the contents.

Red rubber caps are the best; they should be put into a wide-mouthed flask or bottle, which is then plugged with wool and autoclaved and afterwards put in the incubator to dry the plug. A cap can then be taken out with a pair of sterile forceps whenever one is wanted.

In the case of stock cultures, which are to be put away for some time, evaporation may be prevented by pouring a little melted **paraffin** [or **sealing-wax**] over the top of the plug.

[This latter method of sealing tubes or bottles is also of great use in the cultivation of slow-growing organisms such as the tubercle bacillus. After sowing the medium, the top of the plug is carefully sealed with melted paraffin (or sealing-wax), and the culture can then be incubated as long as is necessary without fear of the medium drying up, if the sealing has been efficiently done.]

Method recommended.—[When **paraffin** is used, gently warm the upper $\frac{1}{2}$ cm. of the tube by turning it round in the flame, and then with a pipette or ladle pour a few drops of melted paraffin, kept liquid in a water bath, on to the warm wool and let it soak in to a depth of $\frac{1}{2}$ cm. or so. To unseal the plug, gently warm the upper part of the tube again, stick a needle or pair of forceps into the plug and turn it round at the same time raising it.

[Another simple method of preventing evaporation during cultivation in test-tubes is to place them in a large wide-mouthed ground-glass **stoppered jar,** which has been previously thoroughly washed out with a saturated solution of perchloride of mercury. Place two or three folds of filter paper moistened with perchloride solution at the bottom of the bottle, and after arranging the tubes put a trace of vaseline on the stopper and close the bottle, turning the stopper round to obliterate any air channels. Volatile antiseptics (*e.g.* formalin) are obviously unsuitable for this purpose.]

SECTION I.—LIQUID MEDIA.[1]

1. Media made from animal tissues and fluids.

There is a great variety of these media. A description of those in most general use only will be given here; and the most frequently used of all, peptone beef broth, will be taken as a type and the technique of its preparation given in fullest detail.

Peptone beef broth.

This medium is in everyday use. It will be referred to in future simply as **broth.**

Preparation.—1. Take 500 grams of lean beef. Cut away all fat, tendon and aponeurosis. Mince it, and leave it to macerate in a litre of cold water for 6 to 12 hours.

2. Heat gently to boiling in an enamelled saucepan, stirring constantly, and keep the mixture boiling for 10 minutes.

3. Pour on to a thick clean cloth, express as far as possible all the fluid out of the meat, and while still warm filter the fluid through a thick filter paper (Chardin or Prat-Dumas) moistened with water to keep back the fat.

4. Pour the filtered broth into an enamelled saucepan, and add

Dry peptone (Chapoteaut)	- - -	10 grams, or 1 per cent. of the volume of water used.
Salt,	- - - - - -	5 grams, or 0·5 per cent.
Sodium phosphate,	- - -	about 1 gram.

Boil again, stirring meanwhile to dissolve the peptone

---

[1] The present chapter will be limited to a description of the culture media of general application. Media applicable only to particular organisms will be dealt with when the latter are under consideration.

The addition of the sodium phosphate is not absolutely necessary. Cache states that the addition of magnesium salts increases the value of the culture medium, and advises the addition of 2 grams of magnesium phosphate per litre to ordinary broth, in place of the sodium phosphate. Magnesium phosphate should be added while the meat is macerating (Stage 1 above).

5. The liquid is now strongly acid and must be neutralized, since bacteria grow best in a *neutral or slightly alkaline medium*.

Neutralization.—To neutralize the medium, add normal soda solution to the broth in small quantities at a time with a pipette, testing the reaction at frequent intervals against litmus paper. When a drop of the broth placed on a red litmus paper with the end of the stirring rod turns it slightly blue, sufficient soda has been added. The reaction should be very slightly alkaline to litmus, but acid to phenol-phthaleïn.

Neutralization is the most difficult step in the preparation of broth. The amount of alkali to be added varies considerably with different pieces of meat, and can only be determined by trial. Add the soda solution very slowly stirring carefully after each addition; and as the neutral point is approached, test the broth after the addition of each drop of alkali against both a red and a blue paper. A point is ultimately reached when a drop of the liquid produces no change on either paper; it is then sufficient to add a very small quantity of soda solution to attain the requisite degree of alkalinity. According to Park and Williams, 7 c.c. per litre of normal soda should be added to a neutral broth to obtain the most favourable medium.[1]

[Eyre uses phenol-phthaleïn as the indicator and standardizes after Stage 3 before the addition of peptone and salt.

[Technique.—1. Heat the meat extract in the steamer at 100° C. for 45 minutes.
2. Measure 25 c.c. into a beaker and add about 0·5 c.c. of a 0·5 per cent. solution of phenol-phthaleïn in 50 per cent. alcohol.
3. Immerse the beaker in a water bath and raise to boiling point.
4. Neutralize at the boiling point with deci-normal NaOH solution.

[The *reaction* is expressed by stating the number of cubic centimetres of *normal* alkali required to render *one litre* of the meat extract exactly *neutral* to phenol-phthaleïn.

[For the majority of organisms a medium which requires the addition of 10 c.c. of normal alkali per litre of meat extract is found to be the best. In Eyre's scale the reaction of such a medium is expressed as +10: + indicating that the medium is acid and 10 that it is acid to the extent of 10 c.c. of normal alkali per litre.]

6. Now pour the slightly alkaline broth into a glass flask, or better into an enamelled vessel, and autoclave at 115°–117° C. for 5 minutes. The liquid becomes cloudy and deposits crystals of earthy phosphates.

On taking out of the autoclave, filter while hot through a Chardin paper moistened with water: the filtrate should be absolutely clear. The object of this procedure is to remove any excess of earthy phosphates, and if omitted, the broth is likely to become cloudy when sterilized.

7. Add sufficient distilled water to the filtrate to make the total volume up to 1 litre.

8. This completes the preparation of the broth, which has now only to be distributed into suitable vessels and sterilized.

Sterilization.—**(A)** If the broth is to be kept for future use, it may be sterilized in a large flask the neck of which is either plugged with wool or drawn out in the flame and sealed. The medium can thus be kept indefinitely, and when required for use can be distributed into suitable vessels.

(B) It is however usually more convenient to distribute the broth at once into test-tubes or small flasks.

---

[1] Normal soda contains 40 grams of NaOH per litre of distilled water.

1. Into each test-tube put 10–15 c.c. of broth, and into each flask 25 c.c.

A small glass funnel should be utilized for distributing the broth, because if a drop of it come in contact with the wool plug it will when dry cause the wool to stick to the mouth of the tube, so that the plug can only be removed with difficulty. (The use of a funnel is even more important when solid media, such as agar or gelatin, are being tubed.)

2. Plug the tubes and flasks with wool.

3. Place the vessels containing the medium in a wire basket, and autoclave for 20 minutes at 110°–115° C., taking care that the latter temperature is not exceeded, and that the precautions noted in Chapter I. are observed.[1][2]

Veal broth.

For the preparation of veal broth proceed as for beef broth, using instead of beef 500 grams of lean veal.

Chicken broth.

Chicken broth is prepared in a similar manner, using 500 grams of chicken meat. All skin, tendon, and bone must be removed, otherwise the broth will be of a gelatinous consistency.

Giblet broth.

Liver, spleen, etc., may be used for making broth. The technique is the same as in the preceding cases, substituting 500 grams of the solid organs for beef or other meat. Very often these broths exhibit a slight cloudiness, but this cannot be avoided.

Meat extract.

1. To 500 grams of well-minced lean beef or veal, add 1000 grams of water, and leave in the ice chest to macerate for 12 hours.

2. Shake the mixture, filter through a cloth and squeeze all the fluid out of the meat, then filter through a Chardin paper.

3. Add 5 grams of salt to the filtrate, and heat to boiling.

4. Neutralize as in the case of peptone beef broth.

5. Autoclave for 5 minutes at 115°–117° C.

6. While warm, filter through a moistened Chardin filter paper.

7. Add sufficient distilled water to make up the volume to a litre.

8. Distribute in tubes and sterilize at 110°–115° C.

Martin's peptone solution.
Pigs' stomach broth.

1. Wash, clean and mince finely 4 or 5 pigs' stomachs.

It is better to use a number (5) of stomachs in order to neutralize variations in their pepsin content; if this be done, the broths will have an almost constant average composition.

2. Mix the following :

Minced stomachs, -	200 grams.
Hydrochloric acid (pure),	10 „
Water at 50° C., -	1000 „

and keep the mixture at a temperature of 50° C. for 20–24 hours.

---

[1] Sometimes it will be noticed that the medium when taken out of the autoclave is very slightly cloudy; this will vanish on cooling. But if during sterilization the temperature exceed the temperature of the first autoclaving (par. 6, above), the broth remains permanently cloudy.

[2] In England it is more usual to sterilize media by steaming for 20 minutes on each of three successive days in a Koch's steamer, or some suitable modification of it (Chap. I.).]

3. Then heat to boiling to destroy the excess of pepsin, and pass through a sieve or a thin layer of loosely packed absorbent cotton-wool.

4. Heat the filtrate to 80° C., and neutralize at this temperature : large flocculent masses are precipitated ; filter the clear supernatant fluid through a Chardin paper.

5. Autoclave the filtrate for five minutes at 117°–120° C., and again filter through a Chardin paper.

6. Distribute the clear filtrate in tubes, and sterilize for 15 to 20 minutes at 115° C.

Martin's peptone broth.

1. Mince 500 grams of lean veal, and macerate in 1000 grams of water for 15 to 20 hours at a temperature of 35° C. to get rid of the sugars.

2. Filter through a cloth, squeeze out as much of the fluid as possible, and add 5 grams of salt.

3. Mix this liquid with an equal volume of Martin's peptone solution (see above, Stages 1, 2, 3, 4).

4. Heat to 70° C. to coagulate the albuminoid compounds, and make exactly neutral. Then add 7 c.c. of normal soda solution per litre. Filter through a Chardin paper.

5. Sterilize by filtering through a Chamberland bougie (Chap. I.), and distribute in sterile culture vessels.

Note.—A broth prepared in this way is particularly useful for the preparation of diphtheria toxin.[1] For everyday use, it is simpler to sterilize by heat. After Stage 4 proceed thus :

5a. Pour the broth into an enamelled vessel or flask, autoclave for 5 minutes at 115°–117° C., and filter in the warm through a Chardin paper.

5b. Distribute the filtered liquid into culture vessels and sterilize for 20 minutes at 110°–115° C. The broth is often slightly and permanently cloudy.

Koch's peptone solution.

1. Dissolve in the warm, 10 grams of peptone (Witte or Chapoteaut) and 5 grams of salt in 1000 grams of water.

It is unnecessary to neutralize : peptone itself is sufficiently alkaline.

2. Boil. Filter.

3. Distribute in tubes or flasks. Sterilize at 115° C.

Metchnikoff's peptone-gelatin medium.

1. Dissolve in the warm, in 1000 grams of water,

Peptone (Chapoteaut),	10 grams.
Salt,	5 ,,
Gelatin (extra quality white),	20 ,,

2. Make very slightly alkaline with normal soda solution.

3. Autoclave for 5 minutes at 115° C. Filter through Chardin paper.

4. Distribute in suitable vessels. Sterilize at 110°–115° C.

Miquel's peptone solution.

1. Dissolve at a moderate heat in a litre of water,

Peptone (Chapoteaut),	20 grams.
Salt,	5 ,,

2. Add 0·10 gram of wood ash. Boil. Filter through Chardin paper.

3. The liquid is now generally markedly alkaline. Neutralize very carefully with a solution of tartaric acid, watching the reaction meanwhile by testing against litmus paper.

---

[1] Numerous formulæ for the preparation of broth suitable for the study of diphtheria toxin have been published (Chap. XV.).

C

4. Boil for 5 minutes. Filter. Add sufficient water to make the volume up to a litre.

5. Distribute in tubes. Sterilize at 115° C.

Liebig's broth.

1. Dissolve 5 grams of Liebig's extract of meat ["Lemco"] in 1000 grams of water, warming gently. Neutralize if necessary.

2. Autoclave for 5 minutes at 115°–117° C. Filter through a moistened paper in the warm.

3. Distribute in tubes. Sterilize at 110°–115° C.

Peptone (Chapoteaut) (10 grams) and salt (5 grams) may be added to the medium before neutralization.

In the same way a nutrient broth may be prepared with **Cibils' extract**. In that case 20 grams of the extract is used instead of the Liebig.

These media are used chiefly in German laboratories, [and the former to a large extent also in England].

Thymus broth (Brieger).

1. Obtain the thymus from two or three calves directly after they are slaughtered. Mince the glands finely and add an equal weight of distilled water. Mix, and macerate for 12 hours.

2. Filter through muslin and squeeze out as much of the fluid as possible. Add an equal volume of water to the cloudy viscous filtrate.

3. Make feebly alkaline with a 10 per cent. solution of sodium carbonate.

4. Heat to 100° C. for 15 minutes in the autoclave or steamer (a higher temperature interferes with the properties of the medium). Filter through a piece of fine linen.

5. Distribute in sterile tubes. Sterilize at 100° C. for 15 minutes on each of two successive days.

Some micro-organisms, such as the cholera vibrio, will only grow satisfactorily on this medium provided that 5 or 6 times its volume of sterile water be added just before use.

Serum broth. Blood broth.

These media are prepared by adding to tubes of ordinary sterile broth, one-half, one-third or one-quarter their volume of blood-serum, ascitic fluid, or blood collected under aseptic precautions (p. 45 and Chap. XII.).

Achalme, in the preparation of blood broth, advises the use of a 1 per cent. solution of commercial hæmoglobin instead of blood. The hæmoglobin must first be sterilized by filtration through a Kitasato's filter (p. 25).

The preparation of these media will be more fully considered when dealing with the *Gonococcus* and Pfeiffer's *bacillus*.

On account of the difficulty of obtaining sterile blood, Bernstein and Epstein recommend the following procedure for the preparation of blood broth: collect 400 c.c. of ox blood in a flask containing 30 c.c. of a 1 per cent. solution of ammonium oxalate in distilled water and 0·5 c.c. of formalin; shake, and in half an hour dilute the mixture with 3 volumes of normal saline solution. After standing for a day or two at the temperature of the laboratory, distribute in agar or broth in the proportion of 1 part to 15 of medium. The formalin is thus so highly diluted that it does not interfere with the growth of micro-organisms. The *Pneumococcus*, *Gonococcus* and *Meningococcus* all grow very well in this medium.

Carbohydrate broths.

These media are prepared by adding to beef broth, at the same time as the peptone and salt, 2–4 per cent. of one or other of the following carbohydrates: glucose, saccharose, lactose, galactose, mannite, dulcite, maltose, lævulose; the preparation is completed as in the case of broth.

[The use of carbohydrate media has been considerably extended in recent years in connexion with the identification and differentiation of the various members of the same group of micro-organisms, *e.g.* the differentiation of members of the typhoid-colon group, the differentiation of the streptococci, etc. (Gordon, Andrewes and Horder and others). For this purpose a medium differing somewhat from that given above, and having the following composition, is in general use.

Peptone. - - - - - - - - - - 1–2 grams.
Water, - - - - - - - - - - 100 c.c.
Test substance, - - - - - - - - - 1 gram.
Kahlbaum's litmus solution. - - - - - - - *Q.S.*

[The test substance may be either a sugar, *e.g.* glucose, lactose, lævulose, saccharose, etc.—an alcohol, *e.g.* mannite, dulcite—or a glucoside, *e.g.* salicin, coniferin, etc.

[Care must be taken to obtain guaranteed pure chemicals from reliable firms, and an equal amount of care must be bestowed upon the sterilization of the media, since it is well known that in the presence of water and under the influence of heat many of these highly complex compounds undergo decomposition, often of the nature of an hydrolysis. Filtration through a porcelain filter would seem to be the best method of sterilization. After distribution into sterile tubes the latter must be incubated for a few days and those showing any change rejected.]

Glycerin broth.

Add 5 per cent. or 50 grams per litre of pure glycerin to peptone-beef-broth before distribution into tubes (Stage 7, p. 31).

Glycerin in the same proportion may also be added to the carbohydrate broths prepared as above.

Carbonated broth.

Add calcium carbonate (2 per cent.) [1] to lactose-, mannite-, glucose-, etc., broth before distributing into tubes (Stage 7, p 31).

Calcium carbonate is most frequently added to lactose-broth. When an organism which ferments a given sugar is grown in a carbonated broth containing that sugar, the acids formed by decomposition of the carbohydrate act on the chalk with the formation of CO_2 and the evolution of a considerable quantity of gas. [2]

Milk.

Milk is used as a culture medium in several ways.

(**A**) Fresh milk, alkaline in reaction, is distributed in tubes (15–20 c.c. per tube).

The tubes are plugged with wool, and sterilized at 115° C. for 20 minutes. This is the most simple method of preparation and suffices in the great majority of cases ; it is the method ordinarily employed.

[In England it is usual to add sufficient litmus solution to tint the milk blue (p. 57), and to sterilize by steaming at 100° C. (Chap. I.). In our experience it is a very difficult matter to sterilize milk by steam at 100° C. in bulk ; it is much safer to tube the milk and then sterilize it.]

(**B**) Since a temperature of 115° C. alters to some extent the properties of milk, it may be desirable for some purposes to sterilize at a lower temperature.

In that case, after washing the cow's udder with an antiseptic, the milker

[1 In our experience 0·5 per cent., or even 0·25 per cent. of calcium carbonate is sufficient.]

[2 It sometimes happens, however, that when an organism is grown in a litmus-sugar-carbonate-broth, acid is formed as shown by the change in colour of the litmus but no gas is evolved.]

should sterilize his hands and then collect the milk as it leaves the udder in sterile flasks. (For further details, see Chap. XII.)

Each flask is about three-parts filled, sealed in the flame, [or plugged with sterile wool and covered with an india-rubber cap,] and heated in a water bath at 60°–65° C. for eight days in the manner described on p. 12.

When sterilization is completed, the milk can be tubed into sterile tubes, as described in connexion with the preparation of serum (p. 45).

(C) If the technique of the milking process can be relied upon, it will be sufficient to fill as many tubes as are required, and to incubate them at 30° C. for some days before using the milk as a culture medium. In spite of every precaution some of the tubes will be contaminated, and any tube in which the milk has clotted or which on microscopical examination shows the presence of organisms must be rejected.

Urine.

Though urine was widely employed in the early days of bacteriology, it has now almost ceased to be used as a culture medium.

(*a*) 1. Boil some recently passed urine.

2. If the reaction be markedly alkaline after boiling, add a little tartaric acid solution, testing the reaction with litmus paper.

3. Filter, tube and sterilize at 115° C.

The composition of the urine is distinctly altered by this proceeding, the urea in solution being decomposed at the temperature of boiling water.

(*b*) It is better to sterilize by filtering through a Chamberland bougie (Chap. I.).

(*c*) To collect urine in a sterile manner, and so avoid the necessity for sterilization with the attendant alteration in composition, proceed as in Chap. XII. ("Urine").

The urine which has been collected in a flask may be tubed by any of the methods described for tubing serum (p. 45). Incubate the tubes at 37° C. for 48 hours, and reject any which are then cloudy.

Serum.

Serum is obtained by allowing blood to clot spontaneously or from the fluid of pleural effusions. It is used sometimes as a liquid but much more commonly as a solid medium after being coagulated by heat.

The technique for the collection of serum will be studied under the head of solid media (pp. 45 *et seq.*).

Blood.

Blood is frequently used as a culture medium.

To use it as a liquid medium coagulation must be prevented, and this may best be done by defibrinating the blood. The blood is collected aseptically (pp. 45 and 48 and Chap. XII.) in a sterile flask containing glass beads and shaken for about 10 minutes, then aspirated into a [Cobbett's bulb or] Chamberland flask (pp. 45 and 47) and tubed.

Among the many substances which it has been suggested might be added to blood to prevent coagulation, neutral sodium citrate and extract of leeches' heads may be mentioned.

By the sodium citrate method the blood is collected as it leaves the vein in a flask or tube containing a certain quantity of the following sterile solution: water, 1000 c.c.; sodium chloride, 8 grams; sodium citrate, 15 grams.

Extract of leech heads is obtained by placing the heads in 75 per cent. alcohol for 5 or 6 days. When hardened, the heads are dried and ground up in a mortar. The powder is dissolved in distilled water (100 c.c. per head), boiled, filtered and sterilized at 105° C. for 5 to 10 minutes. The extract is then introduced into the tubes in which the blood is to be collected.

These last two methods are not so good as defibrination.

2. Media made from vegetable tissues.

Vegetable infusions are but seldom used in practical bacteriology. The most important of them, however, may be mentioned.

Malt extract.

1. Grind up 100 grams of germinated barley (malt), and add 1000 grams of water.

2. Heat the mixture to 55°–58° C. for one hour : the starch is converted into maltose by the diastase and a true beer wort obtained ; the temperature must not exceed 58° C., otherwise the diastase will be destroyed.

3. Boil. Filter through Chardin paper.

4. Tube. Sterilize at 115° C.

Yeast extract.

Mix 100 grams of yeast with 1000 grams of water. Boil and filter through Chardin paper.

Tube the slightly acid filtrate or pour it into a flask, and sterilize at 115° C.

The filtrate may be neutralized or made slightly alkaline by the careful addition of normal soda solution before filtering. The addition of 5 per cent. cane sugar or glucose before filtration increases the nutritive value of the extract. If the extract is not clear when filtered, a little phosphoric acid [1] may be added, and the reaction brought back with lime water. Heat to 116°–117° C. for 5 minutes. Filter. Tube. Sterilize at 115° C.

Spronck's peptone yeast extract.

1. To 5 litres of water add 1000 grams of commercial yeast (not brewers' yeast).

2. Boil the mixture for 20 minutes, stirring frequently, pour into cylindrical vessels and leave for 24 hours.

3. Decant the cloudy liquid and add 5 grams of salt and 10 grams of Witte's peptone for each litre.

4. Neutralize exactly, and then make alkaline to the extent of 7 c.c. of normal soda per litre. Boil. Filter through Chardin paper.[2]

5. Pour into flasks. Sterilize at 115°–120° C.

Hay infusion. Straw infusion.

Macerate 15–20 grams of finely chopped hay or straw in 1000 grams of water for 1 or 2 hours. Boil for a few minutes, filter, tube and sterilize at 115° C.

The infusion which is sometimes a little acid may be neutralized in the ordinary way.

Potato infusion.

Clean and scrape a few potatoes ; add a litre of water to each 20–30 grams of pulp. Leave to stand for 3 or 4 hours. Decant. Boil the supernatant fluid. Filter. Tube. Sterilize. The infusion is often acid and can be neutralized before filtration.

Infusion of carrot is prepared in a similar manner.

Haricot decoction.

1. Macerate 50–60 grams of white haricot beans in a litre of water for several hours in the cold.

---

[1] Specific gravity 1·349, and containing 39·4 grams of anhydrous acid per cent.

[2] If the yeast contain meal the filtered liquid remains slightly cloudy, but this is of no consequence.

2. Boil for half an hour.

3. Pour on to a coarse sieve, collect the liquid, and add to it 1 per cent. salt, 2 per cent. ordinary sugar and a pinch of sodium bicarbonate. Boil. Filter through paper.

4. Tube. Sterilize at 115° C.

This medium is used by Mazé for the cultivation of the micro-organism found in the nodules of leguminous plants.

Decoction of dried fruits.

1. Macerate 50–100 grams of dried fruits (prunes or raisins) in a litre of water for several hours. Then stew them in the water.

2. Pass through a coarse sieve.

3. Boil. Filter.

4. Tube. Sterilize at 115° C.

The liquid is slightly acid and is useful for cultivating moulds. For other purposes neutralize with soda solution before boiling (Stage 3).

Wine.

Wine was much used by Pasteur in his early work, but is now hardly ever seen in the laboratory. Before sterilizing, neutralize or make slightly alkaline with soda solution in the ordinary way.

3. Synthetic media.

These media, though seldom used in everyday work, have been employed for the study of certain problems in the biology of micro-organisms.

The formulæ of the best known are given below. Some others will be described in connexion with the organisms in the study of which they have been employed.

Pasteur's medium.

Water,	100 grams.
Candied sugar,	10 ,,
Ammonium tartrate.	0·10 gram.
Ash of yeast,	0·075 ,,

Boil. Filter. Tube. Sterilize. The reaction is alkaline.

Raulin's medium.

Water,	1500 grams.
Candied sugar,	70 ,,
Tartaric acid,	4 ,,
Ammonium nitrate,	4 ,,
Ammonium phosphate,	0·6 gram.
Potassium carbonate,	0·6 ,,
Magnesium carbonate,	0·4 ,,
Ammonium sulphate,	0·25 ,,
Zinc sulphate,	0·07 ,,
Sulphate of iron,	0·07 ,,
Potassium silicate,	0·07 ,,

Prepare as in the case of Pasteur's medium.

The reaction is acid. This medium was used by Raulin in his well-known work on *Aspergillus niger*.

Cohn's medium.

Distilled water,	200 grams.
Ammonium tartrate,	2 ,,
Potassium phosphate,	1 gram.
Magnesium sulphate,	1 ,,
Tricalcium phosphate,	0·10 ,,

Prepare as in the case of Pasteur's medium. The reaction is alkaline.

Nægeli's medium.

Water, - - - - - - - - - - 1000 grams.
Ammonium tartrate, - - - - - - - 10 ,,
Potassium phosphate, - - - - - - 1 gram.
Magnesium sulphate, - - - - - - 0·2 ,,
Calcium chloride, - - - - - - - 0·12 ,,

Prepare as above.

Uschinsky's medium.

Distilled water, - - - - - - - 1000 c.c.
Glycerin, - - - - - - - - - 30 grams.
Sodium chloride, - - - - - - - 5 ,,
Calcium chloride, - - - - - - - 0·1 gram.
Magnesium sulphate, - - - - - - 0·2 ,,
Di-potassium phosphate, - - - - - 2 grams.
Ammonium lactate, - - - - - - 6 ,,
Potassium aspartate, - - - - - - 3 ,,

The method of preparation is the same as in the other cases. This medium was used by Uschinsky in his work on diphtheria toxin.

SECTION II.—SOLID MEDIA.

The introduction of solid media into practical bacteriology is due to Schrœter and especially to Koch. The commonest are transparent media, prepared by adding to broth substances capable of making it solid at ordinary temperatures ; but albumins coagulated by heat (serum, egg, etc.), meat and certain vegetable media are also used.

1. Gelatin media.

Gelatin media are in very general use, and several different sorts are prepared.

General rules.—1. Use extra quality French gelatin, which is sold in thin rectangular sheets weighing about 2·5 grams each. (*Ordinary commercial gelatin* loses its property of solidifying if heated above 102°–105° C. and sterilization must therefore be effected at 100° C.; this introduces an unnecessary complication into the preparation of the medium.)

2. Gelatin is very acid, and the medium must be neutralized after adding it to the other constituents, but the addition of alkali must be stopped at the neutral point or when the reaction is very slightly alkaline, because gelatin will not solidify after being heated in alkaline solution.

3. Ordinary gelatin media liquefy at 25° C., and can therefore only be used when the temperature of incubation is not to exceed 20°–23° C.

Ordinary gelatin.

This medium is generally known simply as **gelatin**.

Method recommended.—Proceed as in the preparation of broth.

1, 2 and 3. Macerate 500 grams of lean beef in a litre of water, heat, express the fluid, filter while hot and make up the volume to a litre.

4. To this broth add

Peptone (Chapoteaut), - - - - - 10 grams.
Salt, - - - - - - - - - 5 ,,
Sodium phosphate, - - - - - a pinch (not essential).
Extra quality gelatin, - - - - - 80–150 grams.

The amount of gelatin required varies according to the time of year : in winter 8 per cent. (80 grams per litre) is sufficient, but in summer as much as 10 to 15 per cent. is necessary—say 120 grams per litre.

Warm the mixture at a gentle heat in an enamelled saucepan, stirring constantly to prevent the gelatin sticking to the bottom. When the gelatin is dissolved, boil for two or three minutes.

5. The medium is now very acid ; add soda solution carefully, testing the reaction with litmus paper after each addition. The end reaction should be neutral or very slightly alkaline.

6. Autoclave for 5 minutes at 115° C. in a flask or enamelled vessel, to precipitate earthy phosphates.

7. On taking out of the autoclave pour the hot fluid on to a moistened Chardin paper fixed in a hot water funnel : the filtration must be done in the warm, otherwise the gelatin will solidify before it has filtered.

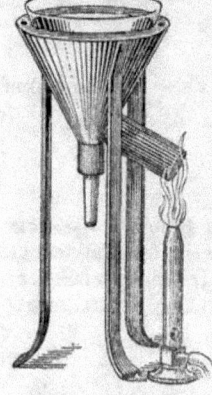

FIG. 32.—Hot water funnel.

A more simple method consists in filtering through a glass funnel fitted into a flat-bottomed flask, the flask and funnel being placed in the autoclave [or steamer], which is heated to 100° C. Filtration is quite easy under these conditions.

Hot water funnel.—This piece of apparatus consists of a copper funnel mounted on legs (fig. 32), and lined by a second —glass—funnel, the delivery tube of which, passing through the neck of the metal funnel, is made to fit it closely with an india-rubber bung. Pour water into the space between the two funnels through a small lateral opening provided for the purpose, and heat the apparatus by means of a Bunsen burner placed beneath a side tube projecting from the lower part of the metal funnel. The temperature should not be raised to boiling point, otherwise the water will be driven out through the inlet tube. Heat the funnel before pouring the gelatin on it. Several patterns of this apparatus are made; one useful form is that in which the metal funnel has two metal walls, and the glass funnel fits inside the inner wall.

8. Collect the filtered liquid in a flask and tube it at once—before it has had time to solidify—in quantities of 10–15 c.c. in each tube.

A small glass funnel should always be used for tubing to avoid soiling the mouth of the tubes, as has been already explained (p. 32).

The medium should be perfectly clear.

9. Plug the tubes : sterilize at 110° C. for 20 minutes, taking care that the temperature does not reach 115° C. [Sterilization may be effected equally by heating in the steamer to 100° C. for 20 minutes on each of three successive days.]

Notes.—(*a*) Gelatin prepared in this way is quite clear and absolutely transparent. If the liquid be slightly cloudy after filtering, add the white of an egg beaten up in 50 to 100 c.c. of water, mix thoroughly and autoclave for 5 minutes at 115° C. ; then after filtering through Chardin paper the mixture is perfectly clear. It is advised that this method be employed as seldom as possible as it is not without influence on the melting point of the gelatin.

(*b*) The low temperature at which gelatin melts (23°–25° C.) being a disadvantage in the use of the medium, bacteriologists have attempted to raise the melting point by modifying the method of preparation. Many of these modifications seem to be of no practical value : the author has never found any advantage in using carbonate of soda for neutralization as recommended by Bertarelli, and similarly Roux's method of sterilization at 100° C. on several successive occasions appears uselessly to complicate the preparation of the medium without offering any corresponding advantage.

By following the instructions given above, and provided that (1) a good quality of gelatin be used, (2) neutralization be stopped at the neutral point or when the liquid is very slightly alkaline, and (3) a temperature of 115° C. be never exceeded, a 10 per cent. gelatin is easily obtained which does not melt below 25° C. ; and by raising the gelatin content to 15 per cent. the medium will remain solid at summer temperature.

An important point is to cool the tubes rapidly on taking them out of the auto-clave, and to store them in a cold place as recommended by Abba.

Fischer's gelatin.

For the cultivation of phosphorescent bacteria Fischer recommends a gelatin very rich in sodium chloride. To a litre of meat extract prepared as above add

Peptone (Chapoteaut),	10 grams.
Salt,	30 ,,
Gelatin (extra quality),	80-120 ,,

Dissolve, and proceed as in the preparation of ordinary gelatin.

Liebig's gelatin.

Dissolve 5 grams of Liebig's extract [Lemco] in 1000 grams of water (add if necessary 10 grams of peptone, and 5 grams of salt), then add 100 grams of gelatin, dissolve, and boil for 2 or 3 minutes. Neutralize and complete the preparation as for ordinary gelatin.

Buchner's gelatin.

1. Dissolve with heat in a litre of water

Gelatin (extra quality),	100 grams.
Cane sugar,	20 ,,
Liebig's extract [Lemco],	5 ,,
Dry peptone,	5 ,,

2. To the solution add

Tricalcium phosphate,	5 grams.

3. Boil for a few minutes, heat to 115° C., filter and proceed in the ordinary way.

Raisin gelatin.

1. Make a decoction as described on p. 38, consisting of 250 grams of dried raisins in a litre of water.

2. Filter, then add 100 grams of gelatin and a pinch of sodium phosphate. Boil for 2 or 3 minutes. Neutralize and finish as usual.

Elsner's potato medium

1. Take 500 grams of potatoes, peel and grate them.

2. Macerate the pulp in a litre of water for 3 or 4 hours.

3. Strain. Stand overnight. Decant the fluid.

4. Make up the volume of fluid to a litre, and dissolve in it with gentle heat 15 to 20 per cent. (150 to 200 grams) of gelatin. Boil for a few minutes.

5. The fluid is now very acid. Add normal soda solution until the reaction is feebly but still *distinctly acid*.

6. Heat to 115° C. for 5 minutes. Filter and complete the preparation in the ordinary way.

Choquet's gelatin.

Choquet recommends the two following media for the cultivation of the micro-organisms concerned in dental caries.

1. Meat extract,	500 c.c.
Gelatin (extra quality white),	35 grams.
Peptone (Chapoteaut),	5 ,,
Calcium glycero-phosphate,	5 ,,

2. Meat extract,										500 c.c.
Peptone,										5 grams.
Gelatin,										35 ,,
Calcium phosphate.										50 ,,
Magnesium phosphate,										5 ,,
Calcium carbonate,										10 ,,

Dissolve the gelatin and peptone in the meat extract : heat in the auto-clave and filter in the ordinary way. After filtration and before distributing in tubes add the calcium and magnesium salts.

2. Agar media.

Agar-agar is derived from a sea-weed growing in the Indian Ocean, and in commerce occurs as dried fibrous strands.

When agar is boiled with water it forms a firm jelly which does not melt below 90° C. Agar is therefore substituted for gelatin whenever a solid medium is required for incubation above 25° C.

The preparation of agar media is tedious because agar readily forms with water a thick jelly which is difficult to filter. This difficulty is overcome by altering the properties of the agar by prolonged boiling or by chemical action, *e.g.* the addition of acid.

Another difficulty arises from the fact that agar is always cloudy if not cleared with albumen, and even then it is sometimes opalescent.

Ordinary Agar.

An agar medium prepared according to the method now to be described is generally spoken of as **agar**, and the word will be so used in this book.

Preparation of Agar.

[**A. Method recommended.—1.** Weigh out 30 grams of agar fibre, turn it into a 2-litre flask and fill the flask nearly full of tap water, then add 10 c.c. of a 2 per cent. solution of acetic acid and stir well with a glass rod.

[**2.** Leave the agar to soak for 10 minutes, then put a large funnel into the flask, stand the funnel under the cold water tap and wash the agar in running water until the washings are neutral to litmus paper (10 minutes).

[**3.** While preparing the agar, stand a flask containing a litre of broth in the steamer and heat to 100° C.

[**4.** Add the washed agar to the hot broth.

[**5.** Heat the mixture in the steamer at 100° C. until the agar is dissolved (20 minutes).

[**6.** The medium is now a little acid. Neutralize with a 10 per cent. solution of caustic soda (about 1 c.c.) and allow the contents of the flask to cool to 50°–60° C.

[**7.** Beat up the white of two eggs in a beaker, and add to the cooled agar. Mix thoroughly.

[**8.** Heat the mixture in the steamer at 100° C. until the egg-albumin is coagulated, and until on holding up the flask to the light the agar is clear ($\frac{3}{4}$–1 hour). At the same time put into the steamer a Chardin filter paper arranged in a funnel, the latter standing in a sterile flask.

[**9.** When the medium is clear—there will nearly always be lumps of coagu-lated albumin floating about in the agar—pour it on to the hot filter in the steamer. The filter must not be taken out of the steamer, and the medium should be poured down the sides of the filter paper.

[**10.** Place the lid on the steamer, and maintain the heat until the medium has all filtered through (15 minutes).

[**11.** Tube, and sterilize for 30 minutes at 100° C. on two successive days. Slope (see B. 9 *infra*).

[This method always gives a perfectly clear, transparent and very slightly opalescent medium. The yield is very approximately 100 per cent.—that is to say that from a litre of broth rather more than a litre of agar is obtained. No trouble is ever experienced in getting the agar to adhere to the walls of the tubes.

[If the autoclave be used the medium is generally of a brownish colour from over-heating, and it is sometimes difficult to get the medium to stand up in the tubes.]

B. Another method, which is also recommended, is as follows :

1. Prepare a peptone-beef-broth, according to the instructions given on p. 30, up to and including Stage 5.

2. To this broth add 20 grams (2 per cent.) of agar cut up into small pieces. The agar should be swollen by soaking in cold water for an hour or two, and then wrung out in a cloth before being added to the broth.

3. Heat the mixture to 100° C. in an enamelled saucepan, and keep it at this temperature until the agar is dissolved (about half an hour), stirring all the time.

4. Test the reaction, which should be neutral or faintly alkaline. If heated in presence of acid, agar becomes converted into sugar.

5. Cool to 55° or 60° C. and add the white of an egg beaten up in 100 grams of water. Mix thoroughly.

6. Autoclave for an hour at 120° C. The albumin is coagulated and carries down the impurities with it.

7. Pour the liquid while still hot on to a moistened Chardin paper arranged in a hot water funnel. Cover the funnel with a glass plate.

8. Collect the liquid as it filters in a previously sterilized flask, and tube at once. This must be done as quickly as possible as agar sets about 40° C., and a funnel as usual should be used in tubing it to prevent the medium soiling the mouths of the tubes. Each tube should contain 8 to 10 c.c.

9. Sterilize at 115° C. for 20 minutes. After sterilization and while the medium is still hot, slope the tubes on some such piece of apparatus as that pictured on p. 52, so that the agar solidifies with a sloped surface. Leave the tubes in this position for 36 hours.

Some bacteriologists recommend the addition of a small quantity of an aqueous solution of gum arabic to the agar, to prevent the thin upper part becoming detached from the wall of the tube when it is placed vertically. But gum arabic makes the medium distinctly cloudy, and does not appear to effect the purpose for which it is added.

If the method of preparation described be followed step by step, the agar will be found to adhere sufficiently well. Gelatin to the amount of 20 grams per litre may be added as recommended by Nicolle.

Modification.—Filtration may be accomplished in the following manner, even more easily than by the above method.

Before adding the agar to the broth (Stage 2) leave it to soak in 6 per cent. hydrochloric acid (water, 500 : HCl, 30) for 24 hours, and wash in a large quantity of water. Then soak in a 5 per cent. solution of ammonia (water, 500 : ammonia, 25) for some hours, wash in a large quantity of water, and squeeze the agar dry in a cloth. The agar is now ready to add to the broth, and the further steps are as described. The resulting jelly does not adhere well to the walls of the tubes, and the process cannot be recommended.

Karlinski's filter.—To facilitate filtration, Karlinski has devised an apparatus in which the agar is filtered under pressure. A water-jacketed copper vessel heated below by a ring Bunsen is fitted with a copper cylinder, the bottom of which shaped like a funnel and terminating in a delivery tube fitted with a tap is covered with a layer of absorbent wool. The agar is poured on to the wool, and the upper opening is hermetically sealed by means of a cover through which the tube of an india-rubber syringe passes. When the ball of the syringe is squeezed, the air in the apparatus above the agar is compressed and forces the agar through the wool. This ingenious piece of apparatus does not appear essential because if the agar be prepared in either of the ways described no difficulty will be experienced in filtering through Chardin paper.

Fischer's method.—Fischer proposes to overcome the difficulty of filtration as follows. Plug the narrow end of a funnel with an ordinary cork, and pour the agar into the funnel at once on taking it out of the autoclave. Allow to cool. The solid particles which cause the cloudiness settle to the bottom of the funnel. When the agar is set the jelly is turned out whole and the opaque conical part cut off with a knife. Cut up the remainder into small pieces and put into tubes. Plug and sterilize the tubes. The resulting agar is always opaque.

Malm's Agar.

Add 2 per cent. of agar to Liebig's or Cibils' broth (p. 34). Proceed as for ordinary agar.

Peptone-agar (Salomonsen).

1. Make a broth with

Water,	1000 grams.
Liebig's extract,	5 ,,
Peptone,	30 ,,
Cane sugar,	5 ,,

If necessary add a little alkali.

2. Dissolve 15 grams of agar in the broth, and proceed as above.

Glycerin-agar.

Add 2 per cent. of agar to glycerin broth (p. 35), and proceed in the ordinary way.

Glucose-glycerin-agar.

Prepare a glucose broth (p. 34), and after neutralization add 5 per cent. neutral glycerin and 2 per cent. agar. Complete the preparation in the usual manner.

Gelatin-agar.

By mixing agar and gelatin a medium is obtained the melting point of which lies between that of agar and that of gelatin. In warm climates, in the summer, agar-gelatin may be used in place of gelatin. But it must be borne in mind that the cultural characteristics of micro-organisms are far from being identical on the two media. Gelatin-agar is prepared as follows.

1. To 1000 grams of peptone-broth, add

Gelatin,	80 grams.
Agar,	5 ,,

or

Gelatin,	50 grams.
Agar,	8 ,,

Dissolve the gelatin in the broth, neutralize, and then add the agar.

2. Complete the preparation as in the case of ordinary agar, but at Stage 5 do not let the temperature exceed 115° C.

Iceland moss.

Some workers use Iceland moss (*Lichen crispus*) in place of agar, but this substitution cannot be recommended.

3. Media made from albuminous fluids and tissues.

Serum.

Serum is the liquid which separates when blood has clotted. In bacteriology bovine, sheep and horse serum are principally used. Serum is most frequently used after coagulation by heat, very rarely in the liquid condition.

An important point about serum media is that they should be almost transparent, hence they cannot be heated to a high temperature because they coagulate *en masse* and become opaque. Liquid serum ought not to be heated above 56° or 58° C., and to preserve its transparency solidified serum should be coagulated at about 70° C.

Serum cannot therefore be sterilized in the ordinary way.

Either (*a*) it must be sterilized by pasteurization combined with tyndallization (Koch's method) or by filtration through a bougie : or (*b*) since the blood in the body is sterile, a sterile medium can be obtained if care be taken to avoid introducing contaminations while collecting the blood and drawing off the serum (Roux and Nocard's method).

Collection of serum.

1. *In the slaughter-house.*

[**A. Method recommended.**—When the blood is collected in the slaughter-house, the following is a simple method of proceeding.

[**1.** When the carotid is severed discard the first spurt of blood, then take the plug out of a sterile 2-litre flask and hold it so that the blood pours into the open mouth : collect enough blood to three-parts fill the flask : replace the plug. Collect as many flasks of blood as are required.

[**2.** On reaching the laboratory place the flask on a cork ring, inclining it as much as possible. Take out the wool plug, burn the mouth of the flask with a Bunsen burner both inside and outside and plug the flask at once with clean wool ready sterilized and wrapped in paper. Then stand the flask vertically, shaking it as little as possible.

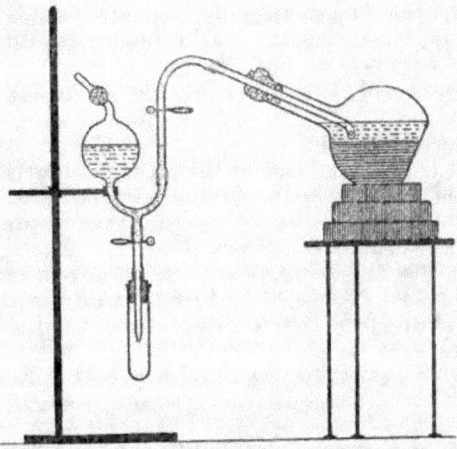

FIG. 33.—Cobbett's bulb as used for decanting serum.

[**3.** When the clot has formed and the serum separated, tilt the flask again —the clot should adhere to the bottom—and introduce a piece of glass tubing connected to a Cobbett's bulb (fig. 33).

[The glass tubing is a piece of ordinary-sized glass tubing bent near the end into an obtuse angle and sealed. About half a centimetre from the sealed end a small hole is blown into it through which the serum enters. This is connected with the Cobbett's bulb by a fairly long piece of india-rubber tubing.

[4. By aspirating through the plugged tubulure the serum can be drawn into the bulb. When the bulb is nearly full, clip the tubing between the flask and the bulb.

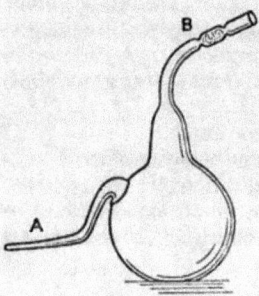

[5. Remove the test-tube enclosing the delivery-tube, and by releasing the clip above it draw off the serum into test-tubes.

[6. Coagulate the serum (p. 51).

[7. Incubate the tubes for 48 hours and reject those which show any growth.]

[For many purposes, e.g. the cultivation of the diphtheria bacillus, the serum may be further sterilized after coagulation by heating it to 100° C. in the steamer (Chap. I.).

[If the serum be wanted in the liquid form for future use, it is best to distribute it in small quantities into tubes, to heat it to 55°-60° C. in the water

FIG. 34.- Chamberland distributing flask.

bath for some time, and then to add a drop or two of chloroform to each tube with a sterile pipette. The chloroform is readily driven off subsequently by heating the tubes to 40-45° C.]

B. Koch's method. Apparatus required.—Prepare beforehand :

1. Three or four large covered glass dishes each consisting of two halves fitting one into the other, and capable of holding 2 litres.

Wrap the dishes in paper and sterilize them in the hot air sterilizer at 180° C. The temperature must be raised slowly to avoid cracking the glass.

2. Chamberland distributing flasks (fig. 34).

Wash and dry the flasks. carefully seal the pointed tubulure A in the flame, plug the other tubulure B with wool between the constrictions, and sterilize at 180° C.

3. Half-litre flasks with long necks (fig. 35). Plug and sterilize at 180° C.

4. Sterile plugged test-tubes.

Technique.—**1.** Collect the blood at the slaughter-house, preferably in cool weather, in the sterilized glass dishes. To collect the blood remove the dishes from their paper wrappings, and when the beast is being bled, after letting the blood which first issues flow away, raise the cover of one of them and collect enough blood to fill it three-parts full : then replace the cover. Several dishes should be filled in the same way.

2. Put the dishes containing the blood in a cool place, but not in the ice chest, because hæmolysis may occur and so impart a red colour to the serum.

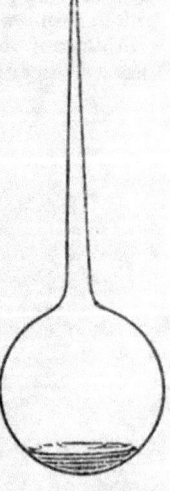

FIG. 35.—Flask with neck drawn out and sealed.

3. After about 36 hours the clot will have formed and shrunk leaving the serum as a clear fluid on top. Break off the fine point of a Chamberland flask (fig. 34), pass it through the flame of a spirit lamp and avoiding all sources of contamination as far as possible aspirate the serum into the flask. Seal the point in the flame.

Modification recommended.—A better yield of serum is obtained if instead of collecting the blood in glass dishes Latapie's apparatus be used (p. 50). For the present purpose the tube *a* (fig. 40) is replaced by a sterile funnel by means of which the blood is collected as it spurts from the severed vessel. When the bottle is half-filled, the funnel is taken out and an india-rubber plug put in its place. The further steps are described at p. 50.

The serum is sterilized in the following manner :

4. Take the Chamberland flasks containing the serum to the laboratory. In spite of the precautions taken the serum will be certain to be more or less contaminated, and must therefore be sterilized. Distribute it first into flasks with long necks thus : flame the mouth of the flask in a Bunsen burner and remove the cotton-wool plug : break off the capillary point of the tubulure on the Chamberland flask, pass the broken end through the flame, and introduce it some distance into the neck of the other flask ; then by blowing through the tube B the serum can be transferred to the other flask. Meanwhile the wool plug of this flask is held between the thumb and index finger of the left hand.

5. The flask being about three-parts filled, its wool plug is removed and the neck heated in the blow-pipe and sealed a few centimetres from the bulb. As many flasks are used as are necessary to contain the serum collected.

6. The flasks after filling and sealing are heated in a water bath, as already described (p. 12), to 56° or 58° C. for one hour on eight consecutive days.

7. When sterilized the serum has to be distributed. A mark is made on the neck of the flask with a glass cutter near the sealed end and to this scratch the end of a very hot glass rod is applied : this cracks the glass and the crack is extended by touching the end of it with the heated glass rod. The two ends of the fracture soon join, and with a gentle tap the end of the neck which was sealed in the flame can be easily separated and the flask opened.

Place the flask B on a cork ring E so that the neck is as nearly horizontal as possible.

Flame the pointed tubulure of a sterile Chamberland flask A in a Bunsen, break off the end with sterile forceps and insert the tube into the flask so as to almost touch the clot D, and aspirate the serum C. Discard the top layer of serum because having been in contact with the air, it may possibly be contaminated by dust (fig. 36).

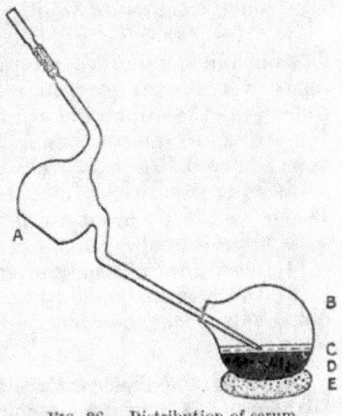

FIG. 36.—Distribution of serum.

8. By means of the Chamberland flask, distribute the serum in sterile tubes, passing the mouth of each tube rapidly through the flame before taking out the wool plug, and flaming also the pointed tubulure of the Chamberland. Pass the tapering end well into the test-tube, and pour about 10 c.c. into each. Replace the plug in the tube.

The serum is now ready either for coagulation or for use in the liquid state (in the latter case it should be first incubated for 48 hours at 30° C.).

If the serum is to be set this should be done as soon as possible, for should any organisms have gained access to the tubes during filling, the heat of coagulation will very probably destroy them.

Note.—The serum may be sterilized by filtration instead of by heat ; but filtration of serum is a long and tedious operation, and usually a troublesome one, as the serum froths considerably as it comes through the bougie. If filtration be decided upon, a Berkefeld bougie through which the serum passes much more readily than through the finer Chamberland F type should be used. The technique is described at pp. 18 *et seq.*

To facilitate the filtration of serum Miquel has devised an apparatus which works at 40° C. The serum is poured into a cylindrical vessel containing a filtering bougie. The cylinder with its contained bougie is placed in a double-walled vessel heated below by a gas burner on which a regulator is placed. The bougie is connected by means of india-rubber tubing to a conical flask with a tubulure attached to a water pump. Before use the flask is sterilized in the hot air sterilizer and the bougie and rubber connexions in the autoclave. The side tube of the flask should be plugged with wool.

2. *From a living animal.*

[A. Method recommended.—When a **living animal**—horse or bovine—is available, the technique will be as follows.

[**1.** Take three or four large (2–3-litre) sterile Jena flasks.

[**2.** Prepare a trocar—Sivori's pattern (fig. 38, p. 49) is very suitable—thus: to the side tube attach a fairly long piece of india-rubber tubing, and to the other end of the latter connect a piece of straight glass tubing long enough to reach from the mouth to the bottom of the flask. Boil the apparatus for an hour.

[**3.** Take the trocar out of the water with a pair of sterile forceps, pass it between the neck and the plug of one of the sterile flasks, and then wrap the plug well round it. Pass the glass tube well down into the flask in the same way.

[**4.** Cleanse the skin of the neck of the animal, and pass the trocar into the jugular vein as described at p. 49.

[**5.** When the flask is about two-thirds filled, pinch the rubber tubing, get an assistant to withdraw the glass tubing and pass it as in (**3**) into a second flask. Release the pressure on the tubing, and fill the second, and in the same way the third and fourth flasks. Care must be taken when withdrawing the trocar in the first instance, and the glass tube later, that the wool plug is so arranged that no air channel is left.

[**6.** Take the flasks of blood to the laboratory and stand them vertically. If any of the plugs be soiled with blood or do not fit well replace them with sterile wool (see **A.** 2 p. 45).

[**7.** When the clot has formed and the serum separated, proceed as in **A** p. 45.

[**8.** The serum should be sterile, but as a precautionary measure a little chloroform may be added to it or it may be heated to 55°–60° C. for an hour.]

B. Roux and Nocard's method. Recommended.—This method has the advantage of furnishing a much clearer serum and a medium more favourable

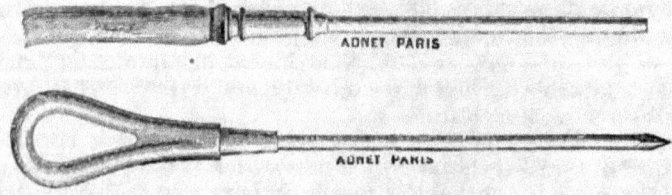

FIG. 37.—Nocard's trocar.

for growth than Koch's, and should therefore be adopted in preference to the latter whenever possible.

Instruments required.—**1.** A Nocard's trocar (fig. 37) on to the cannula of which, when the trocar is withdrawn, a metal adjustment can be fitted; this carries a piece of red rubber tubing about 50 cm. long, to which is attached a piece of glass tubing 15 cm. long bevelled at its free end.

The trocar and the rubber tube with its appendices are wrapped separately in filter paper and autoclaved.

Sivori's trocar (fig. 38) may with advantage be employed instead of Nocard's; it is provided with a lateral tube E to which the rubber tubing is attached. The trocar with tubing attached can be sterilized as a single piece of apparatus, and the blood collected directly, thus avoiding any risk of contamination.

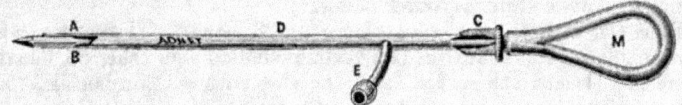

FIG. 38.—Sivori's trocar.

M, handle; C, shoulder into which the conical part AB fits hermetically, so that the blood flows up the cannula D, and passes out through E.

2. A pair of sterile scissors curved on the flat and a sterile bistoury.

3. One or two wide-mouthed bottles of 3 litres capacity.

The bottles are washed and dried, and the mouth of each covered with two or three folds of paper which is tied down round the neck with string; over this another similar but larger covering is fastened with string in a similar manner, so that it can be removed without interfering with the cover beneath (fig. 39). The bottles must be sterilized in the hot air sterilizer.

4. Some sterile Chamberland flasks and test-tubes.

Technique.—As a rule, blood is taken from an horse or an ass while the animal is standing. If necessary its eyes can be covered and the animal can be held with a twitch. If it is proposed to bleed a bovine animal, it will be best to throw it on a table such as is used for vaccination inoculations.[1] The animal, whichever species is used, should be fasting.

1. Proceed as for bleeding from the jugular vein. Sterilize the skin. Press the vein at the root of the neck to render it prominent, and make a small longitudinal incision through the skin with a bistoury along the line of the vessel on the distal side of the point of compression.

2. Introduce the trocar through the incision and push it through the sub-cutaneous tissues for a distance of about 2 cm., then pierce the vein and push the trocar into it in the direction of its long axis.

3. The cannula is now in place; withdraw the trocar, and attach the metal adjustment carrying the rubber tube. Meanwhile an assistant compresses the vein above to prevent blood entering the cannula. This must be done quickly.

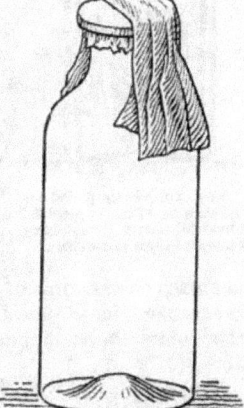

FIG. 39. – Bottle for collecting blood.

Part of the outer cover has been removed to show the one beneath.

4. The rubber tube being attached, pinch it firmly between the thumb and index finger of the left hand. The assistant releases the pressure on the vein above the cannula, but maintains the pressure on the cardiac side.

[1 In our experience it has never been necessary to throw a bovine animal: adults usually stand quite quietly.]

D

5. A second assistant hands the sterile bottle, loosens and removes the outer covering, and perforates the inner cover with the glass tube attached to the rubber tube. Now release the pressure on the rubber tubing, and blood will flow into the bottle.

When the first bottle is three-parts filled, stop the flow of blood by pinching the rubber tubing; the assistant then withdraws the glass tubing from the bottle and covers the mouth as quickly as possible with the paper cap and fastens it round the neck. Fill a second bottle in the same way.

5 to 6 litres of blood can be taken from an horse without harm, but 3 litres is sufficient to take from a young heifer.

6. Place the bottles in a cool place for 36 hours. The serum, which is transparent and of a beautiful pale yellow colour, will then be floating on the surface. Decant the serum from the clot with a Chamberland flask [or Cobbett's bulb], being careful not to contaminate it, and distribute it at once in sterile tubes as described above (pp. 45 and 47).

C. Latapie's apparatus. Recommended.—The technique of Roux and Nocard's method is simplified by using this apparatus. All contamination is avoided, and a yield of about 700 c.c. of serum per litre of blood is obtained instead of 400–450 c.c. by the ordinary method.

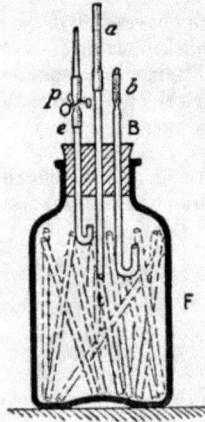

FIG. 40.—Latapie's apparatus, in which to collect blood for serum from large animals (horse or bovine).

Description.—The apparatus (fig. 40) consists of a wide-mouthed bottle F capable of holding several litres, and plugged with an india-rubber plug B perforated with three holes. A number of glass tubes (t) open at both ends and perforated with several holes laterally, is put into the bottle. Three pieces of glass tubing are passed through the india-rubber plug. Through one (a) the blood enters the bottle, a piece of rubber tubing connecting it with the cannula. The tube b is simply to allow access of air to the interior; it is plugged above with wool, while the other end extends some distance into the bottle and is bent in the form of an U. Lastly, the tube e serves for the collection of the serum : its lower end also bent in an U-shape reaches a few cm. below the plug, while its upper end is attached by means of a piece of rubber tubing to a piece of glass tube drawn out and sealed in the flame. A clip can be placed on the rubber connexion p to disconnect the two pieces of glass tubing. Finally, the apparatus is arranged in a special support (not pictured in the figure) which allows the bottle to be inclined so that the neck points upwards or downwards at will.[1]

Technique.—**1.** Sterilize the apparatus in the autoclave. Moisten the wool in the tube b, wrap the bottle in filter paper, and raise the temperature slowly. After sterilization allow to cool, and then lute the plug with paraffin.

2. Puncture the vein as in Roux and Nocard's method, and connect the cannula to the tube a. The bottle must not be more than half-filled and the blood must not reach to the level of the air tube b. The flow of blood is stopped by clipping the tube a. (The bottle is of course held with the neck up during this part of the operation.)

3. Leave for 12 hours or more until the blood has clotted.

[1] In Chapter XII. an apparatus, designed by the same observer, for the collection of blood from small animals will be described.

4. When the clot has formed and shrunk from the pieces of glass tubing, invert the bottle gently in its support so that the neck is the lowest part. The serum will then run down into the neck.

5. Clip the tubing at *p*, break off the pointed end of the glass tubing, and plunge it into the sterile flask in which the serum is to be collected : loosen the clip, and the serum will flow out. By means of the clip the rate of flow can be altered at will or entirely stopped.

The opening of *e* within the bottle being a little distance from the plug, a small quantity of serum containing red cells remains in the bottle ; the bend in the tube prevents these cells from being drawn off with the serum.

Coagulation of serum.

Serum is coagulated by heat. In order that it may retain its transparency, the temperature during coagulation must not exceed 68°–70° C., and to completely solidify the serum it must be kept at this temperature for 2 or 3 hours.

The tubes containing the liquid serum are sloped as in the case of agar. Coagulation is generally effected in a modified form of the apparatus devised by Koch [*e.g.* in an Hearson's serum coagulator.]

[**Hearson's serum coagulator.**—In construction and in some models in appearance Hearson's serum coagulator (fig. 41) is the same as Hearson's

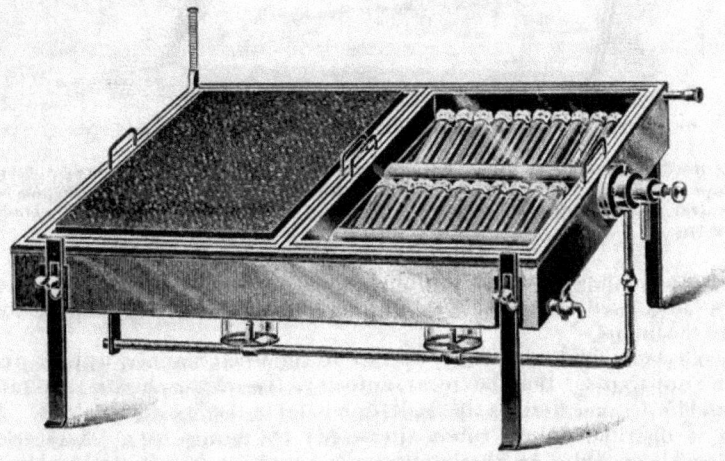

FIG. 41.—Hearson's apparatus for the coagulation of serum.

warm (37° C.) incubator (p. 61), but in the former the capsule is constructed to work at higher temperatures such as are suitable for the coagulation of serum. Special holders are also supplied which retain the tubes in a slanting position. To maintain a saturated atmosphere, dishes of water or wet cloths can be placed on the floor of the coagulator.]

Koch's apparatus consists of a double-walled rectangular copper box supported on legs, by means of which the angle which it subtends with the horizon can be altered at will. The space between the walls is filled with water, and the floor of the apparatus is covered with a thin layer of sand on which the tubes are laid. A thermometer is placed alongside the tubes. The apparatus is closed above with a moveable cover consisting of two sheets of glass mounted in a metal frame with a thin layer of air between them. The apparatus is heated by gas, which passes through a Roux's regulator immersed in the water between the two walls. The technique is as follows :

1. Lay the tubes, each containing about 10 c.c. of serum, in the sand. Incline the apparatus so that the serum does not touch the plugs.

2. Light the gas, and when the temperature as shown by the thermometer inside reaches 68° C., adjust the regulator (Chap. III.), so that the temperature remains constant.

3. The length of time required to solidify different samples of serum varies (from 2 to 3 hours). A tube must be taken out from time to time to ascertain the condition of the serum, which will be sufficiently set when holding the tube upright it retains its slope. [It is perhaps better to hold the tube by the upper end and tap the lower end firmly against the thumb nail: if the serum quiver, the heating has not been continued long enough.] Stop the heating at this stage. The serum when set should still be transparent and of an amber-yellow colour.

To obtain a more transparent product, Vagedes advises coagulating in an atmosphere of water vapour, and this can be effected by placing Petri dishes filled with water alongside the tubes, [or by placing folded cloths which have been wrung out in warm water over the tubes, taking care that they do not touch the wool plugs.]

4. Incubate at 30° [or 37°] C. for 36 hours, to ensure their sterility before using them as culture media.

If only a few tubes of serum have to be coagulated, Koch's apparatus can be dispensed with, and the tubes dealt with as follows. Arrange the tubes in a small flat copper tray ·(fig. 42), about 12 cm. wide, one side of which is notched to receive the upper plugged

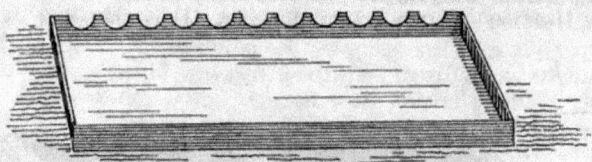

FIG 42.—Copper tray on which to slope tubes of culture media.

ends of the tubes, while the other ends resting against the opposite side keep the tubes in a sloped position. Cover the tray with a sheet of glass, stand it on a saucepan filled with water, and slowly heat the water to boiling. The tubes will be set in about an hour or two.

Other serum media.

In cases of simple pleurisy the fluid which can be drawn off often yields a very clear, easily coagulable serum, very suitable for the purposes of a culture medium.

To collect the fluid aseptically, operate in the usual manner, using a sterile Potain's apparatus. Boil the trocar, autoclave the rubber plug and aspirating tube at 115° C., and sterilize the flask [or bottle] in the hot air sterilizer. The serum is distributed into tubes afterwards by means of a Chamberland flask (see also p. 45). An absolutely sterile serum can frequently be obtained in this way, but it is nevertheless often necessary to tyndallize the fluid before coagulating it (p. 47).

Ascitic fluid as a rule yields only a poorly coagulable serum, which is not of great value when a solid serum is wanted.

Lœffler's serum.

1. Prepare a broth in the ordinary way, using

Water, - - - - - - - - - -	1000 grams.
Beef, - - - - - - - - -	500 ,,
Peptone, - - - - - - - -	20 ,,
Salt, - - - - - - - - -	5 ,,
Glucose, - - - - - - - -	10 ,,
Normal soda solution, - - - - - -	Q.S. to make slightly alkaline.

2. Aspirate 1 part of this broth and 3 parts of sterile liquid serum into a Chamberland flask.

[The quantities can, of course, be measured out with sterile pipettes into a sterile vessel.]

3. Tube. Coagulate at 70°–75° C.

Glycerin-serum.

An excellent culture medium for the tubercle bacillus is obtained by mixing 6 to 8 per cent. of pure glycerin with serum.

1. Aspirate 6 to 8 grams of pure glycerin, previously sterilized in the autoclave, into a Chamberland flask:

2. Then 100 c.c. of sterile liquid serum into the same flask. (To facilitate measurement the flask can be graduated beforehand.)

3. Tube and set at 75° C. This mixture requires a somewhat higher temperature than ordinary serum.

Serum-agar. Ascitic-agar.

1. Dissolve 1·5 grams of agar in 100 c.c. of water. Filter. Tube (about 5 c.c. in each tube). Sterilize at 120° C.

2. Cool to 40° C. To each tube add an equal volume of sterile serum or sterile ascitic fluid. Mix gently by rotating the tubes in the hands. Cool in the sloping position.

Blood-agar.[1]

(*Bezançon and Griffon.*)

1. Take a number of tubes of glycerin-agar, melt in a water bath, and cool to 40° C.

2. Add to each tube a small quantity (about 1 c.c.) of blood from a rabbit's artery (Chap. XII.). Mix without shaking the tubes, and cool in a sloping position.

A solution of hæmoglobin (p. 34) may be used instead of blood in this case.

Serum-agar.

(*Tochtermann.*)

1. Dissolve in 500 c.c. of boiling water

Peptone (Chapoteaut, or Witte),	5 grams.
Salt,	2·50 ,,
Glucose,	2·50 ,,
Chopped and washed agar,	10 ,,

2. Mix with the above solution 500 c.c. of sheep serum, and autoclave for 30 minutes at 115°–120° C.

3. Filter in the warm through moistened Chardin paper. Tube and sterilize at 115° C.

Egg.

Eggs can be used for the cultivation of micro-organisms in several ways.

A. Take a fresh egg, shake it vigorously to mix the white and the yolk : wash the shell in perchloride of mercury and dry with sterile filter paper. Flame the narrow end until the shell blackens. Make a hole with a sterile metal point. Pass a platinum wire or pipette charged with the material to be sown through the hole, then close the latter with a little melted Golaz's wax. It is well to coat the egg with a layer of collodion.

B. Take a fresh egg, flame the pointed end, make a hole as described above : aspirate the white into a sterile pipette. Tube in sterile tubes. Coagulate at 70° C. as in the case of serum.

[1] See also under Pfeiffer's bacillus and Gonococcus.

C. Boil an egg hard. Remove the shell, cut up the egg into pieces, and place in small Petri, or other covered glass, dishes. Sterilize the watch-glasses or dishes at 115° C.

D. Lubenau recommends yolk of egg media for growing tubercle and diphtheria bacilli. The procedure is as follows:

1. Prepare a neutral broth (p. 30) containing 1 per cent. of glucose (for the diphtheria bacillus) or 3 per cent. of glycerin (for the tubercle bacillus). Distribute in quantities of 100 c.c. in 2½-litre flasks. Sterilize.

2. Wash an egg with warm water and soap. Lay the egg in a Petri dish, pour a little alcohol over it, and set light to the alcohol.

Make a hole in the shell with a sterile instrument, and pour the yolk into one of the flasks of broth. Add the yolks of five eggs to each flask. Shake the flasks well.

3. Distribute the medium into tubes. Slope the tubes in a serum coagulator (p. 51). Heat for 2 or 3 hours at 90° C. on three successive days.

[**E. Dorset's egg medium.**—" The eggs are thoroughly cleansed with water of any adherent dirt, and then washed with 5 per cent. carbolic solution, and allowed to partially dry. The ends of the eggs are then gently dried in the flame, and pierced with a burned sharp forceps. The hole at one end should be about ⅜ in. in diameter, and the membrane broken; the other end which is to be blown into should be smaller, and the membrane left unbroken if possible. The eggs are then blown into a sterile Erlenmeyer flask, the blowing being done from the cheeks, which will help to avoid spilling saliva and leakage of air around the outside of the egg. To the egg is then added 10 per cent. of water by volume of the weight of the eggs. The mixing is done by a twirling motion of the flask or by gently stirring with a glass rod. Bubbling is to be sedulously avoided. The mixture is then strained through cheese-cloth by gravity and tubed. The tubes are then inspissated at 70° C. for 2–2¼ hours in a moist chamber " (Park and Krumweide).]

Meat.

Into a litre flask put 500–600 grams of finely chopped lean beef. Add sufficient normal soda solution to make the reaction neutral or slightly alkaline. Plug the flask with wool. Sterilize at 115° C.

Internal organs.

The placenta, liver, spleen, kidneys, etc., can be used as culture media. The organs must be removed with the usual aseptic precautions from healthy animals which have been recently killed.

The technique recommended by Guéniot for the preparation of placenta will serve as an example of the method of preparing these culture media.

1. Lay the placenta (if possible receive it) in a sterile basin with the uterine surface uppermost. Scorch this surface with a large heated metal plate.

2. Cut off a number of pieces with a sterile forceps and scalpel, and place them with the scorched surface downwards in sterile Petri dishes, or better in large sterile tubes.

Place the dishes and tubes in the incubator at 37° C. for a day or two to control the technique. Those that remain sterile (at least 60 per cent.) can then be used as culture media.

4. Vegetable media.

Potato.

A. Petri dish method.—1. Select a number of perfectly sound potatoes, scrub away the soil adhering to them in running water, then dry and peel them.

2. Cut them into slices about 10–15 mm. thick parallel to their long axes, and drop the slices into a dish of distilled water.

The slices should not be touched with the fingers, and it is best to use a silver blade as steel often turns potatoes black.

3. Dry the pieces between folds of white filter paper.

4. Then lay them in Petri dishes (fig. 43) or other suitable covered glass dish.

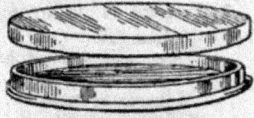

FIG. 43.—Petri dish.

5. Sterilize the potato in the dishes at 120° C. for 20 to 30 minutes.

Potatoes must be sterilized at 120° C., because a highly resistant organism (the potato bacillus), which is often present on the surface, may in slicing the potatoes be carried by the knife on to the cut surface.

B. Method recommended.—1. Wash and scrub the potatoes as above.

2. Cut the potatoes, not into slices, but into elongated parallelopipeds or semi-cylindrical pieces 4 to 5 cm. long, so that they can be put in special potato-tubes also known as Roux's tubes.

These tubes (fig. 44) are rather wider than ordinary culture-tubes and the potato rests on a constriction situated about the lower one-fourth; the bulb below collects the condensation water.

A special cutter may conveniently be used for slicing the potatoes, but the only advantage to be gained is that the pieces are more neatly and regularly cut. The slices should not be too long, otherwise they will curl when boiled.

3. Wash the pieces in distilled water : dry between blotting paper.

4. Put them into tubes. Plug with wool.

5. Sterilize as above.

Note.—Potatoes, though generally neutral in reaction, are sometimes strongly acid, in which case they are not suitable for the cultivation of bacteria. If it be necessary to use these acid potatoes, they must be soaked for some hours before being sterilized in a 0·5 per cent. solution of soda.

FIG. 44.— Potato-tube.

[**C. Glycerin-potato.—1.** After cutting the potatoes into suitably shaped pieces as above (**B**), soak them in a dilute (1–1000) solution of sodium carbonate for 24 hours.

[**2.** Transfer the pieces to a 5 per cent. solution of glycerin in water for a further 24 hours.

[**3.** Tube in ordinary test-tubes, which should have a pledget of wool at the bottom. Fill up the tubes with the 5 per cent. glycerin solution.

[**4.** Sterilize at 100° C. on three successive days.

[**5.** When required for use pour off nearly all the glycerin solution, and sow the surface of the medium.]

D. Potato mash.—1. Peel the potatoes, cut them into large pieces, and boil them in water.

2. Pass them through a sieve.

3. Distribute the mash in layers 1–2 cm. thick in Petri, or other covered glass, dishes.

4. Sterilize at 120° C. for 20 minutes.

Starch jelly.

To 180 grams of water add 10 grams of potato meal and 5 grams precipitated calcium carbonate. Distribute in Erlenmeyer flasks or Petri dishes and sterilize at 115° C. When the starch cools it forms an homogeneous whitish layer on the bottom of the vessel.

Heinemann's jelly.—Heinemann recommends the following jelly in place of potato. The artificial medium has the advantage of being of constant composition and reaction.

1. Prepare the following solution:

Asparagin,	5 grams.
Di-potassium phosphate,	2 ,,
Di-sodium phosphate,	2 ,,
Magnesium sulphate,	2 ,,
Calcium chloride,	2 ,,
Ammonium lactate,	2 ,,
Water,	200 ,,

2. Dissolve 15 grams of agar and 10 grams of peptone in the warm in 600 grams of water.
3. Mix the two solutions. Make neutral to phenol-phthalein. Filter.
4. After filtering, and while still hot, add 30 grams of starch made into an homogeneous suspension with a little water. Boil the mixture for several minutes.
5. Distribute in tubes. Sterilize at 120° C. for 5 minutes. Slope the tubes and allow them to cool.

Bread.

A. Soak some slices of white bread in distilled water, place them in covered glass dishes, and sterilize at 115° C. for 20 minutes.

B. 1. Crumble some bread and dry it in the air between sheets of filter paper.
2. When dry, grind it up in a coffee mill.
3. Put the powder in layers 1–2 cm. thick in Petri dishes or in Erlenmeyer flasks, and add sufficient distilled water to soak all the bread (about 2½ parts of water to 1 part of bread by weight).
4. Sterilize at 115° C. for 20 minutes.

Rice milk.

1. Mix intimately

Milk,	150 grams.
Peptone broth,	50 ,,
Powdered rice,	100 ,,

2. Distribute the mixture in Petri dishes in layers 1–2 cm. deep.
3. Heat to 115° C. for 20 minutes. The mixture solidifies and forms an opaque white layer.

5. Coloured media.

Coloured media are used for the recognition of particular micro-organisms which produce changes of colour in them. Only a few formulæ are given here, the majority being reserved for description later (*vide* the typhoid bacillus, the colon bacillus, etc.).

Media tinted with blue litmus [or neutral-red] and containing a carbohydrate are the most generally used : organisms which ferment carbohydrates with formation of acid, when grown in a litmus medium change the colour of the litmus from blue to red [and in a neutral red-medium produce a bright red colour].

Preparation of litmus solution.—Granulated litmus is ground up, 85 per cent. alcohol is poured on to the powder, and the mixture boiled. 6 to 8 parts of water are added to the residue, and the resulting liquid mixture is heated and filtered through paper. The filtrate is kept in a flask plugged with wool. To one-half of this liquid sulphuric acid is added until the colour is nearly red, and the other half then added to it ; a sensitive indicator is thus obtained. The solution is distributed in tubes which are plugged and sterilized at 115° C.

Litmus-lactose-gelatin.

1. Prepare and sterilize a number of tubes of gelatin in the ordinary way (p. 39), adding (at stage 4) 2 to 4 per cent. of lactose.

2. Prepare a number of tubes of sterilized litmus solution.

3. Just before use, melt the lactose-gelatin in a water bath and to each tube add with a sterile pipette sufficient sterile litmus solution to impart a distinctly blue colour to the medium.

Never sterilize a medium after adding litmus : the subsequent heating is liable to discharge the blue colour.

Litmus-glucose-gelatin, litmus-mannite-gelatin, etc., and various **litmus-agars** are prepared in a similar manner.

Barsiekow's medium.

Prepare separately the two following solutions and sterilize them :

A. Sodium chloride, - - - - - - 0·5 gram.
   Nutrose, - - - - - - - 1 ,,
   Water, - - - - - - - - 75 grams.
B. The carbohydrate (lactose, mannite, etc.), - - 1 gram.
   Water, - - - - - - - - 25 grams.
   Litmus solution, - - - - - - Q.S. to give an amethyst tint to the solution.

After cooling, mix the two solutions and distribute in tubes.

Litmus milk.

Add a sufficient quantity of sterile litmus solution to sterile milk. [It is highly important that the reaction of the milk should be neutral. Milk bought in shops is often acid, in which case a sufficiency of sodium carbonate must be added to neutralize the medium.]

Nœggerath's medium.

1. Mix in the following proportions saturated aqueous solutions of the dyes mentioned :

Methyl-blue, - - - - - - - - 2 c.c.
Gentian-violet, - - - - - - - 4 ,,
Methyl-green, - - - - - - - 1 ,,
Chrysoidine, - - - - - - - - 4 ,,
Fuchsin, - - - - - - - - 3 ,,

2. Add 200 c.c. distilled water.

3. The mixture now has a neutral greyish-blue tint : leave it to stand for a fortnight, then if the colour has altered bring it back to the neutral tint by the addition of any colour that is required. Sterilize at 100° C.

4. Immediately before use add 7 to 10 drops of the sterile mixture to a tube of agar or gelatin previously melted in the water bath.

In place of the above mixture, Gasser prefers to add 20 drops of a saturated aqueous solution of fuchsin to each tube of melted agar.

These media were recommended by their authors for the diagnosis of the typhoid bacillus, but have now fallen into disuse (*vide* the typhoid bacillus).

CHAPTER III.

INCUBATORS.

Introduction.

AN incubator is a piece of apparatus designed to maintain cultures of organisms constantly at any temperature which may be best suited to their growth.

The **shape** of the incubator, provided it be adapted to the size of the tubes, flasks, etc., which it will have to receive, is generally speaking of little consequence, though on the whole the rectangular form is the most convenient because the space can be most completely utilized.

Any metal box which had a door and could be heated by a convenient source of heat could in an emergency be made to serve as an incubator. A rectangular tin or copper box for instance, raised on a suitable stand, and heated by a [Bunsen burner or] small oil lamp, placed below it at such a distance as to keep the temperature within the box at the level required, would do quite well. But with such a rough and ready piece of apparatus the difficulty would be to keep the temperature constant, for quite apart from the question of the control of the heat supply the temperature would be influenced by the temperature of the outside air. And in practice these difficulties are so considerable that it has been found necessary to design special forms of apparatus in which the temperature can be kept more fully under control. These of course are more complicated and more expensive than the simple arrangement just referred to, but are nevertheless indispensable if satisfactory results are to be obtained.

There are two essential points for which provision must be made in the construction of an incubator.

Firstly, the instrument must be protected as far as possible from variations in the atmospheric temperature, and from loss of heat by radiation and convection.

Secondly, it must have some form of automatically acting regulator, which will readily respond to variations of temperature.

The former condition is satisfied by surrounding the outer surface with some non-conductor of heat, *e.g.* wood or felt or a water jacket; or, since polished metal surfaces radiate heat very feebly, a veneer of brightly polished copper will serve the purpose equally well.

To get the most satisfactory results, the temperature throughout the incubator must be as uniform as possible. If the incubator were an hermetically sealed box, there would be very marked differences of temperature at different levels, and the larger the incubator the more noticeable would this be. [In very large incubators, such as incubator rooms, it is even necessary to have thermometers on each shelf and at different places on each shelf, because the differences of temperature in different parts of the room are so considerable.] Hence to maintain as uniform a temperature as possible in an incubator of the ordinary size the shelves are perforated with holes to allow of free circulation of the air, and in some forms ventilation holes are provided in the floor and roof. [With incubators surrounded by a water jacket however this is not necessary.]

SECTION I.—DEVICES FOR AUTOMATICALLY REGULATING THE TEMPERATURE OF INCUBATORS.

Various ingenious pieces of mechanism have been devised for the purpose of automatically regulating the temperature of incubators. Some of these regulators are intended to be used when coal gas is the source of heat, others are constructed for use with paraffin oil, etc., but the former are the most satisfactory, and are the only ones that can be recommended.[1] Those that are in most general use are described below.

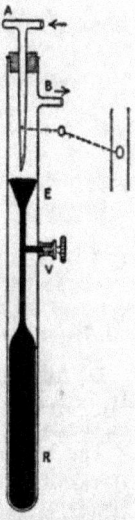

A. Electric regulators.—Of these Babès' may be mentioned as a type. They are very complicated, uncertain in action, and have no advantage over the following.

B. Mercury regulators.—To explain the principle of the mercury regulators, Chancel's may be described. The gas enters through the glass tube A (fig. 45), which terminates within the regulator in an oblique opening through which the gas issues and passes to the burner through the side tube B. When the regulator is placed within the incubator, the mercury contained in the lower part R expands as the temperature rises, so that in time it obliterates the oblique opening of the tube A, and consequently diminishes the volume of gas passing to the burner (in the vertical limb of A there is a small safety opening O which permits of a very small supply of gas to the burner, just sufficient to prevent the light going out altogether when the opening below is completely obstructed by the mercury).

FIG. 45.—Mercury regulator.

When the temperature within the incubator drops, the level of the mercury falls and the supply of gas to the burner is increased. The regulator is controlled by the screw V: when this is turned clockwise the level of the mercury E stands at a higher level for any given temperature, while by turning it contra-clockwise, the volume of mercury in the tube is diminished. This regulator is cheap but not very sensitive, and is only correct within about 3° C.; an improved form of it has been devised by Arloing.

C. Ether regulators.—Rohrbeck's (fig. 46) may be taken as an example of this form of regulator, the working of which depends upon alterations in the vapour tension of ether at different temperatures.

[1] To ensure as constant a temperature as possible with gas, a pressure-regulator should be affixed to the main, so that the gas always reaches the incubator at a constant pressure.

Entering the smaller tube A (fig. 46), the gas passes by way of its lower obliquely cut end through B to the burner. The outer tube is divided towards its lower part by means of a funnel-shaped glass partition E into an upper and lower part. The lower part R is filled below with mercury, and above with ether vapour. When the surrounding temperature rises the pressure of ether vapour increases, and the mercury rises in the funnel and gradually more or less occludes the lower opening of A, thus cutting off the supply of gas to the burner. A pilot opening O is provided, so that the gas shall not be completely extinguished. The apparatus is regulated by raising or lowering the tube A. The regulator is sensitive but fragile.

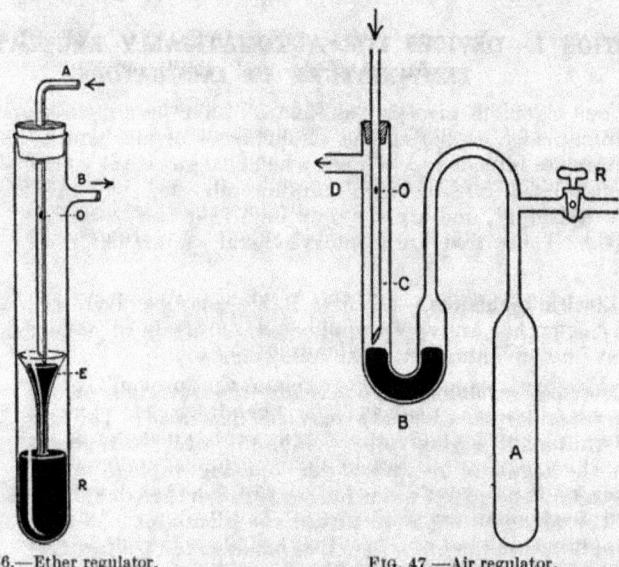

FIG. 46.—Ether regulator.   FIG. 47.—Air regulator.

D. Air regulators.—Of these, Bohr's may be taken as an example (fig. 47). In principle they are similar to the ether regulators, air replacing the ether in the latter.

The regulator is fitted to the incubator with the tap R open, and the reservoir A full of air. When the temperature of the incubator has reached the temperature required, the tap R is closed. If now the temperature of the incubator be raised the air expands, presses on the mercury contained in the U-tube B and forces it upwards thus partially or completely occluding the oblique opening C of the gas supply tube, and so cutting off more or less completely the gas passing to the burner through D. The gas delivery tube is provided with a safety opening O as usual. The regulator is sensitive, but being affected by alterations in the atmospheric pressure requires a certain amount of supervision.

E. Roux's metal regulator.—This regulator is composed entirely of metal (fig. 48). It consists of a strip of zinc and a strip of steel soldered together, and then bent in the form of an U. The metal with the greater co-efficient of expansion, zinc, being on the outside, it follows that any increase of temperature will cause the open ends of the U to approach each other; and conversely if the temperature of the metals be lowered the gap between the limbs is widened.

The left limb of the U is fixed while the right limb R is free, and therefore any change of shape resulting from a rise or fall of temperature in the incubator is integrated on the free limb R, and transmitted by means of a rigid horizontal bar T to a piston placed outside the incubator which controls the supply of gas.

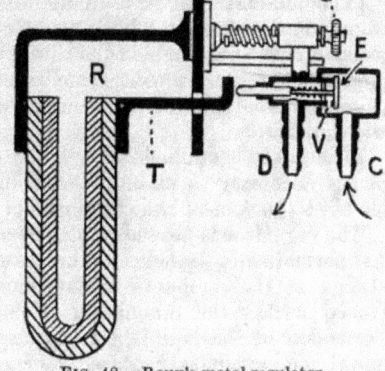

The tube C being connected to a gas tap, the gas must pass under E to reach the chamber to which the tube D leading to the burner is connected.

As the temperature in the incubator rises, the free limb R is drawn towards the other limb taking the rigid bar T with it; the piston, controlled by a spring, closes (this is shown in the figure), only leaving a small safety hole or by-pass V by which the gas can pass to the burners, and accordingly the temperature is lowered. When the temperature is too low, the changes

FIG. 48.—Roux's metal regulator.

described are reversed; the bar is pressed upon by the right limb of the U, and this in turn forces the piston E out so that more gas passes to the burners and the flames are larger. After a few oscillations, the temperature in the incubator will become constant.

By altering the length of the rigid bar T by means of a screw, the temperature can be raised or lowered as required; or as in the pattern shown in the figure, which gives more delicate control, the length of the rigid bar is fixed while the length of the piston can be altered by means of screws.

In some cases the zinc and iron bars are straight, and the apparatus assumes the form of a metal tube. This modification is useful when the regulator has to be immersed in water, as for instance in a water bath or a water-jacketed incubator.

Roux's regulator can be utilized for controlling the temperature of gas stoves used for heating incubating rooms and stoves in laboratories.

FIG. 49.—Hearson's "warm" incubator arranged to work with gas.

SECTION II.—INCUBATORS HEATED BY COAL GAS.

A. Hearson's incubators (fig. 49).—[These are most satisfactory incubators, and are almost if not quite the only ones used in this country. There are two forms, a " hot incubator " for temperatures of about 37° C. and a " cool incubator " for 20° C. or thereabouts. Incubators on the same principle can however be obtained to work at any temperature above 16° C.

[1. **The " warm " incubator.**—This, when once set, will work perfectly for months together without adjustment of any part and without any attention beyond the occasional addition of a little water to replace the small amount lost by evaporation. In actual practice it is found that the temperature can be maintained within half-a-degree centigrade in spite of great changes of gas pressure and of air temperature in the room in which the incubator is working.

[The incubator is rectangular in shape, and consists of a chamber surrounded on five sides by a stout copper water jacket enclosed in an outer wooden case with panels of uralite, the space between the case and the water jacket being packed with some non-conducting material. The sixth side is closed by a double door, the inner of glass, the outer of wood also panelled with uralite.

[The heat is supplied below by an ordinary fish-tail gas burner, the supply of gas necessary to maintain the temperature being controlled by a capsule let in to the roof of the incubator.]

The regulator is based upon the same principle as that of Rohrbeck (p. 59). An hermetically sealed metal capsule containing a few drops of a liquid boiling at the temperature at which the apparatus is required to work is placed within the incubator. When the liquid in the capsule boils, the expansion of the liquid lifts the upper part of the capsule, and so raises a metal rod which actuates a lever controlling the supply of gas. A small safety tube prevents the gas being extinguished when the main supply is cut off by the lever.

[*Technique.*—As full instructions are attached to every incubator, it is unnecessary here to describe the details.]

[2. **Hearson's "cool" incubator.**—The incubating chamber is similar in construction to the "hot" incubator, but on the top is a metal box of the same size as the incubator, surrounded by a thick layer of non-conducting material or wood, and with a large hole in the top through which ice can be introduced. The temperature is controlled by a capsule similar to that used for the hot incubator, but designed to work at about 20° C.

FIG. 50.—Hearson's "cool" incubator.

[The source of heat is a small bath of water placed at the side but near the top of the incubator, and heated by a Bunsen burner. The capsule controls a small pipe connected with a supply of water in such a way that when the temperature is that for which the capsule is set, the water runs to waste; when the temperature has fallen below that required the pipe conveys the water to the little bath, and hot water runs into the incubator jacket, displacing some of the cooler water. When the temperature is too high the cold water runs directly into the water jacket, and if this be insufficient to reduce the temperature then ice must be put in the box. In this country it is very rarely, if ever, that ice is required.

[The only difficulty likely to be experienced with this incubator is in connexion with its water supply. If the supply be taken directly from the main, the water may be cut off or the pressure for one reason or another be so reduced from time to time that the automatic regulation breaks down : so that for satisfactory working it is desirable to have a tank from which a supply of water under constant pressure may be derived.]

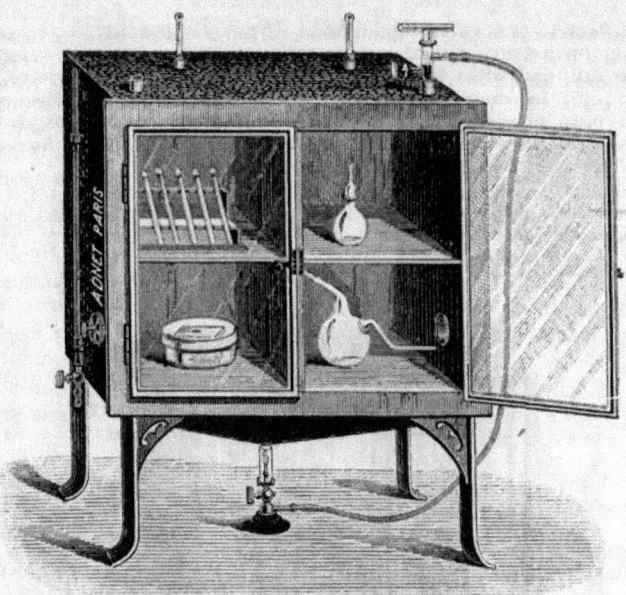

FIG. 51.—Babès' incubator, fitted with a Chancel regulator.

B. Babès' incubator.—One of the simplest forms of incubator is that known as Babès'. It consists of a metal box covered with a layer of felt, and heated by a gas flame the height of which is controlled by one or other of the many forms of regulators devised for the purpose (p. 59).

C. D'Arsonval's incubator.—The control of the temperature in d'Arsonval's incubator depends upon the changes of shape which an elastic lamina undergoes under the influence of changes of pressure.

The incubator is surrounded by a water jacket (2, fig. 52), of which the outer wall is closed below by a flexible sheet of steel, (3, fig. 52), which also forms the roof of a chamber, 10, into which a tube, 12, connected with the gas supply passes. From this chamber the gas reaches the burners through two tubes, 13 and 13′. The end of the tube, 12, can be made to approach or recede from the steel lamina, 3, by means of a screw. When the tube touches the steel the gas is completely cut off, and conversely, when they are separated the gas can pass freely to the burners.

The space between the two walls is hermetically sealed everywhere, except above where an opening, 5, is left for the purpose of filling the space with water. Suppose it be required to regulate the apparatus for a temperature of 37° C. The tube, 12, is screwed down away from the steel lamina, 3, so that the gas burns with a full flame. When the thermometer in the incubator registers 36° C., the tube is raised towards the lamina so that the

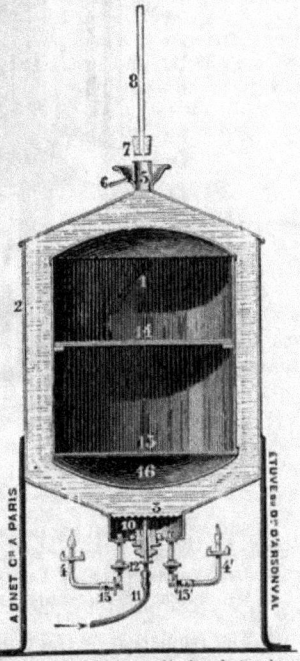

FIG. 52.—D'Arsonval's incubator in section.

size of the flames is somewhat diminished. If now the opening, 5, be hermetically plugged, any further increase of temperature causes the water to expand within the double wall, and so to force the lamina downwards, with the result that the gas is cut off. In practice, instead of completely sealing the opening a plug carrying a piece of glass tubing is inserted, and as the water expands it rises in this tube and at the same time the pressure on the bottom of the wall of the incubator increases, and the lamina is pressed down.

Note.—Recently-boiled water should be used for re-filling the incubator: if tap water be used the bubbles of air which are driven off when the water is heated will alter the water level, and the mechanism will be disturbed.

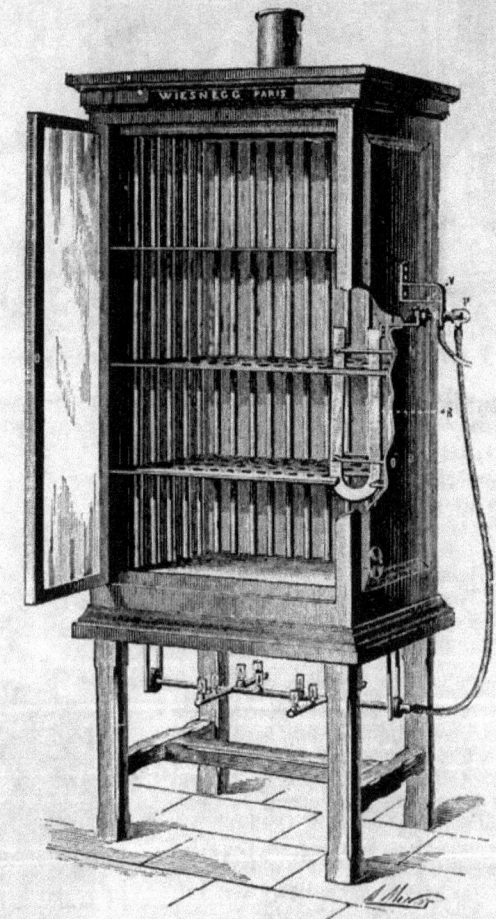

FIG. 53.—Roux's incubator.

In another form the regulator is placed in the side of the incubator, and a rubber membrane which is more sensitive is substituted for the steel lamina.

D'Arsonval's incubator has several disadvantages.

1. By reason of its shape only a small part of its total capacity is available for use.

2. The regulation of the temperature being a function of the level of the water in the tube, 8, it follows that the temperature within the incubator rises as the height of the water falls. Since this may happen as a result of evaporation, leaking

from joints and other causes, it is apparent that the apparatus requires careful supervision.

3. In time the elasticity of the lamina diminishes, with the result that the temperature is not controlled.

D. Roux's incubator.—Roux's incubator meets all practical requirements, and has none of the disadvantages of Babès' and d'Arsonval's, and for these reasons is preferable to those instruments.

The incubator consists of a rectangular wooden box, closed in front by a single or double glass door raised on feet and heated below by a gas burner. The inside walls are lined by a series of vertical copper tubes. The air in the incubator is maintained at a constant temperature by radiation from these tubes, which are heated by the gases from the burners below passing up through them. Ventilation is provided through holes in the floor and roof.

For details of the regulator, see p. 60.

Method of use.—**1.** Before using the incubator, it is well to paste black paper over the doors to shield the cultures from light, which may have an adverse influence on their growth.

2. Lay a thermometer on each shelf to watch the rise of temperature. When the incubator is finally regulated, each shelf has an absolutely constant temperature, which differs slightly from that of the shelf above and below.

3. The tube C (fig. 48, p. 61) is connected to a gas tap, and the tube D to the burners beneath the incubator. Adjust the screw S controlling the piston E, and turn it until the latter is widely open.

4. Light the gas.

5. When the temperature on the middle shelf registers half-a-degree below the temperature required (36·5° C. if a temperature of 37° C. is required), turn the screw S until the piston E is closed, and the burner is fed only by the by-pass V. But when the temperature in the incubator falls, the limbs of the metal U separate, the rod T presses on the piston rod, with the result that the piston E is opened, and a larger volume of gas reaches the burner. The apparatus is now regulated, and the temperature will remain constant without further supervision. If the gas be turned off at the main and then relit, the temperature will be regulated at the height at which it stood when the gas was turned off. A little vaseline must be applied to the piston chamber from time to time to lubricate the piston rod.

SECTION III.—INCUBATORS HEATED BY ELECTRICITY.

When a laboratory has electric power laid on, it may be convenient to use incubators such as those of d'Arsonval, Regaud, Fouilliaud [or Hearson] which can be heated with electricity.

A. D'Arsonval's electrical incubator is similar in appearance to Roux's incubator. It is fitted below with a drawer in which the special form of lamp used for heating the incubator is placed. A metallic regulator is interposed in the circuit: as the temperature rises it causes the expansion of a metal rod, and this breaks the circuit and cuts off the current. When on the other hand the temperature falls the bar returns to its normal position in contact with a platinum point, and the circuit is re-established. To regulate the incubator a screw is turned until the current passes. When a thermometer placed in the incubator registers a few degrees below the temperature required, the screw is slightly reversed. The temperature is noted again in about half-an-hour's time, and after a few trials the regulation is quite perfect.

B. Hearson's electrical incubators.—Hearson's regulator (p. 62) can be applied to incubators heated by electricity. The circuit is broken and the current cut off when the lever is raised by the expansion of the fluid in the capsule.

SECTION IV.—INCUBATORS HEATED BY PETROL, GASOLINE, OR PETROLEUM OIL.

When neither coal gas nor electricity are available as sources of heat, it is difficult to regulate the temperature of an incubator satisfactorily.

In these cases, incubators heated by petrol (Lion's, d'Arsonval-Adnet's, Hearson's) or by gas manufactured on the spot from gasoline (Roux's, Hearson's), [or by petroleum oil (Hearson's)] should be used, but they all require a good deal of supervision.

CHAPTER IV.

THE METHODS OF SOWING AND CULTIVATING AËROBIC ORGANISMS.

AËROBIC organisms should be grown in vessels which while allowing free access of air at the same time protect them from dust.

Various types of culture vessels are in use : for instance test-tubes, circular flat-bottomed flasks, flasks of various other shapes, Petri dishes, Soyka dishes, etc.

Cotton-wool is generally used as a protection from dust. More rarely paper caps and occasionally glass covers are also employed.

In sowing a culture, the following rules must be observed :

1. *The instrument used must be sterile.*

2. *The material to be sown must be collected without introducing extraneous organisms :*

3. *And must be transferred uncontaminated to the medium it is proposed to sow.*

SECTION I.—INSTRUMENTS USED FOR SOWING CULTURES.

Cultures can be sown with a *Pasteur pipette, platinum wire,* or *glass needle.*

A. Pasteur pipettes.

This consists of a piece of glass tubing drawn out in the blow-pipe, and sealed at the pointed end, the other end being plugged with cotton-wool. The pipette should be about 20 to 25 cm. long.

A number of these pipettes ready sterilized should always be at hand.

To make a Pasteur pipette.—**1.** Take a piece of glass tubing of 5 to 7 mm. calibre (the size of a lead pencil), and with a file mark it off into lengths of 25 cm. or thereabouts. (Glass tubing is sold in lengths of about 1 metre, and each length should therefore cut up into four pieces.)

2. Break the tubing by holding it in both hands and pressing the thumbs against the glass, one on each side of a file mark.

3. Round off the cut ends in the blow-pipe.

4. Plug the two ends of each piece with cotton-wool, which should pass some distance into the tubing and should also project a few millimetres from the open end (fig. 54). This is conveniently done by gently pressing the wool in with some blunt-pointed instrument (the thin end of a three-cornered file will do very well).

5. Hold the middle of the tube in a slightly inclined position and heat it in a medium-sized blow-pipe flame, turning it round and round between the thumb and index fingers until the glass is soft. Withdraw the tube from the flame and draw it out quickly into a fine tube about 30 cm. long (fig. 54, A). Divide it into two in the middle by melting it in the tip of the flame. This will give two pipettes with the capillary end of each sealed.

A certain amount of skill is required to make these pipettes. Care should be taken that the tube is drawn out straight, and this can best be done by the operator resting his elbows on the table. The tube should always be taken out of the flame before attempting to draw it, and it should be held horizontally while being drawn. The tube should not be drawn too fine, otherwise it will be too fragile for use.

6. Place the pipettes with their plugged ends downwards in a wire basket

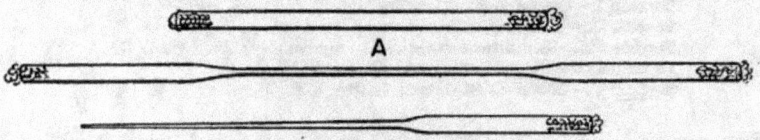

A

FIG. 54—The method of making pipettes.

[or copper cylinder, fig. 1], and sterilize them at 180° C. in the hot air sterilizer. They are then ready for use.

Method of using a Pasteur pipette.—**1.** Break off the fine sealed end of the pipette with a pair of dissecting forceps or between the thumb nail and the pulp of the index finger [it is better to make a light scratch with a carburundum pencil before breaking off the point].

2. Pass the broken end through the flame of a Bunsen burner or spirit lamp to destroy any organisms which may happen to have been deposited on the surface.

3. Dip the sterile end into the fluid which is to be used for sowing the medium. The fluid will rise in the tube by capillarity, or it can be aspirated by slightly withdrawing the wool in the other end of the tube. [A convenient practice consists in slipping an india-rubber teat over the wool-plugged end. By pressing on the teat, air is expelled: if the capillary end be now dipped in the fluid and the pressure on the teat lightly relaxed, as much or as little of the fluid as is required can be drawn up into the tube.]

In doing this, care must be taken that the aspirated liquid does not soil the wool plug; and it is necessary also to watch that bubbles of air are not drawn in.

4. Transfer the fine end of the pipette as rapidly as possible to the medium to be sown, and let one or more drops of the fluid fall on to the medium, either by its own weight or by blowing gently through the other end [or by compressing the teat.]

5. The aspirated fluid can be kept free from contamination for an indefinite time by sealing the end of the pipette. Thus, tilt the pipette gently so that the fluid runs up the tube; heat the point in a small flame (the pilot flame of a Bunsen), and when the glass is soft, draw it out with a pair of forceps, and the tube is completely closed.

B. Platinum wires.

Platinum is to be preferred to all other metallic wires because it does not oxidize after being heated to redness. The wire must be suitably mounted in a glass or metal handle, since on account of its high conductivity it cannot otherwise be held in the fingers.

A platinum wire (German, *öse*) so mounted meets all requirements. It is convenient to have three sizes of wire, stout, medium and fine : each will serve a special purpose (fig. 55). The fine wire is the most generally useful, because it cools more quickly than the stouter wires, and this is an important consideration in the successful sowing of cultures. At the same time it has very little rigidity and is easily bent, so that it cannot be used for instance, to sow cultures which adhere firmly to solid media, nor for sowing a rough-surfaced medium such as potato.

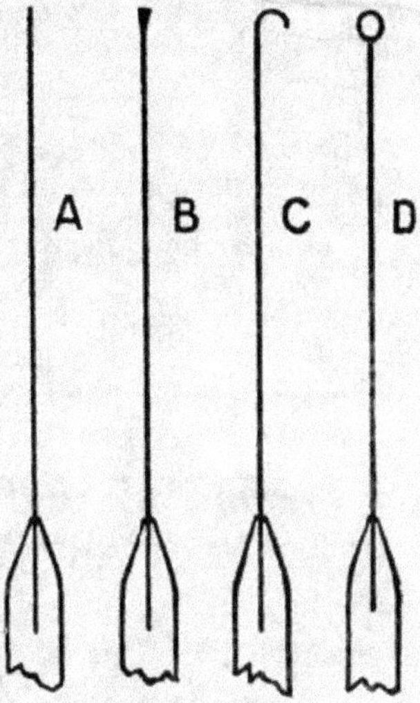

In the laboratory there should always be at hand :

A fine straight wire for sowing stab cultures (fig. 55, A).

A stout wire whose point is flattened in the form of a spatula (fig. 55, B).

A medium-sized wire, which can be bent to any desired angle near its end (fig. 55, C).

A fine wire bent into a loop at the end for picking up a drop of fluid (fig. 55, D).

FIG. 55.—Platinum wires.

Method of mounting platinum wires.—1. Take a glass rod 5–7 mm. in diameter, and divide it into lengths of 20–25 cm. by making a mark with a file and then breaking the rod at this mark between the thumbs.

2. Cut the requisite number of lengths of platinum wire with a pair of strong scissors, making each 5–7 cm. long.

3. Take one of the pieces of glass rod in the left hand, soften one end in the blow-pipe, rotating it between the fingers meanwhile. With forceps in the right hand pick up one of the pieces of platinum wire, holding it about 15 mm. from one end, and heat this end to a white heat in the flame.

4. When the heated end of the glass rod is softened, push the hot end of the platinum wire into it, so that a centimetre or more is embedded in the rod. Heat for a few moments, [pull out slightly] and then allow to cool.

5. Round off the rough edge of the other end of the glass rod in the flame.

6. Then with a pair of dissecting forceps bend the projecting end of the wire into a loop or at a right angle, or flatten it with a hammer, as the case may be.

Technique.—1. Hold the glass rod by its upper one-third, and pass the other end to which the platinum wire is fused rapidly through the Bunsen to destroy any organisms which may be present on the surface.

This sterilization of the glass rod should be done rapidly, because the smooth surface of the glass is quickly sterilized and moreover does not come into immediate contact with the culture : if the glass be overheated there is a risk that it may crack at the point where the wire is fused into it.

2. Heat the wire to redness, and on taking it out of the flame let it cool for a few seconds in the air.

The wire must not be exposed to the air any longer than is necessary for it to cool otherwise it may be contaminated by dust, and it is because it cools more quickly that a fine wire is most generally used.

3. Pick up the material to be sown on the needle, and transfer it to the medium to be inoculated.

4. When the culture is sown, heat the platinum wire to redness to destroy any organisms which may still be present on it.

This is particularly important when dealing with pathogenic micro-organisms. If the needle be not sterilized immediately, it will soil the bench and anything else it may touch.

C. Glass needles.

Draw out a piece of glass rod in the same way as the tubing was drawn out when making Pasteur pipettes. Then with a glass-cutter, [file, or piece of carburundum pencil] divide the fine part squarely in the middle. In this way needles of any degree of fineness can be made.

These needles are not so easy to handle as a platinum wire, but have the advantage of being rigid. They are useful for sowing deep stab cultures in gelatin.

Flame the needles immediately before using them.

SECTION II.—THE METHODS OF SOWING CULTURES.

A culture medium may be sown from another culture, or with water, dust, blood or other material. The method of collecting material differs of course according to the source whence it is derived—and this will be dealt with later (Chap. XII.)—but the technique of sowing cultures is not affected by these variations. Assume for the moment that a sub-culture is to be sown from an already existing culture, and take as an example a broth culture of the anthrax bacillus.

The process may be divided into three stages.

(i) The opening of the tube from which the culture material is to be taken.

(ii) The removal of the material.

(iii) The sowing of the new medium. Here several alternatives present themselves. It may be required to sow—

(a) Broth, or other liquid medium.

(b) Stroke cultures on agar, gelatin, serum or potato.

(c) Gelatin stab cultures.

(It is sometimes required to sow single colonies—for isolating organisms in pure culture: this will be dealt with separately in a later chapter.)

These various problems will now be considered *seriatim*.

A. Method of sowing a liquid medium.—Broth may be taken as a type of liquid media.

1. Take a tube of sterile broth and the tube containing the organism. Flame the plugs of both tubes to burn off the dust which has collected on them. Loosen the plugs by screwing them round with the thumb and index finger of the right hand, at the same time slightly withdrawing them.

2. Place both tubes side by side in the left hand, holding them as nearly horizontally as possible. The bottom of the tubes should rest in the hollow of the hand, their upper parts being grasped between the thumb, index and middle fingers.

3. Take a platinum loop between the index and middle finger of the right hand, and sterilize it as already described.

4. While the needle is cooling, take the plug, which has already been loosened, out of the tube containing the organism with the thumb and index finger of the right hand. Hold the plug between the thumb and first finger.

5. Introduce the platinum wire into the tube, being careful not to let it touch the sides of the mouth. Take up a drop of the culture fluid in the loop, and withdraw the latter from the tube (fig. 56). Flame the mouth of the

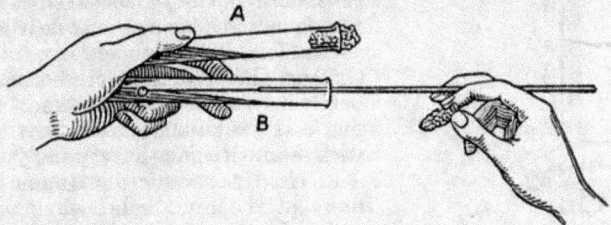

FIG. 56.—Method of sowing a liquid medium.

tube at once to destroy any organisms which may have settled on it during the process, and replace the plug as quickly as possible.

6. Take the plug out of the other tube, dip the loop into the broth, and after withdrawing it flame the mouth of the tube and replace the plug as before.

7. Before laying the platinum loop on the bench, heat it to redness to destroy any organisms which may still be adhering to it (in this particular case the anthrax bacillus, an organism pathogenic to man).

8. See that the tubes have been securely plugged.

Write on the tube which has been sown the nature of the organism and the date when it was sown.

It is often convenient to cover the wool plug with a small cap. This is readily done by twisting over the top a small strip of paper, which has been rolled round its upper part. The details of the culture can be written on this (fig. 57): paper caps, however, are liable to be interchanged, so that it is at least a wise precaution to label the tube as well. The cap has the advantage that it protects the wool from all liability to contamination.

Notes.—It is of the utmost importance that culture-tubes which have to be opened should be held in an oblique, nearly horizontal position so that dust may not fall into them.

No time should be wasted during the sowing of cultures, in order to minimize the chances of them becoming contaminated.

Wool plugs must never be laid on the bench. The part of the plug which goes into the tube ought to be prevented from touching anything.

The handle of the platinum needle should never touch the medium.

B. Method of sowing stroke surface cultures.—As an example of this, the sowing of a sloped agar tube may be described.

1. Proceed as under **A**, substituting a tube of sterile agar for the sterile broth.

2, 3, 4, 5. As under **A**.

6. Remove the plug from the agar tube, place the end of the wire on the lowest part of the surface of the agar, and draw it in a straight or zig-zag line over the medium.

7, 8. As under **A**.

Note.—In sowing potato the technique is the same as above, but more pressure must be used in drawing the needle over the surface and a medium or stout wire is desirable.

FIG. 57.— Culture-tube protected with paper cap.

C. The sowing of stab cultures.—The method of sowing a stab culture in gelatin will be described.

1. Proceed as under **A**, substituting a tube of sterile gelatin for the broth tube.

2. Hold the two tubes in the left hand thus : place the culture from which the material is to be taken in the hollow of the hand, supporting it between the thumb and first finger, and keeping it as nearly horizontal as possible. Place the gelatin tube between the first and second fingers, so that it is held firmly between the dorsal surface of the index finger and the palmar surface of the second, with the mouth pointing vertically downwards.

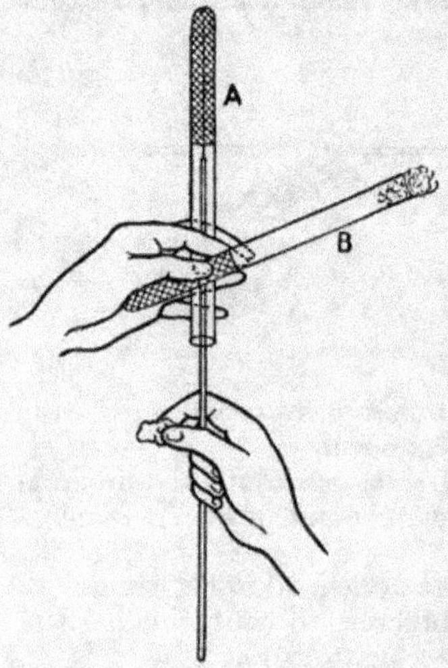

3. Hold the *straight* platinum wire in the palm of the right hand, leaving the thumb and index finger free. Sterilize the needle.

4 and 5. As in **A**.

6. With the thumb and first finger of the right hand remove the plug from the gelatin tube. Hold the already charged platinum wire vertically below the mouth of the tube, pass it into the tube until it touches the surface of the gelatin, let the gelatin tube fall by its own weight on to the wire until the latter touches the bottom, then withdraw it sharply (fig. 58).

7, 8, 9. Complete the operation as in **A**.

FIG. 58.—Method of sowing a stab culture.

Notes.—It is difficult by *forcing* the needle into the gelatin to get a straight stab which reaches to the bottom without touching the sides. A satisfactory stab will be more easily secured by allowing the gelatin to impale itself on the needle. The tube must therefore be held vertically, and not obliquely.

Gelatin which has been made some time is often cracked : in that case stand the tube in a water bath until the medium is liquefied, then let it set, and the gelatin will be found to be quite homogeneous again.

SECTION III.—CONDITIONS ESSENTIAL TO SATISFACTORY GROWTH.

To ensure growth taking place after the medium has been sown, the following conditions must be fulfilled :

(*a*) The cultures must be freely exposed to the air but at the same time be protected from dust. This condition is readily satisfied by the use of the ordinary wool plug, paper cap, etc.

(*b*) The temperature must be kept constant.

(*c*) As far as possible light must be excluded.

The two latter conditions are met by keeping the cultures in an incubator (Chap. III.).

Some micro-organisms grow only at temperatures above 30° C.,—generally 37° or 38° C.,—while others only grow well at temperatures below 30° C. Gelatin cultures of course must not be exposed to a temperature above 20°–22° C.

In the laboratory it is useful to have three incubators :

1. One in which the temperature is maintained at 20°–22° C. (the cool or gelatin incubator).

2. A second in which the temperature is 37°–38° C. (the warm incubator).

3. A third in which the temperature can be altered to meet special cases. Such an incubator is required sometimes for cultures which need a temperature above 38° C. (39°–41° C.), and at other times for growing organisms at temperatures between 20° and 37° C.

(d) The medium must be suitable to the needs of the organism to be grown. All organisms cannot be grown indifferently on any medium; for while some require a medium rich in albuminoid matter, others prefer sugars, glycerin, etc., and others again will not grow on serum, or potato, and so on.

In a later part of the book, when discussing individual organisms, mention will be made of the media most suitable for the growth of each species.

SECTION IV.—THE EXAMINATION OF CULTURES.

Cultures should be examined daily or even two or three times a day, and the character of the growth, which is of great importance in determining the species to which an organism belongs, noted.

Attention should be particularly directed to the following points :

A. In the case of micro-organisms growing in artificial cultivation, no matter what the medium, note :

1. (a) *The optimum temperature of growth,* (b) *the limits of temperature within which growth takes place.*

2. *The time when growth first makes its appearance.*

B. When cultures are growing in liquid media, note :

1. *The mode of growth.* Growth may produce :

(a) A distinct cloudiness of the medium, which may be either an uniform cloudiness or a cloudiness with a watered silk appearance, or sometimes a cloudiness with a surface pellicle. In these different cases flocculent deposits may ultimately form, and if so their occurrence should be noted.

(b) No distinct turbidity of the medium. Under these conditions the growth may show : (a) a surface pellicle, which may be either thin or thick, fatty or wrinkled ; (β) a ring of growth round the wall of the tube at the surface of the liquid ; (γ) flocculent deposits in the liquid, which may subsequently precipitate ; (δ) fine granular deposits, which in some cases adhere to the walls of the tube and in others fall to the bottom of the medium.

2. *The colour of the growth.*

3. *The production of any smell during growth.*

4. *The development of any new substances* in the medium (toxins, indol, acid, ammonia compounds, etc.).

5. The presence or absence of *clot* when grown *in milk.*

C. In the case of stroke cultures :

(i) **On agar, potato or serum,** note :

1. *The mode of growth.*

(a) The growth may remain limited to the line of sowing, and in this case it should be further noted whether (a) the culture takes the form of a delicate homogeneous and transparent streak, or occurs as discrete colonies ; or (β) whether the streak be thick, and if so if it be moist, greasy, viscous, dry or wrinkled.

(b) The growth on the other hand may spread widely over the surface of the medium, and the nature of the growth, that is to say whether it be moist, greasy, viscous, dry or wrinkled, is to be noted.

2. *The colour of the growth.* Whether the line of growth or the surrounding medium is pigmented.

3. *The production of any odour.*

(ii) **On gelatin,** note :

1, 2, 3. The *form, colour* and *smell* of the growth as in the preceding cases.

4. Whether the organism *liquefies the medium,* and if so, the time at which liquefaction begins.

D. In the case of stab cultures in gelatin, note :

1. *The mode of growth.*

The following appearances are observed with different species :

(*a*) A straight line along the line of sowing (fig. 59).

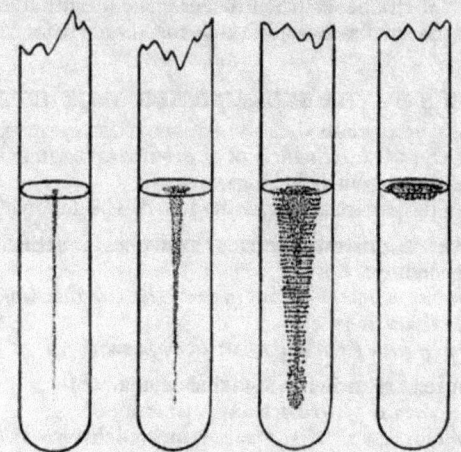

FIG. 59.—Rectilineal growth.
FIG. 60.—Tylotate growth.
FIG. 61.—Arborescent growth.
FIG. 62.—Surface growth.
Stab cultures in gelatin without liquefaction.

(*b*) Growth in the form of a nail, which may be abundant or scanty, and more or less marked at the head of the nail (fig. 60).

(*c*) An arborescent, ramifying growth (fig. 61).

(*d*) A growth strictly limited to the surface (fig. 62).

2. Whether *liquefaction* occurs ; if so, note :

(*a*) The time when it is first observed.

(*b*) The form which the liquefaction takes, whether it be cylindrical, funnel-

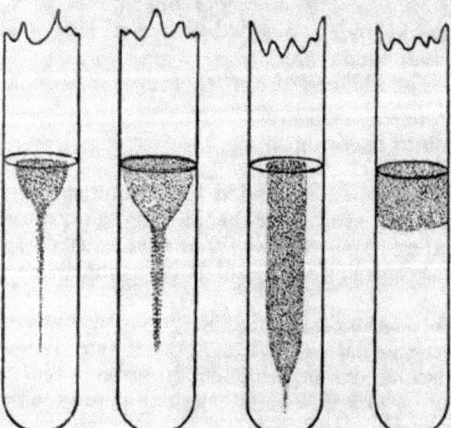

FIG. 63.—Cup-shaped liquefaction.
FIG. 64.—Funnel-shaped liquefaction.
FIG. 65.—Glove finger liquefaction.
FIG. 66.—Cylindrical liquefaction.
Gelatin stab cultures with liquefaction.

shaped, in the form of a glove finger or of a small cup (figs. 63 to 66). Note if an *air bubble* appears at the top of the growth.

3. The *colour of the growth*, and whether the growth itself or the medium around it is pigmented.

4. The *smell* of the culture.

SECTION V.—THE METHODS OF STORING CULTURES.

When growth has ceased, the organisms retain their vitality for a certain length of time varying according to the species from a few days to several months and even years, but ultimately they die and sub-cultures sown from them remain sterile.

The weakening and ultimate disappearance of vitality are in a large measure the result of the prolonged action of the oxygen of the atmosphere on organisms in an old culture medium, and which are not actively multiplying; to keep organisms alive therefore it is necessary to sow them from time to time on a new medium. But the same result is obtained by removing the organism, once growth has finished, from the action of the air; this may be done as follows :

1. Sow a broth culture, incubate it at the optimum temperature of growth until no further development of the organism takes place (the time required will obviously vary with different organisms).

2. Take a Pasteur pipette, and make a constriction just below the wool plug by heating it in the flame and drawing it out a little (*a*, fig. 67).

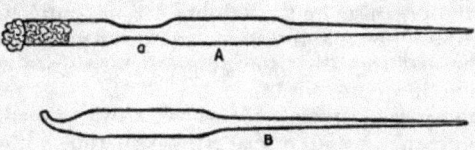

FIG. 67.—Method of sealing up a culture in a pipette.

3. When the pipette has cooled, dip the narrow end with the usual precautions into the culture and suck up the broth until it reaches the constriction *a*.

4. Seal the pipette both at *a* and at the other end as quickly as possible in the blow-pipe.

In this way a small tube is obtained, which is filled with the growth and sealed at both ends. This should be put away in the dark.

[In many cases when the tubes or flasks containing the culture are plugged with wool, it will be quite sufficient to pour melted paraffin wax over the wool and the lips of the opening in order to preserve the organisms alive for an indefinite period.]

CHAPTER V

THE ISOLATION OF AËROBIC MICRO-ORGANISMS IN PURE CULTURE.

BEFORE a study of the morphology and biology of any micro-organism can be undertaken the organism must be obtained in pure culture, a culture, that is, free from all other organisms or as they are technically called *contaminations*. The first step, therefore, in a bacteriological investigation will be the preparation of a pure culture.

It is obviously impossible in view of their exceedingly small size to pick out individual micro-organisms and transfer them to tubes of culture media, so that resort has to be had to more complicated methods. There are numerous processes in everyday use for the isolation of organisms in pure culture ; for convenience of description these may be divided into two groups according as to whether in attempting to isolate an organism a purely mechanical method depending upon *dilution* and *dissemination* is relied upon, or whether advantage is taken of the *biological properties* of the organism.

The former, the **mechanical methods**, will be more useful when every species of organism present in a given material has to be isolated while the latter, the **biological methods**, are more especially applicable when a particular organism of which the chief characteristics are known beforehand has to be isolated from material in which it is suspected to be present.

Above all in attempting to isolate micro-organisms it is of the first importance to distinguish between *aërobic* and *anaërobic* species, for according as to whether the one or the other has to be isolated so the cultures will have to be grown in the presence or absence of air. In the case of anaërobic organisms the methods of isolation will be dealt with later (Chap. VI.). The present chapter is devoted entirely to the methods available for the isolation of aërobic micro-organisms.

SECTION I.—MECHANICAL METHODS.

1. Isolation by dilution.

This method was originally devised by Lister and extensively adopted by Nægeli and by Miquel, but is now of very limited application. It gives very exact results but occupies much time and is exceedingly tedious.

Suppose it be required to isolate the organisms present in a drop of water. Add the water to a tube A containing 10 c.c. of sterile broth. Thoroughly mix the water with the broth by shaking the tube. The organisms present in the drop of water are now diluted in 10 c.c. of broth, and since 1 c.c. corresponds to 20 drops, each drop of broth contains 20×10, i.e. 200 times fewer organisms than the drop of water under investigation. Now transfer one drop of the mixture from the tube A to each of a series of tubes (B, B', B'', ...) containing broth. If the original drop of water contained 200 organisms, every drop of fluid in tube A will contain $\frac{200}{200} = 1$ organism, so that every drop transferred from A to the series B, B', B'', etc., will carry one organism, and that organism will grow in the tube B, B', or B'' to which it has been transferred and will give rise to a pure culture. But if the original drop of water contained only 50 organisms, then only one tube in four of the series B, B', B'', etc. will give rise to a pure culture. On the other hand, suppose the drop of water contained a larger number than 200 organisms, it will then be necessary to dilute further until in fact one drop contains not more than one organism. Thus 10 drops from A will be transferred to a broth tube B, and a series of sub-cultures C, C', C'', etc., will be sown, each with one drop of broth from B.

2. Isolation by dissemination.

The method of isolation by dissemination is due to Koch.

For its application the use of solid media is necessary. It may be carried out in one of two ways : either the medium may be liquefied and then sown, or the organisms may be distributed directly over the surface of the medium.

1. Dissemination in liquefied solid media.

If it be required to isolate all the organisms present in a drop of water, the method would be as follows : Transfer the drop of water to a tube of sterile gelatin previously liquefied in the water bath, and mix the water and gelatin thoroughly by rolling the tube between the hands. The organisms present in the water will now be distributed through the gelatin. Pour the gelatin in a thin layer on to a sterile glass plate, and cool it rapidly. The organisms will be scattered and held in the layer of gelatin like the almonds in a piece of nougat. If the plate be kept at a suitable temperature, each organism will grow in an isolated position, and will give rise to a colony composed of a number of micro-organisms all derived from the one organism which originally settled in that position, and therefore to a pure culture (fig. 68). It will then be easy to pick out each colony separately and sow it on a new medium.

There are in practice several ways of carrying this out, but the following rules must always be observed :

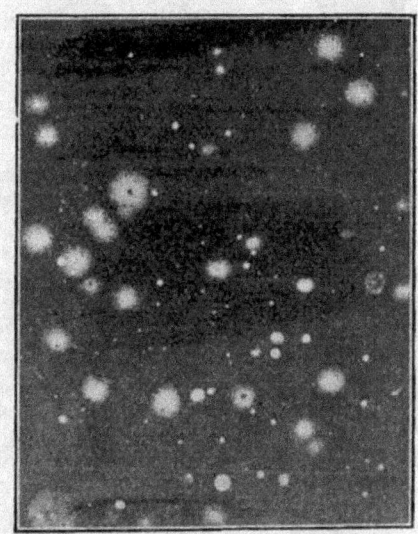

FIG. 68.—Isolated colonies of micro-organisms on a gelatin plate (two-thirds natural size).

1. *After liquefying the gelatin or agar and before sowing it, let it cool sufficiently (to $30°$–$40°$ C.), to prevent the organisms being killed by the temperature of the medium.*

2. *Avoid contaminating the culture.*

3. *Protect the plates from dust.*

A. Petri dishes.   Method recommended.

(i) Using gelatin as the culture medium.

Apparatus required.—(*a*) Three Petri dishes (fig. 43) wrapped in filter paper [or packed in a copper cylinder (Chap. I.)] and sterilized in the hot air sterilizer (a number of these dishes should always be kept ready sterilized).

(*b*) Three sterile Pasteur pipettes.

(*c*) Three tubes of sterile gelatin.

Technique.—**1.** Melt the gelatin by standing the tubes in the water bath.

Gelatin should never be liquefied by holding it in a gas flame, because the air dissolved in the gelatin will appear as bubbles in the substance of the medium and will interfere with the subsequent examination of the plates.

2. Take up a drop of the liquid to be examined in a Pasteur pipette and add it to one of the gelatin tubes (dilution 1), taking the necessary precautions to prevent contamination. Replace the wool plug and mix thoroughly by rolling the tube between the hands.

Never shake tubes of media as is done in chemical investigations, because it gives rise to frothing and this is highly inconvenient.

3. With another pipette transfer three drops from the first tube to another tube of gelatin (dilution 2).   Mix as before.

4. Transfer three drops of dilution 2 to the third gelatin tube (dilution 3).

The three tubes of gelatin will now contain each a different number of organisms. and, according as to whether the original material contained many or few organisms, dilution 3 or dilution 1 will give the best results.   Thus if the number of organisms be large, the colonies will be confluent in the plate poured with tube 1 and isolation will be impracticable; in that case dilutions 2 and 3 will be available.

5. Take out a Petri dish from its envelope.   Take the plug out of the first gelatin tube, and flame the mouth.   Then lift the cover of the Petri dish, pour the gelatin into it and cover the dish again as quickly as possible.

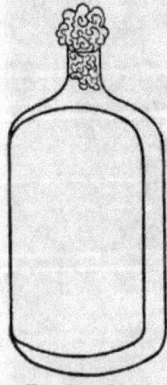

Spread the gelatin in an uniform layer over the surface of the dish by tilting it backwards and forwards, put it on a cold and level surface and allow it to set.   Then label it and put it in the cool incubator (20° C.).

6. Pour plates with the gelatin in tubes 2 and 3 in the same way.

7. Examine the plates every day, and without lifting the cover note the appearance of the colonies and their characteristics (both with the naked eye and with the aid of a lens).   Remove a portion of each colony for the purpose of making sub-cultures and for microscopical examination.

Roux bottles.—It is often more convenient to use a flat flask, such as Roux's (fig. 69), instead of Petri dishes.   The flasks are perhaps better than the Petri dishes because they effectually prevent contamination of the medium and evaporation is reduced to a minimum.

Fig. 69.—Roux bottle.

Note.—The gelatin plate method has some disadvantages, and is not available in all cases.   Thus :

(*a*) Some organisms rapidly liquefy gelatin, and if such are present in the material under investigation the experiment is liable to be a failure.

(*b*) It is only applicable in cases of organisms growing at temperatures below 25° C.   Above this temperature gelatin ceases to be a solid medium.

(ii) Using agar as the culture medium.

Consequently, agar plates are sometimes used, especially when pathogenic micro-organisms are being investigated. The technique is essentially the same as in the case of gelatin plates, but the following points must receive attention.

1. Agar only melts between 90° and 100° C. and does not set again until it cools to 40° C. The agar tubes will therefore have to be melted in boiling water and then allowed to cool until they can be comfortably held in the hand.

2. The tubes must be sown as above, but the experiment must be done quickly otherwise the agar will begin to solidify and the plates will be lumpy.

It is a good plan to have the Petri dishes standing on a levelling apparatus filled with water at 40°–45° C. (see below) before pouring the agar, and to cool the plates slowly in order to prevent the formation of lumps at the time of cooling in the dishes.

3. Incubate the plates at 37° C. The plates should be packed into a large glass dish containing a few pieces of filter paper soaked in water [or perchloride solution] to prevent the medium drying up.

Agar gelatin may be used for cultures which are to be incubated between 25° and 35° C.

This agar plate method never gives very good results, and when agar has to be used for isolating organisms it is much better to employ surface cultures (*vide infra*).

B. Koch's plates.

The use of Koch's plates constitutes an ingenious method of isolating micro-organisms, but the technique is complicated and difficult to carry out under strictly aseptic conditions for the following reasons:

1. The plates must necessarily be exposed to the air for a few seconds while being manipulated, and so are liable to contamination; but if they be prepared quickly and in a still atmosphere, with no dust blowing about, this exposure is not of much moment.

2. In examining the plates, it is also necessary to lift the cover of the moist chamber and so again expose the medium to contamination from the air; the experiment is thus open to error.

The technique of the method is as follows:

Apparatus required.—1. Three glass plates (9 × 12 cm.) each wrapped up separately in paper and sterilized in the hot air sterilizer (a number of these plates should

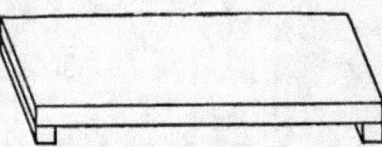

Fig. 70.—Glass support for plate cultures.

always be at hand ready for use). [As in the case of Petri dishes, some bacteriologists prefer to sterilize the plates in metal cases.]

2. Three glass supports on which to rest the plates (fig. 70).

3. Two large circular glass dishes, each about 20 cm. in diameter but one rather larger than the other so that they can be fitted together to form a box.

4. A cooling table consisting of a flat metal box the top of which is well polished and covered by a bell jar (fig. 71). The table rests on screws which enable it to be levelled with the aid of a small spirit level. Two lateral tubes fitted to the box allow a stream of cold water (or if agar is being used, warm water) to be passed through it, and ice can also be put into the box if necessary through a large opening in the bottom closed with a screw cap.

5. Three tubes of melted gelatin and three Pasteur pipettes.

Technique.—1. Pour a little perchloride of mercury solution into the large glass dish, and by rotating the dish wash every part of its interior with the antiseptic.

Lay two or three thicknesses of filter paper in the bottom of the dish and saturate

FIG. 71.—Cooling stand for plate cultures.

them with perchloride solution. (This constitutes a *moist chamber*, the object being to prevent the gelatin plates drying up.)

Wash the glass stands (fig. 70) with perchloride, and place one of them in the bottom of the dish.

2. Place the cooling stand on the operator's left, level it and fill it with cold or iced water. Wipe the top carefully to remove any dust that may be on it, and wash the inside of the glass cover with perchloride solution.

3. Sow the three tubes of gelatin, 1, 2, and 3, as in the gelatin plate method (p. 78 A (i)).

4. Take one of the glass plates, tear off the paper cover along one of the edges, hold it by one of its corners between the thumb and first finger of the right hand [or better in a pair of sterile forceps], slightly raise the glass cover with the left hand, and lay the plate on the glass support already placed there. Replace the glass cover.

5. Take the plug out of tube 3, flame the upper 2 or 3 cm. of the tube, and then while not raising the glass cover more than is necessary introduce the mouth of the tube beneath it, pour the gelatin on to the centre of the glass plate and spread it with the upper flamed part of the tube. Withdraw the tube, replace the bell jar, and allow the gelatin to set.

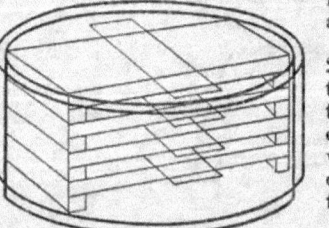

FIG. 72.—Arrangement of the glass plates in a moist chamber.

6. When the gelatin has set, raise the bell jar again, take hold of the glass plate by one of its corners, transfer it as quickly as possible to the moist chamber (the cover of which is raised with the left hand, after replacing the bell jar), and lay it on the glass stand.

Bridge the glass plate with the second glass stand, and replace the cover of the moist chamber.

It will be noticed that the gelatin has not come in contact either with the walls of the moist chamber or with the glass stands, and this explains why perchloride can be used for sterilizing these pieces of the apparatus.

7. In the same way pour a plate with tube No. 2, place it in the moist chamber, and put the third glass support in position.

8. Repeat the process with tube No. 1, and place it on the third plate-rest.

9. Put the moist chamber in the incubator at 20° C. The plates have been arranged so that the one containing the largest number of organisms is nearest the top of the chamber. Colonies will appear on it earlier than on the other plates, and it can be examined and studied without touching the latter, which should not be interfered with until growth appears on them.

C. Esmarch's tubes.

Apparatus required.—1. Three Pasteur pipettes.

2. Three tubes of gelatin. The tubes should be rather longer and wider than ordinary culture-tubes, and each should contain 10 c.c. of sterile gelatin.

3. Three sterile india-rubber caps.

Technique.—1. Sow the tubes 1, 2, and 3 as before (p. 78 **A** (i)).

2. Slip an india-rubber cap over the wool plug of each.

3. Cool each tube in turn under the cold water tap : hold it as nearly horizontally as possible, so that the gelatin coats the whole of the inner surface of the tube below the plug (but without touching the wool), and rotate it between the index finger and thumb of each hand. When the gelatin sets it is thus spread in a thin layer over the whole of the inner surface of the tube and forms a " roll " tube.

4. When the gelatin has set, take off the india-rubber cap and incubate the tubes at 20° C.

This method has the great advantage of absolutely preventing any contamination of the medium, but the investigation of the colonies which develop is rendered more difficult by the cylindrical shape of the gelatin surface.

2. Dissemination on the surface of a solid medium.

When it is necessary to isolate the organisms present in a non-liquid product such as a false membrane, viscous sputum, etc., a small portion of the material is smeared over the surface of some solid medium contained in a Petri dish or sloped in a tube. This, which is the method now universally adopted for the isolation of diphtheria bacilli from false membranes, is available when the media which it is proposed to use cannot be liquefied by heat, *e.g.* potato or serum.

If there be reason to suppose that the material under investigation is very rich in organisms (excreta, for example), a small portion is diluted in a few cubic centimetres of broth or sterile water and a trace of the dilution used for sowing cultures.

Two methods are available.

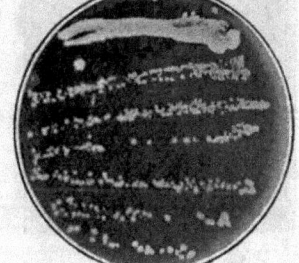

A. Stroke cultures.

The method of isolation on agar plates will be taken as an illustration (fig. 73).

Apparatus required.—1. A medium or stout platinum wire.

2. A tube of agar.

3. A sterile Petri dish.

FIG. 73.—Isolation of organisms by parallel stroke culture on Petri dishes. $\times \frac{1}{2\frac{1}{2}}$.

Technique.—1. Melt the agar and pour it with the usual precautions into the Petri dish.

F

Let the agar set firmly.

2. Take up a trace of the material under investigation on the wire, raise the cover of the Petri dish, and without recharging the needle make a series of parallel strokes on the agar each a few millimetres distant from the other.

As the needle is drawn over the agar the material on it is transferred to the latter, and it is obvious that after the wire has been drawn across the agar three or four times the number of organisms left along the line of any stroke will be but few in number.

3. Incubate the plate at 37° C. The colonies which develop along the first strokes will be very numerous, but will be fewer and fewer along the later ones.

B. Surface cultures.

1. Classical method.—This method of isolation may be illustrated by describing it as it would be used with sloped serum, but the method is the same for agar, potato, etc.

Apparatus required.—1. A stout platinum wire flattened at the end.

2. Three tubes of sloped solidified serum.

Technique.—1. Take up a trace of the material under investigation on the wire.

2. Remove the plug from one of the serum tubes, dip the needle into the tube and smear the whole surface of the medium, commencing below and working towards the mouth (tube 1).

3. Without recharging the needle sow a second tube of serum in the same way (tube 2).

4. Sow the third tube similarly, again without recharging the needle (tube 3).

5. Incubate the tubes at 37° C.

As the result of drawing it over the surface of the serum, the needle is gradually wiped clean of the organisms with which it was charged and which have been deposited on the serum. Tube No. 1 will grow numerous confluent colonies, but tubes No. 2 and No. 3 will grow fewer colonies and some of them will be well isolated. The discrete and isolated colonies on the latter tubes can be used for further investigation.

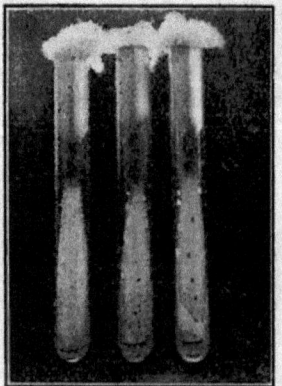

FIG. 74.—Isolation of organisms by Veillon's method.

2. Veillon's method.—1. Take a trace of the material on a platinum wire.

2. Without recharging it, dip it into the water of condensation at the bottom of 4 or 6 agar tubes.

3. Replug the tubes and sow the surfaces of the agar by running the water of condensation over them. Incubate the tubes in the vertical position.

3. Chantemesse's method.—This is useful for the purpose of isolating organisms present in stools.

1. Dilute a trace of the material in several cubic centimetres of distilled water.

2. Dip a sterile badger-hair pencil into this highly diluted material.

3. Brush the surface of a series of 5 or 6 agar plates (prepared in Petri dishes as described above under **A**, p. 81) without recharging the brush.

4. Incubate the plates at 37° C.

Chantemesse adopts this method in isolating the typhoid bacillus, using a special medium instead of agar (Chap. XXI.).

[**4. Burri's method.**—The aim of the method is to grow a colony from a single organism. A dilute emulsion of the organism is made and further diluted in an emulsion of indian ink; small drops of the latter are then laid on the surface of gelatin, covered with a cover-glass and examined under the microscope. Those cover-glasses which cover only a single organism are then transferred to other media and incubated.

[**Apparatus required.**—**1.** A sterile emulsion in water of commercial indian ink,[1] or better a 1 in 9 emulsion in water of a colloidal compound known as Pelikan Tusche No. 54.[2]

2. A number of sterile Petri dishes, slides and cover-glasses.

3. Half a dozen ready poured gelatin plates in large Petri dishes.

4. Sterile dissecting forceps.

5. Two or three fine drawing pens also sterilized.

[**Technique.**—**1.** Prepare a dilute and homogenous emulsion in normal saline solution of the culture or material under examination.

2. Place a sterile slide in one of the Petri dishes; with a small platinum loop put four drops of the indian ink emulsion in a row on the slide and replace the cover of the dish.

3. With a straight platinum wire transfer a small drop of the bacterial emulsion to the left-hand drop (No. 1) of the indian ink on the slide and mix intimately. Transfer similarly a small drop from No. 1 to No. 2 and mix; from No. 2 to No. 3 and so on.

4. With one of the sterile drawing pens take up the right-hand drop of bacterial-indian ink emulsion and lay it in a series of very minute droplets on the surface of one of the gelatin plates. Cover each drop separately with a sterile cover-glass.

5. Disseminate similarly drop No. 3 on another plate and cover as before.

6. Examine the droplets under the microscope using a dry lens and an high eyepiece. If necessary an oil immersion lens may be used; in that case place a drop of oil on the upper surface of the cover-glass. The organisms will be seen as bright objects on a dark background.

7. When a droplet is found in which only a single organism is suspended, raise the cover-glass with a pair of sterile forceps—the indian ink and organism will be found to adhere to the cover-glass—and transfer it to another plate of gelatin or some other suitable medium laying the cover-glass drop side downwards. Incubate.]

SECTION II.—BIOLOGICAL METHODS.

The methods now to be described are only available when the detection and isolation of a given organism is in question, and depend upon a knowledge of one or more properties of the organism; this knowledge is applied to facilitate the growth of that organism while at the same time hindering the growth of any others which may be present.

The separation of anaërobic from aërobic organisms may be quoted as an example of the principles involved. Aërobic organisms cannot grow in the absence of free oxygen; so that by sowing the material in an atmosphere free from oxygen cultures of anaërobic organisms alone are obtained.

The methods most generally in use will be described.

[1] Günther, Vienna.   [2] Grübler, Leipzig.

1. The application of heat to the isolation of micro-organisms.

Spore-forming organisms can resist temperatures of 80° to 100° C. and even 105° C. for several minutes, but non-spore-bearing organisms are soon destroyed when heated to about 60° C. Hence it will be easy to separate a spore-bearing from a non-spore-bearing organism in a mixture containing both; it will only be necessary to heat the mixture for a few minutes to a temperature between 80° and 105° C., according to the resistance of the spore, and subsequently to sow it in a tube of broth. Thus a pure culture of the anthrax bacillus can be obtained by heating an impure culture to 80°–85° C. for 5 minutes.

An infusion of hay if heated to 100° C. for 10 minutes will give a pure culture of the *Bacillus subtilis*. Similarly an infusion of potato chips incubated for two or three days and then heated to 105° C. for 5 minutes will give a pure culture of the potato bacillus, and so on.

In carrying out the above experiments, it is necessary to work with fluid cultures or suspensions, since organisms when dried or mixed with solid matter are much more resistant to heat. It is further essential to the success of the method that all parts of the culture fluid be raised to the required temperature, otherwise some of the non-sporing forms will escape destruction and the experiment will be only a partial success.

The **technique** is as follows :

1. Prepare a very fine Pasteur pipette with a constriction below the wool (p. 75).

2. Fill the pipette with the culture up to the constriction and seal both ends in the flame.

3. Immerse the tube in a water bath heated to the temperature required, and leave it for 5 or 10 minutes. If the temperature required be above 100° C. the tube must be heated in the autoclave.

4. Dry the tube and then break off one end with a pair of sterile forceps after passing it through the flame. Withdraw a little of the fluid into another sterile pipette, being careful to avoid contaminating it, and sow sub-cultures.

2. Isolation by cultivation at the optimum temperature.
Fractional cultivation.

While some organisms will grow at any temperature between 10° and 40° C., the limits of temperature within which growth takes place are in the majority of cases much more restricted. Thus a large number of saprophytes grow slowly and poorly above 30° C. ; many of the pathogenic bacteria attain their maximum development between 30° and 40° C., others will not grow below 30° C., while yet another group (the typhoid-colon group) grows at 43° C.—a temperature which is too high for the multiplication of most micro-organisms. These facts with regard to differences in the optimum temperature at which micro-organisms grow are applied for the purpose of isolating organisms in pure culture.

For example, the colon bacillus can be isolated from stools by sowing a trace of the material in broth and incubating at 43° C. Incubation at this temperature however does not at once yield a pure culture, for the organisms which were present with the colon bacillus in the original material have not been destroyed but their growth merely arrested ; so that were a sub-culture to be sown from this first broth culture and incubated at 37° C. these co-existing organisms would multiply under the more favourable conditions and contaminate the culture of the colon bacillus. To eliminate them the method of **fractional cultivation** may conveniently be adopted ; thus when the first broth culture incubated at 43° C. has become cloudy a trace of it is sown in another tube of broth which is then similarly incubated at 43° C., and from the second tube a third is sown, and so on, until after several sub-cultures in series, each incubated at 43° C., a pure culture of the colon bacillus is ultimately obtained.

A method analogous to this is employed when it is required to isolate the cholera vibrio from specifically infected stools, only in this particular case the action of temperature (37°–38° C.) is combined with that of a special medium (*vide infra*) in which fractional cultivation is effected. This will be found to be in most cases quite a useful method for eliminating saprophytic organisms.

3. Isolation by cultivation on special media.

The growth of any given organism to the exclusion of that of others which are present with it may be effected by sowing the material on a medium which is designed to meet the requirements of the organism to be isolated.

The diphtheria bacillus, for instance, can be isolated in pure culture by smearing the surface of a number of serum tubes with a piece of membrane. Isolation in this case is favoured by the fact that serum is very well adapted to the growth of the diphtheria bacillus, but more or less unfavourable to the multiplication of organisms which are generally found associated with this bacillus.

For the isolation of the cholera vibrio, Koch and Metchnikoff recommend special media which though of poor nutritive value happen to meet its particular requirements. Thus a trace of the " rice water " stool is sown in a tube of Metchnikoff's liquid peptone-gelatin medium (p. 33) and incubated at 38° C. Under these circumstances the growth of the cholera vibrio is much more rapid than that of the other organisms present. The vibrio being a very strictly aërobic organism forms a pellicle on the surface of the liquid, and if after the culture has been incubating for 12 hours or so, a trace of the film be examined, it will be found to consist of an almost pure culture of the cholera vibrio. To further purify the culture recourse must be had to fractional cultivation [sowing the sub-culture with a trace of the pellicle taken up on the point of a fine wire], and three passages will be all that is necessary before finally plating out on gelatin as described on p. 78.

[For the isolation of bacilli of the typhoid-colon group MacConkey has introduced bile-salt media. The material suspected to contain the organism is sown in a liquid bile-salt medium, and after incubation, preferably at 42° C., a trace of the culture is plated out on a bile-salt-agar and suspicious colonies picked off for further examination (for fuller details of the method, see Chaps. XXI. and XXIII.).]

Finally, in some cases, the growth of associated organisms may be arrested by adding to the medium some antiseptic which is not injurious to the organism to be isolated. Chantemesse for instance advises the use of media containing carbolic when attempting the isolation of the colon or typhoid bacillus, and Elsner suggests the use of potassium-iodide-gelatin for the same purpose.

As has been indicated above, this and the method of cultivation at the optimum temperature may be combined; Vincent for instance adopted a combination of the two methods in his attempts to isolate the typhoid bacillus (Chap. XXI.).

4. Isolation by animal inoculation.

In some cases the simplest, and perhaps the only, method of isolating a pathogenic organism in pure culture from material in which it is mixed with non-pathogenic species will be to inoculate the material into a suitable animal.

For example, to isolate the pneumococcus from pneumonic sputum a little of the latter may be inoculated beneath the skin of a mouse ; the animal will soon die and its blood will be found to contain a pure culture of the pneumococcus.

Similarly, to isolate the bacillus of malignant œdema from soil in which there is also present a large number of other organisms, a little of the earth is rubbed up into a thin emulsion in a few drops of sterile water and inoculated beneath the skin of the abdomen of a guinea-pig. The animal dies from an infection known as Pasteur's septicæmia, and the serous peritoneal exudate will contain the bacillus in pure culture.

Many opportunities of studying this method of isolating organisms will occur later.

CHAPTER VI.

THE CULTIVATION AND ISOLATION OF ANAËROBIC MICRO-ORGANISMS.

SOME organisms grow equally well under both aërobic and anaërobic conditions, others grow only when the medium in which they are sown contains no trace of free oxygen ; the former are known as the facultative anaërobes, the latter as the strict anaërobes.   The cultivation of the strictly anaërobic organisms is accompanied by certain technical difficulties arising out of the necessity for removing all traces of air from the culture medium in which they are sown.   The culture media are the same for the two classes, but for the strictly anaërobic organisms special forms of culture apparatus and special methods are required, and it is to a description of these that the present chapter is devoted.

The recent investigations of Tarrozzi, which have been confirmed by Wrzosek, Guillemot, Ori and others, seem to show that oxygen does not directly exert any harmful influence on anaërobic organisms, but that the presence of free oxygen prevents the media furnishing the nutritive substances necessary for anaërobic life.

Anaërobic organisms can in fact, as Tarrozzi has shown, be grown in presence of the oxygen of the atmosphere by simply adding pieces of animal tissue or some reducing agent to the culture media (*vide infra*).

SECTION I.—METHODS OF ABSTRACTING AIR FROM CULTURE MEDIA.

1. By boiling.

Gases dissolved in a liquid can be expelled by boiling.   To expel all the air from a culture medium it must be boiled for 20 minutes to half an hour, and then be cooled rapidly away from the air.

2. By displacing the oxygen of the atmosphere by an inert gas.

The air in a liquid can be displaced by passing a current of an inert gas through it. Hydrogen, carbonic oxide, nitrogen and ordinary coal gas have all been suggested for the present purpose.

A. Hydrogen.—For the growth of anaërobes, hydrogen is preferable to the other gases mentioned. Not only is it easily prepared, but it has no injurious effect on the organisms.

A convenient form of apparatus for readily obtaining a continuous supply of hydrogen is that illustrated in fig. 75. The bottle A contains a 1 in 6

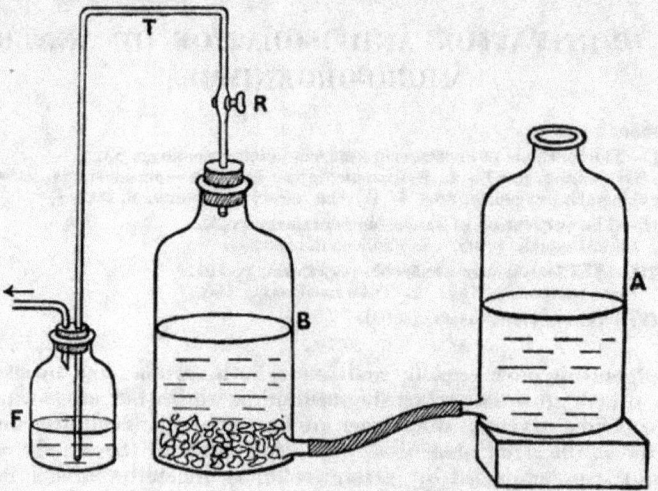

FIG. 75.—Apparatus for yielding a continuous supply of hydrogen.

solution of pure sulphuric acid in water. The bottle B contains some pieces of broken glass at the bottom, and above this a layer of granulated zinc. By simply raising the bottle A and opening the tap R a stream of hydrogen will issue from the tube T ; similarly, by closing the tap R and lowering A to the level of B the supply is stopped. To remove impurities, especially oxygen, it is desirable to wash the hydrogen as it issues from the bottle B by passing it through the following solution :

Caustic potash, 50 per cent. in water, - - - -	50 c.c.
Pyrogallol, - - - - - - - - - -	1 gram

contained in a wash-bottle F.

It is even better to have two wash-bottles, one containing a solution of potassium permanganate slightly acidified with sulphuric acid, the other a solution of potassium permanganate made slightly alkaline with caustic soda. These solutions must be frequently renewed. The method is particularly useful for removing traces of hydrocarbons and phosphides and arsenides of hydrogen.

The hydrogen before being passed through the culture medium should be tested by means of indigo white, to ascertain that it is quite free from traces of oxygen (p. 92).

[Hydrogen is, however, most conveniently obtained by keeping a cylinder of the compressed gas in the laboratory. Cylinders of the gas can be obtained in commerce guaranteed to contain 99·6 per cent. of hydrogen, the remaining 0·4 per cent. being almost if not entirely composed of air, which represents 0·08 per cent. of oxygen. When used for the cultivation of anaërobic organisms in a Bulloch's apparatus (pp. 96 and 100)—which is the method usually adopted in England—the gas requires no preliminary washing, but is passed direct from the cylinder into the bell jar containing the cultures.]

B. Carbonic anhydride.—Carbonic anhydride is harmful to a large number of organisms, and its use for that reason is not to be recommended in the present connexion. The apparatus described above for the preparation of hydrogen can be utilized for the preparation of the gas, if pieces of white marble be substituted for the zinc, and hydrochloric acid for the sulphuric acid. The gas should be washed by passing it through a solution of sodium hydrosulphite contained in the wash-bottle F (fig. 75).

C. Nitrogen.—The preparation of this gas is so difficult that its use should be abandoned in practical bacteriology. [Nitrogen can however now be obtained as a commercial product in the form of cylinders of the compressed gas, which on analysis is found to contain very little oxygen. In our experience the results obtained with this compressed gas in the growth of anaërobic organisms have been quite satisfactory.]

D. Coal gas.—The use of coal gas is not to be recommended in anaërobic methods, because many of the component gases comprising the mixture are inimical to micro-organisms.

Note.—Before passing any gas into a culture medium it must be sterilized by filtration through a sterile cotton-wool plug. The technique of this operation will be referred to later.

3. By absorbing the oxygen.

A. Advantage may be taken of the affinity possessed by some substances for combining with oxygen to remove the latter from culture media. In practice oxygen is generally absorbed by resting the culture-tube on a glass, or metal, support inside a much larger tube (about 20 to 25 cm. in length), and then pouring the following solution into the latter :

Pyrogallol,	1 gram.
Alcoholic potash,	1 „
Water,	10 c.c.

Plug the outer tube with a tightly-fitting india-rubber bung. Under these conditions oxygen diffuses through the wool plug of the inner culture-tube and, being absorbed by the pyrogallol, turns the solution brown.

Sellards, using a similar apparatus, substitutes fragments of phosphorus for the potassium pyrogallate solution.

B. In some cases it will be found convenient to add to the medium some easily oxidizable substance, which does not interfere with the growth of the organism; *e.g.* glucose (2 per cent.), formate of soda (0·5 per cent.), sodium sulphindigotate (0·1 per cent.), fragments of tissue, etc. This method is generally adopted in the case of deep stab cultures in agar (*vide infra*).

C. By sowing the surface of an anaërobic culture in a solid medium with some aërobic organism which absorbs a good deal of oxygen, air can be prevented from reaching the anaërobic culture, the growth of the latter taking place beneath the growth of the aërobic organism (Roux). This method will be described in detail when dealing with stab cultures.

4. By the use of a vacuum.

The use of apparatus by means of which a vacuum can be produced simplifies the methods of cultivating anaërobes and at the same time renders them more exact; and moreover, as both a mercury pump and water pump are in everyday use in the laboratory, the essentials are ready to hand. The use of a vacuum is generally supplemented by washing with an inert gas; by the combination of the two methods it should be possible to remove all trace of oxygen from the culture vessels.

In many laboratories the further precaution is taken of adding some oxygen-absorbing solution, generally pyrogallol and potash, to absorb any traces of oxygen which might still remain.

The reason for washing with an inert gas lies in the physical fact that two gases, which do not enter into chemical combination, rapidly diffuse when brought in contact and form an uniform and constant mixture. The rate of diffusion varies directly as the differences in density of the gases; the greater the difference the more rapid the diffusion.

In practice it is impossible to obtain a perfect vacuum, so that after exhausting a vessel full of air a residuum of air remains. Now if the vessel be filled with hydrogen and exhausted again the residuum will consist of a mixture of air and hydrogen; by repeating the process several times, the amount of air ultimately present will be infinitesimal in amount.

Suppose that after exhausting a vessel of 2 litres' capacity there remains 1 c.c. of air measured at atmospheric temperature and pressure; fill the vessel with hydrogen, and the 1 c.c. of air will be diluted 1 in 2000; exhaust again until only 1 c.c. remains, and the residual gas will contain $\frac{1}{2000}$ c.c. of air and $\frac{1999}{2000}$ c.c. of hydrogen; after a second washing with hydrogen the volume of air will not exceed $\frac{1}{4,000,000}$ c.c.

A. Mercury pump.—With this apparatus an almost perfect vacuum can be obtained, but it is expensive and being delicate is liable to be easily damaged; moreover time and skill are required to use it to the best advantage. Its use is limited in practice to very delicate investigations and to vessels of small capacity. Without going into the details of the working of the pump the following points of importance in connexion with its use may be noted.

1. The pump must always be tested to see that it is working properly and that the taps fit well. Any taps not fitting tightly must be lubricated.

2. Connect the vessel containing the culture to the pump, and exhaust until there is a wide difference between the levels of the mercury in the two limbs of the manometer.

3. Then open the tap connected to the hydrogen supply just a little, and let the hydrogen pass slowly into the receiver until the mercury has reached its original position.

4. Turn off the supply of hydrogen. Exhaust again, and repeat the process two or three times.

5. Seal the neck of the culture vessel in the flame *in vacuo*.

B. Water pump.—On account of its moderate price and of the ease with which it is worked, a water pump is much more often used for producing a vacuum than a mercury pump. The vacuum is only approximate, and exhaustion with a water pump must therefore be combined with washing with an inert gas.

The pump, which is best made of metal (d'Alvergniat's pattern), should consist of a copper pipe fitted with a manometer M and joining the pump proper at a right angle T, as shown in the figure (fig. 76).

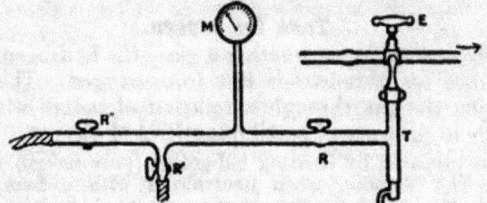

FIG. 76.—A water pump with its fittings.

A water-pressure of about two atmospheres is necessary. The method of exhausting and washing is as follows :

1. By means of pieces of thick-walled rubber tubing (pressure-tubing), connect the vessel containing the culture to the tap R″, and the hydrogen-generating apparatus to the tap R′.

Close the taps R and R′, leaving R″ open throughout the experiment.

2. Turn on the water tap E, gradually open R, and watch the manometer needle.

3. When the vessel has been exhausted as completely as possible, close R, and by gradually opening R′, fill the culture vessel with hydrogen.

4. When the manometer needle has fallen to zero, close R′, open R again, and exhaust the vessel a second time.

5. After exhausting and washing with hydrogen two or three times, seal the neck of the culture vessel in the flame *in vacuo.*

It is not always necessary to wash with hydrogen, but if exhaustion alone be relied upon the culture liquid should be boiled ; this can easily be done by very slowly raising the temperature to 30°–35° C. either by holding the vessel in the hand or by standing it in a vessel of luke-warm water or by heating it with a small flame.

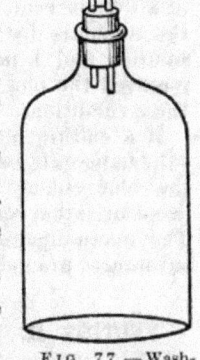

F I G. 77.—Wash-bottle fitted up for use with a water pump to prevent backflow of water into the culture medium.

Note.—In using a water pump the tap R must be closed, so as to cut off all connexion between the water and the culture, before turning off the water tap at the end of an experiment. If this precaution be omitted, the vacuum will induce a violent rush of water into the culture vessel.

A similar inrush of water will also occur if from any cause whatever the pressure in the main is suddenly lowered during the process of exhausting ; consequently a bottle of 2 or 3 litres' capacity, and fitted up as shown in fig. 77, should always be interposed between the pump and the vessel to be exhausted.

If then there be a rush of water, it will collect in the bottle and will not contaminate the culture. It is even better to use a pump fitted with a brass reservoir (fig. 78),

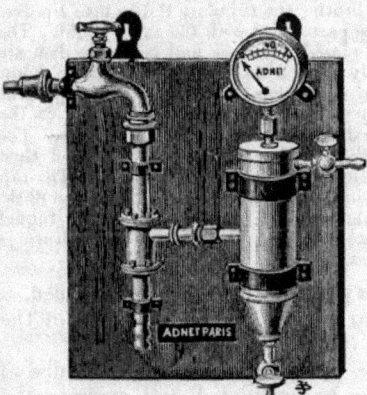

FIG. 78.—Water pump with safety reservoir.

which will act in the same way as the bottle and prevent a rush of water into the culture. The only ground of objection to this piece of apparatus is that of cost.

Tests for oxygen.

It is often necessary to know whether a gas—the hydrogen used in washing anaërobic cultures, for instance—is free from oxygen. This may be determined by passing the gas through a solution of indigo white, a substance which turns blue in presence of small quantities of oxygen.

Indigo white is prepared by treating indigotine (pure indigo) with concentrated sulphuric acid. This solution when neutralized with sodium carbonate gives sodium sulphindigotate, which in presence of an excess of alkali is easily decolourized by reducing agents. Sodium sulphindigotate is generally reduced with sodium hydrosulphite, obtained by adding to powdered zinc a concentrated solution of sodium bisulphite saturated with sulphurous anhydride. Sodium hydrosulphite is a powerful reducing agent and decolourizes the indigo, combining with the oxygen of the atmosphere to form bisulphite.

The gas may therefore be tested for oxygen by bubbling it, away from air, through a solution of indigo white.

To make sure that a culture medium contains no free oxygen, a few drops of a 0·2 per cent. solution of sulphindigotate of sodium may be added until the colour is distinctly blue, then 1 per cent. by weight of a normal soda solution and 1 per cent. of glucose. When all the free oxygen has been removed the blue colour disappears, the glucose reducing the indigo under these conditions.

If a culture medium tinted with a few drops of a solution of sodium sulphindigotate be sown with an anaërobic organism and freed from oxygen, the blue colour will be destroyed as growth of the organism proceeds, decolourization commencing in the immediate neighbourhood of the colonies. The micro-organism takes the oxygen necessary for its growth from the substances around it, and acts therefore as a reducing agent.

SECTION II.—THE CULTIVATION OF ANAËROBIC ORGANISMS.

1. Liquid media.

A. Pasteur's method.—This is the method originally employed in growing anaërobic organisms. It is now only of historical interest.

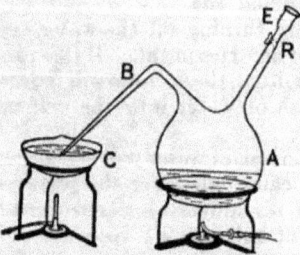

A large round flask A (fig. 79) with two tubulures is filled with broth: the tubulure B dips into a porcelain dish three-parts filled with the same liquid. The tap R being closed, the flask and porcelain dish are simultaneously heated to boiling for half an hour. The dissolved air is thus driven off. The apparatus is allowed to cool *in situ*, and then the end of the tube B is transferred to a vessel full of mercury. The funnel E is filled with carbonic acid gas, and then (away from air) with the fluid to be sown. The tap R is next opened and the fluid runs into the flask, care being taken that a little remains in the funnel to prevent access of air to the flask. The culture is then incubated.

FIG. 79.—Pasteur's original method for the cultivation of anaërobic organisms.

B. Roux's pipette.   Method recommended.—

1. Make a constriction in a sterile Pasteur pipette in a small flame of the blow-pipe just below the cotton-wool plug (fig. 67 *a*, p. 75).

2. After flaming the point of the pipette, break it off, dip it into a tube of broth already sown with the organism to be cultivated and aspirate the broth into the pipette until the latter is three-parts full.

3. Tilt the pipette so as to raise the point and seal the latter in a small flame.

4. Connect the other end to an exhaust pump. Exhaust and wash with hydrogen alternately.

It is often sufficient when a vacuum is established to boil the liquid as described at p. 91. When the pipette is heated even very slightly the liquid will boil violently and will tend to pass into the aspirating tube; this may be prevented by first heating the upper part of the tube above the liquid.

5. Seal the pipette at the constriction *a*, *in vacuo*. Dip the ends of the pipette into Golaz's wax to strengthen them. Incubate.

6. When the culture has grown, flame and break the end *a* of the pipette and withdraw the culture by means of a Pasteur pipette.

C. Pasteur, Joubert and Chamberland's tube.—With this apparatus two successive cultures can be sown without exposing the medium to the air while sowing the second culture.

It consists (fig. 80) of an inverted U-tube, each limb of which is provided with a lateral tubulure terminating in a fine point. A third tubulure originates from the convexity of the U, and this is constricted in two places a short distance apart, and plugged with wool between the two constrictions.

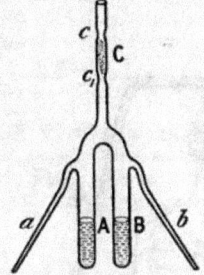
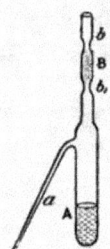
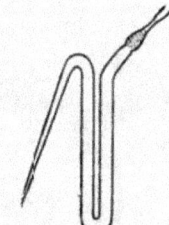

FIG. 80.—Pasteur, Joubert and Chamberland's tube for the cultivation of anaërobic organisms.

FIG. 81.—Pasteur's tube for the cultivation of anaërobic organisms.

FIG. 82.—Lacomme's tube for the cultivation of anaërobic organisms.

1. Plug the vertical part C with wool between the constrictions *c* and *c₁*, seal the points of the lateral tubulures *a* and *b* and sterilize the tube in the hot air sterilizer.

2. When it has cooled, flame the lateral tube *a*, break off its point and dip the latter into the broth, which has been sown beforehand. Aspirate the liquid into the limb A by applying suction to C. Seal up the end of *a* in the flame again.

3. Flame the lateral tube *b*, break off its point, dip the end into a tube of sterile broth, and aspirate the broth into the limb B. Seal the end of *b* in the flame.

Note.—In carrying out the second and third operations, be careful that the liquids in the two limbs do not mix. The limbs should not be more than one-third filled.

4. Attach the upper end of C to the exhaust pump. Exhaust the air, and wash two or three times with hydrogen. Seal the tube at the constriction *c in vacuo*.

5. Incubate the tube in the vertical position. Growth will occur in A, while the broth in B will remain clear and serve as a control.

6. When growth in A has ceased, tilt the apparatus so that a drop or two of the culture passes from A into the sterile broth contained in B. Incubate again, and growth will now take place in B.

D. Pasteur's tube.—This is a more simple form of the preceding, and consists of a single limb of the U-tube just described (fig. 81).

After sterilizing the tube aspirate the broth, already sown with the organism, through the narrow tube *a*, seal the point of *a* in the flame, exhaust through B, seal this tube at *b* and incubate.

Lacomme's tube (fig. 82) is a modification of Pasteur's; it is used in an exactly similar manner, and is cheaper.

E. Long-necked flask method.—This is a useful method when large quantities of culture are required.

1. Take a flask with a long neck, fill it one-third full of broth, plug with wool and autoclave (fig. 83).

2. When cool take out the wool plug and sow the broth with a long Pasteur pipette, taking every care to avoid introducing contaminations. Replace the plug and push it half-way down the neck.

3. Make a shallow constriction below the plug at A, and draw out the upper end B.

4. Connect the upper end B with the water pump: exhaust, and wash with hydrogen: seal the neck above the plug in the flame *in vacuo*, and incubate.

5. To withdraw the culture after incubation, cut the neck above the level of the plug (p. 47), take the plug out, and aspirate the fluid into a pipette or into a sterile distributing flask. [The culture may equally well be drawn up into a Cobbett's bulb by the method used in the preparation of serum (p. 45).]

F. Bottle method. Method recommended.—The advantages of the method are that (1) large quantities of broth can be used, and (2) the culture can be very easily removed.

1. Select a bottle of 1 or 2 litres' capacity with a mouth large enough to take an india-rubber bung perforated with two medium-sized holes. Fill the bottle two-thirds full of broth (fig. 84).

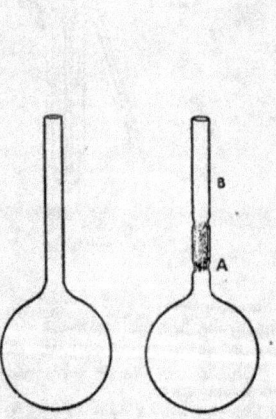

FIG. 83.—Flasks with long necks for the cultivation of anaërobic organisms.

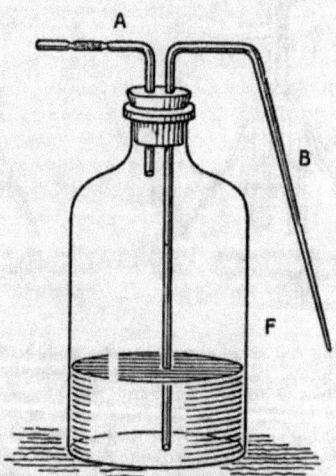

FIG. 84.—Bottle arranged for anaërobic cultivation.

2. Take a piece of glass tubing of the same diameter as the holes in the india-rubber bung, and bend it at right angles about its centre. In one limb make two constrictions a short distance apart, and plug this space with wool. Pass the other limb through one of the holes in the bung so that its lower end projects a distance of 3 or 4 cm. below the bung.

Take another piece of glass tubing and bend it as shown in the figure. Pass the straight limb through the other hole so that it almost reaches the bottom of the bottle, while the other limb terminates outside the bottle in a solid point sealed in a flame.

3. Fit the india-rubber bung firmly into the neck of the bottle and sterilize at 115° C. for 20 minutes, but let the temperature rise gradually for fear of cracking the bottle.

4. When the apparatus has cooled, ascertain that the bung fits firmly, and then lute the joints between the bung and the neck of the bottle and between the tubes and the bung with Golaz's [or paraffin] wax. Dry the wool plug in A by gently heating the glass tube in a Bunsen burner.

5. In order to sow the broth, flame the external limb of B, and break off the point with sterile forceps. Dip the end into the tube containing the organism to be cultivated, aspirate a few drops into the bottle through A. and seal the point of B in the flame.

6. Connect A to the water pump. Exhaust and wash with hydrogen several times, keeping the lower two-thirds of the bottle in a bath of water at 35°–40° C.

7. Seal A in the flame *in vacuo* at the constriction beyond the wool plug. Incubate.

After incubating for 2 or 3 days the gas produced as the result of the growth of the organism accumulates to such an extent as to prevent further multiplication. At this stage it is well to break off the sealed end of A (leaving the cotton-wool plug in position, of course) to allow the pent-up gases to escape ; the pressure of the gases remaining in the bottle and continuously generated by the growth of the organism is sufficient to prevent the entrance of air.

It is as well to add a little calcium carbonate or tricalcium phosphate to the medium before sterilizing it, because with some organisms the amount of acid produced is so considerable as very soon to interfere with and perhaps altogether check the growth. If these salts be added, the acids will be neutralized as they are formed.

8. To withdraw the culture from the bottle, flame the end of B and break off the point. Blow through A and collect the culture in a sterile flask.

G. Pyrogallol method. Buchner's tube.—**1.** Boil a tube of sterile broth, cool rapidly, and sow.

2. Place this tube as already described at p. 89 inside a larger tube containing a solution of potassium pyrogallate, and incubate (fig. 85).

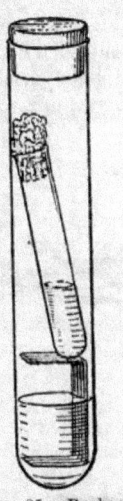

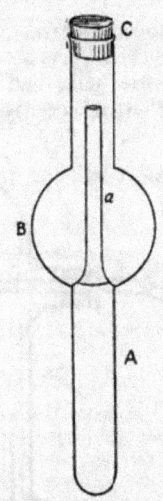

FIG. 85.—Buchner's tube for growing anaë-robic organisms.

FIG. 86.—Turrô's tube.

Turrô's tube.—This method has advantages over Buchner's in that the oxygen is much more rapidly absorbed and the culture is visible during incubation (fig. 86).

1. Pour the medium (broth, agar or gelatin) into A through the narrow tube *a*. Plug the upper end of the apparatus with an india-rubber stopper C and autoclave.

2. When it has cooled, sow the medium through *a* ; then with a pipette fill the bulb B one-third full of potassium pyrogallate.

3. Replace the stopper C and lute it with paraffin or Golaz's wax. Tilt and rotate the apparatus so that the pyrogallate runs all over the surface of the bulb to accelerate the absorption of oxygen. Incubate.

[**H. Bulloch's method. Method recommended.**—Bulloch's method is a modification of the preceding, designed to allow of the incubation of a number

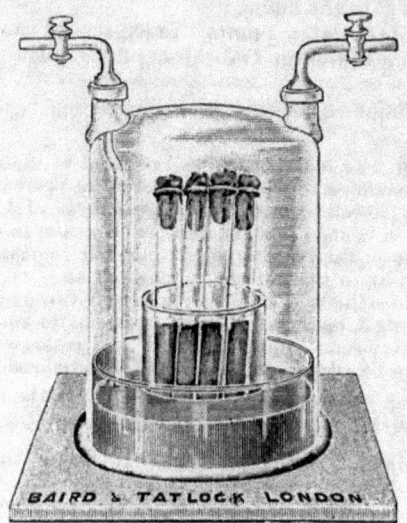

FIG. 87.—Bulloch's anaërobic apparatus.

of plates or tubes at one time. The principle is the same and depends upon the absorption of oxygen by pyrogallate of potassium. The apparatus consists of a circular glass bell jar (fig. 87) flanged below, with two openings above, each of which is fitted with a ground-glass stopper prolonged into a

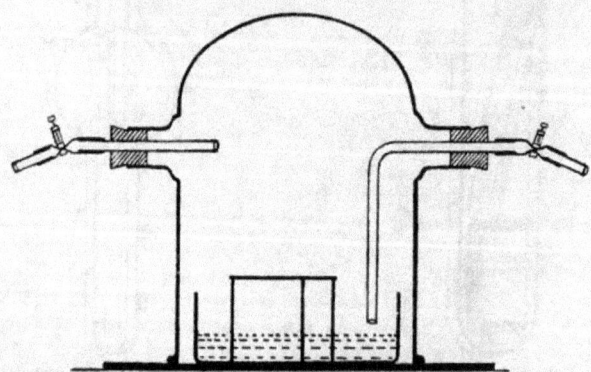

FIG. 88.—A modified form of Bulloch's apparatus.

glass tube bent at right angles and fitted with a closely fitting tap. One of these tubes only passes a few centimetres into the bell jar while the other reaches nearly to the bottom.

[Fig. 88 shows a slightly modified and less fragile form of Bulloch's apparatus. The openings are laterally situated and are plugged with india-rubber corks, each of which is perforated by a piece of glass tubing. To each of the latter a piece of red rubber pressure-tubing is attached at its outer end. When the apparatus is to be closed, the tubing is compressed by screw clips and a piece of glass rod tightly fitted into its distal end.

[1. On the ground-glass plate stand a shallow glass vessel 3–4 inches deep, but having as large a diameter as will permit of the bell jar being placed over it.

[2. Place about ¼ oz. of pyrogallol in the bottom of the vessel.

[3. Place the tubes in a suitable receptacle, and stand the latter on a glass tripod in the vessel containing the pyrogallol.

[4. Grease the lower flanged end of the bell jar with *unguentum resinae*, and press it firmly down on to the ground-glass plate in such a way that the long tube passes into the shallow glass vessel.

[5. Aspirate about 40 c.c. of strong potash solution (30–40 per cent.) into the vessel. Then screw up the clips as tightly as possible, and plug the distal end of the tubing with glass rod.

[6. Incubate.

[7. To remove the tubes, withdraw the pieces of glass rod, gently unscrew the clips, slide the bell jar off the glass plate and lift out the receptacle containing the tubes.]

[I. *Bulloch's method modified*.—**Method recommended**.—The use of pyrogallol and potash is as a rule supplemented by exhaustion and washing with hydrogen.

[1. Proceed as in 1, 2, 3 and 4 above.

[2. Attach the glass tube which passes just inside the apparatus to a water pump connected with a manometer, and the other tube which dips into the vessel to a cylinder of hydrogen.

[3. Exhaust the vessel.

[4. Turn on the hydrogen tap and pass a slow stream of gas until the manometer falls to zero. Unless this be carefully done the pressure of hydrogen will lift the bell jar.

[5. Turn off the hydrogen. Exhaust again.

[6. Wash with hydrogen again, and again exhaust.

[7. Screw up both clips as tightly as possible.

[8. Disconnect the bell jar from the water pump and also from the cylinder of hydrogen.

[9. Connect the tubing that dips into the vessel to a beaker containing a 40 per cent. solution of potash in water. Gently loosen the clip and allow 40 c.c. or so of the solution to enter the bell jar, being careful to allow no air to enter. Screw up the clip.

[10. Insert a tightly-fitting piece of glass rod into each piece of india-rubber tubing on the distal side of the clip.

[11. The vessel is now ready to be placed in the incubator.]

[The security of the joints should be tested on the following day, or even later on the same day. To do this, attach the same piece of tubing as before to the manometer, turn on the water pump to exhaust the rubber connexions, etc., and then loosen the screw clip. If the apparatus is securely fastened the mercury should remain at the same level as when the bell jar was exhausted.]

J. Legros' method. Method recommended.—By this method the air is excluded from the medium by means of a layer of vaseline oil. Pour sufficient oil into the culture-tube to form a layer 5–10 mm. deep on the surface of the medium. Plug with wool, and autoclave. Sow in the ordinary way through the layer of oil.

G

In the case of media which cannot be heated strongly, Ch. Nicolle recommends the following modification of the method: Pour sufficient sterile vaseline oil into the flask or tube containing the medium to form a thin layer on the surface. Stand the culture vessel in a water bath at 40° C. and connect the mouth of the vessel to a water pump. After exhausting the whole of the dissolved air the culture medium is protected from air by the layer of oil.

K. Rosenthal's method. Method recommended.—1. Distribute the medium (broth, milk, etc.) into tubes. Pour lanolin previously liquefied by heat into each tube so that it forms a layer 15 mm. thick on the surface. Plug the tubes with wool and autoclave at 120° C. After sterilization, cool the tubes rapidly in a vertical position.

Tubes prepared by this method can be kept for about two months. If kept longer than this, it is well to heat them to 100° C. for 15 minutes before sowing them. It is an advantage to use tubes slightly constricted about the middle; the medium occupies the lower part of the tube up to the constriction, while the lanolin fills the constricted part. Any gas which may be formed easily escapes by pushing the plug of lanolin into the upper non-constricted portion of the tube.

2. When required for use, melt the layer of lanolin in the flame (it liquefies at 42° C.), and sow the organism in the ordinary way through the melted lanolin. Cool rapidly to solidify the fat. Growth takes place in what is practically a sealed tube (Rosenthal).

L. Tarozzi's method.—To grow the strictly anaërobic organisms (*Bacillus tetani*, *Bacillus maligni œdematis*, etc.) by this method it is only necessary to add to broth contained in ordinary tubes a fragment of tissue freshly removed from a rabbit, mouse or guinea-pig, and to proceed as in the case of aërobic organisms.

Pieces of liver, spleen, kidney or lymphatic glands may be used with success, but blood, milk, or the connective tissues are useless for the purpose. To tubes of broth add a small piece of one of the above-mentioned internal organs which has recently been excised with the usual aseptic precautions. Incubate the tubes for a day or two at 37° C., and they are then ready for use. They may be heated to 100°–107° C. for a minute or two, but if the heating be prolonged for more than 5 minutes growth will fail. Cultures will grow even if the piece of tissue be removed before sowing.

A number of other substances have a similar action in facilitating the growth of anaërobic organisms. Wrzosek, Ori, for example, were able to obtain cultures under ordinary conditions in broth by simply adding pieces of vegetable tissue (potato, elder pith, mushrooms, etc.) to the medium. Tarozzi used a slightly alkaline glucose-broth, which had been heated under a pressure of two atmospheres in the autoclave, with successful results. Aperlo was able to grow strictly anaërobic organisms in a simple peptone-broth by sterilizing the medium for half an hour under a pressure of half an atmosphere, and using it within 24 hours of its preparation.

Kata also succeeded with ordinary broth containing a small piece of agar and 0·3–0·7 per cent. of sodium sulphite, and even better with the same amount of sulphite and a little fresh serum. The latter medium would appear to be very useful for toxin production.

Pfuhl recommends a broth made with liver instead of ordinary meat, and sterilized in the autoclave. Satisfactory results were also obtained with the following technique: To a tube containing 10 c.c. of ordinary broth add 1 gram of spongy platinum, boil for 10 minutes, sow as soon as cool and put in the incubator without shaking the tube.

The vitality of anaërobic organisms is exhausted much more quickly on media prepared on these principles than on media under anaërobic conditions (Jungano and Distaso).

2. Solid media.

(i) Stab cultures.

A. In test-tubes. Method recommended. (*a*) **Gelatin.**—1. Heat a tube of sterile gelatin to boiling, taking care not to let the medium froth and boil over. Boil for several minutes. [This is best done in a water bath.]

A few drops of sodium sulphindigotate solution may be added to the gelatin before boiling it and if this be done the medium will be decolourized by the growth of the organism.

2. Cool the gelatin rapidly, and when it is set sow a stab culture with a fine platinum wire.

A little air would ordinarily be introduced with the needle, and the following arrangement is devised to obviate this. Mount the wire on the wall of a piece of glass tubing, and connect the other end of the latter to a hydrogen-generating apparatus by means of a piece of india-rubber tubing (fig. 89). To use the needle, after flaming it, take up the material to be sown, then turn on the hydrogen and sow in a current of the gas. In this way the oxygen of the atmosphere is prevented from reaching the needle track.

3. After sowing, dip the gelatin tube into very cold water and pour a layer of agar over the surface with a Pasteur pipette. Replace the plug. The object of this procedure is to form a plug impervious to air on top of the gelatin. Sterilized oil or liquid vaseline, etc., may be used instead of agar.

Note.—The agar plug may be omitted if some very oxidizable substance capable of absorbing oxygen be added to the culture medium (Liborius, Kitasato). The best substances for the purpose are glucose (2 per cent.), sulphindigotate of sodium (0·1 per cent.), sodium formate (0·5 per cent.).

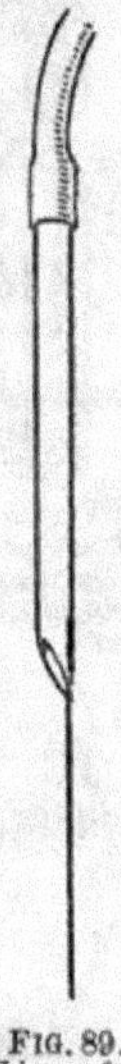

Liborius recommends the following medium :

Fig. 89.—Wire for sowing an-aërobic stab cultures.

Ordinary agar,	1000 grams.
Glucose,	20 ,,
Sodium sulphindigotate,	1 gram.

Nearly fill the tubes with the medium, and sow deep stab cultures as described above.

(*b*) **Agar.**—The method is the same as in the case of gelatin.

B. Absorption of oxygen by an aërobic organism (Roux).—Proceed as above, and when the agar plug has set sow the surface with *B. subtilis*. This organism is strictly aërobic and absorbs the oxygen present in the tube, while growth below takes place under anaërobic conditions in an atmosphere free from oxygen.

To reach the anaërobic organism without contaminating it with the *B. subtilis*, wash the outside of the tube with perchloride of mercury, cut it across about the level of the middle of the growth, break off the lower part of the tube, and the anaërobic organism can then be removed without contaminating it.

C. Roux's pipette.—1. Flame and break off the point of a Roux's pipette. Dip the end into a tube of sterile gelatin which has just been boiled. Draw the gelatin into the tube until it reaches the constriction *a*, fig. 67, p. 75. Seal the narrow end of the pipette and the upper end at the constriction. Dip the whole tube into cold water to cool it quickly.

2. When the gelatin has set, pass the upper part rapidly through the flame, and then break off the point *a* with a pair of forceps. Through the opening sow a stab culture with a fine wire. Seal the opening in a flame.

3. To open the tube when growth has taken place, break off the lower end over a sterile glass plate. If the upper end were opened first the pressure of the gases formed during the growth of the organism would be sufficient to forcibly expel the contents of the tube.

D. Hydrogen method (Roux).—This method is more difficult than those just described.

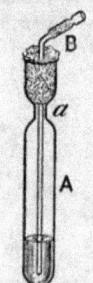

1. Take a tube of sterile gelatin, and constrict in the blow-pipe just below the plug (a, fig. 90).

2. Select a sterile Pasteur pipette the smaller end of which will easily pass through the constriction, and bend it at a right angle below the wool plug. Connect the plugged end of the pipette with a hydrogen apparatus.

3. Melt the gelatin in a water bath. Flame the narrow part of the pipette to sterilize it, and after breaking off the point pass it between the wool and the side of the tube down to the bottom of the gelatin.

4. Pass a stream of hydrogen through the medium for some minutes, and then withdraw the pipette a little so that the hydrogen passes over the surface of the gelatin and prevents air gaining access to the

FIG. 90.— Roux's method of growing anaërobic organisms.

medium while it is being cooled.

5. Take out the wool plug and sow a stab culture with a fine wire, the current of hydrogen being maintained meanwhile.

6. When the tube is sown, take out the pipette and seal the tube as quickly as possible at the constricted part a.

[**E. Bulloch's apparatus** can be used equally well with solid as with liquid media (p. 96).]

(ii) Surface cultures.

Gelatin and agar.

[**Bulloch's apparatus** is available for the growth of anaërobic organisms in surface culture (pp. 96 and 102).]

Roux's tube.—Roux's tube for stroke cultures of anaërobic organisms consists of an ordinary test-tube T drawn out above (A) and provided with a lateral branch B (fig. 91).

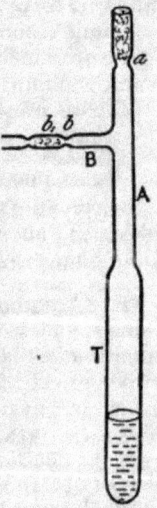

1. Pour some gelatin into the lower, wider part of the tube T, using a narrow-stemmed funnel for the purpose. Seal the tube at the constriction a in its upper part. Plug the side tube B with wool between the two constrictions b and b'. Sterilize in the autoclave.

2. Attach B to the water pump, stand the tube in a water bath at a temperature just sufficient to keep the gelatin melted while the tube is exhausted and washed two or three times with hydrogen.

3. When the air has been displaced by hydrogen, leave the tube in a slanting position while the gelatin sets.

4. Then flame the upper part of A, break off the point a, and sow a stroke culture through the opening; the tube B must at the same time be connected with the hydrogen-generating apparatus and a stream of hydrogen passed

FIG. 91.—Roux's tube for stroke cultures of anaërobic organisms.

into the tube to prevent the access of air. Seal the top of the tube A again.

5. It now only remains to seal B at the constriction b'. Growth then takes place in an atmosphere of hydrogen. If necessary the tube can be again exhausted after sowing and sealing a, before sealing b'.

Potato.

Roux's tube.—1. Blow on to an ordinary potato-tube below the constriction, a lateral tube B, and plug the latter with wool between two constrictions (fig. 92). (These tubes can be bought ready made.) Place a piece of potato in the tube, and sterilize it in the autoclave at 120° C.

2. Sow the potato in the ordinary way, and then seal the upper end of the tube below the wool plug in the flame (fig. 92).

3. Attach B to the pump. Exhaust and wash with hydrogen.

4. Seal the side tube B at the constriction b' under a vacuum, and incubate.

SECTION III.—THE ISOLATION OF ANAËROBIC ORGANISMS.

1. Plate method.

A. On sheets of glass.—The method is similar to that described for aërobic organisms. The technique is difficult but has the advantage that the colonies can be examined under the microscope.

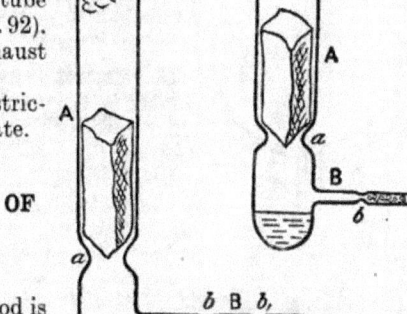

FIG. 92.—Roux's tubes for sowing cultures of anaërobic organisms on potato.

1.—(1) Sow three tubes of liquefied sterile gelatin with the organism under investigation, and pour three plates as in isolating aërobes (p. 78).

(2) Have a vacuum desiccator (previously washed inside with perchloride) at hand, and pour some potassium pyrogallate (p. 89) into the sulphuric acid vessel. A vacuum incubator (p. 104) can also be used.

(3) Arrange the plates on the shelves as they are poured.

(4) Lute the bell jar, exhaust, and wash with hydrogen. Disconnect the bell jar by closing the tap connecting it to the pump.

2. Turrô has simplified the method by arranging the plates on glass benches in a large glass dish into the bottom of which some potassium pyrogallate is poured. The ground glass cover of the glass dish is then sealed with paraffin. Agar can be used for the plates, and the whole incubated.

B. Kitasato's dish.—A circular flat glass dish of the size of a Petri dish is fitted with two tubes A and B on opposite sides. The tube B is drawn out and sealed. A is plugged with wool (fig. 93).

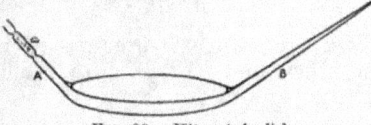

FIG. 93.—Kitasato's dish.

This is a satisfactory though fragile and rather expensive piece of apparatus.

1. Sterilize the apparatus in the hot air sterilizer.

2. After flaming the end of B, break off the point and dip the end into a gelatin tube already sown, and aspirate the medium into the dish through A. Seal B, and let the gelatin solidify.

3. Attach A to the pump. Exhaust, and wash with hydrogen. Seal A at the constriction a.

C. Bombicci's apparatus.—This vessel is cheaper than Kitasato's. It consists of a circular flat glass dish with a cylindrical appendix of about 10 c.c. capacity.

1. Pour the medium, agar or gelatin, into the appendix, plug the neck with wool and sterilize in the autoclave.

2. Select an india-rubber plug with two holes which fits the neck of the dish. Fit it with two tubes as shown in the figure and plug the horizontal limb of each with wool between two constrictions. Wrap the plug with the tubes in position in paper, and sterilize separately from but at the same time as the dish.

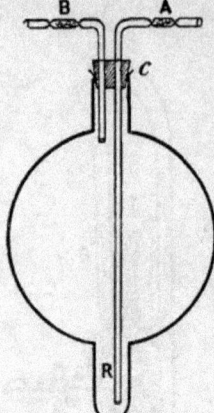

FIG. 94.—Bombicci's dish.

3. Keep the gelatin or agar liquefied in a water bath at 30°–40° C. as the case may be, while sowing the medium. Take the india-rubber plug out of its wrapper and fit it into the neck of the dish as quickly as possible.

4. Lute the plug with Golaz's wax, attach A to the hydrogen-generating apparatus, keeping the medium liquefied in the water bath, and pass a stream of hydrogen through the medium for a few minutes. Lay the apparatus horizontally so that the medium flows into the dish and continue the current of hydrogen for several minutes. Exhaustion may be combined with washing with hydrogen if it be thought necessary.

5. Seal the ends of A and B beyond the wool plugs.

D. Zinsser's method.—Zinsser uses an apparatus similar to a Petri dish, but deeper, and having an annular space of 5–6 mm. between the dish and the cover. The agar or gelatin, as the case may be, is sown and poured into the dish, and after it is set is inverted on to the cover, into which a little alkaline pyrogallol is poured (p. 89). A layer of oil is poured on the surface of the pyrogallate in the annular space.

E. Tarozzi's method.—Tarozzi uses an alkaline glucose agar which has been heated under a pressure of two atmospheres (p. 98). The medium is poured to a depth of 1 cm. into Petri dishes with ground-glass covers, which are luted with paraffin.

F. Marino's method.—1. Take a number of Petri dishes, remove the lids, and place the dishes, cavity upwards, over (and therefore partly within) them. Wrap in paper and sterilize in the hot air sterilizer.

2. Take a number of large test-tubes, and into each pour 30 c.c. of 0·5 per cent. glucose-agar. Sterilize in the autoclave.

3. Cool the agar to 40°–42° C. in a water bath. Add to the contents of each tube 1 c.c. of rabbit- or horse-serum previously heated at 55° C. for 20 minutes.

4. Sow the tubes by the dilution method.

5. Pour the contents of each tube into the lid of one of the sterile Petri dishes, and cover with the other part of the dish in such a way that the agar is contained between and compressed by two sterile glass surfaces, the cavity of the dish being obviously upwards. Cover with a sterile glass plate large enough to project beyond the edges of the Petri dish to protect it from contamination. Incubate.

6. After the colonies have grown, gently separate the agar from one of the glass surfaces, leaving it adhering to the other, and pick off with a fine-pointed pipette, any colonies it is thought desirable to examine.

When separating the medium from one of the glass surfaces it often happens that the agar is torn ; so it may be that the colony which was wanted cannot be found, or else that it has become contaminated, by rubbing up against another colony or by contact with the water of condensation.

Liefman, Fehrs and Sachs-Müke's modification of Marino's method obviates this defect. Instead of the lower part of the Petri dish a plate of sterile glass is used as a cover, and sufficient medium is poured into the lid to slightly overflow the edges. In covering with the sheet of glass care must be taken that no air bubbles are included.

[**G. Bulloch's apparatus.** The technique as now generally adopted has been explained at p. 96. The only modification required in the present connexion is the use of Petri dishes containing a solid medium instead of tubes of a liquid medium. It will be necessary, of course, to have a glass tripod or a thick sheet of cork on which to stand the dishes, in order to prevent the lower ones being flooded with the pyrogallate solution. It is advisable also to stand some water in a Petri dish on the top of the uppermost plate.]

2. Tube method.

A. Esmarch's tubes.—Frænkel, Roux have adapted the Esmarch tube method of isolating aërobic organisms to the isolation of anaërobic species. The technique recommended by these authors is somewhat complicated, and is now very rarely used.

Frænkel's method.—Frænkel prepares an Esmarch tube (p. 81), and after sowing the medium in air, displaces the latter by hydrogen by means of an arrangement similar to that described in Bombicci's method (p. 101). After passing the hydrogen through the medium for 5 or 10 minutes the tube is rolled as in the ordinary Esmarch method.

Roux's method.—Roux sterilizes the medium in a test-tube the upper part of which has been narrowed by drawing it out in the flame (fig. 95, left-hand figure). When cool but still liquid the medium is sown. The narrowed part of the tube is then constricted at two points and the wool plug pushed down between them (fig. 95, right-hand figure.) Attach the tube to a water pump, exhaust and wash with hydrogen, seal at the upper constriction, and roll the gelatin. To remove the colonies, cut off the upper part of the tube and pass a platinum wire through the opening.

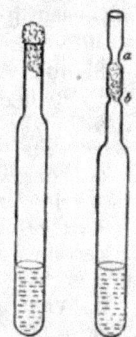

FIG. 95.—Esmarch's tubes applied to anaërobic cultivation.

B. Vignal's tube. Method recommended.—1. Take a piece of glass tubing about 1 metre long and 3 or 4 mm. in diameter. Draw out one end in the flame and plug the other with wool. Make a constriction in the tube 3 or 4 cm. below the wool plug (fig. 96). Heat the tube thoroughly in the flame to sterilize it.

2. Heat a tube of sterile gelatin to boiling point (the medium may, if desired, be coloured with sulphindigotate of sodium). Let the gelatin cool in a current of hydrogen (p. 100), but sow it before it sets, also in a current of hydrogen. Rotate the tube between the hands to distribute the organisms through the medium.

3. Flame the sealed end of the tube, and break off the point. Dip the end into the gelatin, and aspirate the medium into the tube up to the constriction A. (It is necessary to take care that no bubbles of gas enter the tube.) Seal the pointed end and then close the tube at A (fig. 96, A').

Colonies soon appear scattered through the gelatin. The growth can be removed by carefully flaming the tube or washing it, first in perchloride then in alcohol, in the neighbourhood of the colony which it is desired to examine. Cut the tube at the sterilized part and remove the growth with a needle.

FIG. 96.—Vignal's tube.

C. Method of Liborius-Veillon. Method recommended.—Liborius' agar (p. 99) which is used for deep stab culture is also available for the isolation of anaërobic micro-organisms. The tubes are sown by the dilution method (p. 79) and cooled rapidly. For the examination and sub-cultivation of the colonies, Liborius recommended turning out the agar on to the inside of the lid of a sterile Petri dish and cutting out the colonies with a sterile knife, a process which was not only rather difficult but exposed the colonies to contamination. To overcome these disadvantages, the method has been modified by Veillon and as modified by him is now [one of] the best methods of isolating anaërobic organisms.

1. Fill a number of large test-tubes (22 cm. ×15 mm.) to a depth of 10 to 15 cm. with some quite transparent agar containing 1·5 per cent. of glucose. Sterilize in the ordinary way but do not allow the temperature to exceed 120° C.

2. When ready to sow the tubes, heat five to ten of them to 100° C. in a water bath, and boil them for 20 minutes or so to liquefy the agar and drive off the air dissolved in the medium. Then transfer the tubes to a water bath at 40° C. to keep the agar liquid until sown.

3. Add one drop of the material to be sown to the first tube, and disseminate it by rolling the tube between the hands.

4. Sow the second tube with a few drops from the first, the third from the second and so on, as previously described.

5. Immediately the tubes are sown, cool them rapidly in the upright position. Incubate.

Aërobic organisms grow in the upper part of the medium which contains a certain amount of air in solution, while the anaërobes multiply in the deeper layer.

6. When carefully examined it will be found that growth soon makes its appearance, the number of colonies depending upon the extent to which the material with which they were sown was diluted. Examine the different colonies with the naked eye and with a lens and select the tubes containing the smallest number of colonies for the purposes of sub-cultivation.

To sub-cultivate, take a Pasteur pipette with a fine point, break off the end, and holding the culture-tube horizontally remove the wool plug and pass the fine end of the pipette into the agar towards the colony to be removed : as the pipette is passed through the colony some of the growth is forced into it ; withdraw the pipette and sow the colony in a fresh tube of medium.

It facilitates the process of sub-cultivation to put, as Guillemot advises, an india-rubber teat on the plugged end of the pipette ; the colony can then be more easily drawn into the pipette by aspiration, and forced out into the new medium by compression of the teat.

Great care must be taken that the pipette does not touch any colony other than that to be sub-cultivated ; the only way of avoiding such an accident is to work with cultures in which the colonies are few in number and sufficiently well isolated one from another.

Note.—It is often an advantage to use a medium containing serum, since many anaërobic bacteria grow better in albuminous media. Prepare the agar as above, melt the contents of the tubes, and cool to 40° C. in a water bath, then to each tube for two parts of agar add one part of sterile liquid serum also heated to 40° C. : mix the agar and serum thoroughly, keeping the medium at 40° C. to prevent the contents solidifying, and sow with the material.

SECTION IV.—VACUUM INCUBATORS.

Anaërobic organisms can be cultivated in ordinary culture vessels, provided that these are incubated in a special form of incubator which can be exhausted and the vacuum maintained. A little water, or, better, solution of potassium pyrogallate which absorbs oxygen, should always be placed in these incubators to prevent desiccation of the medium.

In discussing isolation of anaërobic organisms by the plate method, Roux's

vacuum bell jar has already been described.   Tretrôp's (fig. 97), or Baginski's
apparatus, or Adnet's vacuum incubator are also available.   The last is a

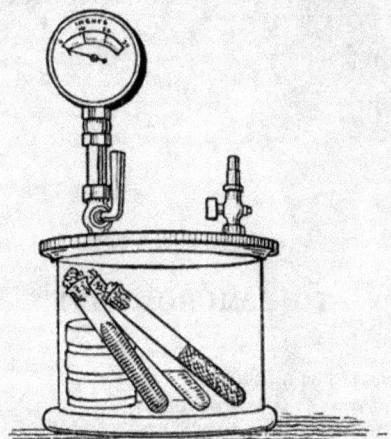

FIG. 97.—Tretrôp's apparatus, in which to grow anaërobic cultures.

stout-walled incubator which can be hermetically closed by means of a door
with an india-rubber washer and is provided with a Roux's regulator and a
gas burner.

CHAPTER VII.

THE MICROSCOPE.

For bacteriological work a good microscope, which will magnify from 600 to 1200 diameters, is necessary. It is seldom that a higher magnification than 1200 diameters is required, though for a few micro-organisms, *e.g.* the organism of pleuro-pneumonia, a magnification of 2000 diameters may be useful. It is to be remembered however that with the very best instruments it is impossible to see organisms measuring less than 0·0001 mm. (0·1 μ) in diameter (p. 113).

A microscope may for purposes of description be regarded as consisting of two parts, the mechanical (the microscope stand) and the optical (the lenses) portions respectively.

SECTION I.—THE MICROSCOPE STAND.

The stand of the microscope must be firm and rigid, and it is desirable that the base be hinged to the body so that the latter can be tilted. The tube should have a rack and pinion mechanism and a micrometer screw adjustment for the grosser and more delicate movements respectively in focussing. The stage, of ebonite or metal, should be large, and it is an advantage if it can be centred and mechanically moved. The mirror, by means of which the light is transmitted to the object, should be concave on one side and flat on the other. The stand, moreover, should be so constructed that an Abbe condenser can be fitted below the stage. A diaphragm either of the cylindrical or iris pattern is also essential. It will be found a great advantage to have a triple nosepiece capable of carrying three objectives, so that one lens can be readily and quickly substituted for another.

SECTION II.—THE OPTICAL PARTS OF THE MICROSCOPE.

The great difficulty in selecting a microscope is the choice of the lenses. For ordinary work two eyepieces, and four objectives including a $\frac{1}{12}$-in. oil-immersion lens, are all that is necessary.[1]

A $\frac{1}{16}$-in. or $\frac{1}{18}$-in. homogeneous immersion lens may be of use occasionally.

In addition to the microscope and its lenses, it is convenient to have a camera lucida and a stage and ocular micrometer.

A. The objectives.[2]

The use of a microscope is to magnify the details of an object, so that those invisible to the naked eye may with its aid be easily seen. The essential requisites then in good lenses are definition and magnification.

1. Magnification.

The apparent linear size of an object AB (fig. 99) varies inversely as its distance BK from the eye of the observer, and depends

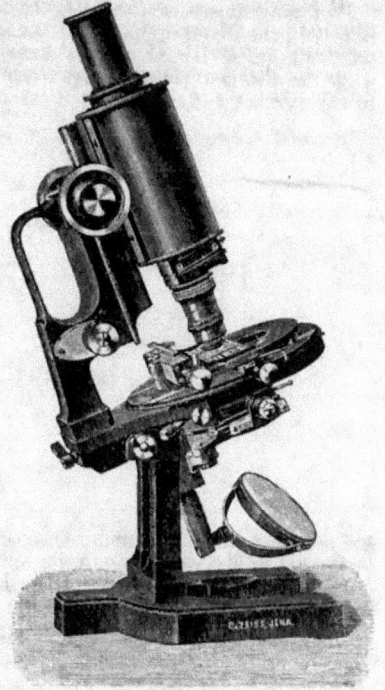

FIG. 98.—A microscope.

upon the tangent of the visual angle α which it subtends at the nodal point K of the eye.

$$\tan \alpha = \tan \text{BKA} = \frac{\text{BA}}{\text{KB}}.$$

Now, let B'K denote the least distance of distinct vision (10 inches), and let it be denoted by l.

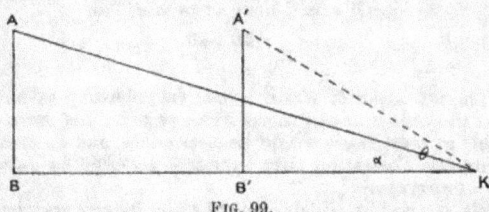

FIG. 99.

Then the greatest apparent size of BA to the unaided eye is when it is in the position B'A', and its apparent size is then

$$\tan \theta = \frac{\text{B'A'}}{\text{KB'}} = \frac{\text{B'A'}}{-l}.$$

It follows from this that the larger the angle subtended by the object at K, the larger will the object appear to be. And a microscope is nothing more than an instrument with which to increase the size of this angle.

[1] In French makes, and also in Reichert's and Leitz' lenses a No. I. or No. II. and a No. III. eyepiece, and a 2, 6, and 8 or 9 dry objective. In Zeiss' list the corresponding objectives are AA, DD, and E in the dry series, and eyepieces, 2, 4 and 8.

[2] The remainder of this section, dealing with the theory of the microscope, has been rewritten and considerably extended.—H. J. H.]

If a convex lens of focal length 2 in. be placed 2 in. in front of an object AB (fig. 100), the divergent pencil of light from A, HAO, will emerge from the lens as a parallel beam F''H, LO, as if it came from an object A' at an infinite distance off, while the divergent pencil of rays from B will emerge from the lens as a beam parallel to the axis, as if it came from B' at an infinite distance off. The image is virtual,

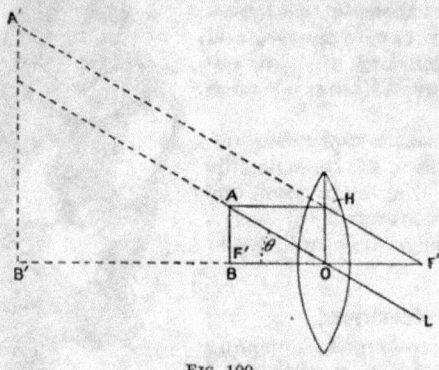

FIG. 100.

not real, and cannot therefore be received on a ground-glass screen, but it appears to the eye as if it came from a big object at an enormous distance away.

The visual angle which this huge image at an infinite distance subtends at the eye will obviously be BOA or θ', and

$$\tan \theta' = \frac{BA}{OB} = \frac{BA}{-F''O} = \frac{BA}{-f''}$$

But

$$\tan \theta = \frac{BA}{-l};$$

$$\therefore M = \frac{\tan \theta'}{\tan \theta} = \frac{l}{f'} = \frac{10}{2} = 5.$$

The object is therefore magnified 5 times by the convex lens of 2 in. focal length placed 2 in. from it. The positive sign shows that the image is erect and virtual.

Now suppose it be required to magnify an object 10 times (linear) it is clear that a lens of 1 in. focal length would have to be used, for

$$M = \frac{10}{f'} = \frac{10}{1},$$

and to obtain a magnification of 400 it would be necessary to have a lens of focal length $\frac{1}{40}$ in., and the object would therefore have to be not more than $\frac{1}{40}$ in. away from the lens; this in most cases would be impossible, not to speak of the extreme spherical and chromatic aberration that would be induced by using a single lens of that high degree of curvature.

There is a simple method by which some of these defects may be overcome, which may be illustrated by a consideration of the simple magnifying glass of 1 in. focal length mentioned above. In this case the lens must be placed 1 in. away from the object to induce a magnification of ten diameters. Now if this biconvex lens be split down the middle, two plano-convex lenses will be formed each of 2 in. focal length. On placing one of these lenses $2\frac{1}{4}$ in. away from the object AB, an enlarged inverted image ab will be formed at a distance of 18 in. from the lens (fig. 101).

(For $$\frac{1}{p} - \frac{1}{q} = \frac{1}{f'}; \quad \therefore \frac{1}{q} = \frac{1}{p} - \frac{1}{f'} = \frac{4}{9} - \frac{1}{2} = -\frac{1}{18}; \quad \therefore q = -18 \text{ in.,}$$

where p is the distance from the object and q the distance from the image to the lens, and f the focal length of the lens.)

On now placing the second plano-convex lens 2 in. beyond this image (i.e. with the image at its focal distance, it will be magnified again. This is the fundamental principle of the **compound microscope**. The first lens, or objective, forms an inverted image 8 times the size of the object.

(For $\dfrac{i}{o}$ or $\dfrac{ba}{BA} = \dfrac{IH}{BA} = \dfrac{F'I}{F'B} = \dfrac{F'I}{FI - BI} = \dfrac{f'}{f' - p} = \dfrac{2}{2 - 2\frac{1}{4}} = -8.$

The negative sign shows that the image is inverted.)

The second lens or eyepiece is now placed 2 in. from this image, *i.e.* 20 in. from

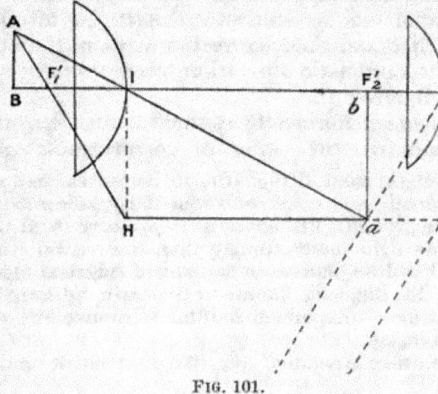

FIG. 101.

the objective, so that the inverted image ba is in its first focal plane. Consequently the image ba will be seen under a magnification of $\dfrac{l}{f'} = \dfrac{10}{2}$ or 5. The total magnification will be therefore $-8 \times \dfrac{10}{2} = -40$.

It follows, then, that by this arrangement the working distance is increased from 1 in. to $2\frac{1}{4}$ in., that the magnification is increased from 10 to -40, while the errors from spherical and chromatic aberration are rather less than with a single biconvex lens of 1 in. focus. Indeed, with the compound instrument, what is called " pincushion distortion " would be almost entirely obviated. If an object were a network of squares, it would be found that, on using a simple magnifying lens, it would present the appearance shown in fig. 102 (pincushion distortion), owing to the fact that the peripheral parts of the object would be more magnified than the central parts. When however a compound instrument is used such as that just described, the objective forms a real image showing " barrel-shaped distortion " as in fig. 103, because the peripheral parts are less magnified than

FIG. 102.—
Pincushion distortion.

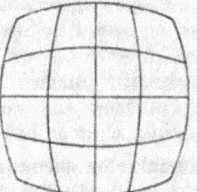

FIG. 103.—
Barrel-shaped distortion.

the central parts. On now viewing this through the second lens, or eyepiece, the barrel-shaped distortion will be completely corrected by the tendency of the virtual image to suffer from pincushion distortion, so that the final image will be rectangular. The fact that it is inverted is no inconvenience.

The enormous advantage of a compound instrument is sufficiently obvious from this simple illustration, but it must be remembered that the eyepiece only magnifies the detail that has already been defined in the real image formed by the objective, and that any defects in this image are exaggerated by the magnification of the eyepiece.

2. Spherical aberration—Coma.

As it is of supreme importance to obtain the most perfect objective possible, some of the defects in the image formed by a simple convex lens when homogeneous light (light of one specific wave-length, *i.e.* of one colour) is used will be considered first and their correction explained, and then the defects in the image due to chromatic aberration when ordinary white light is used will be very briefly referred to.

The defects in the image formed by a simple convex lens when homogeneous light is the illuminant are two :—*spherical aberration and coma.*

Suppose a small bright point P (fig. 104) to lie on the axis of a biconvex lens ; now, although the small axial cone converges fairly accurately to the conjugate focus Q, the eccentric rays PL, PL′ converge to a nearer focus Q′. The peripheral parts of a lens refract light more strongly than the central parts, and hence the image of the point P will be blurred on account of **spherical aberration**. The only way of getting over this difficulty known to the early opticians was to cut off the peripheral rays by means of a diaphragm, but this, of course, very seriously diminished the brightness of the image.

Now take the case where a point P′ (fig. 105) does not lie on the axis of the lens ;

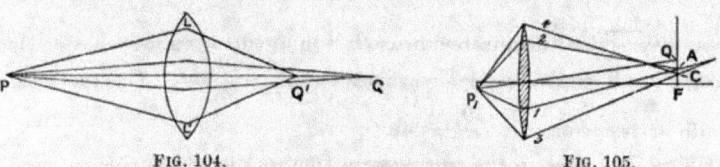

FIG. 104.   FIG. 105.

the image of P′ will be indistinct for a reason which is somewhat similar to that given in the former case, as will be evident from a glance at the figure. The centric pencil will form a well-defined image at Q, but while the rays 1 and 2 will intersect at A the rays 3 and 4 will intersect at C. Hence if a screen be placed in the position QF a bright point will be seen at Q, which becomes an ill-defined flare of light towards F[1] (fig. 106). The image somewhat resembles the tail of a comet and the defect is therefore known as **coma** (κόμη, hair of the head, tail of a comet), and may be regarded as the spherical aberration for object points not on the axis.

These two defects, spherical aberration and coma, must therefore be corrected before any definite distinct image can be formed by an objective. An image free from these defects is known as an **aplanatic image** (ἀπλανής, not wandering). The condition for aplanatism can only be obtained in one way—the lenses must satisfy what is known as Abbe's sine law.

FIG. 106.— Coma.[2]

The sine condition for aplanatism.—Let C be the centre of the circle KAK′ (fig. 107), and let P be a point situated at a distance CP from the centre of the circle, such that

$$\frac{CK}{PC} = \frac{\mu}{\mu'},$$

i.e. the radius of the circle : the distance of the object to its centre
   :: the index of refraction of the first medium : the index of the second medium.

---

[1] F, in this figure, is arbitrary and not the focus.

[2] In "primary" coma the angle formed by the tangents to the series of circles from the point Q in this figure is said to be 60°.

Now find the point Q on the axis such that

$$\frac{QC}{CK} = \frac{\mu}{\mu'}$$

By construction

$$\frac{CK}{PC} = \frac{\mu}{\mu'} = \frac{QC}{CK},$$

i.e. the sides of the triangles PCK, CKQ about the common angle C are proportional.

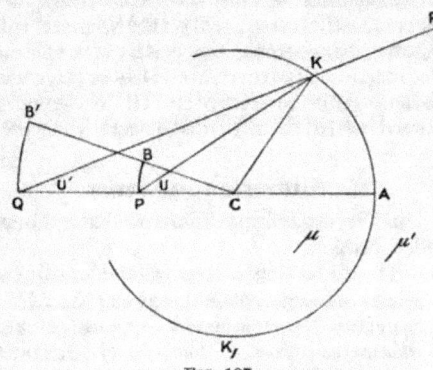

FIG. 107.

∴ by Euclid vi. 6, the triangles are equiangular, viz. CKQ = KPC and KQC = CKP.

Now

$$\frac{\sin CKP}{\sin CKQ} = \frac{\sin CKP}{\sin KPC} = \frac{PC}{CK} = \frac{\mu'}{\mu} = h,$$

where h is the relative index of refraction of the second medium to the first medium.

But if P be a source of light,

$$\sin CKP = \sin \phi = h \sin \phi' = h \sin CKQ,$$

where ϕ = angle of incidence, and ϕ' angle of emergence.

∴ KR must be the refracted emergent ray.

Now the position of K is arbitrary; therefore all the light diverging from P, however great the angle U, must after refraction appear to come from Q. (The angle U, or CPK, represents of course merely one-half the cone of light that falls on the lens.)

This then satisfies the first condition of aplanatism; in other words, the spherical aberration is corrected.

With an oil-immersion lens, seeing that the oil and the glass have practically the same index of refraction, the object may for all practical purposes be regarded as in the glass, which is exactly the condition required to satisfy the sine condition.

For the second condition,—the correction of coma:

From centre C at distances CP and CQ draw arcs PB and QB' (fig. 107). Join CBB'. Then CBB' may be regarded as the axis of the lens, and the image of B will be formed at B'.

Regard PB as the object *o*, and QB' as the image *i*.

Then

$$\frac{i}{o} = \frac{QB'}{PB} = \frac{QC}{PC} = \frac{\frac{\mu}{\mu'}CK}{PC} = \frac{\mu \sin KPC}{\mu' \sin CKP}$$

$$= \frac{\mu \sin KPC}{\mu' \sin KQC} = \frac{\mu \sin U}{\mu' \sin U'};$$

that is,

$$\frac{\text{dimensions of the image}}{\text{dimensions of the object}} = \frac{\text{index of refraction of first medium} \times \sin \text{of half the angle of the rays diverging from the object}}{\text{index of refraction of final medium} \times \sin \text{of half the angle of convergence of the rays forming the image}},$$

and this is the one and necessary test for aplanatism.

3. Angular aperture.

The angle U (fig. 107) is the semi-aperture of the lens; and the total aperture (2U) is the angle formed by the two extreme rays, which starting from the same point on the object ultimately reach the eye of the observer. And obviously, the greater the angle of aperture the greater will be the number of rays of light which leaving the same point on the object reach the eye of the observer, and consequently the brighter will be the image of a given size. So that it is important, especially with the more highly magnifying lenses, that the angle of aperture should be as large as possible. Lenses are now made whose angular aperture is 140° or even more. The angle of aperture is measured with the aid of a special piece of apparatus known as an *apertometer*.

4. Numerical aperture.

The expression $\mu \sin U$ is commonly known as the numerical aperture of the lens, and is denoted by N.A.

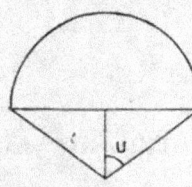

It will be easily seen that the brightness of the image varies amongst other things as $(N.A.)^2$. For the numerical aperture determines the amount of light entering in one diametral plane of the objective, and therefore the total amount of light entering the circular objective must vary as $(N.A.)^2$.

Lenses are more commonly described by their numerical aperture than by their angular aperture. The

FIG. 108.—Numerical aperture.

N.A. of a $\frac{1}{6}$-in. dry lens should not be less than ·82, and of a $\frac{1}{12}$-in. oil-immersion lens not less than 1·3.

The N.A. is determined thus: Suppose a dry objective has an angular aperture of 60°.

Let $U = \frac{1}{2}$ angular aperture, and since the refractive index of air is 1,

$$N.A. = \mu \sin U$$
$$= 1 \sin 30°$$
$$= ·5.$$

Again, suppose an oil-immersion objective has a total angular aperture of 135°.

$$U = \tfrac{1}{2}(135°) = 67\tfrac{1}{2}° \quad \text{and} \quad \sin 67\tfrac{1}{2}° = ·9238795.$$
$$\mu \text{ for cedar-wood oil} = 1·52;$$

then
$$N.A. = \mu \sin U,$$
$$\eqsim 1·52 \times ·924,$$
$$\eqsim 1·404;$$

and if the same lens be used dry, since the critical angle for glass is 41°, the total effective angular aperture would be 82°.

$$\mu \text{ for air} = 1, \quad \text{and} \quad \sin 41° = ·656059;$$
$$\therefore N.A. \eqsim ·656.$$

5. Resolving power.

The resolving power of a lens is the capacity of the lens to optically separate two closely adjacent points on an image which the unaided eye is unable to distinguish as separate, and must be carefully distinguished from magnifying power.

It is found [1] that two objects at a distance d apart can be separated by oblique illumination if

$$d = \frac{·61\lambda}{2\mu \sin U} = \frac{·61\lambda}{2 N.A.};$$

[1] *The Theory of Optical Instruments*, by E. T. Whittaker, M.A., F.R.S.; Camb. Univ. Press; 2s. 6d.

and by direct illumination if

$$d = \frac{\cdot 61\lambda}{\text{N.A.}},$$

where $\lambda =$ wave length of light used.

Thus, in the middle of the spectrum,

$$\lambda = \cdot 00054 \text{ mm.},$$

and N.A. in the very best lenses $= 1\cdot 6$;

$$\therefore \quad d = \frac{\cdot 61 \times \cdot 00054}{2 \times 1\cdot 6} \approx \cdot 000103 \text{ mm.},$$

from which it is apparent that it is impossible to distinguish, *i.e.* to resolve, any two points less than $\cdot 000103$ mm. (approximately $0\cdot 1\mu$) apart, or to see the details of an object of smaller dimensions than $0\cdot 1\mu$.

Limit of effective magnification.—Now the eye can only easily distinguish two objects as separate whose distance apart subtends an angle of $2'$ at its nodal point. This is the angle which a distance of $\cdot 1477$ mm. subtends at 10 inches. Then the necessary magnification is $\frac{\cdot 1477}{\cdot 0001} \approx 1477$. Consequently, the limit of resolution of the microscope is attained when the total magnification is about 1450. With an high eyepiece a further magnification may be obtained up to 1600 or 2000, or even 3000, but no more detail will be discoverable. The effect of the higher magnification will merely make the detail larger: it will add no new detail but will still further diminish the brightness of the image.

Resolving power $\propto \frac{1}{d}$, for it varies inversely as the least distance between separable points. Hence, *whatever the focus*, the resolving power $\propto$ N.A., and *comparing lenses of the same focus*, the brightness of the image $\propto \frac{(\text{N.A.})^2}{\text{M}^2}$.

6. Brightness of image.

Suppose M and M' be the magnifications obtained, using the same objective but different eyepieces,

and let $\qquad \text{M}' = 2500, \text{ and M} = 1450.$

The relative brightness $\frac{B'}{B}$ of the image in the two cases will be

$$\frac{B'}{B} \propto \frac{(\text{N.A.})^2}{\text{M}'^2} \div \frac{(\text{N.A.})^2}{\text{M}^2} \propto \frac{(1450)^2}{(2500)^2}, \qquad \propto \approx \frac{1}{3};$$

so that the penalty of increasing the magnification from 1450 to 2500 is to make the brightness of the image $\frac{1}{3}$ what it was with the lower magnification.

The **penetrating power** $\propto \frac{1}{\text{N.A.}}$, so that

		Resolving power $\propto$ N.A.	Brightness of image $\propto (\text{N.A.})^2$	Penetrating power $\propto \frac{1}{\text{N.A.}}$
Ratio	Oil	1·404	1·972	·7121
	Dry	·656	·43041	1·5242
or	Oil	2·14	4·58	1
	Dry	1	1	2·14

H

That is to say the resolving power—the capacity to recognize as distinct two closely adjacent points—is more than twice as great, and the brightness of the image more than $4\frac{1}{2}$ times as great when the same lens is used with oil as when used dry ; while as regards penetrating power a dry lens is more than twice as efficient as an oil lens, so that when thick sections have to be examined a dry objective of low N.A. should be selected.

And it is clear that the N.A. is of fundamental importance in determining the efficiency of a microscope.

7. Definition.

The *definition* of a lens is its capacity to render the outline of an object or image distinct to the eye, and depends partly upon the sufficiency of the correction for aplanatism, which can be assisted by the use of diaphragms, and partly upon the sufficiency of the correction for chromatic aberration : from which it follows that an achromatic lens has a better definition than a lens not so corrected.

The definition and resolving power of a lens are in practice tested by means of preparations of diatoms, those generally used for the purpose being *Pleurosigma angulatum, Grammatophora subtilissima, Navicula crassinervis, Surdrella gemina,* etc. With a good objective a very distinct image with sharply defined outlines will be obtained ; in the case of *Pleurosigma angulatum* it should be possible to make out, under a magnification of 500–600 diameters, a central venule on to which two systems of oblique lines abut, crossing each other at an acute angle and forming a reticulated system of fine lines.

It is also well when testing an objective, to examine some small organism such as the *Bacillus tuberculosis*, in order to ascertain the magnification produced as well as the sharpness of the image.

8. Chromatic aberration—achromatism and apochromatism.

So far the conditions which must be fulfilled by a lens when homogeneous light is the illuminant have been considered. But in practice white or non-homogeneous light, *i.e.* light of different wave lengths, is used. And with white light a series of images will be formed of different colours, in different places, and of different sizes. Further, only one of these images corresponding to one definite wave length will be aplanatic. As will be readily appreciated, the calculations required for the correction of chromatic errors are of necessity extraordinarily complex ; it must therefore suffice here to say that in practice chromatic aberration is corrected by the use of lenses combined in pairs (or triplets), one lens being concave the other convex. The convergent convex lens is made of *crown* glass, which has a low dispersive power, while the divergent concave lens is made of *flint* glass, which has a high dispersive power. By making these two lenses of a suitable curvature the chromatic aberration for two colours is corrected, and the lenses are said to be achromatized for those colours.

If ϖ denote the dispersive power of the glass between the F and C lines of the spectrum, and f denote the focal length of the lens, then, in order to achromatize the blue and red colours F and C, the couplet must be such that $\varpi_2 f_1 = -\varpi_1 f_2$.

In apochromatic ($\dot{a}\pi o$, apart from ; $\chi\rho\hat{\omega}\mu a$, colour) couplets, fluorite takes the place of crown glass. Fluorite has a similar relative dispersion to flint, so that with these couplets 3 (not 2) different colours can be achromatized. Suppose that in a given apochromatic system the focal lengths for the red, the yellow, and the green rays are the same : then the magnifications will be the same, but the images will not all lie in one plane. Again, suppose a system were so constructed that all the different colours should come to a focus in the same plane : then the images would, though superposed, be of different sizes. And in order to correct as far as possible these defects, an under correction in one couplet is compensated by an over-correction in the next. Abbe's apochromatic oil-immersion objective is made up of ten lenses, as illustrated in the figure. Even then the achromatism is only carried out with regard to the position of the image, not to its size. Without entering

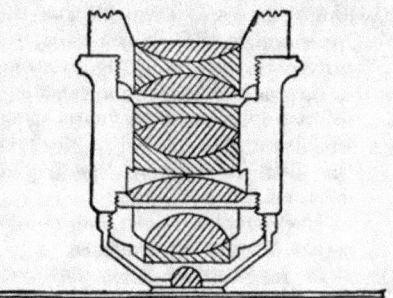

FIG. 109.—Abbe's apochromatic oil-immersion objective.

upon a discussion of the calculations necessary to determine the curvatures of the different lenses, it may be said that in practice if an objective be over-corrected, that is if the power of the flint glass be too great in proportion to that of the crown glass, the error may be rectified by slightly separating the lenses by means of a **compensating collar**: this has practically the effect of decreasing the power of the flint lens. Now if an object be examined under a microscope with an achromatic objective it will be found that the image has either a bluish outline or a yellowish outline. The former is the more common defect, and is due to over-correction, while the latter is a result of under-correction. Abbe's apochromatic objective being achromatic for three colours, is free from secondary spectra ; since however the achromatism has regard to the position of the images, not to their sizes, the blue image though formed in the same plane as the red image is larger than the latter. This error is subsequently corrected by the **compensating ocular**, which produces larger red images than blue. Moreover the sine condition is attained for two colours so that each of these images is aplanatic, *i.e.* the image formed by these two colours is free from spherical aberration and coma. It will therefore be seen that no lens is, in the strict sense of the word, achromatic.

Apochromatic lenses with the necessary compensating eyepieces are very expensive, and are only necessary for special work. Ordinary achromatic objectives are quite sufficient for general purposes.

9. Flatness of image.

Even now all the defects of the image formed by a simple lens have not been studied, for it will always be found that the image is curved. To secure flatness of the image, a condition known as Petzval's condition must be satisfied.

Petzval's condition for flatness of image.—A couplet must satisfy the condition that

$$\mu_1 f_1 = -\mu_2 f_2.$$

But the essential condition for achromatism is

$$\varpi_2 f_1 = -\varpi_1 f_2.$$

It is necessary therefore to use glass with a high refractive index but low dispersive power to obtain an achromatic flat image. It is only recently that Messrs. Schott of Jena have succeeded in making such a glass—barium silicate glass—which produces a greater refraction and a smaller dispersion than crown glass.

B. The eyepieces.

An eyepiece is a system of lenses so arranged that the real image produced by the objective in the tube is magnified and transmitted to the eye of the observer.

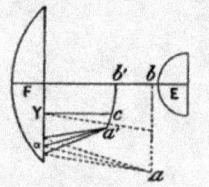

FIG. 110.—Theory of Huygenian eyepiece.

(To render the figure less complicated than it would otherwise appear, the principal plane is taken as on the flat side of the lens.)

In practice it is found expedient to form the eyepiece of two plano-convex lenses separated by an interval; the lower lens is called the field-lens, for it increases the field of view of the instrument, the upper lens is called the eye-lens.

The eyepiece most commonly used is that known as the **Huygenian eyepiece.**

In its simplest form this eyepiece consists of two plano-convex lenses—a field-lens F, and an eye-lens E. The focal length of the former f_1 is three times that of the latter f_2. The curved surface of each lens faces the incident light, and the lenses are separated by an interval d which is twice the focal length of the eye-lens.

Thus $$f_1 = 3f_2, \text{ and } d = -2f_2.$$

Let ba (fig. 110) represent the aplanatic image of the object formed by the objective. The field-lens of the eyepiece is placed below it, indeed half its focal length below it. Consequently the image ba is not actually formed, but the converging pencils proceeding towards the separate points of the image ba are made by the field-lens to converge towards the separate points of the image $b'a'$. Now since the cone of light that corresponds to any point of the image ba meets only an exceedingly small portion of the field-lens, we may neglect the aberrations which occur within each of these cones. Therefore we may regard each point of the image $b'a'$ as being fairly distinct; but it is necessary to consider in what way the spherical aberration of the field-lens will affect their relative positions. Now the field-lens may be regarded as consisting of several annular zones, the refracting power of each zone increasing with its distance from the centre. The axial ray of the peripheral pencil aa' will consequently undergo a greater deviation than that of the intermediate pencil such as γc. The consequence of this will be that the image $b'a'$ will be—

1. Curved, because the refracting power of the peripheral portion of the field-lens being greater than the more central portion, the focus a' of the peripheral pencil will be nearer the lens than the focus c of the intermediate pencil.

2. Distorted, for the peripheral parts of the image ca' will be smaller than the more central part $b'c$, *i.e.* the distortion is barrel-shaped (p. 109).

3. Smaller than the image ba.

In the majority of text-books the Huygenian eyepiece is also proved to be achromatic. But the proof is only applicable for incident *parallel* rays. The eyepiece does not give a strictly achromatic image of the objective's image, so that the proof is not worth considering. It has already been said that the apochromatic objective forms larger blue images than red. The compensating eyepiece having the eye-lens of a flint and crown glass combination forms larger red images than blue, and consequently the final image is completely achromatic.

It will be obvious from what has been said that there are so many errors to correct that it would appear well nigh impossible to correct them all. So would it be were it necessary to form a point image on the retina of each point of the object; fortunately the structure of the retina obviates this necessity. The smallest visual area of the retina is a retinal cone. In the region of most distinct vision, the fovea, the cross section of a cone is a circular area of diameter ·002 mm.   Now if not more than one cone is stimulated the resulting impression will be that of a single point of light.[1]   Therefore errors will be absolutely negligible if they give rise to such small confusion circles that they do not extend over more than one foveal cone.

SECTION III.—THE CARE OF THE MICROSCOPE.

The microscope must be kept at as uniform a temperature as possible and away from direct sunlight and all other sources of heat, because the lenses are held in position by Canada balsam, and they would be displaced and the instrument put out of order if the balsam were to be melted.   It is also essential to protect the microscope from dust, which may best be done by standing it on a piece of thick felt or india-rubber on the bench, and covering it when not in use with a glass shade.

Objectives and eyepieces should always be wiped with a piece of soft linen before use, to ensure their being absolutely clean.   If on looking down the microscope a speck of dust be seen in the field, one must find out where it is in order to wipe it off.   To determine the position of the speck, first rotate the eyepiece ; if the dust be on one or other of these lenses it will of course alter its position ; and if rotation of the eyepiece do not alter its position, then it is on the objective.   By holding the lenses up to the light some distance from the eye, it can be seen if they are cloudy or if specks of dust adhere to them.

To clean the front lens of the objective, rub it with an absolutely clean piece of fine linen ; if this fail to clean it, take a piece of elder pith, strip off a thin layer, and with the clean surface so exposed gently rub the lens.

If cedar-wood oil, Canada balsam, or dammar varnish be sticking to the lens, moisten the cloth with a drop of xylol, and gently wipe the lens.   An excess of xylol must not be used nor should xylol be poured on to the objective, for fear that it should penetrate between the lenses and their mountings and dissolve the balsam holding them in position.

When it is necessary to examine preparations in caustic potash, acids or other chemical reagent, great care must be taken to keep the lenses from coming in contact with the reagent : but if by accident the lens should be soiled, wash it at once in distilled water and dry with a soft linen rag.

If the objective be cloudy, and cleaning the outer lens does not remove the cloudiness, it must not be unscrewed to clean the inner lens, but should be sent to the maker, who is the only person capable of putting it right.

Objectives should be carefully protected against the slightest shocks or falls.

The eyepiece and the Abbe condenser can be cleaned in the same way as the objective, but these lenses are much more accessible and infinitely less delicate.   The mirror can also be cleaned in the same way.

Before putting the microscope away, always wipe the eyepiece and objectives, and remove every trace of oil from the immersion lens.

The stand should be wiped frequently with a chamois leather, and rubbed in the direction in which the lacquer has been applied.   Should the stand be accidentally soiled with balsam or cedar-wood oil, apply a little xylol on a soft cloth, and remove it at once with a chamois leather ; if too much xylol be used or if it be not carefully wiped off it will dissolve the lacquer from the metal.

[1] For simplicity of explanation the question of diffraction is ignored in this case.

A little xylol can also be used to clean the stage.

The coarse and fine adjustments should be lubricated from time to time by the application of a trace of vaseline.

SECTION IV.—METHOD OF USING THE MICROSCOPE.

1. The source of light.

The microscope when in use should rest on a firm table in front of a window. The best light for microscope work is that reflected from a white cloud, but the light may be taken also from a clear sky or a white wall. Direct sunlight is totally unsuitable.

In default of a satisfactory natural light, a good petrol-air or albo-carbon lamp may be used, though a lamp such as Ranvier's with an Auer burner is better. With these lamps it is sometimes necessary to interpose a sheet of ground glass between the source of light and the microscope to moderate the intensity of the former.

[Many observers prefer a small oil lamp, but for general use a very satisfactory artificial light is to be obtained by the use of an inverted incandescent gas mantle, the light from which is passed through a large flask filled with distilled water before reaching the mirror (fig. 111).]

FIG. 111.—Illumination with an inverted incandescent gas burner.

Turn the microscope towards the source of light, look down the tube, and taking hold of the sides of the mirror move the latter about until the field is brightly illuminated.

1. With dry lenses use a concave mirror, which throws a convergent pencil of light on to the object.

2. When using an immersion lens, it is necessary to have an Abbe condenser fitted below the stage. *With a condenser a flat mirror must always be employed*; the rays reflected from the flat mirror are converged by the condenser and brought to a focus on the object, so that by the use of a condenser a considerable amount of light is obtained.

Every microscope should be provided with a diaphragm below the stage. The size of the opening in the diaphragm will be determined by the magnification employed; the greater the magnification the smaller should be the opening in the diaphragm. By cutting off the marginal rays—which are not only useless but actually detract from the sharpness of the image—the diaphragm assists in the correction of spherical aberration, and produces a sharper definition of the object.

2. Arrangement of the object.

The object to be examined under the microscope must be mounted on a microscope slide—a thin very transparent piece of glass free from bubbles of air—and may be covered with a cover-glass—a much thinner and smaller piece of glass, square or circular in shape, and measuring 18–25 mm. in diameter, but not exceeding 0·15–0·20 mm. in thickness.

The rays of light coming from the object as they pass through the cover-glass will be displaced to a greater or less extent, depending upon the thickness of the glass. Fig. 112 shows this. Given any point A on the object, its image on account of displacement will appear along the line DE, and will be diffuse; with [dry] high-power lenses especially, much of the brightness and sharpness of the image will be lost.

[To obtain perfect definition with the higher powers of the microscope, the thickness of the cover-glass is important, and for two reasons :

1. "If the cover-glass be very thick, there may not be room enough to bring the front lens sufficiently near to focus the specimen.

2. "The varying thickness of the actual glass introduces errors in the adjustment of the components of the lens system" (Spitta).]

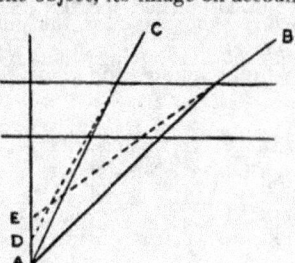

FIG. 112.—Displacement of the rays of light as they pass through the cover-glass.

To overcome this difficulty, it is only necessary to use cover-glasses of the thickness indicated on the objective, each objective being corrected to work for a given thickness. Or, since cover-glasses of exactly the same thickness cannot always be obtained, one may have objectives of certain magnifications, which can be corrected by altering the distance between the component lenses: the thicker the cover-glass the nearer must the lenses be together.

But now that all microscopes have a draw tube this correction is really not of vital importance, because by altering the length of the tube the effect of the thickness of the cover-glass can within certain limits be counteracted. The thicker the cover-glass the shorter must the tube be. With the draw tube right down in its socket, cover-glasses 0·25 mm. thick can be used, but with a normal length of tube (160–170 mm.) one must have cover-glasses no thicker than 0·15 to 0·18 mm.

3. Homogeneous immersion lenses.

Immersion lenses are used in order to counteract the refraction of rays of light in passing from glass to air. In using an immersion lens a drop of some liquid, the refractive index of which is as nearly as possible the same as that of glass, is placed on the cover-glass, and the lens lowered into it. Cedar-wood oil has a refractive index of 1·515 to 1·520, a mixture of castor oil and essence of anise about 1·510, and monobromonaphthaline 1·66. Homogeneous immersion objectives do not need to be corrected.

When rays of light pass from the cover-glass into air, their direction is altered in such a manner that all rays making with the surface of the cover-glass a smaller angle than [48° 12′] are totally reflected and are lost to the objective. By substituting a substance of the same refractive index as glass for air, this loss of light is avoided. An immersion lens makes the image very much [brighter and] sharper ; so that an homogeneous immersion objective whose angle of aperture measures 82° has the same value (*i.e* numerical aperture) as a dry lens whose angle of aperture is 180° (μ sin U) (p. 112). Moreover for the same magnification an immersion objective has a greater focal length than a dry lens.

It is necessary to use an Abbe condenser with an immersion lens, and perfect results can only be obtained with a given length of tube (generally 160–170 mm.). A drop of cedar-wood oil is placed on the cover-glass, and the objective is lowered until its front lens touches the oil.

Immersion lenses should be used only with stained preparations. They are not suitable for the examination of unstained preparations, because the light focussed by the condenser is so intense that it drowns unstained objects and renders their outline very indistinct.

4. The nosepiece.

The nosepiece in most general use is constructed to carry three objectives—usually a No. 2, and a No. 8 or No. 9 dry, and a $\frac{1}{12}$th homogeneous immersion lenses. Each objective is screwed into its proper place in the nosepiece, which is marked for the purpose; it is necessary that this be done in order to get the centering true. By simply rotating the nosepiece, it is thus possible without unscrewing them to use any of the objectives.

5. Eyepieces.

In the great majority of cases a low-power eyepiece should be used. A high-power eyepiece only magnifies at the expense of brightness and sharpness (p. 113). Eyepieces I. or II. are generally used, III. and IV. only when delicate work requiring considerable magnification is in hand.

6. Focussing.

Focussing is done in two stages. The object is first brought approximately into focus with the coarse adjustment, and then sharply focussed by means of the fine adjustment.

The focal length varies with the different objectives, being in inverse ratio to the magnification. The approximate focus for each objective is soon learnt with a little practice, so that the first stage of the process is quickly done.

The object having been brought more or less into focus with the aid of the coarse adjustment, is exactly focussed by means of the fine adjustment working on a micrometer screw.

When high powers are used the objective will be close to the cover-glass, and a rough movement downwards of the lens will most certainly break the slide. [Microscopes are now made so that it is impossible to force the objectives through the cover-glasses.] In any case, to avoid this possibility proceed as follows:

1. Before looking down the microscope, fix the eye on the preparation, and lower the tube slowly with the coarse adjustment until the front lens touches the cover-glass.

2. Now look down the tube, and raise the coarse adjustment until the preparation is approximately focussed.

3. Then get the exact focus by gently rotating the fine adjustment.

The fine adjustment should never be used for large alterations of focus; it is a very sensitive and delicate screw, acting on the microscope tube through a spiral spring, which would soon be put out of gear if used for large excursions.

While the fine adjustment is being used, the thumb and index finger of the right hand should not be taken off the micrometer screw, but should continually move it backwards and forwards gently, until, without any effort of accommodation, the different parts of the preparation are brought into focus and seen in succession, and the shape of the object distinctly made out.

While examining a preparation, the slide should be held between the thumb and index finger of the left hand, and moved about on the stage, so that the different parts can be brought within the field as required. [It is a great advantage to have a " mechanical stage " fitted to the microscope;

this enables the observer by turning a milled screw to place with ease and accuracy any portion of the slide under the objective, and by continuous rotation, rapidly to examine the whole preparation.]

SECTION V.—THE MEASUREMENT OF MICROSCOPICAL OBJECTS.

1. The experimental determination of the magnification produced by a system of lenses.

Magnification produced by an optical system is of course magnification in diameters.

Microscope makers supply a table with each of their instruments, showing with a given tube length the magnification produced with every combination of objective and ocular. This table may be verified roughly by one or other of the two following methods :

A. With a camera lucida.—For this purpose a camera lucida and a stage micrometer are necessary. A stage micrometer is a thin glass slide, on which a scale mechanically divided by parallel lines into $\frac{1}{100}$ths of a millimetre has been engraved.

1. Select the eyepiece and the objective of which the magnification produced by the combination is to be determined. Lengthen the tube to 160 mm., or whatever is the proper working length. Place the micrometer slide on the stage and get it into focus, so that the divisions on the scale are sharply defined.

2. Place a sheet of paper, bluish for choice, on a small drawing table level with the stage of the microscope on the right-hand side of the instrument. Fit the camera lucida to the eyepiece.

3. On looking down the tube of the microscope, two images of the scale on the micrometer will be seen—one formed directly by rays passing through the camera lucida, the other projected by reflection at the prism on to the paper. If the image projected on to the paper be approached with the point of a pencil, the latter will also come into view, and it will be easy to outline on the paper the position of the image of the scale on the micrometer. Trace the position of a few of the divisions of the scale.

4. With a millimetre scale measure the distance between any two of the lines sketched.

Let n =the distance in millimetres of two adjacent divisions, and M =the magnification of the optical system employed.

Since the scale on the micrometer slide is divided into $\frac{1}{100}$ mm., it follows that

$$n = \tfrac{1}{100} M$$
$$M = 100n.$$

Suppose, for instance, the distance between two adjacent lines on the paper be 5 mm. The magnification produced will be 100×5. This is expressed by saying that the magnification is 500, or to be more accurate 500 diameters.

The magnification produced by an optical system can also be determined by simply projecting the magnified divisions of the micrometer directly on to a millimetre scale arranged on the same level as the microscope stage.

Then, if n denote the number of divisions on the scale occupied by m divisions on the micrometer,

$$\text{the magnification is} = 100\frac{n}{m}.$$

Suppose for example that three divisions of the micrometer occupy fifteen divisions on the millimetre scale, then the magnification is $=100 \times \frac{15}{3} = 500$.

This is a simple and convenient method but the results are only approximate, the magnification being somewhat exaggerated.

B. With the ocular micrometer.[1]—An ocular micrometer consists of a small circle of glass on which a scale divided into $\frac{1}{10}$ mm. is engraved. The ocular micrometer is placed between the eye and field lenses of the eyepiece.

The magnification of the eyepiece being known (generally 10 diameters), each division of the scale as seen through the eyepiece is equal to $\frac{1}{10} \times 10$ mm. $=1$ mm.

1. Place the stage micrometer on the stage of the microscope, drop the ocular micrometer into the eyepiece, and turn on the objective to be examined. Adjust the tube to the proper working distance and focus the scale on the stage micrometer, then arrange the latter so that any two lines on it coincide with any two lines on the ocular micrometer.

2. Determine how many divisions of the ocular micrometer are covered by one division of the stage micrometer, and let n be the number.

The magnification produced is given by

$$M = 100n \, ;$$

and if five divisions of the ocular micrometer are covered by one division of the stage micrometer, the magnification is $=5 \times 100$.

2. The measurements of objects under the microscope.

The standard adopted for microscopical measurements is the one-thousandth part of a millimetre, which is designated by the Greek letter μ; the *Bacillus tuberculosis* for example is said to measure $1 \cdot 7$ to $3 \cdot 5 \mu$ long by $0 \cdot 2$ to $0 \cdot 5 \mu$ broad.

Two different methods may be employed for measuring microscopical objects.

A. Camera lucida method.—**1.** First ascertain the magnifying power of the system of lenses to be used by means of the objective micrometer and camera lucida (p. 121).

2. Substitute the slide on which the object to be measured is mounted for the stage micrometer, and an outline of the object will be thrown on a sheet of paper arranged as for the preceding determination.

3. Measure the length of the outline in millimetres, and let n be the length.

4. Then, the magnifying power m of the combination of lenses being known, the diameter D of the object is easily determined from the equation

$$D = \frac{n}{M}.$$

Let us suppose the magnifying power of the optical system to be 500 diameters,

[1 Hermann, Whittaker, and Young warn against the use of an Huygenian eyepiece with a micrometer. Ramsden's is the only eyepiece that can be relied upon. Clearly a micrometer scale put between the lenses of an Huygenian eyepiece will only be magnified by the eye lens and will therefore undergo "pincushion" distortion. But the image of the stage micrometer at that place has "barrel-shaped" distortion which is rectified by the "pincushion" distortion produced by the eye lens.

[The object must be in the centre of the field, as towards the periphery "pincushion" distortion would be more marked.

[Using the same microscope measurements made with the same combination of lenses are comparable among themselves but are not comparable with measurements made with any other microscope nor with the same microscope and any other combination of lenses.]

and the greatest diameter of the outline of the *Bacillus tuberculosis* as sketched with the camera lucida to be 1·5 mm. : then, from the formula

$$D = \frac{1 \cdot 5}{500} = 0 \cdot 003 \text{ mm.} = 3\mu,$$

we find the length of the tubercle bacillus to be 3μ.

A table can be readily drawn up showing the magnification obtained with any combination of lenses, and such a table will save considerable time in the measurement of microscopical objects.

B. Measurement with the ocular micrometer [see footnote p. 122].—**1.** The stage micrometer is examined through the ocular micrometer, and the number of divisions on the ocular micrometer corresponding to one on the stage micrometer determined for each objective. For example, supposing that with objective No. 8 one division of the stage micrometer cover five divisions on the ocular micrometer, then five divisions on the ocular micrometer are equal to $\frac{1}{100}$ mm., and one division to $\frac{1}{500}$ mm., that is to 2μ.

2. Replace the stage micrometer by the object to be measured. Suppose it occupies n divisions on the scale.

3. Now, knowing that one division is equal to 2μ, and using D to denote the diameter of the object,

$$D = n \times 2\mu.$$

If the object cover for example two divisions, then

$$D = 4\mu.$$

Note.—A table giving the value of each division of the ocular micrometer when used with any objective can be drawn up. It is then only necessary to multiply this figure by the number of divisions occupied by an object. For example, using Reichert's lenses—

With objective No. 2 one division on the ocular micrometer scale $= 27\mu$.

,,	No. 4,	,,	,,	,,	$= 11\mu$.
,,	No. 8,	,,	,,	,,	$= 2 \cdot 2\mu$.
,,	No. 9,	,,	,,	,,	$= 1 \cdot 9\mu$.
,,	$\frac{1}{12}$th,	,,	,,	,,	$= 1 \cdot 8\mu$.

Thus : Suppose, using objective No. 8 (Reichert), an object covers two divisions on the ocular micrometer ; then

$$D = 2 \cdot 2\mu \times 2 = 4 \cdot 4\mu.$$

Similarly, an object seen through a $\frac{1}{12}$th immersion lens covers three divisions ; then $\qquad D = 1 \cdot 8\mu \times 3 = 5 \cdot 4\mu.$

It will be readily understood that the higher the magnification the more exact the measurement. With high powers the errors of observation are reduced.

SECTION VI.—DARK-GROUND ILLUMINATION.

It has already been shown (p. 113) that it is impossible even with the best microscopes to distinguish, *i.e.* to resolve, any two points less than about $0 \cdot 1\mu$ apart, or to see any details of smaller dimensions than $0 \cdot 1\mu$.

To render small delicate objects more readily visible under the microscope, Siedentopf and Zsigmondy have utilized the fact that very fine particles placed on a dark back-ground and powerfully illuminated are rendered much more easily visible than when examined on a brightly illuminated surface. Everyone is familiar with this fact in connexion with the stars—the darker the night the brighter the stars. This is the whole principle of [the dark-ground illuminator, or, as it sometimes unfortunately is termed] the ultra-microscope. *The dark-ground illuminator does not increase the resolving power of the system of lenses,* but merely illuminates particles when on a dark background and so renders them more easily visible.

The researches of Siedentopf and Zsigmondy, afterwards extended by Cotton and Mouton, have been taken up by optical instrument makers, who have constructed and are daily improving the apparatus necessary for dark-ground illumination.

1. The application of dark-ground illumination to micro-biology.

Whatever the form of apparatus employed, the dark-ground illuminator does not appear likely to be of assistance in the study of infinitely small things, such as the so-called "invisible micro-organisms," [for the simple reason that objects less than $0·1\mu$ are not resolved. They are seen just as stars are seen, which subtend no appreciable angle, but are visible because their image forms such an intensely bright point of light on part of the apex of one retinal cone that they become visible. Such minute objects appear as bright points in the field of vision surrounded by light and dark diffraction rings; they have neither shape nor form.]

The instrument is, however, of considerable practical value in that it affords more favourable conditions than are obtainable with the ordinarily illuminated microscope stage for the examination of material in the fresh unstained condition. The dark-ground illuminator renders cells and organisms easily visible in the living condition with their natural movements unimpaired. The valuable aid afforded by the instrument in the rapid diagnosis of certain micro-organic diseases, and particularly of syphilis, has been demonstrated by Landsteiner and Mucha, by Gastou and others.

2. The construction of the dark-ground illuminator.

The essential features of the dark-ground illuminator.—The dark background and the powerful illuminant that it is necessary to apply can be realized in several ways.

A. Zeiss' diaphragm.—The simplest and cheapest method—sufficient moreover in the majority of cases for purposes of clinical diagnosis—is to use an

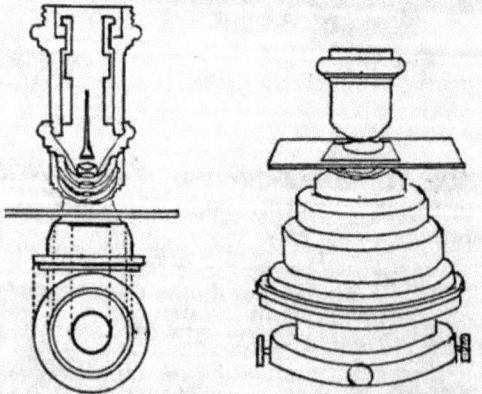

FIG. 113.—Dark-ground illuminator for fixing below the stage.

ordinary microscope fitted with an Abbe condenser (N.A. 1·40), a dry lens (7 or 8) and a high eyepiece (Zeiss' 12 or 18 compensating ocular): the apparatus for dark-ground illumination consists of a special diaphragm which is placed below the condenser. Slides and cover-glasses of a given thickness, varying with every condenser, are essential.

B. Special condensers.—In these cases the ordinary Abbe condenser is replaced by a prismatic condenser (Cotton and Mouton), a parabolic condenser (Zeiss) or a spherical condenser (Leitz) arranged in such a way that the rays reflected by the mirror are deviated, so that they pass obliquely through the film of liquid which is placed between the slide and cover-glass, and cannot enter the objective. Under these conditions any particles held in suspension in the preparation on the stage of the microscope are lighted from the sides while the back-ground is obscure.

In most patterns the dark-ground condenser is placed below the stage

FIG. 114.—Dark-ground illuminator for fixing on the stage.

in the collar generally used for the Abbe condenser, but instruments are now made to fix on the stage of the microscope.

These latter are the better, and they can be used either with a dry lens or with an immersion lens.

3. Method of using the dark-ground illuminator.

To use dark-ground illumination it is necessary to have :

1. A powerful source of light ;

2. A lens to form the image of this source on the mirror ;

3. A firm microscope stage on which to fix the dark-ground illuminator, an objective and an eyepiece.

These are all arranged on a rigid table, and it is an advantage to have an optical bench 1 metre long.

A. The source of light.

The specific intensity of the light increases the visibility of the objects under the microscope. A Nernst lamp, an arc lamp or an inverted incandescent gas burner are the sources of light generally used. Electric light is perhaps better, but an Auer burner (inverted incandescent) (p. 118) is quite good enough for most purposes.

Sometimes it is necessary to use sunlight, and particularly when photographing objects under the ultra-microscope. For this purpose the apparatus is arranged in a dark chamber, and the rays of light falling on an heliostat worked by clock-work pass into the chamber through an opening made in the shutter of the window.

Whatever the pattern of apparatus used, the rays of light must be condensed by a lens on to the flat surface of the microscope mirror.

Sometimes it is better to use instead of a lens a large round flask filled with water lightly tinted with copper sulphate, an arrangement which has the advantage of absorbing the heat rays and so prevents deterioration of the preparation from that cause.

The image of the source of light must be formed on the mirror : to secure this, a sheet of white paper may be placed upon the surface of the mirror

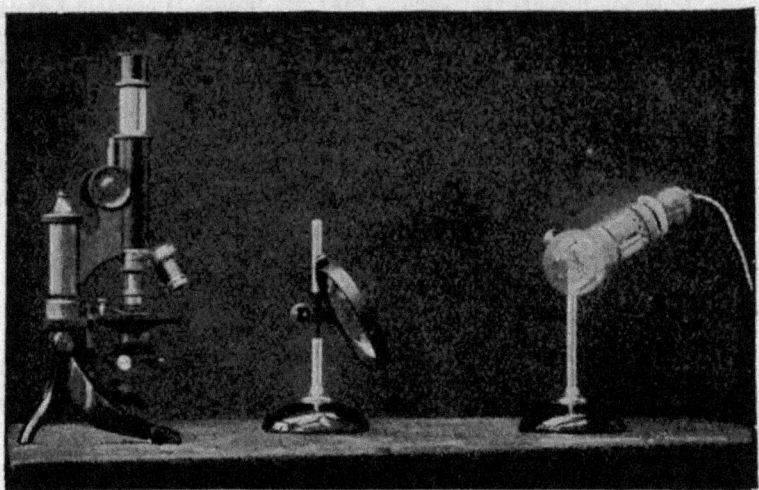

FIG. 115.—Illumination with Nernst lamp and lens.

which is then moved about until the image is clearly defined. The mirror should be uniformly illuminated and the whole surface covered with light.

To get the light arranged satisfactorily requires prolonged manipulation, so that for clinical work where time is an important consideration the apparatus should be arranged beforehand. It is of great advantage in this connexion to have an optical bench, for with it the respective positions of the light, the lens and the microscope can be found once for all. Roughly speaking, the lamp, the lens and the mirror are placed at a distance of 15 to 20 cm. from one another according to the apparatus used. The tube of the microscope should be vertical.

B. Centering.

The dark-ground illuminator, whether placed on the stage of the microscope or arranged in the place of the Abbe condenser, must be centered. The method by which this is done will depend upon whether the apparatus is above or below the stage.

(a) Dark-ground illuminators fixed in the collar ordinarily carrying the Abbe condenser must be so arranged that the lower flange is close up against the collar, and the upper surface just below the upper surface of the stage.

Using a low-power objective and looking down the tube of the microscope, the centre of the apparatus should be brightly illuminated without shadows or halos. If the field be not bright, adjust the lateral screws (fig. 113) until the lighting appears quite uniform.

(b) In those forms which are made for use on the stage, first fix the apparatus with the clips and then, using a low-power objective and looking down the tube, take hold of it on each side with thumb and finger and move it about gently until the centre appears brightly and uniformly illuminated.

C. Arrangement of the preparation to be examined.

1. The preparation to be examined should be mounted on a slide and covered with a cover-glass.

(*a*) The slide should be of crystal glass free from flaws and absolutely clean, because any dust or dirt will seriously interfere with the observation.

Slides and cover-glasses should be washed in acid rinsed in distilled water and kept in alcohol (p. 130). When required for use, it is advisable in order to ensure cleanliness to paint the slide with a layer of collodion, which can be peeled off just before it dries.

Dust which falls on the cover-glass during the examination interferes with the satisfactory lighting of the preparation, and if the observation be prolonged the cover-glass should be washed or dusted from time to time.

(*b*) To secure the most satisfactory illumination the slide should be of a thickness suitable to the particular apparatus in use (all dark-ground condensers are marked with a number indicating the thickness of slide to be used—generally about 1·4 mm.). When working with sunlight it is absolutely necessary that slides of the exact thickness indicated on the condenser should be used ; but with the sources of light ordinarily employed this precision is of less importance, and one-third of a millimetre one way or the other is a matter of no great moment.

The thickness of the cover-glasses should correspond with the correction of the objective (p. 119).

2. There should be continuity between the media through which the light passes, so that refraction may take place under the best conditions ; a large drop of very fluid immersion oil should therefore be placed between the condenser and the slide.

An inferior quality of oil is a frequent cause of failure. The oil should be quite fluid, absolutely homogeneous, contain no air bubbles, and be used in sufficient quantity to completely fill the space between the lens and the condenser.

3. The film to be examined should be as thin as possible, uniform and free from air bubbles. If the material be sufficiently fluid and viscous to keep the slide and cover-glass together the preparation may be examined without any addition. In the contrary case, dilute the material in a drop of blood serum, aqueous humour or ascitic fluid ; water or normal saline solution may be used but these solutions have the disadvantage that they alter the shape and interfere with the vitality of the cells.

If the experiment is to be prolonged it is advisable to lute the edge of the cover-glass with a little vaseline or paraffin to prevent evaporation.

D. Focussing the microscope.

For dark-ground illumination work a dry lens (No. 7, 8, or 9) may be used (though an immersion lens is better) and a high eyepiece (No. IV. or Zeiss' compensating ocular 18).

To obtain a quite black background, special objectives can be employed in the mounting of which a carefully centered diaphragm is suspended to intercept marginal rays : these objectives (Leitz, Zeiss) give remarkably distinct images.

A certain amount of skill which can only be obtained with practice is required to get satisfactory results.

1. With a dry objective.—The lighting being satisfactory, the apparatus centered and the preparation fixed with the clips, the eye is applied to the tube of the microscope which is then slowly lowered. At first there is a certain amount of diffused light, but this soon gives place to complete dark-ness; by continuing carefully to lower the tube, the back-ground will suddenly become lit up in places and dotted with bright points; the preparation is then focussed.

2. With an oil-immersion lens.—Place a drop of cedar-wood oil on the cover-glass and lower the tube until the lens touches the oil. Then with the mechanical adjustment gently raise and lower the tube until the back-ground is illuminated with bright spots.

If the field be unequally lighted or if it be narrowed by shadows, the centering is at fault and must be corrected by careful manipulation of the dark-ground condenser (p. 126).

E. Appearances seen in the field under dark-ground illumination.

When the lighting and centering are satisfactory, and the object focussed, luminous points and spots of different appearances—motile or non-motile—will be seen corresponding to the microscopical objects (micro-organisms, cells, particles of colloid matter, etc.) in the preparation. Certain non-motile

FIG. 116.—Preparation showing spirochætes, leucocytes and red cells (after Gastou).

spots, generally taking the form of rosettes or flocculent masses, may be seen; these are merely flaws in the glass and must not be confused with the objects in the preparation. [This generalization of course only applies when the size of the objects is greater than the resolving power of the com-bination of lenses employed. Any objects in the field which are beyond the resolving power of the combination of lenses will appear as bright spots with light and dark diffraction rings and the size of the objects which will appear as such will depend upon the intensity of the illumination. It has already been pointed out that the so-called ultra-microscope or dark-ground illuminator does not increase the resolving power of the microscope, hence whatever the shape of the object if it be so small as to be below the resolving power of the system of lenses used it will appear as a bright dot surrounded by rings.]

It will be found easy to study the movements (Brownian movements, move-ments of propulsion, etc.) of the different corpuscles. In interpreting these it must not be forgotten that an universal movement of the illuminated elements in the same direction is due to currents set up in the preparation.

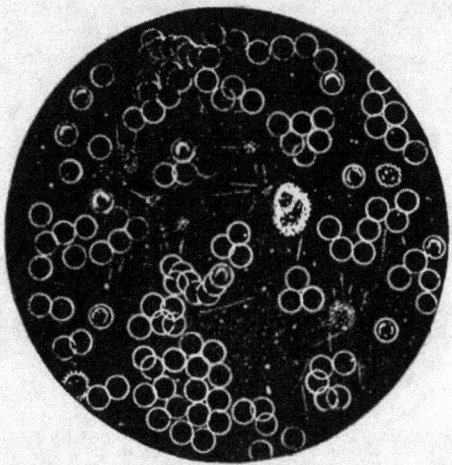

FIG. 117.—Preparation showing red blood cells, hæmatoblasts and strands
of fibrin (after Gastou).

Lastly it cannot be too strongly emphasized that the smallest trace of dust on the slides or cover-glasses interferes materially with the examination of the preparation.

I

CHAPTER VIII.

THE MICROSCOPICAL EXAMINATION OF CULTURES OF MICRO-ORGANISMS.

CULTURES should be examined microscopically in two ways.

(*a*) An unstained preparation of the living organisms should first be examined. By this means not only can the shape of the organisms be determined but also whether they are motile or not, and if motile the nature and rapidity of the movements.

(*b*) Secondly, the morphological study of an organism must be completed by the examination of stained preparations, which will allow a more detailed study of its structure with the higher powers of the microscope.

For the preparation of objects for the microscope a supply of clean slides and cover-glasses is essential, and the methods of preparing these may first be described.

SECTION I.—THE PREPARATION OF COVER-GLASSES AND SLIDES.

The essential qualities of cover-glasses and slides have already been mentioned (p. 119). Before being used they must be carefully cleaned.

1. Cleaning of cover-glasses and slides.

A. New cover-glasses are more or less greasy and cannot be moistened with water. Before using them therefore wash them in 95 per cent. alcohol, and wipe with a piece of soft smooth-surfaced cloth; then to get them perfectly clean they must be passed several times through the heating flame of a Bunsen burner.

In wiping a cover-glass never hold it in both hands because it will certainly be broken, but hold it between the folds of the cloth with the thumb and first finger of the right hand, and rub it gently.

It is convenient to have a wide-mouthed ground-glass stoppered pot on the bench containing 95 per cent. alcohol in which to keep a stock of cover-glasses, so that they can be taken out and dried as wanted.

Slides similarly should be carefully washed in alcohol and dried.

B. Slides and cover-glasses can be used over and over again. They must however be carefully cleaned to remove all traces of material on them ; unless this be properly done mistakes are likely to occur when they are used a second time. The thorough cleaning of soiled slides is therefore of great importance and can be done as follows :

1. Drop all used slides and cover-glasses when they are finished with into a dish containing spirit.

2. When a number have collected put them into a porcelain dish, cover them with a 4 per cent. solution of sodium carbonate and boil for half an hour.

3. Pour off the soda solution, wash in a large volume of water, then drop them into the following solution :

Water,	1000 grams.
Potassium bichromate,	50 ,,
Sulphuric acid,	100 ,,

and boil again for half an hour.

4. Pour off the bichromate solution, wash again in a large volume of tap water, then in distilled water, wipe them dry and drop them one by one into covered pots filled with 95 per cent. alcohol, out of which they can be taken as required.

This method will ensure the glasses being clean.

2. Method of using cover-glasses and slides.

Cover-glasses should be picked up by one of their angles with a pair of Cornet's (fig. 118) or Debrand's (fig. 119) forceps.

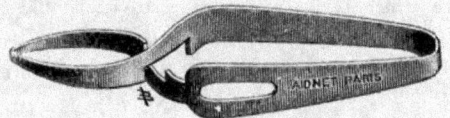

FIG. 118.—Cornet's forceps.

Debrand's forceps, a very useful modification of Cornet's, are well balanced and easily held in the hand : they give a firm hold and do not break the cover-glasses.

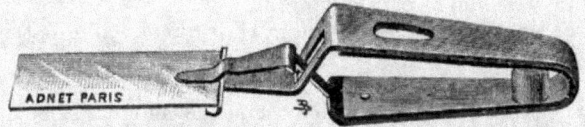

FIG. 119.—Debrand's forceps.

SECTION II.—THE EXAMINATION OF UNSTAINED PREPARATIONS.

A little drop of a culture of a micro-organism may be mounted between a slide and cover-glass and examined. But to keep the organisms alive while they are being examined for the purpose of studying the method of multiplication, etc., special slides having a small concavity or cell ground in their centre, are used. A drop of broth is placed in the cell and sown with the organism ; in this way a living culture is available for the purposes of microscopical examination.

1. Examination of a culture on an ordinary slide.

A. Cultures in fluid media.—**1.** Prepare an absolutely clean slide and cover-glass.

2. Aspirate a few drops of the culture into a Pasteur pipette, taking care of course not to introduce contaminations.

3. Pick up a cover-glass by one of its corners with a pair of Cornet's forceps, and let fall a drop of the liquid from the pipette on to the centre of the cover-glass.

4. Invert the cover-glass on to a slide and the drop will spread out in a thin layer. One must be careful not to introduce any air bubbles as these would interfere with the subsequent examination.

5. Place the preparation on the stage of the microscope and examine with a No. 8 or No. 9 objective and a No. I. or No. II. eyepiece. If the examination is likely to be prolonged the edges of the cover-glass can be luted with paraffin in the following manner : Soak up the excess of culture fluid which has exuded from the edges of the cover-glass with a cigarette paper or piece of filter paper : then apply a heated iron rod—it is better to use a special instrument such as that shown in fig. 120—to a block of paraffin, so as to

FIG. 120.—Instrument for luting with paraffin.

melt a little of it : in doing this some of the paraffin will adhere to the rod and can be transferred to each of the corners of the cover-glass to fix it in position. Then by taking up some more paraffin on the rod the edges can be luted.

The pipette with which the culture was removed should not be used again. Pipettes which have been in contact with a culture must never on any account be laid on the bench. All pipettes after use should be put into a metal vessel, and when the experiment in hand is completed sterilized either in the autoclave or more readily by boiling for a few minutes : only then can they be safely thrown away.

B. Cultures on solid media.—**1.** Take a cover-glass in a pair of forceps, and put a little drop of recently filtered water (Chamberland filter) or sterile broth in its centre.

2. Open the culture-tube in the ordinary way, take up a trace of the culture on a platinum wire and re-plug the tube.

3. Make an emulsion of the culture in the drop of water on the cover-glass with the wire. Flame the wire.

4 and 5. As above.

A common mistake is to remove too much of the culture. If more than a trace be taken, there will be too many organisms in the field of the microscope and the examination of them will be exceedingly difficult. It cannot be too clearly understood that the fewer the organisms the better can their shape, movements, etc., be studied.

2. Hanging drop preparations.

By using a hollow-ground slide any organism under examination can be kept alive for a long time and its development studied.

(i) The technique of the hollow-ground slide.

There are many patterns of slides or cells for use with the microscope.

A. Koch's hollow-ground slide.—This is simply a slide of the ordinary size having a circular cup-shaped hollow about 15 mm. in diameter ground in its centre (fig. 121).   Sterilize the slide as well as the cover-glass with which

FIG. 121.—Koch's hollow-ground slide.

it is to be covered by rapidly passing them through the flame several times just before they are about to be used.

(*a*) In the case of cultures already incubated, take a drop of the culture and place it in the centre of the previously heated and cooled cover-glass, invert the cover-glass over the hollow in the slide and ring the edges with a little vaseline to prevent evaporation.   The drop of culture hangs from the lower surface of the cover-glass into the cavity ground in the slide.

The drop of culture placed on the cover-glass should be small enough to prevent it touching the sides of the cavity otherwise the liquid will run by capillarity between the slide and cover-glass and the hanging drop will disappear.

When examining a hanging drop under the microscope great care must be exercised in lowering the tube, because the cover-glass is only supported at its edges and the least pressure on it will break it.   It is best to use a No. 8 or No. 9 objective and No. I. or No. II. eyepiece (Reichert's lenses).

The small quantity of air contained within the cell is quite sufficient to provide all the oxygen necessary for several days.

(*b*) Most frequently a hanging drop is used to study the development of an organism.   In this case the culture must be sown in the cell.   It can be done thus: Put a drop of sterile broth or sterile aqueous humour on the cover-glass and sow it with the organism under investigation.

It is absolutely essential in doing this that only a very few organisms be sown. A trace of the culture may be picked up on the end of the straight wire and the drop then very lightly touched with the latter, but it is better to adopt the dilution method: thus, sow a broth tube (No. 1) with a loopful of the culture and shake; sow a second broth tube (No. 2) with one or perhaps two drops from tube No. 1, and then transfer a drop of the broth from tube No. 2 to the cover-glass to form the hanging drop.   If tube No. 2 still contain too many organisms, sow a third tube (No. 3) with a few drops from No. 2.   The hanging drop is then made with a drop of broth from No. 3.

The successive steps, then, are as follows:

1. Flame the slide and cover-glass and allow them to cool.

2. Place a drop of sterile broth in the centre of the cover-glass and sow it with a trace of the culture (or, better, take a drop of broth from a tube sown by the dilution method).

3. Invert the cover-glass on the hollow-ground slide and lute the edges with paraffin.

4. Examine the hanging drop on a warm stage (*vide post*), or if a warm stage be not available, incubate it in the ordinary incubator and examine at frequent intervals on the ordinary stage, using a No. 8 or No. 9 objective and a No. I. or No. II. eyepiece.   Make certain that at the time when the hanging drop is made there are not more than two or three organisms in each field of the microscope.

The culture can be kept for examination for 1 to 3 days.   The air present in the cell is generally quite sufficient for the growth of the organism during this period.

To improvize a hollow-ground slide.—A hollow-ground slide may be improvized by taking a rectangular piece of pasteboard about 3 × 2 cm. and 1·5 to 2 mm. thick, and cutting out of its centre a small piece about 15 mm. square. Sterilize the piece of pasteboard in the autoclave at 115° C., take it out with a pair of sterile forceps and lay it on a slide which has been passed through the flame: the cover-glass on which the drop of fluid is placed can be inverted on this to form a hanging drop.

B. Bœttcher's cell.—This cell consists of a glass slide on to which a glass ring (15–20 mm. in diameter and 5 mm. deep) is stuck (fig. 122). The cover-

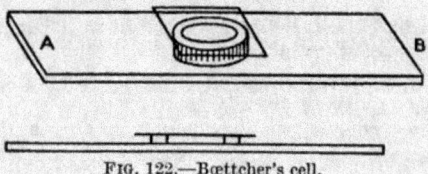

FIG. 122.—Bœttcher's cell.

glass carrying the hanging drop is inverted on to the ring. A little drop of water should be put in the bottom of the cell to prevent evaporation of the culture medium.

C. Ranvier's cell.—In the foregoing cells the hanging drop has a spherical lower surface, with the result that the rays of light passing through it are refracted at points which are not equally distant from the lens, and this to some extent interferes with the examination of the preparation. For delicate work it is better to have the two surfaces of the liquid under examination parallel to each other. This can be attained by using Ranvier's cell (fig. 123),

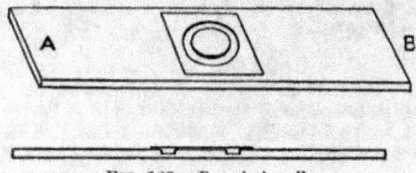

FIG. 123.—Ranvier's cell.

which consists of a rather thick glass slide having a circular groove 15–20 mm. in diameter running round its centre marking off a central elevation which it surrounds on all sides like a moat. The upper surface of this elevated central part is about $\frac{1}{10}$th mm. below the surface of the slide. The drop of liquid, being placed on the central elevation and covered with a cover-glass, is flattened out between the elevated part and the cover-glass, and forms a layer $\frac{1}{10}$th mm. deep surrounded on all sides by the air in the groove ; the edges are luted and the subsequent procedure is the same as in the foregoing cases.

(ii) The cultivation and preservation of hanging drop preparations.

To grow an organism under these conditions it is necessary to keep it at the temperature best suited to its growth, which in the majority of cases is 37° C. This may be done by keeping the slide in the incubator, taking it out when required for microscopical examination ; but it is better to maintain the slide at the temperature required on the stage of the microscope itself, by making use of some form of **warm stage** for the purpose, Vignal's for example (fig. 124) or Malassez's or Ranvier's. These really are small incu-

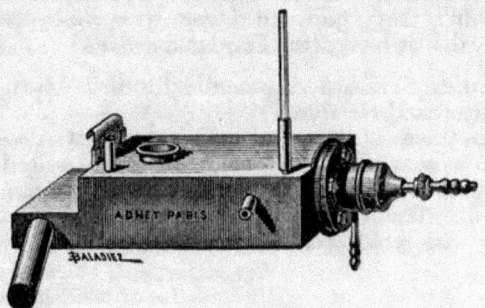

FIG. 124.—Vignal's warm stage.

bators, allowing of the examination of the culture through a circular aperture cut in the apparatus.

Pfeiffer's warm stage is simpler than those already mentioned and serves the same purpose. It consists of a rectangular glass box (fig. 125A), the upper surface of which is hollowed out to form a cell, in which the culture is placed. The box is filled with water and is connected by means of two lateral tubulures to a thermostat. The temperature is indicated by a thermometer placed as shown in the figure.

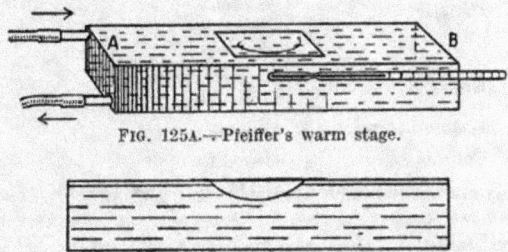

FIG. 125A.—Pfeiffer's warm stage.

FIG. 125B.—Pfeiffer's stage in section.

The apparatus is placed on the stage of the microscope like an ordinary slide.

By another method the lower part of the microscope is enclosed in a box— a small incubator—which entirely surrounds the stand ; the box has a window for lighting purposes and lateral openings to allow of the preparation being moved (Zeiss, Plehn). The apparatus is fitted with a regulator and is heated by a gas burner. The temperature must not exceed 45° C., to avoid injury to the microscope.

SECTION III.—THE EXAMINATION OF STAINED PREPARATIONS.

Staining methods allow a more detailed study of the morphology of micro-organisms than is possible with unstained preparations, and furnish important data for the diagnosis of species. For different species of bacteria do not react in the same way to stains : some are readily stained and cannot be decolourized with alcohol, others which stain with equal readiness lose the stain in alcohol, while a third group stain with difficulty but after being stained resist the action of the most powerful decolourizing agents.

Bacteria are vegetable cells of which the greater part is occupied by the nucleus (Bütschli) : they stain with those dyes which stain the nuclei of vegetable cells, that is to say, **the basic aniline dyes.**

Stains.—Ehrlich divided dyes according to their action on cells into two groups : *basic dyes* and *acid dyes.*

Basic dyes are those in which the staining property depends upon a base combined with a colourless acid. They are called selective dyes, because they exhibit a marked selective affinity for nuclei and especially the nuclei of vegetable cells. The basic dyes are the true micro-organic dyes. Those most commonly used are the following :

Violets, -
- Crystal-violet.
- Thionin (Lauth's violet).
- Gentian-violet.
- Methyl-violet B (Bâle's violet).
- Methyl-violet 6B.
- Paris violet.
- Dahlia.

Blues, -
- Methylene blue.
- Victoria blue.
- Azur.
- Nile blue, or Capri's blue.
- Quinoline blue.
- Unna's polychrome blue.

Reds, -
- Fuchsin.
- Rubin.
- Safranin.
- Neutral-red.

Greens, -
- Methyl-green.
- Malachite green.

Bismarck brown, - Vesuvin.

Colin black, - Indulin.

In **the acid dyes** on the other hand the staining agent is an acid combined with a coloured or colourless base. They are non-selective dyes and stain all tissues indifferently. Fluorescein (phthalic ether of resorcin), eosin (tetrabrom-fluorescein), aurantia, coccinine, acid fuchsin, tropæolin, magenta S, orange G, and picro-carmine are the acid dyes in most common use.

Note.—The aniline dyes have intense staining properties, and should be carefully handled; if the hands be stained accidentally they can be quite easily decolourized with soap. The powders should not be shaken.

Mordants.—In dyeing, an intermediary agent is used to fix the dye more firmly in the fabric. This intermediary agent is known as a mordant, and combining both with the dye and with the tissue unites the two intimately together.

Mordants are also used in staining micro-organisms, and though their mode of action is not as yet thoroughly understood, they undoubtedly increase the affinity of the dyes for the cells and render the staining more rapid and more lasting. The mordants in ordinary use are :

Acids.—Acetic acid.
Phenol.—Creosote.
Tannin.
Iodine in iodine-iodide solution.
Bromine in iodine-bromide and bromine-bromide solutions.
Perchloride of mercury.

Alkalis.—Caustic potash, ammonia, sodium borate, ammonium carbonate, and certain organic alkalis (aniline, phenylamine, toluidine).

Mixtures of two dyes, of which one acts as a mordant towards the other.

Action of heat.—The rapidity and depth of the staining can be increased by heating the preparation in a bath of the stain to 60° or 100° C.

1. Staining solutions.

The staining solutions used in bacteriology are very numerous. Every observer has his own preferences, so that there is a multiplicity of formulæ, making the subject very complicated and embarrassing for the beginner and practical work would gain much by a reduction and simplification of these staining processes. As a matter of fact a few formulæ will meet all ordinary requirements, and if these be thoroughly understood errors which often arise from the use of too complicated and unfamiliar methods will be avoided.

The various formulæ to be found in papers published during recent years must be given, but those methods which in our own experience have given good results will be distinctly indicated and will be found sufficient for practically all purposes. The acid dyes will not be dealt with in this chapter but will be referred to later, and the consideration of some of the staining methods of limited application will be deferred until occasion for their use arises.

To avoid mistakes only good dyes obtained from well-known sources should be used.

A. Simple solutions.

These solutions have only a limited use ; staining solutions containing a mordant are generally better.

(i) Alcoholic solutions.

Alcoholic solutions of the basic aniline dyes are prepared by mixing in a ground-glass stoppered bottle :

Dye, ... 1 gram.
Absolute alcohol, ... 10 c.c.

Shake well and leave the alcohol standing on the dye. Filter before use. Alcoholic solutions keep for a very long time in the dark, and solutions of the following dyes should be kept in the laboratory, viz. **fuchsin, crystal-violet** or **gentian-violet, and methylene blue.**

These solutions are not used for staining, but when diluted with water serve for the preparation of watery alcoholic solutions.

(ii) Watery alcoholic solutions.

Watery alcoholic solutions are prepared by mixing

Filtered alcoholic solution of the dye, ... 1 to 5 c.c.
Distilled water, ... 100 c.c.

Filter immediately before use.

These solutions are seldom used in this form, as they do not keep well : it is simpler to make them up as required by pouring several cubic centimetres of water into a porcelain dish, and adding to it a few drops of the filtered alcoholic solution until an iridescent pellicle with a metallic lustre appears covering the surface.

(iii) Aqueous solutions.

Mix in a small bottle

Dye, ... 0·25 gram.
Distilled water, ... 25 c.c.

Shake and leave the water standing on the dye. Filter before use.

The above proportions give a saturated solution and there should be an excess of the dye at the bottom of the bottle.

These solutions are very little used : they do not keep well and should be prepared as required. They stain slowly but sharply.

Aqueous solutions of **quinoline blue, vesuvin, methyl-green** and **neutral-red** are used for staining living organisms.

B. Staining solutions containing a mordant.

(i) Carbolic acid solutions.

These are more used than any other stains and retain their properties for a very long time.

Ziehl's carbol-fuchsin.

Basic fuchsin,	1 gram.
Carbolic acid crystals,	5 grams.
Absolute alcohol,	10 c.c.
Distilled water,	100 ,,

Rub up the fuchsin and alcohol in a glass mortar, add the carbolic acid and mix ; add two-thirds of the water little by little, stirring all the time ; pour the mixture into a bottle then rinse out the mortar with the remainder of the water and add it to the mixture in the bottle. Leave for 24 hours before filtering into a clean ground-glass stoppered bottle.

A diluted solution prepared as follows is often used :

Dilute carbol-fuchsin.

Mix

Ziehl's carbol-fuchsin,	1 c.c.
Distilled water,	3 to 10 c.c.

Mix and filter just before use.

Carbol-gentian-violet (Nicolle).

Gentian-violet,	1 gram.
Carbolic acid crystals,	2 grams.
Absolute alcohol,	10 c.c.
Distilled water,	100 ,,

Prepare as in the case of carbol-fuchsin. Use as such. This solution is chiefly used for Gram's stain.

Carbol-crystal-violet (Roux).

Substitute crystal-violet for gentian-violet and prepare in the same way as the preceding solution.

Crystal-violet has the advantage over gentian-violet of being a well-defined crystalline compound. Gentian-violet is an amorphous product which varies in composition. Crystal-violet however is not so powerful a dye as gentian-violet.

Carbol-thionin (Nicolle).

Thionin,	0·5 to 1 gram
Carbolic acid crystals,	1 gram
90 per cent. alcohol,	10 c.c.
Distilled water,	100 ,,

Prepare in the same way as carbol-fuchsin. This stain is recommended for sections and films ; it stains rather more slowly but gives better results than crystal-violet and gentian-violet and does not overstain.

Carbol-methylene-blue (Kühne).

Methylene blue,	1·5 to 2 grams.
Carbolic acid crystals,	2 grams.
Absolute alcohol,	10 c.c.
Distilled water,	100 ,,

Prepare in the same way as the foregoing solutions.

Unna's polychrome blue.

Unna's polychrome blue solution (Grübler),	100 c.c.
Carbolic acid crystals,	1 gram.
90 per cent. alcohol,	10 c.c.
Distilled water,	Q.S. to 100 c.c.

Dissolve the carbolic acid in the alcohol, add sufficient water to make up to 80 c.c. and then add the polychrome blue.

(ii) Aniline solutions.

These solutions keep badly and should be freshly prepared every time they are wanted. They have no advantage over carbolic solutions and are gradually dropping out of use.

In preparing them, the following solution must first be made up:

Aniline oil water.

Aniline oil,	5 c.c.
Distilled water,	100 „

Mix the oil and water in a yellow glass bottle, shake vigorously and leave them in contact. Just before use filter the solution through a previously moistened filter paper. See that no fine droplets of oil pass through the filter as this would spoil the results of the staining; should this accident occur, filter the solution again.

Ehrlich's aniline-violet.

Filter into a porcelain dish about 10 c.c. of aniline oil water. To the filtrate add a few drops of a filtered alcoholic solution of gentian-violet until an iridescent pellicle appears. Use the solution at once. It should be freshly prepared every day.

Aniline-fuchsin, aniline-crystal-violet and aniline-methylene-blue are all prepared in the same way.

(iii) Alkaline solutions.

These solutions have been extensively used in Germany. Almost the only alkaline solution now used however is **Lœffler's alkaline methylene blue.** **Borrel's blue** (*vide Hæmatozoa*) is an alkaline dye. [**Borax blue** also is not infrequently used for staining some of the hæmatozoa (*q.v.*).]

Lœffler's alkaline methylene blue.

Alcoholic solution of methylene blue,	30 c.c.
1 in 10,000 aqueous solution of caustic potash,	100 „

Mix in a bottle and filter before use. This solution is rapidly decomposed by the caustic potash combining with the CO_2 of the atmosphere.

Kühne's alkaline blue.

Alcoholic solution of methylene blue,	30 c.c.
1 per cent. aqueous solution of ammonium carbonate,	100 „

Mix and filter before use. This solution keeps better than Lœffler's.

(iv) Perchloride solutions.

Nastikow's violet.

1 in 2000 aqueous solution of perchloride of mercury,	10 c.c.
Alcohol solution of gentian-violet,	1 „

Mix and filter. This stain does not keep well.

(v) Complex stains.

Roux's blue.

SOLUTION A.

Violet dahlia, · · · · · · · · ·	1 gram.
Absolute alcohol, · · · · · · · · ·	10 grams.
Distilled water, · · · · · · · · ·	Q.S. for 100 grams.

SOLUTION B.

Methyl-green, · · · · · · · · ·	2 grams.
Absolute alcohol, · · · · · · · · ·	20 „
Distilled water, · · · · · · · · ·	Q.S. for 200 grams.

1. Prepare each solution separately by rubbing up the dye with the alcohol in a mortar and add the water gradually. Let the mixture stand for 24 hours in a bottle.

2. Then mix the two solutions, filter and store in a well-stoppered bottle.

2. Simple staining.

For purposes of staining there should be at hand—

(*a*) Several small glass funnels and a number of pieces of filter paper folded to fit them. Staining solutions ought always to be filtered before being used and should be dropped from the filter straight on to the preparation.

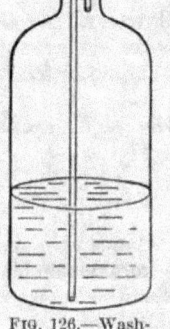

(*b*) A wash-bottle filled with water recently filtered through a Chamberland filter (fig. 126). This bottle is so arranged that by simply tilting it the water runs out through the glass tube.

(*c*) In the absence of a sink, a large glass dish to collect the washings.

(*d*) A number of slides and cover-glasses, a pair of Cornet's or Debrand's forceps, platinum needles, a piece of soft cloth, some small squares of filter paper or a packet of cigarette papers and a few Pasteur pipettes.

(*e*) A Bunsen burner with a pilot flame.

(i) The methods of staining living organisms.

FIG. 126.—Wash-bottle.

The object of staining living organisms is to make them more readily visible for microscopical examination while at the same time preserving their motility.

For this purpose aqueous solutions of dyes which have no toxic action on the organisms are used *e.g.* vesuvin (Metchnikoff), methyl-green (Babès), quinoline blue, fuchsin, neutral-red, etc.

Technique.—Make the preparation in the same way as for the examination of unstained living organisms. Invert the cover-glass on the slide and run a drop of a watery solution of the dye along the edge of the cover-glass ; by capillary action it will be drawn between the cover-glass and slide.

Or if preferred a small drop of the stain can with a very fine pipette be added to the culture on the cover-glass and the two solutions mixed with the end of the pipette ; the cover-glass is then inverted on the slide and the preparation is ready for examination.

(ii) The staining of dried films.

This is the best method of examining the morphology of micro-organisms and it gives moreover preparations which are practically permanent.

Technique.—A. 1. Pick up a cover-glass with a pair of Cornet's forceps, place a drop of a broth culture about its centre and spread the drop with the end of a pipette ; *or*

Place a small drop of filtered water on the cover-glass, mix a trace of the growth from a solid medium with it and spread the mixture with a platinum wire.

2. *Dry* the film gently either by holding it above the pilot flame of a Bunsen or by placing it on a Koch's drying stage (fig. 127) heated to 45° or 50° C.

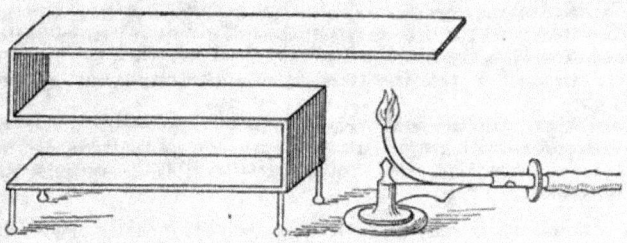

FIG. 127.—Koch's drying stage.

While the film is drying keep the liquid evenly spread over the cover-glass to prevent the formation of concentric circles.

3. Now *fix* the film to prevent the organisms being washed off by the stain, etc. This may be done either (*a*) by passing the cover-glass, film upwards, two or three times through the ordinary Bunsen flame ; the organisms are liable to be distorted and shrivelled by this procedure so that it is better (*b*) to pour two or three drops of alcohol-ether on the film side of the cover-glass and let it evaporate. This method produces no distortion of the organisms.

Alcohol-ether.

Absolute alcohol,	50 c.c.
Ether pur.,	50 „

In special cases it is better to fix the films by immersing them in absolute alcohol for 15 or 20 minutes or by exposing them to the vapour of osmic acid (*vide Treponema pallidum*).

4. Filter two or three drops of *stain* straight on to the film (diluted carbol-fuchsin, carbol-thionin, alkaline blue, etc., may any of them be used). Be careful not to let the stain run on to the under surface of the cover-glass. Stain for $\frac{1}{2}$ to 1 minute.

5. *Wash* off the stain by running a gentle stream of water from the wash-bottle on to a corner of the cover-glass. The water ought not to be poured on to the centre of the film for fear of washing it off.

6. The film may now be examined (with the $\frac{1}{12}$ in. immersion lens and a No. II. eyepiece, for preference) :

(*a*) Provisionally, in water, by inverting the wet cover-glass on to a slide, blotting the upper surface of the cover-glass with a fine cloth [or filter paper] and then placing a drop of immersion oil on the blotted surface.

(*b*) After drying and mounting in Canada balsam. If the film is to be mounted, dry the cover-glass either in the air or by heating it gently, and place a small drop of balsam with the end of a fine glass rod on the film side, invert it on to a slide and press gently to spread the balsam.

To sum up : *Spread a drop of culture on a cover-glass, dry, fix, stain, wash in water, dry, mount in balsam and examine.*

Notes.—(*a*) It is important in staining films to remember which is the film side of the cover-glass. If this should be forgotten, gently scratch the surfaces of the cover-glass near the edge with the point of a needle, and the side on which the film has been spread will be easily distinguished by the little scratches which will remain. But the possibility of losing the film side can be avoided by marking the upper limb of the forceps with a glass pencil, or by the use of a pair of Cornet's forceps on one limb of which a small knob-like depression is impressed in the metal; if the forceps be always used with this knob upwards it will act as a guide.

(b) A small quantity only of culture should be used for making the film. The shape of the organisms can best be made out when there are only a few in the field of the microscope.

(c) The Canada balsam should be dissolved with xylol and the solution should be of such a consistency that it does not tail when a drop is taken out with the glass rod. Balsam should be kept in a bottle stoppered with a glass bell-stopper and having a rim arranged so that the excess of balsam taken up on the glass rod can be drained off.

(d) Alcohol-ether, alcohol and volatile reagents generally are best kept in drop-bottles stoppered with ground-glass stoppers (several patterns can be obtained at the shops): shallow thick glass bottles of 60 to 100 c.c. capacity are perhaps the most convenient.

B. The method which has just been described is especially useful for delicate work and for making films which are to be preserved. But for the provisional examination of cultures and for routine work it is quicker and more economical to work with slides.

1. Take a slide between the fingers or hold it in a pair of Debrand's forceps, and place a little drop of the culture on it.

2. Spread, dry and fix as in the former case (**A**).

3. Stain the film and wash it in the manner described, and then dry the slide. Put a drop of cedar-wood oil straight on to the film without using a cover-glass and examine with the oil-immersion lens.

If after examination it is desired to preserve the preparation the cedar-wood oil can be washed off with a few drops of xylol, and when this has evaporated the slide is put away dry. Or after washing off the cedar oil the film may be mounted with a drop of balsam and cover-glass.

[**C.** Another method which gives excellent results and is commonly adopted by us seems to deserve description here.

[1. The film is spread, dried and fixed on a slide as in **B**.

[2. Wash the film for a moment or two in a 10 per cent. aqueous solution of acetic acid. Wash thoroughly in water. Blot and dry.

[3. Place a drop of the stain on the centre of the film and lower a cover-glass on to the stain, avoiding the introduction of air bubbles. Blot the upper surface of the cover-glass with blotting paper.

[4. Put a drop of oil on the dried surface of the cover-glass and examine with a $\frac{1}{12}$th oil immersion.

[If the preparation is to be preserved, float off the cover-glass from the slide by putting a drop or two of water at the edge of the cover-glass. Wash the slide in water, blot and dry it. The slide may then be kept indefinitely.

[Both in this method and in the preceding, the film may be decolourized in a 10 per cent. aqueous solution of acetic acid or in alcohol, and then restained with another dye; so that using the same film one may first determine the morphology of the organism by examination in a simple stain and then ascertain its reaction to Gram's stain (*vide infra*).]

3. Gram's method of staining.

Gram devised a method of staining which serves to divide bacteria into two large groups.

Some bacteria, when stained with a basic pararosanilin dye in aniline or carbolic solution and treated afterwards with a special mordant containing iodine, are not decolourized by absolute alcohol and similar decolourizing agents. The anthrax bacillus is an example of this group.

On the other hand, other bacteria when treated in the same way are readily decolourized with absolute alcohol, *e.g.* the typhoid bacillus.

Bacteria then are classified with reference to these reactions into two groups, termed **gram-positive** (organisms which retain the stain) and **gram-negative** (organisms which are decolourized). The anthrax bacillus is said to be gram-positive, and the typhoid bacillus gram-negative.

The mordant has the following composition :

Gram's (or Lugol's) solution.

Iodine, - - - - - - - - -	1 gram.
Potassium iodide, - - - - - - - -	2 grams.
Distilled water, - - - - - - - -	300 c.c.

In the original method absolute alcohol was used as the decolourizing agent. But pure aniline oil (Weigert) or acetone alcohol (Nicolle) are now sometimes used in its place.

Acetone alcohol.

Absolute alcohol, - - - - - - -	5 parts.
Acetone, - - - - - - - -	1 part.

According to Nicolle, a bromine-bromide, iodine-bromide, or bromine-iodide solution may any of them be used in place of Gram's solution. They are all prepared in the same proportions as Gram's iodine-iodide solution.

Gram's stain has undergone many modifications, and is used as a double stain for films, sections, etc. These modifications will be dealt with in a special chapter and for the present the use of this classical method as a means of diagnosis will alone be considered.

Technique.—**1.** Prepare a film on a slide or cover-glass.

2. Stain for 30 to 60 seconds with carbol-gentian-violet.

3. Blot up the excess of stain (but do not wash), drop two or three large drops of Gram's solution on the film and let it act for 20 to 30 seconds. The preparation will have now assumed a brown tint.

4. Wash in water and dry.

5. Pour absolute alcohol over the film a drop at a time until no more violet stain comes away—usually 20 to 30 seconds (Notes (*a*) and (*b*) *infra*).

6. Wash in water quickly.

7. Examine the film in water. If the organisms are gram-positive they are stained deep violet, but if gram-negative decolourized : sometimes some of the organisms will be decolourized while others are still stained violet ; in that case a further washing in alcohol will complete the reaction.

[Many bacteriologists prefer to counterstain the film. For this purpose, after washing in water (Stage 6) the film is flooded with some weak staining solution the colour of which is in sharp contrast with violet. Dilute carbol-fuchsin (1–5 or 1–10) or bismarck brown (p. 136) is convenient ; the former is allowed to act for about ½ minute, while bismarck brown requires rather longer (2 minutes). Wash in water, blot and dry. Gram-positive organisms are as in the former case stained violet, while gram-negative organisms being decolourized by the alcohol take the counterstain and appear pink or brown as the case may be.]

To keep the cover-glass preparation, dry and mount in balsam. If the film was made on a slide it merely requires to be dried.

To sum up : *Prepare and fix a film, stain, treat with iodine solution, wash, dry, treat with alcohol, wash [counterstain, wash,] dry and examine.*

Notes.—(*a*) Stage 5, decolourization, is a delicate manipulation. The length of time during which decolourization must be continued varies with the intensity of the stain used, the length of time during which it is allowed to act, the number of organisms, etc.; practice and a certain amount of skill are more than any rules the secrets of success. It is obvious that insufficient decolourization of a gram-negative organism may lead to mistakes; on the other hand the most resistant bacteria can be decolourized by prolonging unduly the action of the alcohol, and such treatment might result in a gram-positive organism being classed with the gram-negative group.

[(*b*) In view of the difficulty as to the time of decolourization we have found it useful, especially for beginners and in dealing with organisms such as the meningococcus, to prepare on the same slide three separate films. At one end there will be a film of a gram-positive organism *e.g. Staphylococcus*, at the other end a gram-negative organism *e.g. Bacillus coli communis*, and in the centre the organism whose reaction is to be tested *e.g. Meningococcus*. The films are stained and decolourized as described above, and then examined in water. If decolourization is sufficient, the staphylococci are all violet and the colon bacilli all pink (or brown). If the films have been under-decolourized some of the bacilli will be stained violet, and if over-decolourized some of the cocci will be pink (or brown). When it is evident that the decolourization has been correctly done the organism whose reaction is being tested is examined.]

(*c*) Films prepared by Gram's method do not keep so well as when stained with ordinary stains and ultimately become decolourized.

4. Claudius' method.

Claudius suggested a method of staining which, while having all the advantages, has both a simpler technique and gives more constant results than Gram's method. Thus the bacillus of malignant œdema and the bacillus of quarter ill are somewhat readily decolourized by Gram's method but retain the stain well by Claudius' method.

The author repeated Claudius' experiments and obtained results which fully confirm that observer's. Claudius' method has many advantages for the student; beginners using Gram's method never know when to stop decolourizing, sometimes they leave the alcohol on too long and sometimes they do not let it act for long enough, and in either case the results are unsatisfactory. These difficulties do not arise in Claudius' method.

The following solutions are required :

(*a*) A 1 per cent. aqueous solution of methyl violet 6B (or a solution of carbol-gentian violet).

(*b*) A solution of picric acid.

Saturated solution of picric acid, - - - - - 1 volume.
Distilled water, - - - - - - - - 1 „

Technique.—1. Prepare and fix a film in the ordinary way.

2. Stain with violet for 1 minute.

3. Wash in water, and blot up the excess.

4. Treat with the picric acid solution for 1 minute and blot.

5. Decolourize with chloroform or clove oil until the decolourizing agent is no longer tinted blue.

6. Examine in clove oil or mount in balsam.

CHAPTER IX.

THE STAINING OF SPORES, CAPSULES AND FLAGELLA. THE STUDY OF THE MOTILITY OF BACTERIA.

SECTION I.—SPORES.

WITHIN the protoplasm of certain micro-organisms, a small bright refractile spot is seen at one period or another of their existence. This refractile body, which does not stain readily with the ordinary aniline dyes, is known as a **spore**, or more strictly an **endospore**. The occurrence of spores was first described by Pasteur.

On the death or destruction of a spore-bearing organism the spores are set free from the protoplasm in which they originated. They are surrounded by a highly resistant membrane, which not only renders them immune to the agents ordinarily destructive of bacteria, but also prevents them becoming stained by the methods generally employed for staining micro-organisms.

Endospore formation does not occur in all bacteria : it is unknown in the micrococci, in which the resistant form is due to a thickening of the enveloping membrane and is known as an *arthrospore*. **Arthrospores** differ from endospores in that they react to stains in the same way as do their corresponding organisms.

It is therefore only necessary to describe the methods of staining endospores. The organisms more commonly used for illustrating the methods are the *Bacillus anthracis, Bacillus megatherium, Bacillus maligni œdematis, Bacillus tetani,* and *Bacillus subtilis.*

1. Examination of unstained preparations.

In unstained preparations the spore appears as a small, refractile, spherical or oval spot within the protoplasm of the cell ; it is surrounded by a bright refractile ring, and is always smaller than the mother cell. The mother cell gives rise to a single spore which becomes free on the disappearance of the cellular protoplasm ; the spore in turn germinates, giving origin to a new bacterium.

All these facts can be observed under the microscope in a hanging drop culture of the anthrax bacillus (p. 134). When it is desired merely to determine the presence of spores in bacteria an ordinary film is made on a slide as described on pp. 140 and 141.

2. The staining of spores.

When spore-bearing organisms are stained with the basic aniline dyes the spores do not take up the dye and appear as unstained spots in the stained bacilli. To stain the spore it is necessary therefore to apply special methods which have been designed to overcome its resistance to staining reagents.

(i) Simple staining.

This method stains both the bacilli and the spores.

A. Method recommended.—1. Prepare a cover-glass film of the culture to be examined and dry it.

2. Pass the cover-glass, film side uppermost, ten times through the heating flame of the Bunsen, but sufficiently quickly to prevent the preparation being scorched.

3. Stain with carbol-violet for 15 to 30 minutes.

4. Wash, dry, mount in balsam. Examine. The bacteria and the spores are both stained violet.

B. Chromic acid method.—1. Make a film on a cover-glass and dry it.

2. Drop a large drop of a 1 in 20 aqueous solution of chromic acid on the film, and leave it for 4 or 5 minutes.

3. Wash in water.

4. Stain in carbol-violet for 15 minutes to half an hour.

5. Wash. Mount. Examine.

(ii) Double staining.

The object of double staining is to differentiate the spore from the bacillus by staining the bacillus one colour and the spore a different colour.

Principle of the method.—Spores stain with difficulty, but once stained they retain the dye with more tenacity than the bacillary protoplasm, so that decolourizing agents will decolourize the latter before they take the stain out of the spores.

A. Method recommended.—1. Prepare a cover-glass film, dry and fix it by passing it rapidly through the flame two or three times.

2. Drop a large drop of carbol-fuchsin on the film and warm it over a small flame until steam just begins to rise, then keep the solution warm for 4 or 5 minutes by moving it about over the flame.[1] Both the bacilli and the spores are now stained an intense red.

3. Wash in water.

4. Decolourize for a few seconds in a solution of nitric acid :

Pure nitric acid,	1 part.
Distilled water,	3 parts.

The bacilli should be decolourized while the spores are still stained red.

5. Wash well in water.

6. Counterstain with a drop of diluted alcoholic solution of methylene blue for 30 to 60 seconds. The decolourized bacilli take up the blue stain.

7. Wash. Dry. Mount in balsam.

The bacilli are stained blue ; the spores red.

[1 This may be done on a warm stage (p. 141, fig. 127) taking care to select a place where the metal is not too hot.]

Note.—This method gives excellent results with *B. megatherium* but is not so good for *B. anthracis* : absolute alcohol is a better decolourizing agent for the latter. Decolourization is in fact the difficult part of double staining, but after a few trials the extent to which decolourization must be pushed to decolourize the bacilli while leaving the spores stained can be determined for different organisms.

B. Mœller's method.—**1.** Make a cover-glass film. Dry. Fix in absolute alcohol for 2 minutes, then in chloroform for 2 minutes. Dry.

2. Drop a few drops of a 1 in 20 aqueous solution of chromic acid on the film and leave for 4 or 5 minutes. Wash in water.

3. Stain in carbol-fuchsin in the warm as described above (**A**). Wash in water.

4. Decolourize for a few seconds in a 5 per cent. solution of sulphuric acid and complete the decolourization in absolute alcohol.

5, 6, 7. Wash. Stain in blue. Mount.

C. Aladar-Aujeszky's method.—**1.** Make a cover-glass film. Dry in the air.

2. Dip the preparation for 2 to 4 minutes into a porcelain capsule containing a 0·5 per cent. solution of pure hydrochloric acid which has been heated but not boiled.

3. Wash freely in water. Dry. Fix in the flame.

4. Stain with carbol-fuchsin in the warm by heating until steam rises : as the stain evaporates add a fresh supply.

5. Decolourize rapidly in a 4 per cent. solution of sulphuric acid.

6. Wash. Stain in blue. Mount.

D. Orszag's method.—**1.** Place a small drop of the following mixture on a cover-glass.

> 0·5 per cent. aqueous solution of sodium salicylate, - - 4 parts.
> 5 per cent. aqueous solution of acetic acid, - - - 1 part.

Make an emulsion of the organisms in this solution. Dry. Fix the film in the flame.

2. Stain with carbol-fuchsin in the warm for 2 minutes.

3. Decolourize with a 1 per cent. aqueous solution of sulphuric acid.

4. Wash. Counterstain with blue. Mount as before.

E. Thesing's method.—**1.** Prepare a cover-glass film. Dry. Fix in the flame.

2. Place a drop of a 1 per cent. aqueous solution of platinum chloride on the film. Heat to boiling.

3. Wash in a large quantity of water. Dry.

4. Stain with carbol-fuchsin as in the foregoing methods.

5. Decolourize with 33 per cent. alcohol.

6. Wash. Dry. Counterstain with blue. Mount.

SECTION II.—THE STAINING OF CAPSULES.

Some micro-organisms are surrounded by a bright hyaline area called a capsule which can be demonstrated by certain staining devices. When these are employed the organism is deeply stained, while the capsule surrounding it is pale with a feebly stained margin.

A. 1. Having dried and fixed a film, stain for 1 minute in carbol-fuchsin.

2. Wash. Treat for 20–30 seconds with water containing 1 per cent. acetic acid.

3. Wash. Dry. Mount in balsam.

In the author's hands this method has given better results than the following.

The method may be modified by treating the film first with a 1 per cent. solution of acetic acid for 1 minute, then drying it and afterwards staining with carbol-violet.

Simple staining with dilute carbol-fuchsin also gives quite good results.

B. 1. Dry and fix a film on a cover-glass.

2. Stain with a drop of the following solution for 30–60 seconds :

Acetic violet.

Acetic acid, - - - - - - - - -	1 gram.
Alcoholic solution of gentian-violet, or crystal-violet, - -	5 c.c.
Distilled water, - - - - - - -	100 grams.

3. Wash. Dry. Mount in balsam.

C. Ræbiger suggests staining the dried but unfixed films in the following solution, which must be filtered :

Gentian-violet, - - - - - - -	15 grams.
Commercial formalin, - - - - - -	100 ,,

After staining, wash, dry and mount in balsam. The bacteria are stained violet and the capsules violet with a pink tint.

D. Nicolle recommends staining with carbol-gentian-violet followed by rapid decolourization in a 1 in 3 solution of acetone-alcohol. Mount and examine in water.

Hiss fixes the films in the flame, stains in a 5 per cent. aqueous solution of gentian-violet or fuchsin in the warm, then washes in a 20 per cent. solution of copper sulphate, dries and mounts in balsam.

He also recommends staining in a half-saturated aqueous solution of gentian-violet followed by washing in a 0·25 per cent. aqueous solution of potassium carbonate. The films should be examined in a drop of the potassium carbonate solution.

The staining of capsules in sections requires special methods which will be studied later (*vide Pneumococcus*).

SECTION III.—THE STAINING OF FLAGELLA.

Flagella are the organs of locomotion of the motile bacteria and are only visible in the living unstained condition in such large organisms as the sulpho-bacteria (*Bacterium photometricum, Beggiatoa roseopersinica*, etc.). To demonstrate flagella in other motile organisms complicated staining methods have to be adopted.

1. The staining of flagella in living organisms.

Straus' method.

1. Place a drop of a broth culture of the organism on a slide.

2. Add a drop of carbol-fuchsin diluted with three or four parts of water and mix the culture with the stain.

3. Cover with a cover-glass and examine at once with an oil-immersion lens.

The bacilli are stained an intense red and the flagella, which will be seen especially well on the living actively-motile bacilli, assume a pale pink colour with deeper red points scattered along their length.

Note.—This is a very rapid method but it only succeeds with certain organisms, and its action on these is uncertain. The best results are obtained with *Vibrio choleræ asiaticæ, V. finkler-prior, V. metchnikowi.* The method fails altogether to stain the flagella of many organisms·such as *B. febris entericæ, B. coli communis, B. subtilis,* etc.

2. The staining of flagella in dried preparations.
General rules.

1. Take a small quantity of a young agar culture and make a perfectly homogeneous, very faintly opalescent emulsion in a watch-glass containing ordinary water, or preferably distilled water.

2. Fix an absolutely clean cover-glass in a pair of Cornet's forceps, pass it through the flame, and when cool place a drop of the emulsion on it with a pipette. Unless the cover-glass be perfectly clean the liquid will not spread uniformly.

3. By tilting the cover-glass run the liquid over the surface, then let the excess gravitate to one corner and aspirate it with a pipette.

4. Dry the film in the air away from dust and, without fixing, stain by one of the methods described below.

By following the above instructions, a dilution is obtained such that each field of the microscope contains only a few organisms, which is an essential condition for good results. By this method, also, the mucoid substances which agglomerate organisms in cultures and form precipitates on the cover-glass (thus obscuring the details) are as far as possible excluded.

(i) Van Ermengem's method.
(Method recommended.)

This method is based on the reduction of nitrate of silver in the flagella and gives very beautiful preparations. It is generally used in preference to any other method. [*Vide* also method (xi) p. 153.]

1. Place the film for 1 minute at 50° C. (or 30 minutes at room temperature) in the following bath which must be freshly prepared :

2 per cent. aqueous solution of osmic acid,	8 c.c.
10 per cent. aqueous solution of tannin,	16 ,,
Glacial acetic acid,	1 drop.

2. Wash in water, then in absolute alcohol.

3. Treat the film with silver for 1 or 2 minutes.

Crystals of silver nitrate,	1 gram.
Distilled water,	200 c.c.

4. Without washing, transfer the film for 1 minute to the reducing solution.

Gallic acid,	5 grams.
Tannin,	3 ,,
Fused sodium acetate,	10 ,,
Distilled water,	350 c.c.

5. Without washing, put the preparation into the silver bath again and keep the liquid moving over the film until the latter assumes a black tint.

6. Wash. Dry. Mount in balsam.

(ii) Lœffler's method.

The method devised by Lœffler, for a long time the classical method of staining flagella, requires a good deal of practice and gives only mediocre results. The films are often covered with an abundant precipitate which obscures the flagella and renders their detection difficult.

The following reagents are required :

Fuchsin ink.

25 per cent. aqueous solution of tannin,	- - - -	10 c.c.
Cold saturated solution of ferrous sulphate,	- - -	5 „
Saturated alcoholic solution of fuchsin,	- - - -	1 „

Alkaline solution.

Alcoholic soda,	- - - - - - - -	1 gram.
Distilled water,	- - - - - - - -	100 c.c.

Acid solution.

Pure sulphuric acid,	- - - - - - -	1 gram.
Distilled water,	- - - - - - - -	100 c.c.

Staining solution.

Aniline water,	- - - - - - - -	100 c.c.
1 per cent. solution of soda,	- - - - -	1 „
Gentian-violet or fuchsin,	- - - - -	4 to 5 grams.

Shake. Leave for a few hours in a bottle. Filter.

Technique.—1. Mordanting. Drop on to a film prepared as above (see general rules) a large drop of the fuchsin ink containing a few drops of the acid or alkaline solution. The amount of the latter depends upon the species of organism under examination. Heat the film over the pilot flame of a Bunsen until steam just begins to rise and continue the heating for 30–50 seconds : be careful not to boil the liquid. This stage of the method is very tricky and is liable to failure.

The amount of the acid or alkaline solution to be added to 16 c.c. of fuchsin ink has been determined by experiment. The following table shows the amounts necessary for the principal ciliated bacteria.

Mordant alone without acid or alkali, - - - -	Spirillum concentricum.
Mordant + ½–1 drop of acid solution, - - - -	V. choleræ asiaticæ.
„ +6 drops „ „ - - - -	B. pyocyaneus.
„ +18–20 „ „ „ - - - -	Micrococcus agilis.
„ +20 „ „ „ - - - -	B. chauvæi.
„ +20–30 drops of alkaline solution, - - -	B. febris entericæ.
„ +28–30 „ „ „ - - - -	B. subtilis.
„ +26–28 „ „ „ - - - -	B. maligni œdematis.
„ + { from 20 drops of acid solution / to 15 drops of alkaline solution, } - -	Bacillus of blue milk.

2. Washing.—Wash in water, then in absolute alcohol.

3. Staining.—Place a drop of the staining solution on the film, heat until steam begins to rise gently, and let the hot stain act for about a minute.

4. Mounting.—Wash in a large volume of water and examine in water. If satisfactory, dry and mount in balsam.

(iii) Remy and Sugg's method.

This method, which is a modification of Lœffler's, is designed to avoid the formation of granular precipitates. The fuchsin solution is used cold, and after staining the film is treated with iodine.

Instead of Lœffler's the following stain is used :

Staining solution.

Phenylamine water,[1]	- - - - - -	20 c.c.
Alcoholic solution of gentian-violet,	- - - -	1 drop.
Distilled water,	- - - - - - -	5 c.c.

Mix the gentian-violet with the water and then add the phenylamine water.

Technique.—1. The mordanting process is the same as in Lœffler's method but the solution is left on the film for 15–30 minutes and is not heated.

2. Pour off the mordant and replace it at once with a drop of Gram's solution.

3. Wash in water, then in absolute alcohol.

4. Place the film in a watch-glass filled with the stain and leave it for half an hour, preferably in the warm (37° C.) incubator.

5. Wash in water. Examine in water. Dry. Mount in balsam.

[1] Prepare in a similar manner to aniline oil water (p. 139).

(iv) Nicolle's and Morax's method.

(Method recommended.)

This method, a simplification of Lœffler's, does away with the use of the acid and alkaline solutions and is a satisfactory stain for the flagella of all motile organisms. Carbol-fuchsin is used instead of Lœffler's stain.

1. Mordanting.—Place a large drop of fuchsin ink (without any addition of acid or alkali) on the film. Heat for 10 seconds over the pilot flame of a Bunsen.

When steam begins to rise, pour off the solution, tilt the cover-glass and run a gentle stream of water from a wash-bottle on to its upper angle so as to wash the film well without washing the organisms away.

Repeat the mordanting and washing two or three times. Wipe the under surface of the cover-glass and the teeth of the forceps each time after washing: otherwise when the mordant is poured on again the solution will run under the cover-glass and along the forceps.

2. Staining.—Put a drop of carbol-fuchsin on the film and heat once or twice until steam has been rising for 15 seconds.

3. Mounting.—Wash and examine in water. If the preparation be satisfactory, dry and mount in balsam.

Bunge's and de Rossi's methods.—These methods differ slightly from that of Nicolle and Morax but have no advantage over the latter.

Bunge's mordant.

Saturated aqueous solution of tannin, - - - -	3 parts.
1 in 20 aqueous solution of perchloride of iron, - - -	1 part.

To ten parts of this mixture add one part of a saturated aqueous solution of fuchsin. The mordant must be exposed to the air for a few weeks before use: while so exposed it acquires a brownish-red colour.

Filter a few drops of the above solution on to the film and leave it for 5 minutes: wash in water and dry: stain with carbol-fuchsin as in Nicolle and Morax's method: wash, dry and mount in balsam.

De Rossi treats the film for 10 minutes with the following solution:

Tannin, - - - - - - - - -	5 grams.
0·1 per cent. aqueous solution of potash, - - - -	100 c.c.

Wash in water. Dry. Stain with carbol-fuchsin as in Nicolle's and Morax's method. Wash. Dry. Mount.

(v) Trenkmann's method.

This method gives satisfactory results but it takes too long to be of use in routine work.

1. Leave the film in the following solution for 6–8 hours:

Tannin, - - - - - - - - -	2 grams.
Distilled water, - - - - - - - -	100 c.c.
Pure hydrochloric acid, - - - - - -	4 drops.

2. Wash in water. Treat the film for 1 hour in a watch-glass containing a saturated solution of metallic iodine in distilled water.

3. Wash in water. Stain in aniline-gentian-violet for half an hour.

4. Wash in water. Examine. Dry. Mount in balsam.

(vi) Cerrito's method.

This method requires a good deal of care, because the mordant frequently gives rise to troublesome deposits.

The mordant consists of the following somewhat complex mixture:

25 per cent. aqueous solution of tannin in ether, -	20 c.c.
5 per cent. aqueous solution of pure iron-alum, - -	10 ,,
Saturated solution of fuchsin in 90 per cent. alcohol, -	1 ,,

Pour these solutions into a well plugged flask: heat the mixture to 100° C. in a water bath, thoroughly mixing the ingredients by shaking and continue to heat until the liquid is pale in colour. The solution must be kept in an hermetically sealed bottle.

1. Mordanting.—Flood the cover-glass with a few drops of the mordant for 2 or 3 minutes at 25° C. or 10 minutes at 15° C. Wash in water. Dry.

2. Staining.—Stain in the following solution for a few seconds, heating the solution until steam just begins to rise:

Fuchsin, - - - - - - - -	0·25 gram.
Absolute alcohol, - - - - - - -	10 c.c.
Carbolic acid crystals, - - - - - -	5 grams.
Distilled water, - - - - - - -	100 c.c.

3. Mounting.—Wash in water. Dry. Mount in balsam.

(vii) Pitfield's method. Benignetti and Gino's method.

In Pitfield's method the mordant and the stain are combined in one solution, thus :

Saturated aqueous solution of alum, - - - -	10 c.c.
10 per cent. solution of tannin in water, - - - -	10 ,,
Saturated alcoholic solution of gentian-violet, - - -	2 ,,

Benignetti and Gino obtain very satisfactory results with the following very simple method which is a modification of the above.

The combined mordanting and staining solution is prepared thus :

A. Zinc sulphate, - - - - - - -	1 gram.
Tannin, - - - - - - - -	10 grams.
Distilled water, - - - - - - -	100 ,,
B. Solution A, - - - - - - -	5 c.c.
Saturated aqueous solution of alum, - - - -	5 ,,
Saturated alcoholic solution of gentian-violet, - - -	3 ,,

Technique.—Fix the film with heat, and when cool flood it with a large drop of solution B, and heat it over the pilot light of a Bunsen until steam just begins to rise. Wash in water. Dry. Mount in balsam.

[R. Muir's modification of Pitfield's method.

[Prepare :

A. Filtered 10 per cent. aqueous solution of tannin, - -	10 c.c.
Saturated aqueous solution of perchloride of mercury, -	5 ,,
Saturated aqueous solution of alum, - - - -	5 ,,
Carbol-fuchsin, - - - - - - -	5 ,,

[Mix thoroughly. Allow the precipitate to settle. Decant off the clear supernatant fluid. The mordant will keep for about a fortnight.

B. Saturated aqueous solution of alum, - - - -	10 c.c.
Saturated alcoholic solution of gentian-violet, - - -	2 ,,

The stain does not keep.

[1. Prepare and fix a film as above (p. 149).

[2. Flood the preparation with the mordant and heat until steam just begins to rise. Let the solution act for 1 minute.

[3. Wash thoroughly in running water. Blot and dry over the flame.

[4. Flood the film with the stain, heat as before for 1 minute.

[5. Wash well in water.

[6. Blot. Dry. Mount in balsam.]

(viii) Bowhill's method.

This method is troublesome and offers no special advantages.

1. Mordanting.—Place the film for 10 minutes at a temperature of 40°–50° C. in a bath consisting of equal parts of the following solutions mixed just before use and filtered :

SOLUTION A.

Orcëin,	1 gram.
Absolute alcohol,	50 c.c.
Distilled water,	40 ,,

SOLUTION B.

Tannin,	8 grams.
Distilled water,	40 c.c.

Use heat to dissolve the tannin.

In staining the flagella of the *Vibrio choleræ asiaticæ* add 1 c.c. of a saturated solution of alum for each 10 c.c. of mordant.

2. Washing.—Wash in water. Dry.

3. Staining.—Flood the film with aniline-gentian-violet and heat until steam just rises from the film for 15–30 seconds.

4. Mounting.—Wash. Dry. Mount in balsam.

(ix) Gemelli's method.

1. Immerse the films in the following solution for 10–20 minutes.

Potassium permanganate,	0·25 gram.
Distilled water,	100 c.c.

2. Wash in distilled water.

3. Stain for 15–30 minutes in the following mixture :

0·75 per cent. aqueous solution of calcium chloride,	20 c.c.
1 per cent. aqueous solution of neutral-red,	1 ,,

4. Wash in water. Dry in the air. Mount in balsam.

(x) Sclavo's method.

Sclavo's method fails to stain the flagella of some micro-organisms especially the flagella of the cholera vibrio. The author's experience has been that it is equally unsuited for the flagella of the colon bacillus.

1. Flood the film with a large drop of the mordant, viz. :

50 per cent. alcohol,	100 c.c.
Tannin,	1 gram.

Leave for 1 minute and then wash in water.

2. Treat for 1 minute with the following solution on the film :

Phospho-tungstic acid,	5 grams.
Water,	100 c.c.

3. Wash quickly in water.

4. Stain for 3 or 5 minutes with a drop of aniline-gentian-violet heating the stain until steam rises gently from the film.

5. Wash and examine in water. Dry. Mount in balsam.

[(xi) Stephens' method.]

(Method recommended.)

[The method worked out by J. W. W. Stephens is a modification of van Ermengem's (p. 149) and depends upon the use of very strong ammonia as the reducing agent. With ordinary care a satisfactory result can be absolutely relied upon.

[*To clean the slides.*—Rub the slides with a clean cloth, place them on a piece of clean wire gauze and heat with a smokeless flame for some minutes (by this means grease is completely removed). Leave the slides until cool.

[*To prepare the film.*—Rub a little of the culture in a small drop of tap-water in a watch-glass. Transfer a drop with a very small platinum loop to a minute drop of water on the slide. Mix. Spread with the loop as quickly as possible. The film should dry immediately if only a small drop of water has been used. A twenty-four hour growth on agar does quite well (a younger one is perhaps better, but flagella can be shown for a week or fortnight or more).

[The following solutions are required:

(a) *The mordant.*

2 per cent. aqueous solution of osmic acid, - - - -	1 part.
20 per cent. aqueous solution of tannin, - - - -	3 or 4 parts.

(b) *Silver nitrate solution.*

Crystals of silver nitrate, - - - - - - -	1 gram.
Distilled water, - - - - - - - -	100 c.c.

(c) *Reducing solution.*

2 per cent. aqueous solution of gallic acid, - - -	1 part.
Ammonia fort.,[1] - - - - - - - -	1 ,,

Mix immediately before use.

[1. Place the mordant on the film for one or two minutes or less (time unimportant).

[2. Wash in tap-water thoroughly. Shake off as much water as possible.

[3. Place a few drops of the silver nitrate solution on the film for a few seconds or longer.

[4. Shake off the excess of silver solution.

[5. Allow one drop of the reducing solution to fall on the *middle* of the film from a pipette. A wave spreads away from the centre to each end of the slide. As soon as the film is seen standing out clearly and black (a few seconds), wash off in tap-water.

[6. Pour another drop or two of the silver solution on to the film and leave for half a minute or so.

[7. Wash in tap-water. Blot. Dry over a flame. The preparations fade rapidly if mounted in balsam or cedar-wood oil.]

SECTION IV.—METHODS OF STUDYING THE MOTILITY OF MICRO-ORGANISMS.

Closely connected with the morphological study of the flagella of micro-organisms is the investigation of their motility.[2] Motile organisms can make their way through porous substances such as sand or a filter composed of porcelain or siliceous earth. The time occupied in traversing a given thickness of sand, etc., will vary according as to whether the organism is actively or feebly motile.

These observations are utilized for determining whether or no an organism is motile, for separating motile from non-motile species, for determining the relative motility of different strains of the same organism, and even for creating, by a process of selection, races which are endowed with exceptional powers of movement.

A.—Cambier has drawn attention to the property possessed by the typhoid bacillus of traversing the walls of porous structures, and has suggested that this property might be made use of in attempting the isolation of the organism.

A porous porcelain bougie is placed in a large test-tube, and both the bougie and the test-tube are half-filled with broth; the tube is plugged with wool and the whole apparatus autoclaved. When cool, the broth in the bougie is sown with a culture of the typhoid bacillus. After incubating for a few hours at 37° C. the broth in the tube surrounding the bougie will be distinctly cloudy, and this is due to the fact that the typhoid bacillus

---

[1] [It is essential that the solution of ammonia be the strongest obtainable.]

[2] *Vide* also Chap. VII., Dark-ground illumination.

has passed through the porous walls of the bougie. Only motile organisms can do this, and of these the typhoid bacillus is one of the first to pass through. In attempting the isolation of the typhoid bacillus from water, the broth in the bougie would be sown with some of the suspected water, and when the broth surrounding it became cloudy a small quantity would be removed for the purposes of further investigation by the usual methods (Chap. XXIII.).

B.—Carnot and Garnier conceived the idea of making motile organisms pass by their own efforts through a layer of sand of known thickness, and then collecting the first organisms to pass through; they were thus able to determine exactly the time required by a given organism to make its way through a given thickness of sand. The degree of motility possessed by any species of microorganism can by these means be exactly measured.

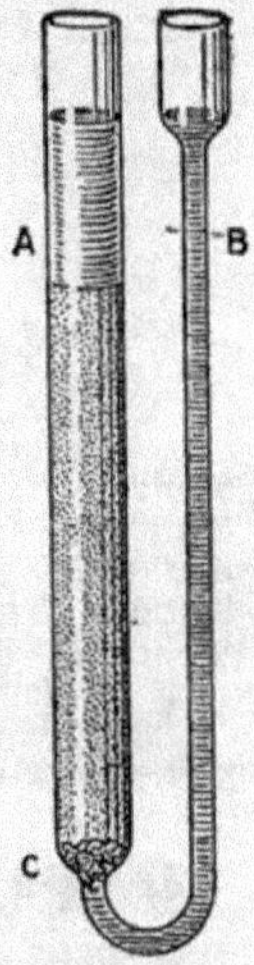

Technique.—**1.** A piece of glass tubing, 7 mm. calibre, is drawn out in the flame about its middle, and bent into an U-shape with the two limbs parallel and closely applied to each other, each being about 25 cm. long (fig. 128).

2. A loosely-packed plug of glass wool C is pushed down the limb A as far as the constriction in the lower part. Broth is then poured in to a depth of about 10 cm. in each tube. Very fine quartz sand (previously washed in hydrochloric acid for 48 hours, and then in water for several days and afterwards calcined in the hot air sterilizer) is slowly dropped down the tube A until it forms a column 10–15 cm. high. A and B are then plugged with wool, and the tube autoclaved.

3. The organism whose motility is to be investigated is then sown in the broth contained in the limb B in which there is no sand, and the tube incubated at 37° C. The passage of organisms through the sand is made manifest by a cloudiness of the broth in A. Only motile organisms reach the broth in A, and the time occupied varies with different species.

FIG. 128.—
Carnot and Garnier's tube.

Carnot and Garnier give the following times for the most motile organisms :
Vibrio choleræ (Massaouah) traverses 1 c.c. of sand in 1 hr. 38 m.

,,	,,	(Dantzig)	,,	,,	,,	2 hrs. 4 m.
,,	,,	(Paris, 1884)	,,	,,	,,	4 hrs.
Bacillus psittacosis			,,	,,	,,	2 hrs.
Bacillus febris entericæ			,,	,,	,,	3–6 hrs.
Bacillus coli communis			,,	,,	,,	(variable, 1 hr. to several days).

Streptococci with feeble undulatory movements,   4 hrs. 50 m.

Bacillus anthracis, Staphylococcus pyogenes, Pneumococcus, etc., do not pass through the sand.

This method like that of Cambier may be utilized for the isolation of motile organisms ; moreover it renders possible, by successive passages of selected organisms, the creation of strains of a given bacillus possessed of exceptional motility. In this way Carnot and Garnier were able after five passages to isolate from a culture of the typhoid bacillus, which originally took 6 hours to traverse a centimetre of sand, a strain which passed through the same thickness in 1 hour and 4 minutes.

CHAPTER X.

ANIMAL INOCULATION.

SECTION I.—THE SELECTION OF ANIMALS FOR INOCULATION.

FOR purposes of experimental inoculation, animals are chosen preferably from among the mammalia, less frequently from among the other vertebrata. In deciding upon what species of animal shall be used for a given experiment there are of course various considerations which must be taken into account.

1. Susceptibility.—In the first place it is obviously necessary to select a species of animal suitable for the experiment in view. To produce a given disease experimentally, an animal susceptible to the virus should, generally speaking, be chosen, though it is sometimes desirable to use an animal immune to the particular disease and to destroy its immunity in some way or another. Some knowledge, therefore, of the diseases to which animals available for experimental purposes are susceptible is more or less indispensable. In subsequent chapters those animals which are susceptible to the action of the principal micro-organisms will be mentioned. When a new organism is under investigation and its pathogenic properties have to be determined it is desirable to inoculate as many different species of animals as possible.

2. Economic considerations.—In the majority of cases small animals are used; they are cheap to buy, and can be kept and fed at small expense, and, if need be, can be bred in the laboratory.

3. General considerations.—Whenever possible animals of quiet habits are chosen because they are easy to handle, and do not require elaborate cages.

Small rodents, such as rabbits, guinea-pigs, white mice, white rats, common brown mice [*Mus musculus*], and house rats [*Mus decumanus*] are, on the whole, more often used than any other animals for experimental inoculation; they are easily obtained, and the first four—to which the term "laboratory animals" generally refers—are susceptible to most of the organisms pathogenic to man.

Cattle, goats, pigs, horses, sheep, asses and birds (fowls and pigeons) are also used for experiment in special cases.

Cats are difficult to handle, and dogs are only slightly susceptible to most of the organisms pathogenic to man.

Frogs are occasionally used, but they are not very susceptible.

Ground squirrels [*Mus citullus*] are not only difficult to get in this country, but they do not breed in captivity.

Monkeys, and especially the anthropoid apes, have for some time been little used for experimental purposes on account of the difficulty of obtaining them, their initial cost, and the great care with which they have to be tended in captivity. Nevertheless, the work of Metchnikoff and his pupils on syphilis, [of the English Commission on tuberculosis, of Levaditi and Landsteiner on acute polio-myelitis, etc.] has shown the value of experiments upon these animals in the investigation of human disease. [*Macacus rhesus* is the most suitable—for most purposes—and at the same time the cheapest monkey.]

SECTION II.—THE KEEPING OF ANIMALS.

A. Small animals.

Accommodation.—The "small animal" house ought to be spacious, well ventilated, floored with concrete or some similar impervious material and have water laid on, so that it can be frequently washed down.

Animals generally, and especially monkeys, rabbits, mice and rats, are very susceptible to cold and damp; the animal house must, therefore, be kept dry, and facilities for warming it in winter should be provided.

Cages, feeding, etc.—The cages should, as far as possible, be made of metal. It is a bad practice to place one cage on top of another, since fluids from the upper cage may soil [and infect] the one beneath. If, from want of space, it becomes absolutely necessary to place one cage on top of another there must be a sheet of metal between them tilted and guttered, so that the urines run off. The bottom of the cages should always be perforated.

[A rectangular cage made entirely of fairly stout galvanized iron wire and resting on a metal tray is both efficient and cheap (fig. 129).

FIG. 129.—Animal cage. The cage is raised on blocks from the tray.

[The cage itself consists merely of a rectangular wire box of which one of the two largest sides is hinged to form a lid allowing access to the interior. A clasp must also be provided so that the top can be securely fastened down. The cage is placed on a sheet of galvanized iron turned up to the extent of 1 or 1½ in. all round to make it water-tight and measuring 1 inch larger in each direction than the cage. To clean, it is only necessary to lift the cage off the tray. It is an advantage to have two trays for each cage so that one may be disinfected and dried while the other is in use.]

The cages must be cleaned daily, and should an animal die or be removed to make room for a new occupant the cage in which it lived must be dipped in some strong antiseptic solution (phenol, lysol) or be sterilized by flaming it with spirit or with a large specially constructed gas burner (fig. 130).

FIG. 130.—Gas burner for flaming cages.

Each cage should carry a label indicating the nature of the experiment, and the day upon which the animal was inoculated.

(i) **Rabbits, guinea-pigs, rats and mice.**—Rabbits and guinea-pigs can be conveniently kept in the cages described above.

Grey rats [*Mus decumanus*], as well as house [*Mus musculus*] and field [*Mus silvaticus*] mice should, as a rule, be caged singly. When several of these rodents are put in one cage they fight, and frequently kill each other. They are best kept in large wide-mouthed jars, the mouth being covered with metal gauze fastened down round the neck with iron wire.

White rats are frequently very tame, and can be kept in small-mesh wire-netting cages or bird cages, or even in wooden boxes fitted with a wire-netting door.

White mice should be kept in glass jars or metal boxes such as Palmer's biscuit boxes, the lid of which must be pierced with a number of holes. The floors of the boxes or jars, whichever be used, should be covered with a layer of sawdust several centimetres deep, and a little wool should also be put in the cage as mice do not like the cold.

There is no need here to discuss the proper feeding of rabbits, guinea-pigs, etc.

Mice and rats should be fed on corn and moistened bread. White rats are very fond of water, and a small dish containing it should, therefore, be put in their cages.

(ii) **Monkeys.**—Monkeys require a great deal of attention. Their cages must be large and be kept scrupulously clean. They are very susceptible to the cold, and the house in which they are kept ought, in these climates, to be artificially heated most of the year. The nature of the food required varies with the different species, but, generally speaking, milk,[1] dried fruits, [cooked rice], bananas and bread constitute the staple articles of diet of monkeys in captivity. They should be given something to drink twice a day, but it is advisable not to leave water [or milk] in their cages.

(iii) **Frogs.**—Frogs can be easily kept at ordinary room temperatures, but at temperatures approximating to that of the human body, such as are sometimes necessary under experimental conditions, they often die in a day or two without any apparent reason.

Ledoux-Lebard suggests the following as a useful method for keeping frogs (*Rana esculenta* is better than *Rana temporaria*) at a temperature of 35° or 37° C. for a month or more. Keep each frog in a bottle containing a few centimetres of water and covered with a piece of stout muslin tied on with string, renew the water daily with a fresh supply at the same temperature, and cram the frog once a week with beef, veal or mud worms.

Isolation of inoculated animals.—Inoculated animals must, of course, be rigidly kept away from the neighbourhood of normal animals.

[1 It must not be forgotten that monkeys and apes are highly susceptible to tuberculosis, so that the milk must either come from an unimpeachable source or be sterilized before being fed to them.]

Breeding of small animals.—Rabbits, guinea-pigs, white rats, and white mice can easily be bred in the laboratory, and because of their inclination to kill the newly-born animals, it is as well to separate the males from pregnant females. This precaution is absolutely necessary in the case of rabbits and mice, but is less important with white rats and perhaps unnecessary for guinea-pigs.

B. Large animals.

[With regard to the housing of large animals such as cattle, sheep, pigs, etc., it is unnecessary to say more than that the houses or pens should be designed on lines similar to those in which they are ordinarily stabled. The structures should be light and well ventilated the floors concreted or cemented and well drained. The walls should be constructed of material which readily lends itself to efficient cleansing with antiseptics. If the stalls be of wood they should be limewashed out with lime mixed with 2 per cent. lysol before a new animal is introduced.]

SECTION III.—THE SPONTANEOUS DISEASES OF EXPERIMENTAL ANIMALS.

Laboratory animals are liable to certain infectious diseases with the more common of which it is important to be familiar because they are sometimes responsible for a heavy mortality among experimental animals.

Abscess.—Large abscesses containing thick, fetid pus not infrequently occur in rabbits in various parts of the body. They lead to a cachectic condition, and ultimately end in death. The disease is contagious.

The infected animal must be isolated and the cage carefully disinfected. Treatment consists in opening the abscess, evacuating the pus and gently curetting the wall, subsequently washing it out at frequent intervals, and dressing it with antiseptic dressings.

Acari.—An *acarus* sometimes develops in the external auditory meatus of the rabbit; it soon invades the middle ear, and causes serious nerve troubles, such as gyratory movements, convulsions and epileptiform seizures which lead to the death of the animal. The disease may be recognized by the yellow crusts which are seen in the rabbit's ear and which, if examined microscopically (Oc. 2, obj. A. Zeiss), are found to be composed of amorphous débris and numerous *acari*. The disease is highly contagious but yields to treatment if taken in hand in the early stages.

Immediately a case is found in the animal house, the infected individual should be killed, unless the experiment be of special interest, in which event it must be isolated and treated. Treatment consists in washing off the crusts formed on the auditory meatus daily with a sponge made by twisting a little piece of wool round a small rod, and dropping a few drops of a 0·5 per cent. solution of polysulphide of potassium (liver of sulphur) into the ear. The infected animal's cage, and those near it, should be disinfected and the ears of all the other rabbits in the house frequently examined.

Septicæmias.—Rabbits and guinea-pigs are subject to epizootic diseases which, only too often, decimate the population of the animal house in a few days.

As a rule, rabbits and guinea-pigs are affected at the same time. The animals curl themselves up, their coats are rough, and they suffer from a running from the nose and diarrhœa. Death soon follows these symptoms. *Post mortem*, lesions of broncho-pneumonia are seen. The disease appears to be due to a small bacillus morphologically similar to Pfeiffer's [influenza] bacillus.

Pasteurellosis.—Another disease, caused by an organism of the *Pasteurella* group (Chap. XXVIII.), is sometimes seen among rabbits. Infection is due to contamination of the food and floors of the cages with infected excreta. The animals are listless, suffer from diarrhœa and succumb rapidly. *Post mortem* : excess of fluid in the pleuræ, pericardium and peritoneum ; congestion of the lungs, intestines, etc.

Phisalix has described a disease caused by a *Pasteurella* similar to the canine pasteurella (Chap. XXVIII.) which may sometimes produce an epizootic among guinea-pigs.

Certain contagious **pneumonias** (Weber, Tartakowsky and others) may also attack laboratory animals.

When a septicæmic infection makes its appearance in the animal house, isolate the animals which are obviously or suspected to be infected and disinfect the house. It is even better, especially if there be any which it is particularly important not to lose (animals undergoing immunization, etc.), to remove the animals which are healthy to some other place altogether, and to transfer them to disinfected cages. Still it will often be difficult, whatever be done, to prevent the spread of the infection.

[Several epizootics resulting in a heavy mortality among the guinea-pigs in the small animal house and due to organisms other than those mentioned have now been recorded. From the internal organs bacilli of **the paratyphoid group** have been isolated (*B. Gaertner*, M'Conkey, Petrie, Bainbridge and O'Brien, *B. aertrycke*, O'Brien, Petrie and O'Brien). There is evidence, however, that these organisms were not, at least in some of the epizootics, the real cause of the disease, which appeared to be a **filter-passing organism** (Petrie and O'Brien), the paratyphoid bacilli being "secondary" infections.]

Coccidiosis.—Rabbits are frequently infected with *Coccidium oviforme* (*vide* Sporozoa). It is important that this fact be kept in mind and it should be noted that the disease is particularly troublesome in young animals.

Numerous other parasites may be found in experimental animals, and reference will be made to these in due course, more particularly when dealing with Tuberculosis, Glanders, Pleuro-pneumonia, the Pasteurelloses, the Hæmatozoa, the Trypanosomata, etc.

[**Pseudo-tuberculosis.**—Pseudo-tuberculosis is a most troublesome disease among guinea-pigs and rabbits not only because the naked-eye lesions so closely resemble the lesions produced by the tubercle bacillus but because so many animals become infected and die once the disease makes its appearance in the animal house. Pseudo-tuberculosis is the result of infection with a short stout bacillus with rounded ends which grows readily on agar at the temperature of the body. A feature which may arouse suspicion is the fact that the mesenteric glands are markedly affected which is, of course, not the case when an animal has been inoculated sub-cutaneously or intra-peritoneally with the tubercle bacillus. Infection apparently takes place through the alimentary canal. When the disease appears all the animals in the animal house must be isolated and as many as possible killed. It is a wise thing to destroy any food or bedding in stock and order a fresh supply.]

SECTION IV.—THE HANDLING OF EXPERIMENTAL ANIMALS.

Most animals struggle when they are caught and try to bite or scratch the person holding them. It is important to avoid these wounds, which may be dangerous, especially when the animal is infected with a disease transmissible to man, *e.g.* hydrophobia. A skilled worker should never be damaged by the animals he handles.

During an experiment the animal may be held either by the person operating or by an assistant. This is quite satisfactory in the case of most animals when a simple sub-cutaneous inoculation has to be done, but for the more difficult operations, such as inoculation into the peritoneum, meninges or veins, or when dangerous viruses, such as those of glanders, hydrophobia, etc., have to be inoculated, it may be better to hold the animal in some suitable form of apparatus designed for the purpose. [But as a matter of fact occasion for the use of such apparatus very, very seldom arises. If there is likely to be any difficulty this may be overcome by administering an anæsthetic. But for sub-cutaneous, intra-peritoneal or intra-venous inoculation not even an anæsthetic is necessary.] In the handling of small animals an assistant should be dispensed with as far as possible.

1. Rabbits.

To catch a rabbit grasp it by the skin of the back, or by one of its ears. These are the only ways to secure the animal without being scratched in the attempt. To hold a rabbit, place the animal on the knees, and hold it there with the left hand, using the right hand for the inoculation. If the animal be troublesome, take hold of the skin of the back with the right hand, and put the rabbit under the left arm, so that the head and fore limbs are fixed between the arm and the chest wall, support the trunk on the forearm, and hold the hind limbs with the left hand. The right hand is then free to make the inoculation.

When an assistant is available he turns the rabbit on its back and takes hold of the four limbs in his left hand, holding the head in his right in such a way that the top of the animal's head rests in the palm and his thumb passes under the lower jaw.

Apparatus for holding rabbits.—A rabbit can be very simply held by wrapping it up to its neck in a duster, or a large strip of cloth, and fastening the limbs beneath the body. Operations on the head and ears can then be performed. To inoculate one of the limbs, take it out of the duster, and hold it extended in the left hand.

To prevent the animal moving at all, several pieces of apparatus are available, for example, Malassez's, Czermak's, Piorkowski's, Latapie's, and Debrand's.[1] The two latter, which may be used for all the smaller animals, are very ingenious, but complicated and expensive. Latapie's apparatus is, moreover, difficult to disinfect.

The simplest and at the same time the best piece of apparatus consists of a rectangular sheet of zinc or copper, the edges of which are turned up and pierced with holes 2 or 3 cm. apart. Place the animal on the metal tray, put a noose (of string, or, better, leather) round one of the hind limbs and fasten it above the wrist, pass one of the ends through a hole near the end of the tray and tie it to the other end. In the same way fasten the fore limb of the opposite side to a hole at the other end of the tray. Then make the other two limbs fast. The animal is now absolutely unable to move. The head can be held by an assistant, or can be fixed with a string passing from a bar introduced behind the incisor teeth, and fastened as before to two holes of the tray.

Or the head can be held with a Ranvier's ring. This device consists of an horizontal iron bar moving on a vertical bar by means of a double joint, which allows it to be fixed in any position. The horizontal bar ends in a ring perpendicular to its axis, and on to the ring two small hooks are adjusted, to which a piece of elastic can be attached. Fit the ring on the animal's nose and attach the elastic to one of the

---

[1] *Annales de l'Institut Pasteur*, 1894 and 1900.

hooks, then pass it between the ears, and fasten the end to the other hook. The apparatus is fixed on to the tray by means of a screw clamp. The animal can be secured to the tray and its head held by the ring, either on its back or belly at will.

Anæsthetics.—Rabbits are very sensitive to anæsthetics. Chloroform should not be given either slowly or in small and repeated doses, because it will thus almost certainly kill the animal; but by giving a large dose to begin with and then stopping the administration after a few moments accidents can almost always be avoided.

Twist two or three thicknesses of filter paper into a cone, pour a teaspoonful of chloroform on to it and hold it over the animal's mouth. Respiration stops after a few seconds but soon begins again; at this moment anæsthesia is complete; the administration of chloroform should now be stopped, and the operation quickly performed.

2. Guinea-pigs.

It is best to catch guinea-pigs by the skin of the back; they are easier to handle than rabbits, and can generally be held in the left hand, leaving the right hand free for inoculation.

If it be desired to hold a guinea-pig with instruments the simplest way is to catch hold of the animal by the skin of its back with a large pair of pressure forceps, the grasping ends of which are ring-shaped (fig. 131). The

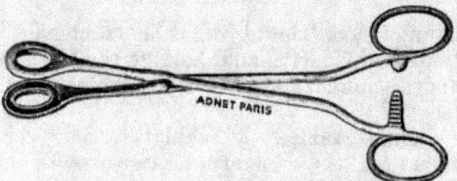

FIG. 131.—Forceps for grasping small animals.

forceps being clipped together are hung by one of the finger holes on a nail in the wall, and the animal being thus suspended is rendered quite motionless.

For holding guinea-pigs still while taking temperatures, making inoculations into the hind limbs, etc., it will be found convenient to secure the anterior part of the animal in a metal cylinder with slits along the sides.

For carrying out delicate operations, it is preferable to fix the animal on the tray described above. Such trays should be kept in two sizes, the larger ones for rabbits and the smaller for guinea-pigs.

Anæsthetics.—Guinea-pigs are less susceptible to chloroform than rabbits. It is very seldom that they have to be anæsthetized, but should it be necessary to give a guinea-pig an anæsthetic, the method of administration is the same as for rabbits.

3. White mice and white rats.

To catch the animals.—These species are perhaps in a general way more used for experimental inoculation than any other " laboratory animal." They can be caught by the tail with the fingers. Sometimes they struggle and may inflict a painful bite; this can be avoided by grasping them by the tail or skin of the neck with a pair of forceps.

To hold them for inoculation, the only method that can be recommended is to catch hold of the tail with the fingers or a pair of forceps and draw it out of the pot, the animal being thus suspended head downwards inside; then, as a precaution against being bitten put a small piece of board over the mouth of the jar so that only the tail projects. Inoculation can now be made at the root of the tail. If it is necessary to inoculate into one of the limbs, pull the limb out of the jar with a pair of forceps. [The inoculation can however be performed in the following manner which is we think simpler than that just described. Let an assistant catch the rat by the tail and hold the hind limbs with one hand and the fore limbs with the other at the same time wrapping a small cloth loosely round the animal's head. The inoculation can now be made by the operator into any part of the animal's body. If the rat bites it does not damage the cloth.]

Anæsthesia.—For all delicate operations it is better to anæsthetize the animal. Rats and mice are easily killed by chloroform, but take ether well.

Put the animal under a bell jar with a small piece of absorbent wool soaked in ether ; or the wool can be put straight into the jar in which the animal is living. When it falls over motionless take it out of the bottle and fix it on a small tray ; in the case of rats, if necessary, put on a Ranvier's rat-bit as well, or use a Debrand's apparatus. Anæsthesia can be prolonged by giving a little ether from time to time.

4. Grey rats.

Gréy rats [*Mus decumanus*] struggle vigorously and may give very nasty bites. They can only be caught with a pair of large strong forceps, such as those described above (fig. 131).

Pass the forceps into the bottle containing the rat and catch hold of the animal quickly wherever it is possible to do so. The rat at once attacks the forceps and bites them ; while the animal is thus engaged, fix a second pair of forceps on to the skin of the neck, clamp the two pairs of forceps firmly and lift the animal out of the bottle.

During the inoculation, an assistant holds the rat securely with the two pairs of forceps, inclining the forceps fastened to the neck towards the ver-tebral column in order to pick up the tail with the same hand. With the other hand he takes the other pair of forceps and pulls on them gently so as to make it impossible for the animal to use his teeth. If this second pair of forceps was badly fixed, the skin over the lower jaw should be held with another pair. When the operation is done place the rat safely inside the bottle again before releasing the forceps.

Anæsthesia.—Grey rats should always be anæsthetized before beginning a difficult or dangerous inoculation. Put a pledget of wool soaked in ether in the bottle, and proceed as already described in the case of white rats.

5. Dogs.

If the dog be a quiet animal catch hold of him firmly by the skin of the neck. When dealing with a surly or snappy dog use a special pair of long iron forceps (*pince à collier*), which, when closed, form a collar round the animal's neck. Alternatively, throw a noose round the animal's neck and fasten the loose end to a bar of the cage or a post ; as the dog pulls the noose tightens, the animal falls over half suffocated, and the opportunity is taken to slip on a muzzle and tie its feet.

Muzzling.—No operation should be performed on a dog without muzzling it beforehand. The simplest method is to pass a piece of stout string into the animal's mouth behind the canine teeth, make a simple knot below the jaw, bring the two ends up and tie them in a double knot on the nose. Or, after passing a stout round iron wire behind the canine teeth, take two turns with a piece of string round the muzzle behind the bit, tighten the ligature and fasten the ends securely.

If it be necessary to gag the animal with the mouth open for the purpose of passing a catheter into the stomach, use a Claude Bernard's bit with a double transverse branch.

Instead of the bit a rectangular wooden gag of a size suitable to the animal, and pierced with a hole in the centre may be used. After placing this gag in the mouth behind the canine teeth fix the jaws with string.

Method of holding dogs.—After muzzling, the dog can generally be held with the hands. For long operations the animal should be held with Claude Bernard's, Malassez's, or Debrand's apparatus, or more simply by fixing it by the feet, as has been described in the case of the rabbit, on to a heavy wooden table perforated with holes or fitted with hooks through which strings can be passed.

Anæsthesia.—It is rarely necessary to anæsthetize dogs in bacteriological work. The animals take chloroform well, provided that large doses are not given, and the liquid does not come in contact with the nasal mucous membrane.

In giving an anæsthetic a long muzzle is generally used, ending in a small perforated box in which a sponge soaked in chloroform is placed. The administration can be suspended or continued at will by taking off or replacing the box on the muzzle. Small doses should be used to begin with. Anæsthesia is complete after 8–15 minutes.

6. Cats.

Cats are very difficult to manage, and are rarely used for bacteriological experiments. It is best to take hold of them firmly by the skin of the back ; or, if the cat be wild, adopt the noose method described above for dogs.

In operating on a cat it is well to anæsthetize it. As soon as it has been caught, put it into a large wide-mouthed jar in which there is a sponge soaked in chloroform, and cover the jar at once. Cats are very sensitive to chloroform, and the animal must be taken out of the jar as soon as it falls : anæsthesia will continue for several minutes without any further administration. The animal may either be fixed on to the table already described, or may be wrapped up in a large duster with the feet under the belly, the head and anterior part of the body being pushed into a sack. This is an excellent method for inoculations into the posterior part of the body, rectal injections, etc.

7. Monkeys.

Monkeys especially *Macacus rhesus* are also difficult to manage [and are so active that it is hard to catch them, if their cage be at all large. Wear rough leather gloves and grasp the animal by the body or limbs, then take a fresh grip, of] the skin at the back of the neck, and treat them in the same way as dogs. It is necessary to chloroform them if the operation is likely to last any length of time.

8. Horses and Asses.

Horses can nearly always be inoculated without adopting any special method for holding them. It is enough for an assistant to hold them with a bridle or halter. If a horse be nervous its eyes may be covered, and should it struggle, a twitch may be used or one of its fore legs flexed and fastened. For longer operations the horse may be shackled and thrown by methods well known to people who have to do with horses. Vinsol's apparatus is to be strongly recommended, but unfortunately it is very expensive.

9. Cattle.

Bovine animals are, as a rule, easily managed. For long operations the animal is thrown on a vaccination inoculation table, or placed in a Vinsol's apparatus.

10. Birds.

Fowls and other birds ordinarily used for inoculation are easily held with the hand. They may be secured by their feet and wings on the metal tray described on p. 161, or on a Debrand's apparatus.

Note.—After every operation, all bits, gags, dishes, etc., which have been used must be carefully cleaned and washed with a solution of carbolic acid, [lysol or] formalin, and the antiseptic washed off, of course, with water before the instrument is used again.

SECTION V.—EXPERIMENTAL INOCULATIONS.

1. Instruments.

There is no need to go into details with regard to instruments in common, every-day use such as are required for making incisions, exposing vessels, etc. Bistouries, scissors, forceps, retractors, inoculation needles, suture needles, etc., must all be sterilized before an operation either in the hot air sterilizer at 180° C. or by boiling in water for 10 minutes, and then transferring to a 0·1 per cent. solution of oxycyanide of mercury [or 2 per cent. lysol]. When the operation is over the instruments must be cleaned, passed through alcohol and dried with a piece of soft cloth.

Sterile absorbent wool, thread and silk should be at hand, ready for use when needed.

Preparation of sterile silk.—(a) The silk may be sterilized just before it is wanted by boiling it for 15 minutes in 3 per cent. carbolic acid. But it is better to keep a quantity in stock prepared by one or other of the following methods:

(b) Cut the silk into lengths of about 12 inches and wrap three or four lengths in two or three pieces of filter paper. Prepare a number of these packets of silk and heat them to 120° C. in the autoclave, dry them in the incubator and keep them in a well-stoppered bottle in a box with a tightly-fitting lid. Open the packets one by one as they are wanted. Any packet which has been opened and not used must be thrown away.

(c) Place the reel of silk and a little water in a small bottle. Pass the end of the silk through a narrow piece of glass tubing which should perforate the cork of the bottle and be plugged with wool (fig. 132). Sterilize in the autoclave. When the silk is wanted it is only necessary to pull on the end of it to unreel it. So much of the silk thread as projected before it was pulled must of course be cut off and thrown away. The silk in the bottle will keep sterile so long as the wool plug remains in position.

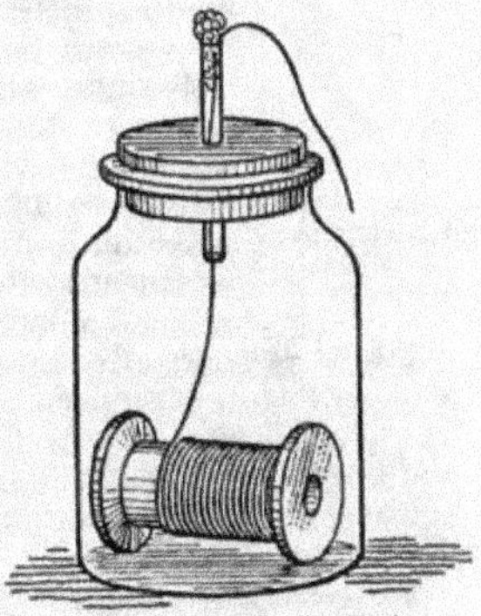

FIG. 132.—Bottle for storing sterile silk.

A solution of some antiseptic (e.g. 0·1 per cent. oxycyanide of mercury) [or 2 per cent. lysol] and some sterile water must also be at hand.

Preparation of sterile water.—A number of test-tubes or small flasks of 50–100 c.c. capacity should be three-quarters filled with water and sterilized at 115° C. When wanted, aspirate the water out of the tubes or flasks with a Pasteur pipette. The contents of any tube or flask opened and not used should be thrown away at once.

Water may also be sterilized in larger quantities in flasks : but it is better to use small vessels containing only sufficient water for one experiment. All risk of contamination is then avoided.

A number of sterile glass dishes, glass rods, platinum wires and Pasteur pipettes must also be ready.

Lastly, *syringes* and *needles* are necessary with which to inoculate the virus into the tissues.

(i) Inoculation syringes.

There are numerous patterns of inoculation syringes, and this fact in itself shows how difficult it is to obtain a syringe which fulfils all the conditions required of it. A satisfactory syringe should—

1. Lend itself to sterilization in boiling water or in steam under pressure.

2. Have perfectly water-tight joints and plunger, and be so constructed as not to need frequent renewal of the parts.

3. Be accurately graduated on the glass barrel or piston rod.

Pasteur pipettes.—A Pasteur pipette may on occasion be used as an inoculation syringe. And for this purpose the drawn-out part is made short and slightly curved, with a sharp point (fig. 133). [The opposite end is plugged with wool and the pipette sterilized before use. A small bulb may be very conveniently blown on the tube.] The liquid to be inoculated is sucked up into the pipette and the pointed end pushed through the skin [or, when required, into the peritoneal cavity], and the material deposited in the tissues by blowing through the plugged end. This is all the apparatus that is necessary in a large number of cases.

A. Older patterns.—Of the older patterns of syringes the use of which is now being given up, the following may be mentioned :

Pravaz's syringe.—One of the oldest and at the same time one of the best from the point of view of the security of its joints is Pravaz's syringe. But, unfortunately, the joints and the leather plunger will not stand the temperature of boiling water. If it be used it must be disinfected by soaking it in a 5 per cent. solution of carbolic acid for several hours and subsequently rinsing well in sterile water.

Syringes with air piston.—Koch, Petri and others eliminated the plunger. The glass cylinder forming the body of the syringe was fitted at one end with a needle and at the other with an india-rubber ball : by squeezing the ball, the liquid was forced out. When used for injection however especially into tense tissues, the liquid either cannot be inoculated or runs out when the needle is withdrawn. These syringes are not now in use.

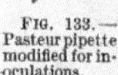

Fig. 133.— Pasteur pipette modified for inoculations.

Straus' syringe.—By substituting compressed elder pith for the leather in Pravaz's plunger, Straus obtained a syringe which stands the temperature of boiling water and of the autoclave very well. The plunger can be changed as often as is necessary, but though it is easily done it takes some time ; moreover it has to be done frequently since the elder pith rapidly loses its elasticity. With these reservations the syringe is a good one.

To renew the plunger.—Take a piece of elder pith with a regular and fine grain, cut off the outer, fibrous layer, and compress it between the fingers so as to flatten it longitudinally as much as possible. Then cut out a small cylindrical piece to fit the barrel of the syringe tightly. Perforate its centre with a needle heated to redness in the flame and fix it on to the end of the piston rod. Then with a very fine file polish its sides and introduce it into the barrel. By soaking in water for a few minutes the elder pith swells and the plunger becomes water-tight. The pith can also be compressed at will by means of the screw at the top of the piston rod.

In Roux's modification the glass barrel is narrowed below and ground so that the needle is fitted directly on to it. The plunger can be freely withdrawn or inserted since the tube is merely fitted above with a plug.

Malassez's syringe.—There are several patterns of this syringe. The only ones which can be recommended are those in which the plunger consists of a mixture of india-rubber and asbestos, or of "fibre," a combination of cellulose and rubber. The lower end is narrowed and ground and the needle fitted on to it by means of a "fibre" washer.

Metal plunger syringes.—In these forms the elastic plunger is replaced by a rigid piston consisting of an accurately calibrated metal rod, the body itself being an hollow metal cylinder. These syringes soon deteriorate and are inconvenient in that the liquid within them is not visible.

B. Patterns recommended for use.

Roux's syringe (fig. 134).—For serum-therapeutic inoculations, Roux devised a syringe of 20 c.c. capacity with a plunger made of some rubber preparation. The needle is connected to the nozzle of the barrel by a piece of rubber tubing about 10 cm. long. This arrangement allows the injection to be forced into the tissues without the risk of detaching the needle. Before sterilizing the syringe be careful to loosen the upper socket to leave the glass barrel room to expand, and so prevent it being cracked.

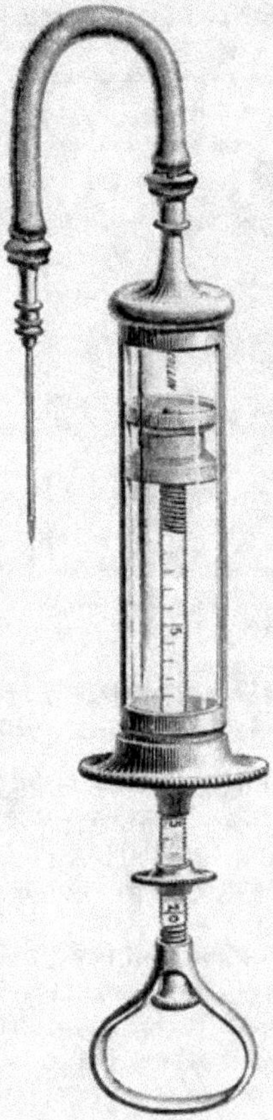

Debove's syringe (fig. 135).—In the author's opinion Debove's syringe is better than Roux's. It is both easy to manipulate and easy to sterilize: it is solid and perfectly water-tight and can be used for all sorts of inoculations.

The syringe consists of a glass tube held between two metal sockets by means of a movable metal armature

FIG. 134.—Roux's syringe.   FIG. 135.—Debove's syringe.

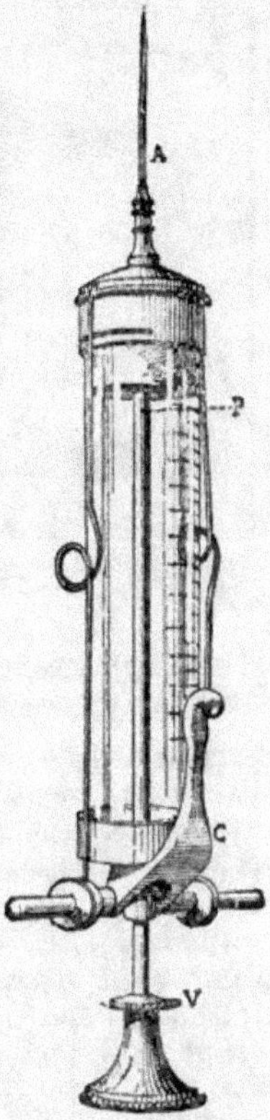

which is entirely distinct and controlled by a lever. The barrel is accurately calibrated and the syringe is made water-tight with asbestos washers.

The lower metal socket has a conical extension on to which the needle fits either directly or through an india-rubber connexion.

The plunger consists of asbestos rings held between two metal discs. The piston rod carries a screw which allows the pressure on the asbestos rings forming the plunger to be varied so that the play of the plunger can be regulated.

The syringe is easily taken to pieces by raising the lever; this relaxes the lateral stays which can then be disconnected and the sockets taken off.

All parts of the syringe are made to a standard pattern, so that broken parts can be replaced without sending it to the maker to be repaired. The syringe is made in several sizes to hold from 2–100 c.c., those ordinarily in use being of 2, 10 and 20 grams' capacity.

Method of sterilizing Debove's syringe.—Withdraw the plunger as far as possible, raise the lever to relax the spring and allow expansion of the glass cylinder. Place the syringe and needle in a vessel of cold water and heat to boiling for 15 or 20 minutes. Let the syringe cool; then take it out of the water with a pair of sterile forceps, let the water above the plunger run out, lower the lever and fit the needle on its socket.

(When the syringe has been used for an injection, rinse it out in cold water to wash out all albuminoid matter—which would coagulate on boiling—and boil it in the same water, so that both the latter and the syringe are sterilized at the same time.)

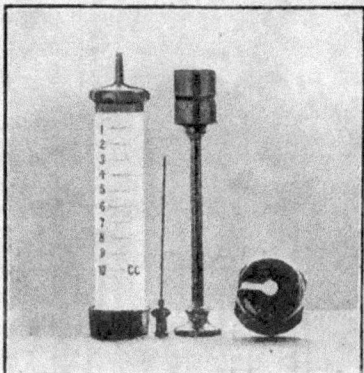

FIG. 136.—Another reliable form of syringe ("The Record").

The syringe may be sterilized in the autoclave if preferred: it is prepared as above and then heated to 115° C. for a quarter of an hour. In most cases, boiling is sufficient to completely sterilize it, but when it has been used for inoculating cultures of spore-bearing bacilli, such as *B. tetani*, *B. maligni œdematis*, etc., it should be autoclaved.

Syringes with glass pistons.—Malassez has had a syringe made by Luer which is entirely of glass. The piston itself consists of a calibrated glass rod. Numerous forms of syringes based on this pattern can now be bought at a low price. These syringes are easily sterilized quite water-tight and are excellent in every way, particularly for small volumes (1–2 c.c.).

Apparatus for injecting large quantities of fluid.—In immunizing animals with toxins when large quantities of filtered cultures are inoculated, syringes are not large enough, and moreover the fluid cannot be injected sufficiently slowly. In these cases the following arrangement is useful (fig. 137).

The liquid to be inoculated is poured into a tall glass vessel graduated on the glass from above downwards and closed with an india-rubber plug which is perforated by two glass tubes, one of which, reaching to the bottom of the vessel, has a needle attached to its upper end with india-rubber tubing. The other tube passes a few centimetres below the stopper, is plugged with wool, and through it the air in the vessel can be compressed. When this is done the liquid flows out of the needle, and the rate of flow can be regulated at will.

The apparatus is sterilized in the autoclave and then the liquid to be inoculated aspirated into the vessel.

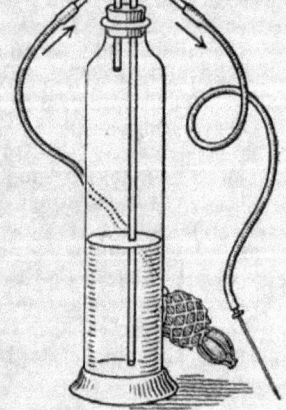

FIG. 137.—Apparatus for inoculating large volumes of liquids.

(ii) Needles.

Steel needles are very generally used for inoculations. The disadvantage of steel is that it so readily rusts with the result that the lumen of the needle soon becomes obstructed. This difficulty is overcome by carefully washing the needles after use, and keeping them—after they have been sterilized by boiling—in a small bottle filled with absolute alcohol [to which a few lumps of calcium chloride may with advantage be added] or in a 3 per cent. solution of sodium borate.

On account of the difficulty of preventing steel from rusting platinum-iridium needles are gradually replacing steel. Platinum-iridium does not rust and the needles can be heated to redness. On the other hand they are expensive and delicate and as they are but little stronger than needles made of pure platinum, it is on the whole better to use steel especially when a thick skin has to be penetrated as is generally the case in animal inoculation.

A selection of needles of different calibre and of different lengths ought to be kept in the laboratory.

2. Preparation of the material for inoculation.

The material to be inoculated may be either a solid or a liquid. The procedure will be different in the two cases.

(i) Of fluids.

Broth cultures are the commonest fluids inoculated but other fluids, such as blood, serum, pleural and peritoneal exudates, have also to be inoculated at times.

(a) **Cultures.**—*Every culture should be examined microscopically before being inoculated*, to test its purity.

When ready to perform the inoculation remove a little of the culture with a Pasteur pipette and transfer it to a sterile watch-glass and cover the latter again with the paper in which it was sterilized. Aspirate the culture into a sterile syringe through the needle either by puncturing the paper with the needle or by slightly raising the paper and passing the needle beneath it. Hold the syringe with the needle pointing upwards and gently press the plunger to expel any air which may have been drawn into the syringe, taking care to hold the piece of sterile paper which covered the watch-glass alongside the needle to catch any drops of culture which may inadvertently be driven out. Burn the paper and dip the watch-glass into a vessel of boiling water to sterilize it.

(b) **Exudates.**—Blood and serous exudates must be collected in the manner to be described in Chaps. XI. and XII. With a pipette transfer the amount required for inoculation to a sterile watch-glass and proceed as above. It is very difficult to inject blood directly because it so readily coagulates and blocks the needle. If the virus pass into the serum, the blood should be allowed to clot and the serum used for inoculation. On the other hand if the virus be retained in the clot this should be dealt with as though it were a solid tissue (*vide infra*).

To facilitate the inoculation of whole blood it is occasionally necessary to have resort to the anti-coagulating action of sodium citrate. The blood is collected in a sterile vessel containing a little of the following solution also sterilized:

Water, - - - - - - - - - -	1000 c.c.
Sodium chloride, - - - - - - -	8 grams.
Sodium citrate, - - - - - - -	15 ,,

Use two to four volumes of the citrate solution to one volume of blood. Mix thoroughly and inoculate without delay.

(ii) Of solid substances.

(a) **Solid substances.**—Fragments of internal organs, splinters, etc., may be inserted directly into the tissues of the animal. After making a small incision separate the cellular tissue with a director and introduce the material into the pocket so formed, suture the wound and cover it with collodion. Material may be similarly introduced into the peritoneal cavity, muscles, etc.

(b) **Cultures on solid media.**—A small incision may be made in the skin and then a wire, charged with the micro-organisms by scraping the surface of the medium, rubbed into the tissues. But it is more usual to make an emulsion by rubbing up some of the material in sterile water or broth and then to inoculate the emulsion with a syringe.

Scrape the surface of the medium with a stout platinum wire and transfer the growth to a sterile watch-glass containing a little sterile water and stir it about until an homogeneous emulsion is obtained. If the culture be difficult to break up—as, for example, a growth of the tubercle bacillus—and does not mix with water, it should be ground up as described below (d).

(c) **Pus.**—As a rule, pus is too thick to be inoculated undiluted. Transfer a few drops of the pus with a pipette to a sterile watch-glass, add a little sterile normal saline solution (0·8 per cent. aqueous solution of sodium chloride) or broth and mix them thoroughly with the end of the pipette.

(d) **Fragments of organs.**—For the method of collecting fragments of internal organs, portions of the central nervous system, etc., see Chap. XI.

Transfer the material to a sterile watch-glass and break it up with a sterile glass rod. When the tissue is reduced to a fine paste add a little sterile normal saline solution drop by drop from a pipette, and mix until a quite homogeneous suspension is obtained.

It is often necessary to filter the suspension through a small piece of previously sterilized fine muslin to get rid of little lumps. This precaution is very necessary when the emulsion has to be inoculated intra-venously, in order to avoid the formation of an embolus.

When a very tough material has to be dealt with such as tuberculous or leprous nodules, it should be cut up into quite small fragments with sterile scissors, ground up in a sterile mortar with some fine sterile sand [1] (p. 155) and a fluid emulsion made by adding sterile normal saline solution drop by drop: the emulsion is then filtered through fine, sterile muslin before being inoculated.

When a larger quantity of material has to be emulsified, Borrel's *broyeur* will be found useful. With this machine fine powders or emulsions can be obtained without contaminating the material and without exposing the operator to the inhalation of dust containing pathogenic organisms.

3. Technique of inoculation.

General rules.—Before inoculating an animal shave the hair and cleanse the skin of the part.

The hair may be cut very short with a pair of curved scissors but it is better to shave the skin. For delicate inoculations it is preferable to epilate the hairs with one of the following solutions:

1. Recently slaked lime, - - - - - - - 2 parts.
   Water, - - - - - - - - - - 3 ,,

Pass a stream of hydrogen sulphide through the emulsion, shaking it frequently, until it is saturated. Apply the paste to the part to be epilated and after a few minutes wash it off with water and a nail brush.

2. Sodium sulphide, - - - - - - - 3 parts.
   Powdered quicklime, - - - - - - - 10 ,,
   Starch, - - - - - - - - - 10 ,,

Mix into a powder. When required for use, add sufficient water to a little of the powder to form a soft paste and apply it to the skin. After 3 or 4 minutes' application the hair is removed.

[1 Sand was used as a triturating agent by Cobbett in his earlier experiments on tuberculosis but was almost immediately given up. A little patience is all that is required to grind up even a tough tuberculous lesion.]

In a large number of cases it is only necessary after cutting the hair, to rub the skin with a [2 per cent. solution of lysol or a] 0·1 per cent. solution of oxy-cyanide to render it aseptic. But for more perfect asepsis the skin should be scrubbed with soap and a nail brush, washed with an antiseptic solution, rinsed with alcohol and wiped with sterile filter paper.

(i) Intra-dermal inoculation.

1. Shave and cleanse the part but use no antiseptics.

2. Scarify the skin very superficially with a bistoury, or pick up the skin in the fingers and shave off the epidermis with the blade of a sharp knife.

3. Rub the material to be inoculated into the prepared part with a sterile glass rod or a piece of sterile wool held in a pair of forceps.

In some cases, it is sufficient merely to rub the skin briskly with a sponge soaked in the material without scarifying or scraping the surface.

As a rule, the material should be inoculated into some part of the skin of the body which the animal cannot reach, such as the dorsal surface of the ear, the skin of the back, or the root of the tail. This rule applies to the inoculation of all species of laboratory animals. For some viruses there are special sites used for inoculation, the eyelids or eyebrows in syphilis, for example.

(ii) Inoculation on the surface of mucous membranes.

Abrade the surface of the mucous membrane, as in the preceding case, by scraping it with the blade of a knife, and spread the virus over the surface thus abraded. Sometimes it is better, before inoculating the mucous membrane, to cauterize it with a moderately hot iron or platinum rod so as to produce a superficial slough.

(iii) Sub-cutaneous inoculation.

A. Of a liquid.—**1.** Shave and cleanse the skin.

2. Pick up a fold of skin between the thumb and index finger of the left hand, insert the needle at the base of the fold, inject the fluid and withdraw the needle. Care must be taken to see that the fluid does not find its way out again through the needle puncture and that the injection is not made into the muscles. [By lightly pressing the fold in which the puncture has been made between the finger and thumb and twisting it gently, exudation through the puncture can almost always be prevented.]

B. Of a solid.—**1.** Shave and cleanse the part.

2. Make a small incision through the skin with a bistoury, separate the sub-cutaneous tissue with a director over a sufficient area and then introduce the material to be inoculated into the pocket with a pair of sterile forceps.

3. Put a stitch or two into the incision and cover it with a little collodion.

Note.—As has already been stated above, inoculations are most satisfactorily made into some part of the body that the animal cannot reach; they are however often made beneath the skin of the abdomen or thigh.

When the material to be inoculated is solid, some part of the body where the skin is very loose should be chosen, as for instance, the flanks or the groin.

(iv) Inoculation into lymphatic spaces.

In the frog, inoculations are frequently made into the sub-cutaneous connective tissue which consists of large inter-communicating lymphatic sacs.

A. Dorsal sac.—The dorsal sac is situated over the posterior part of the back. The animal's hind limbs are wrapped in a cloth so that it cannot move, and by compressing the sides of the back with the fingers the sac is made to stand out. A fine needle is introduced obliquely into the sac from above downwards and the material injected.

B. The posterior limb sacs.—Introduce a needle obliquely from above downwards under the skin below the femoro-tibial joint and inject the material.

(v) Intra-muscular inoculation.

1. Shave and cleanse the part.

2. Push the needle deeply into the muscles, inject the material and withdraw the needle.

Inoculations are made for preference into the muscles of the thigh in the mammalia, and into the pectoral muscles of birds.

(vi) Intra-venous inoculation.

Whenever possible intra-venous inoculation should be made into a superficial vein. The needle may be passed through the skin directly into the vein without first exposing the latter. Intra-venous inoculation cannot be effected in the case of very small animals such as mice.

A. Rabbits.—**1.** One of the dorsal veins of the ear, and, for preference, the external marginal vein should be selected. Avoid the median veins, because, being embedded in a lax cellular tissue, they are liable to slip away from under the needle.

2. Cut the hair over the vein with a pair of curved scissors [or better, shave

FIG. 138.—Pressure forceps for the ear vein.

it], and cleanse the skin. The rabbit is placed on the operator's knee and, if necessary, held by an assistant.

3. Take hold of the margin of the ear between the index finger and thumb of the left hand so as to extend it. Put a pair of pressure forceps on the vein at the base of the ear so as to make the vein prominent (fig. 138).

By rubbing the skin with a sponge soaked in warm carbolic water the vein can be rendered more distinct.

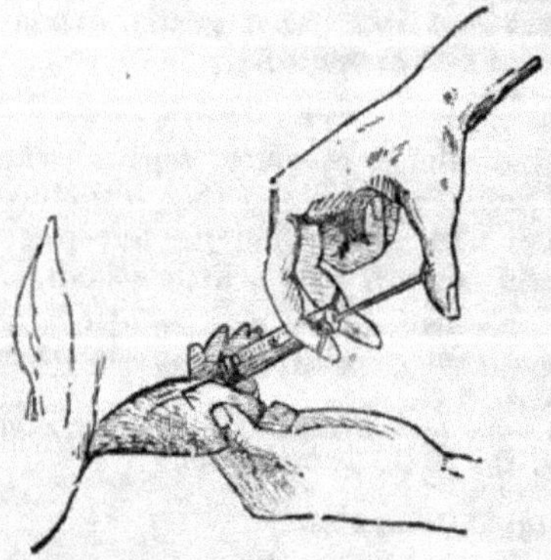

4. Hold the needle very obliquely, almost parallel to the vessel and pierce it in the direction of the blood-stream (fig. 139).

5. When the needle is in the vein, take off the forceps and inject the fluid slowly. It is a good plan to apply the forceps higher up on the needle itself, so as to hold it in the skin. If the needle has missed the vein, the injection will cause a subcutaneous swelling, and the operation must be begun again lower down.

FIG. 139.—Inoculating into the ear vein of a rabbit.

After the fluid is injected withdraw the needle, and if the vessel bleed leave the forceps on the bleeding point for a few minutes, [or pass the vein between the finger and thumb moving them against the blood-stream].

B. Guinea-pigs.—The superficial veins are not large enough in the guinea-pig for intra-venous inoculation, and recourse must be had to the external jugular. This vein is superficially situated, lying beneath the skin the sub-cutaneous muscles and some cellular tissue, and follows a line from the angle of the jaw to a point mid-way between the shoulder and the sternum.

1. Fix the guinea-pig on its back, with its head extended. Shave and cleanse the part.

2. Make an incision through the skin and sub-cutaneous muscles in the middle of the line described above, tear through the cellular tissue with a director, and the vein will be exposed lying to the outer side of the incision.

3. Pass the needle obliquely into the vein (it is very convenient to have a needle with the lower end bent at a right angle) inject the fluid and with-draw the needle.

4. Cleanse the wound with a sponge soaked in carbolic water and make quite sure that there is no bleeding from the prick in the vein. Put two or three stitches in the skin and cover the incision with collodion.

C. Dogs.—For intra-venous inoculation in dogs the external vein of the hind limb—the small saphenous—should be selected.

1. Muzzle the animal, and get an assistant to hold it.

2. Shave the skin on the outer side of the limb where the calf muscles are inserted into the *Tendo achillis*. Compress the limb above, and rub the shaved part with a sponge soaked in carbolic water. The small saphenous vein will thus be made to stand out and is easily accessible at the upper part of the *Tendo achillis*.

3. Avoid, if possible, having to expose the vein, and in performing the inoculation pierce the skin and the vein at one and the same time.

D. Horses and Cattle.—Locate the jugular vein and render it prominent as described on p. 49. Make the injection with the usual precautions.

E. Birds.—Birds are best inoculated intra-venously in the axillary vein.

1. Fasten the bird down, and let an assistant extend the wing, and at the same time compress the base. Pluck the down from the inner surface of the root of the wing, and rub the part with a sponge soaked in carbolic water.

2. When the vein has swelled, inoculate the material.

(vii) Arterial inoculation.

In mammals for purposes of arterial inoculation the femoral or carotid artery is chosen.

A. Femoral artery.—The femoral artery takes the same course in animals as it does in man. In the fold of the groin, the vein is on the inside, the artery next and the crural nerve on the outer side. The artery takes a line from the middle of the fold of the groin to the inner side of the knee.

1. Fix the dog on its back. Rotate the leg outwards and extend it. Shave and cleanse the part.

2. Determine the exact position of the artery by finding the pulse near the middle of the fold of the groin, and make an incision through the skin and sub-cutaneous tissue, a few centimetres long, along the line of the vessel.

3. Divide the aponeurosis on a director and the sheath of the vessels and nerve will be exposed.

4. Having found the artery prick it very obliquely, inject the material and withdraw the needle.

5. Put a few stitches in the skin and paint the wound over with collodion.

B. Carotid.—In all mammals, the carotid artery lies in close relation to the trachea in the middle of the neck, being contained in a sheath common to it, the internal jugular, the pneumogastric and the great sympathetic.

1. With the animal lying on its back and its head extended shave the skin of the middle of the neck and wash it with an antiseptic.

2. Make a longitudinal incision, a few centimetres long, in the middle line, in front of the trachea.

3. Divide the aponeurosis connecting the two sterno-mastoids on a director.

4. Separate the cellular tissue along the trachea with the rounded end of a probe and then, by pulling the sterno-mastoid outwards, the sheath of the vessel will come into view.

5. Open the sheath with a pair of forceps and a director. The artery will be recognized from the vein by its larger size. Make the injection as described above.

(viii) **Intra-peritoneal inoculation.**

A. Of a fluid.—Every precaution should be taken not to perforate the gut.

1. Fix the animal firmly on its back. [An assistant can hold it equally well.] Shave and disinfect a few square centimetres of the skin of the abdomen.

2. Pick up the whole thickness of the abdominal wall between the thumb and index finger of the left hand.

3. Insert the needle of the syringe into the base of the fold in such a way that the point is directed upwards, withdraw it a little, and then, altering the direction of the point, pass it into the cavity of the abdomen. Inject the material and withdraw the needle.

The following method affords greater security. A curved needle is used, in which the opening is situated in the centre of the concavity of the arc (fig. 140).

Insert the needle through the whole thickness of the abdominal wall, including the peritoneum, and bring the point to the surface; the needle-opening is now within the peritoneal cavity, and the material is injected. The point of the needle being outside all the time that the injection is being made no injury can be done by it to the gut.

FIG. 140.—Needle for intra-peritoneal inoculation.

B. Of a solid.—**1.** The animal is fixed on its back, and the skin of the part shaved and cleansed.

2. Make an incision in the median line, the length of the incision varying with the size of the substance to be inoculated.

3. Cut through the aponeurosis, using a director to avoid injuring the intestine.

4. Take hold of the aponeurosis on each side with pressure forceps, and hold it up as much as possible to prevent the intestines protruding. Introduce the material to be inoculated into the wound and push it well under the muscular layer into the peritoneal cavity.

5. Sew up the aponeurosis with silk, stitch the skin and cover the incision with a layer of collodion.

Note.—The most careful asepsis is necessary in performing intra-peritoneal inoculations. These inoculations should always be done with pure cultures or with material which can be obtained free from contaminating organisms for, if sputum, excreta, etc., are injected into the peritoneal cavity the animal will very quickly die of peritonitis. [But see Chap. XVIII. Sect. IV.]

C. Collodion sacs.—In their researches on cholera toxin, Metchnikoff, Roux and Salimbeni devised a method of growing organisms in small closed collodion sacs in the peritoneal cavities of animals.

By this method, organisms can be cultivated in the body, and at the same time be protected from phagocytes. The thin walls of the collodion sacs, while allowing osmotic changes to take place which alter the composition of the medium in the sacs and permit the toxins secreted by the organisms to diffuse into the tissues of the animal, prevent cells (micro-organisms and phagocytes) passing through. This method has many applications in bacteriological work.

All the soluble products of micro-organic metabolism dialyse more or less through the walls of collodion sacs, but they do not pass through *in toto*. The collodion membrane does not act as a perfect filter, so that some toxins pass through only with difficulty and very slowly (Rodet and Guechoff). The immunizing substances seem to pass through first (Crendiropoulo and Ruffer).

(*a*) **Method of preparing collodion sacs.**—Have ready (1) a wide-mouthed bottle containing collodion, free from castor oil, and of a medium syrupy consistence, (2) some small glass tubes of 5–6 mm. internal diameter, (3) some small conical india-rubber plugs, (4) a test-tube carefully calibrated and not enlarged at its lower part, (5) some silk thread.

Collodion containing castor oil [*collodium flexile* B.P.] may be used instead of ordinary collodion but such sacs are not so transparent, though they are more elastic and stronger than the others.

1. Rotate the lower end of the test-tube regularly on the sloping surface of the collodion and prolong the contact according to the thickness which the layer of collodion is to have.

2. Remove the tube from the collodion and continue turning it between the fingers for about a minute. Then let the layer of collodion dry for a few minutes until it is of a semi-liquid consistence.

3. With a scalpel cut round the layer of collodion near its upper end and separate the sac from the tube, thus : Free the upper end of the sac with the thumb nail, turn it back like a glove finger and gradually peel it off. This must be done slowly. [Dipping in methylated spirit softens the collodion and makes it strip easily.]

4. Distend the sac by blowing into it. A sac may be made to hold from one to several cubic centimetres according to the particular requirements of the experiment.

5. Fit a piece of small glass tubing into the open end of the sac, fasten it on with several turns of silk thread and cover the silk with a little collodion. Fill the sac with water and suspend it by a thread in a bottle or test-tube containing a little water. Plug the vessel with wool. Place the india-rubber plugs in a flask plugged with wool and containing a little water. Sterilize at 115° C. in the autoclave.

All these manipulations are delicate and take a long time, and it will often be advantageous to use the collodion sacs which can now be bought in the shops.

6. Take each sac out of the bottle with a pair of sterile forceps and transfer it to a sterile dish covered with paper. Suck up the water out of the sac with a pipette and replace it with broth sown with the organism under investigation.

7. Close the opening of the tube with an india-rubber plug picked up with sterile forceps, and cut it off short close to the tube with a sterile scalpel. Dehydrate the plug in absolute alcohol and cover it with several layers of collodion.

Notes.—(i) Bertarelli has recommended a method which considerably simplifies that just described. In Bertarelli's method, the free end of the piece of glass tubing to which the sac is affixed is drawn out beforehand so that the free end is fine and conical. After the sac has been sterilized the water is withdrawn with a syringe and the culture introduced. The pointed end is then sealed off in a small flame. Bertarelli further suggests using a solution of collodion in ether of the same consistence as is used for embedding, in place of collodion.

(ii) Phisalix has introduced a modification of the method of preparation by which much stronger sacs can be made, and the risk of breakage in the abdominal cavity thereby reduced.

FIG. 141.—Phisalix' guides for collodion sacs.

A collodion sac is prepared as above (Stages 1 to 4) and slipped on to a guide consisting of a perforated glass ampoule (fig. 141, A). The sac which now covers the guide is fastened to the neck of the ampoule with a few turns of silk thread, and this is covered with a layer of collodion.

The sac is then sterilized in the autoclave in the ordinary way. After the sac has been filled with the culture the neck of the ampoule is sealed off in the flame (fig. 141, B).

(iii) Gorsline noted that the principal difficulty in making collodion sacs is the separation of the sac from the tube on which it is moulded. He overcomes this difficulty by using test-tubes perforated with a small hole in the bottom. In making a sac the hole is first obliterated by gently touching the bottom of the tube on the collodion. After the sac has then been made in the ordinary way and dried, the tube is filled with water. By blowing down the open end of the tube, the water forces its way through the small hole at the bottom of the tube and insinuates itself between the tube and the sac, with the result that the latter is easily separated.

(b) **Insertion into the peritoneal cavity.**—Collodion sacs may be used in experiments upon guinea-pigs, rabbits, dogs, sheep, cattle [and birds (fowls and pigeons)]. All these animals tolerate aseptic sacs filled with sterile broth very well. The technique of the operation is described above (p. 174, **B**).

After an interval varying from a few days to several months, the animal is killed and the sac withdrawn and its contents investigated. When the sac has been in the peritoneal cavity several weeks it not infrequently happens that it is found to be broken; [even then it is in the case of birds at least usually covered with a fibrous sheath which prevents dispersal of the contents]. It is well therefore to use several animals, to be sure of finding at least one sac intact. To remove the contents, sterilize the bottom with a hot wire, insert a pipette and aspirate the fluid.

D. Reed sacs.—Roux and Nocard suggested using, in place of collodion, a small piece of the tubular membrane lining the central canal of the bulrush. Sacs made with this membrane are more permeable than those made of collodion.

1. Take a few pieces of common bulrush and, if they are fresh boil them in water for about a quarter of an hour, but if dry autoclave them at 115° C. for an hour instead.

2. After softening them sharpen one end in the same way as a lead pencil, in order to expose the membrane lining the central canal. This membrane is then carefully denuded for a certain length.

3. Tie one end of the separated membrane firmly like a purse, then by pressing gently on this end with a glass rod it can be turned inside out.

4. Tie a small glass tube into the open end and fasten it with a stout ligature, and place another ligature on the sac itself below the end of the tube. Fill the sac with water and sterilize as in the case of collodion sacs.

5. Suck up the water out of the sac and replace it with the culture, tie the ligature on the reed and disconnect the sac from the tube above this ligature. Cover each ligature with a drop of melted gum lac.

Introduce into the peritoneal cavity in the manner already described.

(ix) Inoculation into the biliary passages.

In all animals in common use for experimental purposes the bile is poured into the duodenum through a simple channel—the bile duct—of which the orifice is more or less close to the pylorus. In the dog the opening is 4–12 cm., in the rabbit about 1 cm., beyond the pylorus, and in the guinea-pig about the middle of the duodenum. The operation in the rabbit, guinea-pig and dog will be described. The strictest asepsis must be observed.

A. Guinea-pig. Rabbit.—1. Anæsthetize the animal and fasten it on its back. Shave the hair and cleanse the skin of the abdomen.

2. Make an incision about 6 cm. long in the middle line commencing about 1 cm. below the xiphoid cartilage. Cut through the skin and aponeurosis, stop any bleeding, then incise the peritoneum on a director.

3. Identify the pyloric end of the stomach, then, using the index finger, find the duodenum and bring it to the surface. The opening of the bile duct will be seen about its centre.

4. Having identified the canal, isolate and fix it on a blunt hook. Pass the end of a fine needle bent at a right angle very obliquely through the wall and in the line of its long axis. Inject the material.

5. Withdraw the needle, and touch the point where it penetrated the wall with a sponge soaked in carbolic water. Stitch up the aponeurosis with silk, suture the skin and paint the incision with collodion.

B. Dog.—1. Anæsthetize the animal and fix it on its back. Shave the hair and cleanse the skin of the abdomen.

2. Make a longitudinal incision in the middle line, or a little to the right, about 8 cm. long, commencing a few centimetres below the xiphoid cartilage. Cut through the skin and aponeurotic layer, and stop any hæmorrhage. Incise the peritoneum on a director.

3. With the first finger in the wound, follow the lower surface of the liver, then bend the finger to hook up the duodenum and bring the latter to the surface, and to the left.

4. Find the right edge of the duodenum, and follow it until the finger meets the fold of the lesser omentum in which the bile duct lies with the portal vein, hepatic artery and nerves. The duct lies superficially in the fold and can be recognized by its pearly appearance, its structure, its direction, and by the fact of its opening into the duodenum at a distance of from 4 to 12 cm. beyond the pylorus.

5. Isolate the duct on a small director and fix it in a blunt hook. Pass the bent needle very obliquely through the wall in the line of its long axis. Inject the material, and complete the operation as above.

(x) Inoculation into the portal vein.

The operation is easier in the dog, but can also be done, though with some difficulty, in the guinea-pig and rabbit. The walls of the vein are very thin, and easily torn. The technique described above for finding the bile duct is applicable to the isolation of the portal vein in the dog. It is better to operate as follows :

1. Anæsthetize the animal, and fix it on its left side. Shave and cleanse the skin of the part.

2. Make an oblique incision in the right hypochondrium commencing above at the junction of the last rib with the vertebral column and extending to the outer edge of the rectus muscle at the crest of the ileum. Cut through the skin and muscular layers. Stop the hæmorrhage. Incise the peritoneum on a director.

3. Get an assistant to pass his fingers into the wound and push the intestines to the left as far as possible and hold them in the abdomen.

4. Deep down, in the upper part of the wound below the liver the bend of the duodenum will be seen, and, on a level with it, the principal mesenteric veins converging above towards the portal vein.

5. Having recognized the vein, isolate it, fix it and perforate the wall with a bent needle. Inject the material.

6. Withdraw the needle and wipe the puncture in the vein with a sponge soaked in carbolic water. Put in two layers of sutures and cover the skin incision with collodion.

(xi) Inoculation into the kidneys.

The dog, rabbit, guinea-pig, etc., may be used for this experiment.

1. Lay the animal on the side opposite to that on which it is proposed to operate, and anæsthetize it. Shave and cleanse the skin over the region to be operated upon.

2. Make an incision outside the sacro-lumbar muscles from the anterior end of the last rib to the sacrum.

3. Incise the muscles for the whole length of the skin incision on a level with the external border of the floor of the lumbar region.

4. Retract the margins of the wound widely and the peri-renal adipose tissue will be exposed behind the peritoneum. Then, after tearing through the loose cellular adipose tissue with a director, the kidney appears in the wound.

5. Push the needle into the renal parenchyma and make the injection. Withdraw the needle and touch the needle prick with a sponge soaked in carbolic water. Insert two sets of sutures, and paint the skin incision with collodion.

Ureter.—The ureter can be exposed by operating in the same way. When the kidney is freed from the peri-renal adipose tissue the ureter will be seen lying with the renal vessels and nerves.

(xii) Inoculation into the anterior chamber of the eye.

A. Liquids.—**1.** Fix the animal on its belly, and fasten the head so that the animal cannot move it. It is well to anæsthetize the eye by dropping into it a few drops of a 1·5 per cent. solution of cocaine. In about 5 minutes the eye is completely insensitive.

2. Hold the eyelids apart and fix the eye with the thumb and index finger of the left hand. Insert the needle perpendicularly to the axis of the eye at the margin of the cornea at the corneo-sclerotic junction. Inject a few drops of the fluid and withdraw the needle.

B. Solid.—**1.** As in " **A** " above.

2. Then, the eyelids being held apart and the eye fixed make· an incision a few millimetres long along the upper border of the cornea with a cataract knife or very fine scalpel.

3. Holding the tissue to be inoculated with a fine pair of bent forceps pass it through the incision and force it as far as possible into the anterior chamber by rubbing the cornea lightly with a Daviel's curette or the blunt end of a probe.

Inoculations into the anterior chamber are generally done on rabbits and are practised chiefly for infecting with hydrophobia or syphilis, or for studying the development of tuberculosis or the phenomena of phagocytosis.

(xiii) Inoculation into the respiratory passages.

A. Inoculation into the lungs.—**1.** Shave and cleanse the skin of the thorax in the neighbourhood of the axillary fold.

2. Push the needle perpendicularly through one of the upper intercostal spaces to a depth of from one to several centimetres, according to the size of the animal. Inoculate the material and withdraw the needle.

B. Intra-tracheal inoculations (*mammalia*).—**1.** Fix the animal on its back with the head extended and the neck raised by means of a firm plug of cotton-wool, a large india-rubber cork or a small block, etc. Shave and cleanse the skin in the middle line below the larynx.

2. Make an incision 2–3 cm. long through the skin in the middle line of the neck in front of the trachea.

3. Incise the aponeurosis on a director.

4. Having exposed the trachea, pass the needle obliquely into the lumen between two of the cartilaginous rings. Inject the material. Withdraw the needle and wash the perforation with a sponge soaked in carbolic water.

Notes.—(*a*) In small animals it is convenient as soon as the trachea is exposed to fix it by passing a suture needle threaded with silk through it.

(*b*) To avoid all risk of inoculating the material into the cellular tissue or into the walls of the trachea itself the following precautions may be taken. Use a small, very fine trocar with a cannula which should be shorter than the syringe needle. When the trachea is exposed pass the cannula between two of the cartilaginous rings, withdraw the trocar leaving the cannula in position. Pass the syringe needle through the cannula so that the point of the needle passes beyond the end of the latter. Make the injection and withdraw the needle first, then the cannula.

5. Suture the skin. Cover the wound with collodion.

C. Intra-tracheal inoculation in birds.—The opening of the trachea is behind the base of the tongue.

1. Open the beak and draw the tongue forwards with a pair of forceps.

2. The opening of the trachea will be seen behind the tongue and the material to be inoculated is injected straight into it.

D. Intra-pleural inoculation.—Besson has shown as a result of some observations which he made with Pourrat that it is a very difficult matter to inoculate a fluid into the pleural cavity without, at the same time, injuring and penetrating the lung. Consequently intra-pleural inoculation is not very exact.

In those cases where the syringe needle is passed obliquely from below upwards through an intercostal space (6th or 7th) it often happens either that the pleural cavity has not been reached or that the needle has passed beyond it. To perform a true intra-pleural inoculation the following technique must be followed.

1. After fixing the animal on its left side, shave and cleanse the skin over the middle of the right side of the thorax.

2. Make an incision about 3 cm. long through the sixth space about its centre parallel to the rib and passing through the skin, subcutaneous cellular tissue and, if desired, the external intercostal muscle.

3. Have ready a blunt-pointed needle, laterally perforated and previously sterilized and connected to a syringe by means of an india-rubber tube (p. 167). Pass the needle into the intercostal space, directing it somewhat obliquely. The parietal pleura attached to the outer wall of the cavity allows the needle to pass through : the visceral pleura, with the lung, is driven back by the blunt end of the needle and when the latter has gone a distance of from a few millimetres to 2 cm. according to the species of animal used, the operator can feel that it is moving freely in a cavity. At this stage the fluid is inoculated.

4. Withdraw the needle quickly. Stitch up the skin and paint the wound with collodion.

E. Inhalation.—**1.** Put the animal in a solid-walled metal cage having an observation window on one side and on the other two holes lightly plugged with cotton-wool to allow of interchange of air: the tube from the pulverizer passes through a third hole.

2. Liquid cultures may be pulverized by means of Richardson's apparatus, but when the virus can be used dry without losing its virulence it is better to pour the liquid culture on to lycoperdon spores, lycopodium powder, or on to wood charcoal reduced to an impalpable powder. The culture is intimately mixed with the powder and then dried *in vacuo* over concentrated sulphuric acid. The powder well dried is then pulverized in the cage with a pair of bellows.

Note.—When dealing with micro-organisms pathogenic for man, the operator should be particularly careful to protect himself from the dust, and the experiment should be done, for choice, in the open air.

(xiv) Intra-cranial inoculation.

Intra-cranial inoculation is generally performed on the dog, guinea-pig or rabbit.

A.—**1.** Fasten the animal on its belly, the head being firmly held by an assistant.

2. Shave and cleanse the scalp behind the orbits.

3. Make an incision through the skin and aponeurosis about 3 cm. long in the middle line, commencing at a point level with the upper borders of the orbits. Retract the edges of the incision with a speculum. In the dog the incision is made preferably a few millimetres from the middle line to avoid the superior longitudinal sinus.

4. Place a small trephine about 5 millimetres in diameter (fig. 142) on the skull towards the middle of the incision and a little outside the middle line.

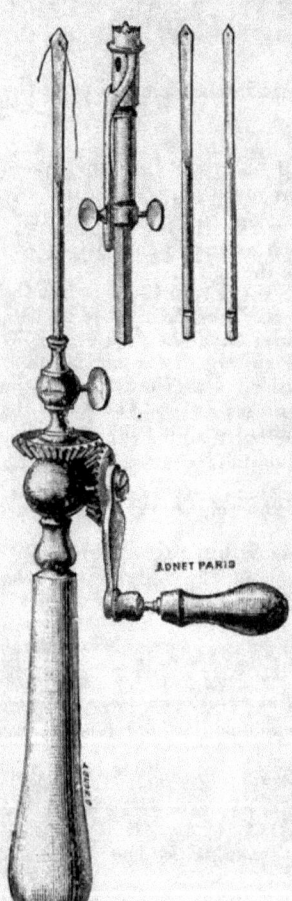

ADNET PARIS

FIG. 142.—Trephine for small animals.

Commence trephining and when the teeth bite raise the axis to prevent wounding the brain. Ascertain frequently to what depth the trephine has reached and when resistance has ceased raise the circle of bone with a pair of forceps or small elevator.

5. The dura mater is now visible at the bottom of the wound. Pass the needle very obliquely through the meninges so that the brain may not be injured, and inject the fluid. It is well to use a needle bent to a right angle at the middle point of its length.

6. Withdraw the needle, touch the puncture with a sponge soaked in carbolic water, suture the skin and apply collodion to the incision.

B.—In injecting small doses of toxin into the cerebral tissues the foregoing technique may be simplified.

After shaving, cleansing and incising the skin make a small hole in the skull with a drill, limiting the depth of the perforation with a shield to avoid damaging the meninges. Then the needle is introduced to the required depth (determined beforehand by means of a probe) the toxin injected, the puncture touched with carbolic and a stitch put through the skin and covered with collodion.

(xv) Intra-spinal inoculation.

Intra-spinal inoculation is effected through the posterior occipito-atloid ligament. With a little practice it is easy to inoculate directly into the spinal canal of a rabbit or dog by forcing a curved needle into the ligament through the skin (which must have been previously cleansed): the needle passes behind the posterior occipital tuberosity just outside the middle line and is then turned and made to follow the occipital bone. Before injecting make sure that the needle has passed well into the canal by aspirating a little of the cerebro-spinal fluid into the syringe (which should be not quite full of the material to be injected). The inoculation must be done very slowly.

It is easier, especially in the guinea-pig, to fix the animal on its belly with the head flexed, then to divide the posterior cervical muscles transversely on a very small director below the posterior occipital tuberosity, avoiding the vertebral veins. When the muscles are divided the ligament will be seen and will be recognized by its pearly-white appearance. Bleeding is easily arrested by plugging the wound with wool soaked in a solution of peroxide of hydrogen. By keeping close to the lower surface of the occipital bone the membrane is easily pierced with a curved needle. After inoculating insert two sets of sutures and paint the wound with collodion.

(xvi) Inoculation into the alimentary canal.

A. Ingestion.—(*a*) The simplest method is to mix the culture with the animal's food, viz. bran in the case of rabbits and guinea-pigs and meal in the case of dogs.

(*b*) In small animals the culture may be sucked up into a Pasteur pipette which is then introduced into the animal's mouth, and the liquid allowed to fall drop by drop while the head is held up. The end of the pipette should be short and stout.

[In the experiments of the Royal Commission on Tuberculosis (1901) it was found that pipette-feeding experiments with liquids were unsatisfactory. The animal may cough or choke, and the fluid find its way to the lungs. Even when the fluid appeared to be swallowed quite satisfactorily disease of the lungs was sometimes found which could not have been of intestinal origin. Experiments conducted on these lines are likely, therefore, to lead to erroneous conclusions.]

(*c*) In the case of birds make small pellets of flour and mix the culture with it. Put a pellet on the base of the tongue and close the beak.

B. Oesophageal catheterization.—This method is more certain than the foregoing and moreover the amount of liquid injected can be measured.

Guinea-pigs and rabbits.—The animal with the head moderately extended is held by an assistant. By pressing the cheeks near the molar teeth the mouth can be opened and a small gag with a hole in the centre, or, better, a piece of iron wire bent into a rectangle, can be placed between the jaws behind the incisor teeth (fig. 143). A very fine gum-elastic catheter can then be easily passed through the hole in the gag into the stomach. By attaching the needle of the syringe to the open end of the catheter, the fluid can be injected.

Dogs.—Fix the animal on its back, and gag it as described at p. 163. Pass a small œsophageal sound or a rather firm piece of ordinary india-rubber

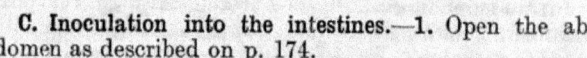

tubing of the size of an ordinary pen-holder into the stomach. Inoculate the culture through the sound or tubing.

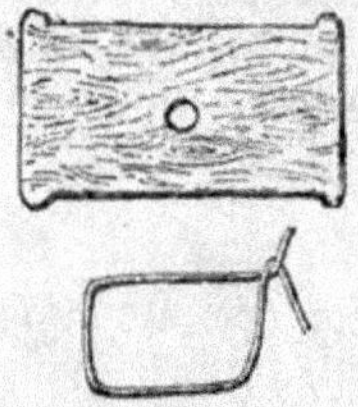

It is often necessary to render the contents of the stomach alkaline before introducing the culture ; this can be done by injecting 1 or 2 grams of sodium bicarbonate dissolved in a little water.

C. Inoculation into the intestines.—1. Open the abdomen as described on p. 174.

FIG. 143.—Gag for œso-phageal catheterization.

2. Pick up and hold a loop of intestine with a pair of forceps.

3. Pierce the wall of the loop obliquely with the syringe needle. Inject the material and withdraw the needle at once.

4. Dab the loop with a sponge soaked in carbolic solution : suture the aponeurosis and then the skin. Paint with collodion.

D. Rectal injection.—The animal must be firmly held by an assistant, then with a stout blunt-pointed needle inject the fluid into the rectum.

SECTION VI.—OBSERVATIONS TO BE MADE ON INOCULATED ANIMALS.

In studying an experimentally-induced disease the symptoms to which it gives rise in the inoculated animal should be observed and recorded day by day.

A note should be made of the following points :

1. The local lesion.—The presence or absence of a local lesion. The time when it appears. Its situation, extent, nature and the changes which it undergoes. Enlargement of the glands.

2. Temperature.—The temperature should be taken at least twice a day in the rectum, with a thermometer graduated in tenths of a degree centigrade and of a size suitable to the species of animal under observation. The temperature must always be taken before inoculating the animal. It is necessary to bear in mind that all animals have not the same normal temperature ; in guinea-pigs, rabbits[, goats, pigs] and cattle [1] the normal temperature varies from 38·5° to 39·5° C., that of horses and asses between 38° and 39° C. and of birds between 41° and 42° C. In small animals, complete immobilization rapidly reduces the temperature which should therefore never be taken with the animal tied to the operating table. A curve of the temperature should be kept.

3. Weight.—Animals must always be weighed before inoculation. A ratio can then be established between the weight of the animal and the quantity of virus which must be inoculated in order to produce sickness or lead to death. In chronic conditions, the animal should be weighed from time

[1 The normal temperature of an adult bovine animal is usually constant in the neighbourhood of 38·5° C.; that of a young calf may vary from 38·9° C. to 40·0° C. with a mean of 39·2°–39·4° C.]

to time : the weight curve furnishes valuable information as to the course of the disease.

4. Auscultation.—The development of pulmonary lesions can be detected and followed by auscultation.

5. Condition of the alimentary canal.—Observation should be kept upon the animal's appetite and it is highly important to notice if it suffers from diarrhœa, etc.

6. Urine.—Does the urine contain pus, blood albumin, etc. ?

7. The appearance of the animal.—Whether the animal is lively and active or dull and quiet, the condition of its coat whether rough and badly kept or smooth and well tended, as well as the position assumed by the animal (whether lying on its side or curled up) are all important facts which should be noted. The appearance of twitchings, convulsions or signs of paralysis should also be carefully watched for.

Observation of the clinical condition must be subsequently supplemented by an examination of the tissues, fluids and exudates for micro-organisms.

CHAPTER XI.

POST MORTEM EXAMINATIONS.

Introduction.
1. Instruments, p. 184.   2. Preliminary operations, p. 185.   3. Examination of the external surface of the carcase, p. 185.   4. Examination of the internal organs, p. 185. 5. Removal of tissues for histological examination, p. 188.

THE objects of *post mortem* examination are two in number.

1. To ascertain as far as possible the nature of the lesions which were the cause of death.

2. To collect material for further investigation.   This will involve the search for micro-organisms in the blood, exudates and internal organs by microscopical, cultural and inoculation methods, as well as the histological examination of portions of the internal organs.   The material therefore will have to be collected under very strict aseptic precautions.

The following **general rules** must be observed in conducting a *post mortem* examination on small animals.

A. To avoid soiling the bench fasten the animal to a sheet of zinc or copper [or pin it by the paws to a sheet of cork or a wooden board covered with cork linoleum, either of which can be washed with antiseptics, preferably 2 per cent. lysol, before and after use].   Lay all the instruments in use on the metal tray [or wooden board] and not on the bench while the examination is in progress.

B. Use sterile forceps and not the fingers for raising the skin, muscles and internal organs.

C. Sterile instruments must be used throughout.

D. Conduct the examination at the earliest moment possible after the death of the animal.

E. As soon as the examination is completed burn the carcase and any wool or paper which may have been used in an incinerator (fig. 13, p. 16) or in a fire with a good draught, boil the instruments, and if a metal tray has been used immerse it in a vessel of boiling water if it is not too large, or wash it with a strong solution of lysol or carbolic acid.

1. Instruments.

Have ready before commencing a *post mortem* examination—

1. Sterile scalpels, bistouries, dissecting forceps and scissors both large and small.

2. A number of sterile Pasteur pipettes.

3. Two or three platinum wires one of which should be stout and flattened at the end in the form of a small spatula.

4. An iron rod of the size of a large goose-quill and 15–20 cm. long, mounted in a wooden handle.

5. A tray of zinc or copper or a sheet of cork.

6. Sterile absorbent wool in a glass bottle plugged with wool and some sterile filter paper.

Cut a sheet of filter paper into pieces about 10 cm. square, wrap them in a piece of ordinary paper and sterilize the packet in the autoclave.

7. An enamelled iron bowl or glass dish containing an antiseptic solution (0·01 per cent. corrosive sublimate, or oxycyanide of mercury [or 2 per cent. lysol]).

8. A Bunsen burner or a spirit lamp.

9. Tubes of agar, broth, etc.

10. [Slides and cover-glasses.]

11. Wide-mouthed, glass-stoppered bottles of 30–50 c.c. capacity.

2. Preliminary operations.

1. Fasten the animal securely to the tray. In the case of rabbits, guinea-pigs, cats, etc., lay them on the back, pass a slip knot round each paw and tie to holes in the sides of the tray [or if a wooden board be used pin the animal's extended paws to it with large drawing pins].

Frogs, mice, sparrows, etc., can be pinned down on their backs to the cork sheet, one pin being passed through the neck the others through the extended paws or wings.

In the case of fowls and pigeons cut off the wings, lay the animal on its back and fasten the neck and legs by cords passed through holes in the sides of the tray.

2. The animal being fastened out, thoroughly wet the surface of the thorax and abdomen with the antiseptic and cut off the hair gathering up the loose hair in a piece of paper which is then burnt. Never cut off the hair without first of all wetting it.

In the case of birds the same precaution must be adopted before plucking the feathers.

3. Examination of the external surface of the carcase.

Before opening the carcase, examine the external surface carefully for lesions of the skin, abscesses, etc. If an abscess be found, cut away the hair, cauterize the surface thoroughly with a red-hot iron rod, flame and break off the point of a Pasteur pipette as quickly as possible, push the pipette through the centre of the cauterized area and aspirate the pus through the plugged end.

If the pus be thick and inspissated—as it often is in the case of rabbits—and cannot be drawn into the pipette, make an incision with a sterile knife after cauterizing the skin and collect the contents either on the point of the knife or with a stout platinum wire.

After sowing two or three tubes of culture media with some of the material prepare films for microscopical examination.

4. Examination of the internal organs.

As a rule it is better to open the thorax first. If the abdomen were opened before the thorax it might happen that contamination of the thoracic organs could not be prevented.

A. Mammalia.

1. Pick up the skin over the manubrium sterni, incise it and prolong the incision to the lower part of the abdomen, then extend the incision outwards to the roots of the four limbs. Dissect the skin from the subjacent tissues and throw the flaps outwards. This incision must involve the skin only.

2. Should there be reason to suspect the presence of excess of fluid in the pleural cavities, cauterize the muscles over one of the intercostal spaces,

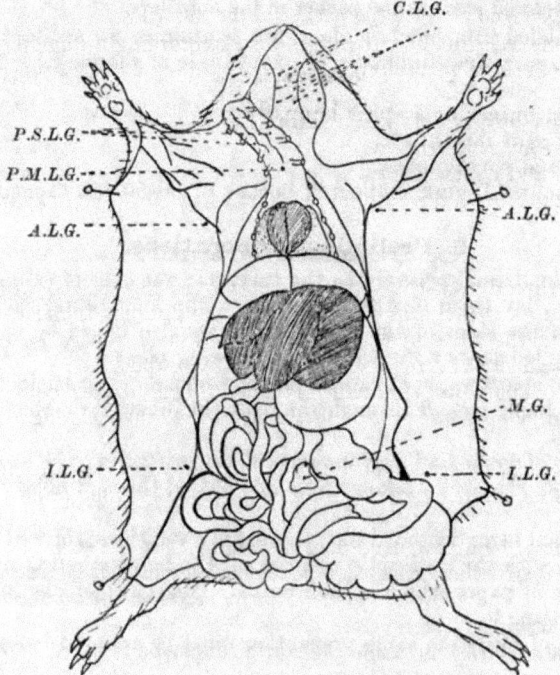

FIG. 144.—Appearance presented by a normal guinea-pig.

C.L.G. Cervical lymphatic glands; A.L.G. Axillary lymphatic glands; M.G. Mesenteric gland; I.L.G. Inguinal lymphatic glands; P.M.L.G. Post-manubrial lymphatic gland; P.S.L.G. Post-sternal lymphatic glands.

introduce the point of a flamed pipette, aspirate some of the fluid, sow cultures and make films.

3. To open the thorax. Raise the tip of the xiphoid cartilage with a pair of sterile forceps, introduce the point of a pair of strong scissors beneath the costal cartilages a little to one side of the sternum, and inclining the scissors slightly outwards cut through the costal cartilages as far as the clavicle and then divide the clavicle. By doing the same on the other side of the sternum a flap is formed which can either be turned upwards or detached. This will expose the heart and lungs.

4. If there be any fluid in the pericardium, take hold of the latter with a pair of sterile forceps, flame the point of a pipette and push it through the membrane close to the forceps. The hot end of the pipette will sterilize the surface of the pericardium as it passes through it. Aspirate the fluid.

5. To collect blood from the heart, tear through the pericardium with two pairs of forceps, or holding it with a pair of forceps slit it up with a pair of fine scissors. Cauterize the surface of the ventricle with a red-hot rod, pass a pipette through the sterilized area and aspirate the blood.

6. To collect material from an hepatized or congested area of the lung, cauterize the surface of the latter and pass a pipette or the bent end of a stout platinum wire into the affected part : or the latter may be exposed by taking hold of the lung with two pairs of sterile forceps and tearing it.

7. When the examination of the thorax is completed, open the abdomen.

To collect the peritoneal fluid lift up the muscular wall with a pair of forceps, make a small slit with a sterile scalpel, introduce a pipette through the opening and aspirate the fluid from the flanks. The pipette should be held parallel to the abdominal wall so as to avoid damaging the intestine.

Complete the incision of the abdominal wall along the middle line and throw the flaps outwards.

8. Note carefully the appearances presented by the internal organs. In taking material from the liver, spleen, kidneys or lymphatic glands, first cauterize the surface, then pass a stout wire bent in the form of a hook through the centre of the cauterized area deeply into the organ, twist it round and round and on withdrawing it sow the material at once on a suitable culture medium. For making films, simply tear off a small piece of the organ with a pair of forceps (Chap. XIII.).

To examine the intestinal contents, cauterize the surface, pass a pipette through the cauterized area and aspirate some of the contents. Urine may be collected from the bladder in a similar manner, a ligature being first tied round the urethra.

9. *Bone marrow.*—To examine the bone marrow expose one of the long bones, divide it across with a pair of strong sterile scissors and collect the medulla in a pipette or platinum loop.

If the bone be divided with non-sterilized scissors the cut end must be cauterized with a heated iron rod before aspirating the medulla.

10. *Examination of the central nervous system.*—Lay the body on its ventral surface and fasten the feet firmly to the tray [or board] as before.

Make an incision through the skin from the root of the nose to the sacrum along the line of the spinous processes of the vertebræ, and reflect the skin ; detach the scapulæ at their humeral articulations and turn them on one side ; then dissect away the masses of muscle from the vertebral laminæ with a strong bistoury, taking care in so doing not to penetrate the abdominal cavity in the lumbar region. With a pair of curved Liston's forceps (fig. 145)

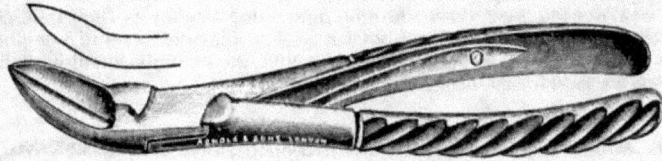

FIG. 145.—Curved Liston's forceps.

open the skull along an horizontal line passing through the superciliary ridges ; free these ridges on each side by an oblique incision ; then raise the frontal bone with an elevator and detach it with the Liston's forceps. This will expose the brain. Having reached the occipital foramen raise the spinous processes with the elevator and cut through the laminæ of the bodies of the vertebræ with the forceps alternately on the right and left sides. This if properly done (a certain amount of skill and patience is required to avoid injury to the spinal cord) will remove the spinous processes in the form of a rosary held together by the ligamenta flava.

If there is any meningeal exudate, cauterize the surface of the membrane, introduce a pipette through the centre of the cauterized area and aspirate the fluid.

To remove portions of the nerve tissue tear through the meninges with two pairs of forceps, cauterize the area (cerebrum, cerebellum, medulla oblongata or spinal cord), and push the point of a strong pipette deeply into the tissue ; then aspirate the material, twisting the pipette about if necessary. Or, after cauterization, portions of tissue may be removed with a platinum loop or with a sterile bistoury.

B. Birds.

To open the thorax in birds it is best to divide the skin along the middle line, and after reflecting it to each side to make a curved incision, extending to the bone, round the sternum ; beginning at the root of the neck, continue along the right margin, round its lower end and up the left margin. Cut through the clavicle on each side with a pair of stout scissors, and following the line of the incision through the soft parts detach the sternum from the ribs, then cut away the muscular attachments and remove the breast plate.

The examination will then be proceeded with as in the former case.

C. Post-mortem examination of human bodies.

The technique to be employed in the collection of material *post mortem* from the human subject does not differ from that already described in the case of animals, and the methods of examination should also be the same, but it must be remembered that if the results of the bacteriological investigation are to be relied upon the examination must be made within a few hours of death ; if it be delayed until the interval required by law has elapsed (24 hours after death) the bacteriological findings must be accepted with caution especially in summer ; the presence of the colon bacillus in particular in the internal organs would be under such circumstances without significance, since this organism multiplies in the tissues of the body immediately after death and sometimes even during the period immediately preceding death.

Note.—The material collected *post mortem* may either be examined and sown at once, or may be put aside for examination at a later stage, provided that both ends of the pipettes containing the material be sealed. In the latter case to reach the contents of a pipette, push down the wool plug almost as far as the top of the fluid, and cut off the part of the pipette above it with a glass-cutter and a point of red hot glass ; the plug can then be taken out and the contents manipulated with a pipette just as in the case of a culture-tube.

5. Removal of tissues for histological examination.

For purposes of subsequent histological examination, small pieces of the internal organs [and other tissues] should be removed at the time of the *post mortem* examination. The pieces should be quite small (cubes of 10–15 mm.), but should be cut off with a sharp sterile bistoury so that the section may be as clean as possible. Place the pieces at once in ground-glass stoppered bottles containing one of the following fixing solutions :

1. Absolute alcohol.—For bacteriological purposes absolute alcohol is the simplest and the most generally useful fixative.

The method of placing the tissue in the first instance in weak alcohol and subsequently transferring to solutions of increasing strength not only takes more time but yields only moderate results.

To fix in alcohol, place the tissue (about 1 cm. cube) in 25–30 c.c. of absolute alcohol: renew the alcohol after 3 hours and again after 24 hours. The tissue is then fixed, but it is found to stain better if left in the alcohol for 3 days. Tissues should not under any circumstances remain in absolute alcohol for more than a week at the outside, and if it is not convenient to use them then they should be transferred to 90 per cent. alcohol.

The tissue should be suspended in the fluid or laid on a piece of wool at the bottom of the bottle to ensure its hardening uniformly.

2. Formalin.—Formalin is an excellent hardening agent; it does not interfere with any of the staining methods and is particularly valuable for tissues which are to be cut by the freezing process. The best solution is the following:

Commercial formalin,	10 c.c.
Distilled water,	90 ,,

The tissue is fixed in about 6–8 hours; but better preparations are obtained if the formalin is allowed to act for 24–48 hours.

If frozen sections are to be cut the tissue is used straight out of the formalin: but before embedding in paraffin, transfer the tissue first to 90 per cent. alcohol, then to absolute alcohol, leaving it for 24 hours in each solution.

3. Corrosive sublimate.—Corrosive sublimate is another most useful hardening agent, and can be used whatever method of staining is to be subsequently employed. It can be used as a cold saturated solution[1] but it penetrates and fixes the tissue better when acidified with acetic acid.

Acid sublimate (Mayer).

Saturated aqueous solution of corrosive sublimate,	100 parts.
Glacial acetic acid,	1–3 ,,

Allow 20–30 c.c. for each piece of tissue. The solution penetrates well and rapidly, so that the pieces may be relatively large (cubes of 2 cm.). Leave in the acid solution for at the most 12 hours. The tissue is then white and opaque. Wash in running water for an hour (this is not absolutely essential). Transfer to 100 c.c. of 70 per cent. alcohol containing xv–xx drops of tincture of iodine and leave for 24 hours to remove the excess of perchloride and prevent the deposition of crystals in the tissue. Transfer to 80 per cent. alcohol for 24 hours and finally to 90 per cent. alcohol for a similar period.

It must be remembered of course that perchloride of mercury acts on metal instruments, so that in removing tissues from perchloride solutions horn, glass or wood spatulas must be used.

4. Flemming's solution.—For purposes of bacteriological examination the diluted solution is better than the concentrated.

(a) **Diluted solution.**

1 per cent. aqueous solution of chromic acid,		25 volumes.
1 per cent. ,, ,, osmic acid,		10 ,,
1 per cent. ,, ,, acetic acid,		10 ,,
Distilled water,		55 ,,

(b) **Strong solution.**

1 per cent. aqueous solution of chromic acid,		15 volumes.
2 per cent. ,, ,, osmic acid,		4 ,,
Glacial acetic acid,		1 volume.

[1] Cold water dissolves about 6·6 per cent. of perchloride of mercury. A saturated solution is easily prepared by dissolving 70 to 75 grams of perchloride in 1 litre of distilled water in the warm: filter while warm, and, as the solution cools, white needles crystallize out at the bottom of the vessel; pour off the supernatant liquid which is then ready for use.

These solutions ought to be prepared just before use.   The use of Flemming's solution should be limited to the hardening of nerve tissues, and the pieces should be very small.   Many stains cannot be used after Flemming's solution; the best to use are hæmatoxylin, safranin and the basic aniline dyes.

Suspend the tissue in the solution and leave it for 36–72 hours (weak solution) or 1–24 hours (strong solution), wash in running water for 24 hours, transfer to distilled water for 1 hour, and then for 24 hours to each of the following solutions successively, viz. 70 per cent., 80 per cent., 90 per cent. alcohol.

5. Flemming-perchloride solution.—A mixture of acid perchloride and Flemming's solution combines the advantages of both.   The mixture is prepared according to the following formula :

Saturated aqueous solution of perchloride of mercury, -	500 c.c.
1 per cent. aqueous solution of chromic acid, - - -	500 ,,
Osmic acid crystals - - - - - - -	1 gram.
Glacial acetic acid, - - - - - -	50 c.c

Harden for 12–24 hours.   Wash and transfer to alcohol as in the case of Flemming's solution.

CHAPTER XII.

THE COLLECTION OF MATERIAL FOR BACTERIO-LOGICAL EXAMINATION.

1. Hair.
Man and animals.

Pull out a few hairs with a pair of sterile forceps, lay them between two sterile microscope slides and wrap up the slides in a piece of sterile paper. They can thus be put aside or be transmitted to the laboratory without fear of contamination.

2. Skin.
Man and animals.

1. Cut off the hair with a pair of sharp scissors.

2. Scrub with a nail brush and soap, wash with boiled water and ·rub briskly with a sponge wrung out in a 1 in 1000 solution of sublimate, wash with absolute alcohol, then with ether and wipe quickly with a piece of sterile filter paper.

3. Pick up a small fold of the skin with a pair of sterile forceps and cut it off through the base with a sharp-pointed sterile bistoury.

If the skin be thick or adherent to the deeper tissues it will be difficult to pick up a piece of the size required.   In that case mark out a small rectangular area with the bistoury, detach one corner and then, taking hold of the latter with a pair of sterile forceps, dissect the piece of skin from the deeper tissues.

4. If the material be collected at the bed-side it can be taken to the laboratory between two sterile watch-glasses or in a sterile glass dish wrapped up in paper.

3. Sputum.
Man.

A. Ordinary method of collection.—For the ordinary microscopical examination of sputum for the tubercle bacillus, it will suffice if the patient cough the sputum into a sterile bottle or clean pocket handkerchief. The material should be examined as soon as possible.

B. Kitasato's method.—This method is much to be preferred when cultures are to be sown or investigations of a more delicate nature are to be made.

1. The patient rinses his mouth and gargles the back of his throat several times with boiled water and then coughs the sputum into a sterile Petri dish.

2. Transfer the sputum immediately to a tube containing several cubic centimetres of sterile water and shake it up well. Remove the sputum from the tube with a sterile platinum loop or a pair of sterile forceps to a second tube of sterile water and wash it in this way three or four times to free it, as far as possible, from contaminating organisms (but note that sputum can only be washed when it is tenacious and lumpy as in influenza, advanced tuberculosis (nummular sputum), etc.).

3. After washing spread the sputum in a thin layer in a sterile Petri dish and cut off a small fragment with a small pair of sterile scissors or platinum needle from as near the centre as possible. Use this for sowing cultures.

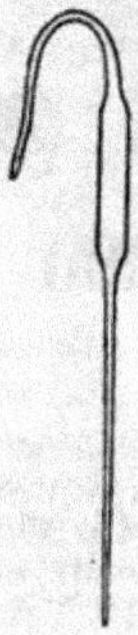

4. Blood.
Man.

A. Pricking the skin.—A small quantity of blood is readily obtained by pricking the distal end of the finger near the nail and collecting the drops in some suitable sterile vessel such as a Pasteur pipette, a small tube [or a Wright's capsule] or on a glass slide. This method is however only applicable when the blood is required for immediate microscopical examination, e.g. for anthrax bacilli, hæmatozoa, etc., as it is liable to contamination during collection. When the blood is required for sowing cultures, it should be taken from a vein.

FIG. 146.— Wright's capsule for collecting blood.

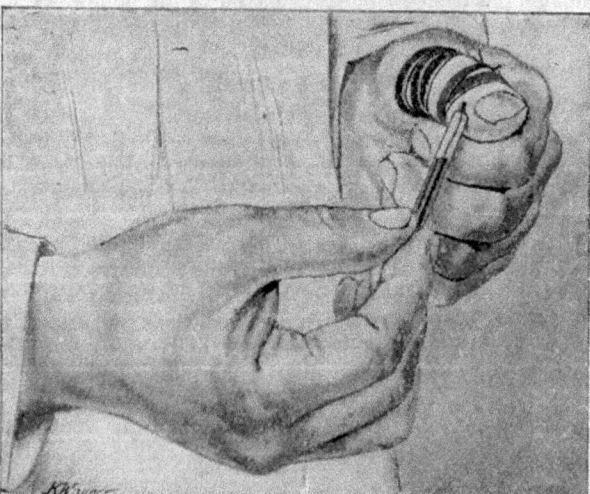

FIG. 147.—Method of collecting blood by pricking the finger.

1. Scrub the ball of the finger with soap and water. Wash it in perchloride, alcohol and ether. Dry with sterile paper.

2. Compress the base of the finger by grasping it with the left hand or by tying a ligature round it (fig. 147).

3. With a sterile pin or small lancet [or a straight surgical triangular needle] prick the skin sharply and deeply.

4. Wipe away the first drop or two of blood which issues with a piece of sterile paper and collect the remainder.

As a further precaution the skin of the finger after being washed and dried may be painted over with a very thin layer of collodion. The finger is then pricked through the collodion and in this way the blood is prevented from coming in contact with the skin.

B. Cupping.—A larger volume of blood can be obtained by cupping, but otherwise the method is open to the same objections as the foregoing.

1. Asepticize about 10 sq. cm. of the skin of the thorax, back or sides.

2. Apply a sterile cupping-glass over the part.

3. When the glass has fixed itself, raise it (the operator's hands having, of course, been already sterilized), scarify the skin with a sterile razor and apply the cupping-glass again.

4. When sufficient blood has collected put the patient in such a position that the blood will not be spilt, then lift the glass and cover it at once with sterile paper.

C. Bleeding from a vein at the bend of the elbow.—By this method all danger of contaminating the blood is avoided, and it should be adopted in all cases when cultures have to be sown. It is attended by no danger and is less painful than the foregoing methods.

1. Procure a sterilizable syringe of 2–20 c.c. capacity according to the amount of blood to be collected, and adjust a sharp and clean-bored needle. Test its efficiency and, if satisfactory, sterilize it with the needle attached, by boiling it in water for 15 minutes or by heating in the autoclave at 115° C.

2. Lay the patient's forearm flat on the bed, and get an assistant to compress the middle of the arm or put a tight bandage round it as in the operation for bleeding.

3. Wash the skin over the bend of the elbow with soap and rinse with sublimate or oxycyanide, then with alcohol and finally with ether. As the result of the combined compression and friction the veins at the bend of the elbow will stand out prominently.

To make certain of asepsis it is sometimes advised to lightly touch the point through which the needle is to be passed with a cautery ; but in the great majority of cases it is sufficient to wash the arm in the manner described.

4. Select the largest vein, push the needle through the skin and then into the vein. The vein lying, as it does, immediately beneath the skin is generally penetrated at the same time as the skin. The needle should be held parallel to the long axis of the vein and at a very acute angle to the surface. When the vein is entered, by gently withdrawing the plunger, blood will flow into the syringe.

Notes.—There is nothing to be gained by pointing the needle towards the hand ; on the contrary, it is easier to point it towards the arm and the calibre of the vein is such that it flows just as easily into the needle whichever way the latter is directed. The alternative method which consists in first making an incision through the skin and exposing the vein should never be practised.

5. When the syringe is full, withdraw the needle from the vein, relieve the pressure and apply a drop of collodion to the needle prick. Be careful not to let the blood clot in the syringe but squirt it at once into a sterile test-tube ; then wash the syringe in cold water and sterilize it.

Horses, asses and cattle.

In these animals the jugular vein is the most convenient from which to bleed. The method has already been described in dealing with the preparation of serum (p. 48). When a small quantity only of blood is wanted a syringe is used instead of a trocar.

Guinea-pigs.

Guinea-pigs may be bled from the jugular vein, from the femoral or carotid arteries, or by cardiac puncture.

A. From the jugular vein.—For the anatomical data see p. 172.

1. Fix the guinea-pig on its back with the head extended. Shave the skin over the front of the neck and cleanse it in the ordinary way.

2. Make an incision through the skin and sub-cutaneous tissue along the line of the vein, dissect away the cellular tissue with a director and the vein will come into view.

3. Pass the needle of a sterile syringe or the end of a pipette (similar to that described at p. 166) very obliquely into the vein. If a slip knot be passed under the vein with a Deschamps needle on the cardiac side of the puncture, the vessel can be compressed and the flow of blood into the pipette facilitated.

4. Having collected the blood, withdraw the needle or pipette and make certain that there is no hæmorrhage from the puncture. If the vein be bleeding, tie a ligature above and below the puncture. Put two or three stitches in the skin and cover the wound with collodion.

Note.—The blood may be collected directly in a sterile tube or flask by passing a fine trocar into the exposed vein. The operation in this case is described at p. 49.

B. Carotid and femoral arteries.—1. Expose the vessel (pp. 173 and 174).

2. Puncture the wall of the artery obliquely with a syringe needle, the end of a bent pipette or a small trocar.

3. Having collected the blood, withdraw the instrument, stitch up the skin and paint the incision with collodion.

Sometimes hæmorrhage occurs when the needle is taken out of the artery. This can be guarded against by placing two ligatures beneath the vessel, one above and the other below the puncture, then, if hæmorrhage occur, the two threads can be tied and the wounded part of the vessel isolated.

C. Cardiac puncture.—Cardiac puncture as practised in physiological laboratories may be usefully applied for bacteriological purposes (Pagniez). It allows a much larger volume of blood to be collected than is possible by other methods, is easily performed and is unattended by danger to the animal ; moreover the blood is not exposed to any risk of contamination. The technique, which is as follows, has been worked out by Raybaud and Hawthorn.

1. Tie down the animal on its back, shave and cleanse the skin over the front of the cardiac area. Have ready a sterile syringe capable of holding 5 c.c. and fitted with a needle of the ordinary pattern but very sharp.

2. At a point on the left margin of the sternum, about 8–10 mm. above the angle formed by the base of the xiphoid cartilage and the last rib cartilage articulating with the sternum, push the needle sharply to a depth of 15–17 mm. above the last but one or last but two chondro-sternal articulations.

The needle will pass into the left ventricle, and by inclining it a little towards the middle line it can be made to enter the right ventricle. This method is to be recommended because the risk of wounding the anterior margin of the left lung is lessened, and if the heart were punctured at a higher level than that described the auricle would be penetrated and ruptured.

3. Fill the syringe slowly with blood, and withdraw the needle sharply and quickly.

Rabbits.

A. The ear veins.—The simplest method of collecting blood from a rabbit is to take it from one of the veins of the ear. An adult rabbit can easily be bled to the extent of 20 c.c. in this way.

1. Prepare a large Pasteur pipette with the pointed end short but strong and bent at an obtuse angle to the shaft (fig. 133, p. 166). The point must be sharp and have thin cutting edges. Sterilize the pipette by passing it through the flame but be careful to allow it to cool before using it. In this particular case a pipette is better than a syringe.

2. Let the animal sit on the knees of the operator or of an assistant. Shave the hair over the line of the marginal vein and cleanse the skin in the ordinary way (p. 172). Compress the vein at the root of the ear between the finger and thumb or with a pair of pressure forceps.

3. Holding the ear in the left hand, thrust the point of the pipette through the skin and then into the lumen of the vein. A flow of blood into the pipette will indicate when the point is in the vein. The point of the pipette should be directed towards the tip of the ear and must be held absolutely parallel to the axis of the vein to avoid penetrating both walls.

The flow of blood into the pipette is slow : sometimes it ceases, owing to the formation of a small clot in the end of the pipette ; this, however, can easily be displaced by aspirating at the plugged end of the pipette.

It is a good practice to puncture the vein near the root of the ear so that if unsuccessful at the first trial another attempt may be made nearer the tip. By bleeding from the two ears in turn, blood may be collected at frequent intervals from the same animal.

4. When sufficient blood is collected remove the pipette and seal the point in the flame. The blood can afterwards be aspirated into other Pasteur pipettes through the plugged end, the plug being well flamed before being taken out.

5. Clip the wound with a pair of pressure forceps for a moment to stop any hæmorrhage. After being bled the animal will be thirsty, and some water should be left in its cage.

B. Jugular vein.—The anatomical data and the technique of the operation are the same as in the case of the guinea-pig (p. 194).

C. Carotid and femoral arteries.—Here again the description given for the guinea-pig is applicable (p. 194).

D. Cardiac puncture.—The technique is similar to that described for cardiac puncture in the guinea-pig. C. Nicolle and Duclaux recommend using a rather large needle, about 2 cm. long, fitted to a sterile syringe of 10–20 c.c. capacity.

1. The animal is held down on its back and the skin over the heart shaved and cleansed.

2. The needle with syringe attached is driven in sharply to a depth of 17–18 mm. to the left of the sternum in the fourth intercostal space counting upwards from the xiphoid cartilage. The needle must be inclined from below upwards and slightly inwards.

3. Aspirate the blood slowly into the syringe and then withdraw the needle quickly.

Dogs.

Dogs are most easily bled from the jugular or external saphenous vein (p. 173), or from the carotid or femoral artery, the ordinary rules of asepsis being observed. It is to be remembered that dogs' blood coagulates very quickly.

Birds.

Bleed from the axillary vein (p. 173) adopting the ordinary aseptic precautions.

Collection of serum.

On account of the importance at present attaching to a study of serum reactions it is often necessary to collect serum from immunized animals.[1] In the case of large animals it is best to bleed by the method described on p. 49. Small animals may be bled preferably from the carotid by the method just described, and after the clot has contracted the serum can be decanted. But by this method much of the serum is lost, being retained in the meshes of the clot and it is better, therefore, when the amount of blood available is strictly limited as is the case with small animals, to bleed into a Latapie's tube. By using this apparatus all chance of contaminating the blood is avoided and a yield of 80 per cent. of the total volume of serum is assured.

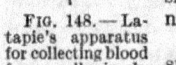

Latapie's apparatus.—This (fig. 148) consists of a large glass tube B constricted about its lower third E, and having a small cup F at its lower end. Below the constriction there are two tubulures, one T, straight and open and plugged with wool between two constrictions; the other D, on the opposite side, bent in the form of an inverted U, and drawn out at its free end to a fine point which is sealed. The upper end of the large tube B is connected by means of a piece of india-rubber tubing with another tube A, known as the trocar tube, consisting of an ordinary test-tube bent in the form of a right angle and drawn out to a fine point at its free end. This second tube A passes well down into the larger tube B, leaving the bent and pointed end projecting. Occupying the centre of the apparatus from top to bottom is a small glass tube H, sealed at one end and the sides perforated with numerous holes.

FIG. 148.—Latapie's apparatus for collecting blood from small animals.

Technique.—Place a few drops of water in the apparatus and sterilize it in the autoclave at 120° C. Expose the carotid in the ordinary way then break off the point of the tube A with a pair of sterile forceps and pass it into the vessel, holding the apparatus so that the broken point is downwards. Blood will now ascend into the tube A. Stop the flow of blood before the latter is quite full, then seal the pointed end of A in the flame, gently aspirating through T to prevent the blood being clotted by the heat. Stand the apparatus on one side with the tube A downwards. The clot forms around the narrow central tube H, and retracts from the walls of A. If the apparatus be now inverted the serum will fall into the collecting bulb R, the red cells precipitating into the cup F. In this way 80 per cent. of the serum can be collected in a few hours, and can be easily withdrawn through the tubulure D by breaking its point and blowing through T. With a little experience and skill a small animal such as a rabbit or guinea-pig can be bled two or three times without killing it.

Stassano's apparatus.—Stassano's apparatus is somewhat similar to Latapie's but is fragile and more expensive.

Lumiere's tube.—This consists of a glass tube (fig. 149) on which two bulbs B and D are blown, the interior of the lower B having a number of projecting points. The tube is plugged with wool at the ends and sterilized in the hot air sterilizer. To use the apparatus the tube A is fitted with a short piece of india-rubber tubing carrying a sterilized platinum-iridium syringe needle. As soon as the vessel is penetrated, blood will flow into the bulb B. When the latter is full, the tubing is pinched and the needle withdrawn from the vessel. The

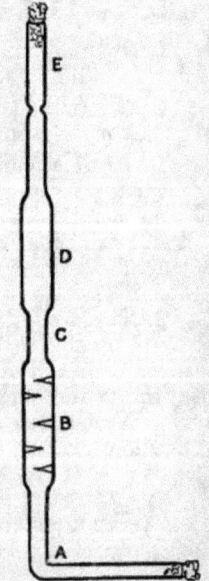

FIG. 149.—Lumiere's tube for collecting blood.

---

[1] The collection of serum for use as a culture medium is described in Chapter II.

tube A is then tilted, the india-rubber tubing detached, and after passing it through the flame the end A is plugged with wool. When the blood is clotted the apparatus is inverted, the clot will be held by the points in B and the serum will run into the bulb D.

Centrifuging.—The maximum yield of serum is obtained in the minimum of time by centrifuging the blood (Camus). Collect the blood, without contaminating it, in a number of sterile centrifuge tubes (*vide infra*), plug the tubes with wool and centrifuge at once. The serum collects in the upper part of the tube and the clot below. If the animal was fasting at the time of bleeding, the serum will be clear and transparent ; on the other hand if digestion was going on the serum will be milky and slightly opaque.

5. Pharyngeal exudates.

Man.

A. Puncture of the tonsil.—**1.** Get the patient to clean the surface of the mucous membrane by thoroughly rinsing out his mouth with boiled water.

2. Make the patient sit up and incline his head at a suitable angle, then press the tongue down.

3. Take a rather long stout-pointed Pasteur pipette with a sharp cutting end, heat it well in the flame and then pass it rapidly and deeply into the tissue of the tonsil. The heated end cauterizes and sterilizes the surface of the gland and is itself cooled before reaching the deeper parts. Aspirate lightly through the plugged end of the pipette and then withdraw the instrument.

4. The small quantity of material which will be obtained should be sown at once into broth and the pipette washed out two or three times by aspirating some of the broth and blowing it out again.

B. False membranes.—After the patient has washed out his mouth with boiled water, press the tongue down and strip off the false membrane with a pair of sterile forceps. If the membrane be friable it may be that the forceps will not pick it up, in which case it can be removed by rubbing it with a plug of sterile wool held in a pair of forceps or affixed to an iron wire.

When the membrane is detached it should be blotted firmly between two pieces of sterile filter paper to remove any contaminating organisms that may be on the surface.

6. Abscesses.

Man.

1. Cleanse and if necessary shave the skin.

2. Puncture the abscess with a needle of large calibre and aspirate the pus into a sterile syringe.

3. If the pus be inspissated and cannot be aspirated in this way, make a small incision through the skin, introduce the end of a large Pasteur pipette and aspirate the pus into the pipette, or collect some of it with a platinum loop.

Animals.

1. Shave the hair and cauterize a small area of the skin over the abscess.

2. Pass a Pasteur pipette through the centre of the eschar and aspirate the pus.

7. Aqueous humour.

Animals.

1. Fix the animal so that it cannot move and keep the eyelids retracted with a speculum. Wash the conjunctiva with warm sterile water.

2. Hold the eye firmly between the thumb and index finger of the left hand, and with a screwing movement pass a fine Pasteur pipette perpendicularly to the axis of the eye at the margin of the cornea into the anterior chamber. The fluid will ascend into the pipette without aspiration.

8. Pleural and pulmonary exudates.
Man and animals.

A small quantity of a pleural effusion can easily be collected with a sterile syringe. Use a long needle (5–7 cm.) with a large bore. When the exudate consists of thick and granular pus it is better to use a small trocar attached to a suitable syringe.

1. Asepticize the skin; to make quite sure of the asepsis the site through which the needle is to pass may be superficially cauterized.

2. Pass the needle mounted on a syringe into one of the intercostal spaces and aspirate the fluid into the syringe.

3. Transfer the material to a sterile test-tube.

These small punctures are quite unattended by danger and may, if necessary, be repeated.

The same technique may be employed, when there is no fluid in the pleura, for puncturing the lung to reach (for example) a pneumonic patch previously delimited by auscultation. In this case a fine needle is passed perpendicularly and more or less deeply through one of the intercostal spaces and a little blood-stained fluid aspirated into the syringe.

9. Ascitic fluid.
Man.

A large volume of ascitic fluid may be collected aseptically by using a trocar with sterile rubber attachment. The fluid is best collected in a sterile bottle covered with paper. The operation must be carried out under the ordinary surgical conditions and the rules for puncture of the abdomen observed.

The fine trocar of a Potain's apparatus with the india-rubber adjustments on its lateral tubulure is very useful for the purpose.

10. Tumours and lymphatic glands.

Tumours and lymphatic glands must be removed in the ordinary surgical manner, strict asepsis being maintained and care being taken that the structure is not touched with the hands.

When the tumour or gland, as the case may be, is enucleated, sterilize a small area of the surface with a well-heated iron wire, pass a sterile platinum needle or bistoury through the cauterized part and remove the material required from the centre.

11. Spleen.
Splenic puncture in man.

The spleen has been punctured for the purpose of recovering the typhoid bacillus from patients suffering from enteric fever and is sometimes practised in the study of certain other infections, e.g. Leishmanioses, etc.

1. Delimit the spleen by percussion and asepticize the skin.

2. Use a long needle (5 cm.) attached to a syringe by a piece of india-rubber tubing (p. 167) and penetrate the skin perpendicularly over the centre of the spleen. Aspirate, withdraw the needle, and paint the puncture with collodion.

3. A few drops of blood generally represent the material collected. This is sown into broth by drawing some sterile broth into the syringe and expelling it again into the culture-tube.

The india-rubber connexion is absolutely necessary : it allows the needle to follow the movement of the spleen and so avoids any risk of tearing the organ.

Splenic puncture is not often performed and is not altogether unaccompanied by danger.

Splenectomy in animals.

The functions of the spleen in the resistance of the body to certain infectious diseases can be studied by observation of the results following the removal of the organ. Dogs and rats are the best animals for the experiment, but the operation can be performed on many other species.

The spleen is situated in the left flank beneath the lower false ribs and near the left curvature of the stomach.

1. Fix the animal on its right flank and anæsthetize it.

2. Shave and asepticize the skin of the left flank. Sterilize all instruments and asepticize the hands carefully.

3. Make an incision a few centimetres long through the skin and subcutaneous tissues immediately below the margin of the last rib, commencing at the angle and continuing parallel to the bone.

4. Cut through the aponeurosis of the external oblique and then of the internal oblique on a director.

5. Separate the fibres of the transversalis with the blunt end of a director.

6. Incise the peritoneum for the whole length of the incision.

7. The spleen will then be exposed or can readily be found by passing the finger along the greater curvature of the stomach : draw it out of the wound, being very careful not to lacerate it.

8. Tear through the gastro-splenic omentum and put a firm silk ligature around the vessel of the hilum. Cut through the pedicle on the distal side of the ligature.

9. Suture the muscles, close the skin wound with a few stitches and cover the incision with collodion.

12. Lumbar puncture.

Man.

By means of lumbar puncture, an operation devised by Essex Wynter, a needle can be passed into the cerebro-spinal canal and the fluid withdrawn. Bacteriological investigation of the cerebro-spinal fluid is of great interest and importance in cases of meningitis.

Anatomical data.—In the adult the spinal cord only reaches to the lower border of the first or upper border of the second lumbar vertebra, but in children twelve months old it reaches to the level of the third. The spinal cord cannot then, be injured by passing a fine trocar into the spinal canal through the third, fourth or fifth lumbar spaces. In these situations the nerves comprising the cauda equina are suspended in the cerebro-spinal fluid and are collected into two lateral fasciculi separated by an interval of 5 mm. The lower down the puncture is made the smaller the chance

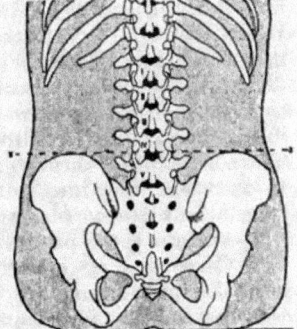

Fig. 150.—Landmarks for the operation of lumbar puncture.

of wounding the nerves since they diminish in number as the canal is descended.

The transverse width of the third and fourth lumbar spaces is from 18–20 mm. and their depth from above downwards 10–15 mm. Their shape varies with age : the fifth space between the last lumbar arch and the upper border of the sacrum is wider than but not quite so deep as the two above and marks the situation of the inferior arachnoidal cul-de-sac, which is a true reservoir of cerebro-spinal fluid.

The operation is generally performed between the fourth and fifth lumbar vertebræ. An horizontal line drawn between the highest points of the two iliac crests passes through the tip of the spinous process of the fourth lumbar vertebra ; by inserting the needle immediately below this process the space between the fourth and fifth lumbar vertebræ is entered.

The depth to which the needle must be inserted will depend upon the age and also upon the state of nutrition of the patient ; in a child it will be 1·5 cm., 2 cm. and sometimes 3 cm. according to its condition ; in the adult, 4–6 cm. If the needle pass too far it may reach the premeningeal venous plexus and cause a slight hæmorrhage, in which case the needle must be withdrawn a little before the cerebro-spinal fluid can be collected.

Operation.—1. Sterilize a platinum-iridium needle with a short bevel and a calibre of 0·8–1 mm. and about 5 cm. long for a child and 8 cm. for an adult.

The needle should have a fine platinum wire passed through it reaching as far as the bevel but not interfering with its cutting edge.

2. Place the patient in the lateral decubitus on the edge of the bed with the thighs strongly flexed on the abdomen and the legs on the thighs, the head being slightly raised on a pillow and flexed on the thorax.

The patient may also sit up with his legs hanging over the edge of the bed, the body being bent forward and the back arched. This position, however, though more convenient for lumbar puncture, is often rendered impossible by illness and has the disadvantages of tiring the patient and stimulating muscular reaction.

3. Asepticize the skin by washing with soap, ether and alcohol. Or, more simply, paint the surface of the skin a few minutes before doing the operation, with tincture of iodine. The surgeon must of course prepare his hands as for any other surgical operation.

4. Determine the position of the line connecting the highest points of the crests of the iliac bones (*vide ante*). This line will pass through the upper border of the spinous process of the fourth lumbar vertebra.

5. Put the tip of the left index finger on the spine of the fourth lumbar vertebra and keep it in that position throughout the operation. Take the needle with the platinum wire in it in the right hand and pass it perpendicularly to the surface immediately below the spinous process and very near (not more than 1 cm. away from) the median line, slowly but deliberately into the spinal canal. Direct the needle forwards and a little upwards. The needle will pass through in order, the lumbo-sacral muscles, the ligamentum flavum, the dura mater and the arachnoid membrane. As soon as it enters the sub-arachnoid space the liquid will issue from the needle.

6. Withdraw the platinum wire and collect the fluid in a sterile test-tube.

7. Collect 5 or 6 c.c. in the case of a child and 10–15 c.c. in the case of an adult. Lumbar puncture is unattended with danger if no more fluid than this be aspirated. Withdraw the needle and paint over the puncture with iodoform and collodion. The patient should remain in bed for 24 hours after the operation.

Notes.—1. No advantage is obtained from local anæsthesia, but in the case of very nervous patients the skin may be sprayed with ethyl chloride.

2. As the needle passes through the skin there is occasionally a reflex muscular contraction of the lumbo-sacral muscles. Should this occur desist for a few moments before continuing to push the needle into the canal.

3. Should the needle be driven against bone its point will be bent and another attempt will have to be made taking a better direction.

4. If, during the operation, the needle becomes obstructed it is easily cleared with the platinum wire.

5. Occasionally the fluid is blood-stained : in that case the needle has wounded some of the small meningeal veins ; this is a matter of no importance and can be remedied by slightly altering the position of the needle.

13. Milk.

Duclaux adopts the following technique which, though delicate, gives a sterile milk without any heat :

1. Take a number of plugged sterile test-tubes.

2. Wash and brush the cow's udder with soap and water, rinse with per-chloride of mercury then with alcohol and finally with sterile water. The milker then sterilizes his hands.

3. Reject the first few drops of milk, which serve to wash the walls of the excretory canals.

4. An assistant takes the plug out of one of the tubes and holds the latter as close to the mouth of the teat as possible without touching it ; when the tube is half-full he replaces the plug. A number of tubes may be filled in the same way.

14. Urine.
Man.

To collect urine in a sterile manner proceed as follows.

1. Take a red rubber catheter, protect the upper end with a small cap of filter paper, then wrap up the whole instrument carefully in several folds of paper and autoclave for 20 minutes at 115° C. On taking it out of the autoclave dry it in the incubator.

2. Put the man on his back and carefully wash the glans and meatus with a 1 in 1000 solution of oxycyanide of mercury, sponge with wool which has been sterilized in the autoclave and wrap the penis in another wool sponge similarly sterilized. The operator now sterilizes his hands.

3. Remove the catheter from its paper covering by taking hold of its upper end ; dip the other end in oil sterilized at 115° C.

4. Lay the catheter for a moment on the paper in which it was sterilized. Hold the penis in the left hand, and pick up the catheter about its middle with the right, introduce it into the meatus and push it along the urethra still resting the upper end on the paper which should be held by an assistant.

5. On reaching the entrance to the bladder pinch the catheter firmly between the thumb and index finger and pass the catheter through the sphincter.

6. The assistant flames the mouth of a flask previously sterilized in the hot air sterilizer, removes the wool plug and holds the mouth to the end of the catheter from which he now removes the paper cap.

7. Relax the pressure on the catheter and the urine will flow into the flask. When the latter is three-parts filled pinch the catheter to stop the flow of urine. The assistant flames the neck of the flask and replaces the wool plug which he has been holding in his left hand during the time the flask has been filling.

A similar technique can be adopted in the case of **large animals.**

Small animals (rabbits, guinea-pigs, etc.).

It is impossible to use a catheter on these small animals and the only way to collect the urine in the male is to let it flow into a sterile tube or Pasteur pipette. The animal should be fixed on its back and the emission of urine is easily provoked by laying towels wrung out in very cold water on the abdomen and loins.

15. Stools.

Stools for bacteriological examination should be collected in a sterile vessel and care must be taken that they are not mixed with urine.

When solid, cauterize the surface with a red-hot iron rod and collect some of the material from the centre. When liquid, take up the quantity required with a Pasteur pipette or platinum loop.

CHAPTER XIII.

THE BACTERIOLOGICAL EXAMINATION OF FLUIDS AND TISSUES.

SECTION I.—FILM PREPARATIONS.

Pathological material whether taken during life or after death, from man or from one of the lower animals, may be examined :

1. either fresh and unstained, or
2. after drying and staining.

1. Unstained preparations.

(a) **Fluids.**—Blood, fluid exudates and pus may be collected in a Pasteur pipette and ought to be examined at once.

The examination of blood may be described in illustration of the method. As soon as the blood is removed from the body a drop is placed on a slide and covered with a cover-glass ; the blood spreads out in a thin layer between the slide and cover-glass, and by pressing lightly on the latter the excess can be squeezed out at the edges and wiped away with a piece of soft linen. In this way a very thin uniform layer is obtained and must be examined immediately (obj. D ; oc. 2 Zeiss).

The slides and cover-glasses must be absolutely clean, because dirt or grease prevents the blood from spreading in a thin and uniform layer, and renders satisfactory examination of it impossible. It is also essential that the red cells should not be heaped one on another, as this would mask the presence of micro-organisms.

Should the examination be very prolonged the edges of the cover-glass may be luted with paraffin, but in the majority of cases this is unnecessary, because the blood at the edges of the cover-glass, being in contact with the air, coagulates and thus affords sufficient protection to the central parts of the preparation.

Serous exudates, liquid pus, etc., should be treated in the same way ; but if the pus be inspissated it must be treated as though it were a scraping from an organ.

A warm stage can be used to maintain the preparations at the temperature of the body (p. 135).

(b) Scrapings of organs.—Scrapings of the internal organs are to be collected in the manner already described (Chap. XI.), and transferred with a platinum loop to a slide on which they are rubbed up in a drop of filtered water or, better, in a drop of normal saline solution (water 1000, NaCl 8, filter, distribute in tubes, sterilize in the autoclave) ; then spread the material with a platinum loop, cover with a cover-glass, and examine at once (obj. D ; oc. 2 Zeiss).

2. Stained preparations.

Before being stained fluids and scrapings of organs should be spread in a thin layer on a slide or cover-glass, and dried and fixed to preserve the form of the cells and to make them adhere to the surface of the glass.

A. Preparation of films.

(a) Fluids.

The treatment of fluids such as blood, serous exudates, pus, etc., will first be described.

1. Spreading of films. (*a*) **On cover-glasses.**—**1.** Hold a perfectly clean cover-glass by one of its angles, A, and place a drop of the fluid to be examined in the centre.

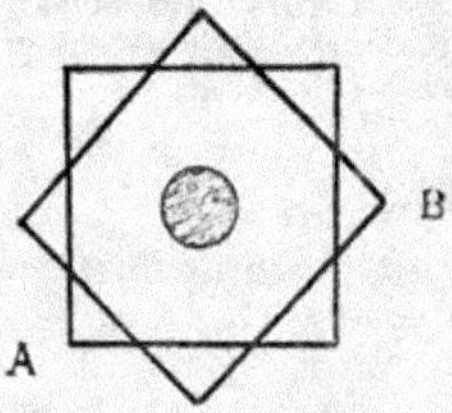

2. Cover with a second cover-glass laying the latter across the former as shown in the figure (fig. 151).

3. Take hold of the second cover-glass at the angle B, opposite to A, and slide them apart so that the liquid is spread out in a thin, uniform layer.

4. Allow the films to dry either in the air or by placing them on a drying stand heated to 40° or 45° C. (fig. 127, p. 141).

FIG. 151.—Preparation of films on cover-glasses.

(*β*) **On slides.**—For pus, serous exudates, etc., slides may be used in a similar way to cover-glasses: place the drop of fluid near one end of the slide, lay another slide over it and then draw the two slides apart.

For blood the following method is better :

1. Take a perfectly clean slide and lightly touch the drop of blood as it oozes from the prick, taking care that the blood is drawn up by the slide and not the slide pressed down on to the drop.

2. Hold the slide in the left hand, apply the edge of a cover-glass to the drop of blood and the latter will spread along the edge of the cover-glass by capillary action. (The end of a slide or a visiting card or even a small glass stirring rod will serve equally as well as a cover-glass.)

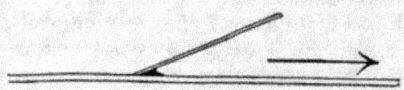

FIG. 152.—Preparation of a blood film on a slide.

3. Draw the cover-glass slowly and without pressing upon it towards the other end of the slide. In this way a very thin and uniform layer of blood is left on the slide which dries as fast as the cover-glass passes over it (fig. 152).

The preparation of satisfactory blood-films requires a certain amount of practice, so that if the first attempts fail one must not be discouraged ; remember always that absolutely clean and flat slides and cover-glasses are indispensable.

2. **Fixation.**—Several methods are available for fixing films on slides and cover-glasses.

(α) **Heat.**—The slide or cover-glass with the film upwards is held in a pair of Cornet's forceps (fig. 118, p. 131) and passed three times through the heating flame of a Bunsen burner or spirit lamp. The shape of the cells is somewhat distorted by this procedure and it cannot therefore be used (for example) for fixing blood-films.

(β) **Alcohol-ether.**—Pour two or three drops of alcohol-ether on the cover-glass (p. 141) and allow it to dry in the air. This method is preferable to the preceding as it preserves absolutely the shape of the cells. It is occasionally necessary to allow the solution to act for several minutes.

(γ) **Absolute alcohol.**—In many cases absolute alcohol can be used in place of alcohol-ether for fixing films. The technique is the same as that described in the preceding paragraph. With many dyes staining is facilitated by allowing the alcohol to act for 10–30 minutes.

(δ) Other solutions are occasionally used for fixing films, e.g. osmic acid vapour, absolute methyl alcohol, etc. These will be referred to when occasion for their use arises.

(b) Scrapings of organs.

Films of the internal organs are prepared as follows :

1. Transfer to a slide with a platinum loop or pipette a small piece of tissue from the organ, and spread it by rubbing it on the slide so as to cover a rectangular area about 15–20 mm. square. Or a piece of the tissue (liver, spleen, etc.) may be taken up in a pair of dissecting forceps and lightly rubbed over the surface of the slide.

The film in any case should be thin and uniform and any lumps which would interfere with the application of a cover-glass must be removed.

2. Dry as above.

3. Fix by heat or with alcohol-ether.

Films of the brain or spinal cord should always be washed several times in the alcohol-ether mixture after fixing to remove fatty matters, as these would interfere with the subsequent staining processes.

(c) Sputum.

When the sputum is fluid it can be treated as a fluid exudate, but should it be tough or inspissated it should be spread with a platinum loop on a slide ; it will facilitate the preparation of a thin and uniform film if the slide be gently heated while the sputum is being spread. Dry and fix.

B. Staining methods.

Films whether of fluids or scrapings of organs contain structures of two different kinds.

1. The groundwork, which is formed of tissues of animal origin—cells and amorphous elements.

2. Bacteria, which are of vegetable origin.

Such films may be stained in one of two ways,

(a) With a *simple stain* by which at a single operation the groundwork and the micro-organisms are stained the same colour.

(b) With a *double stain* by means of which the micro-organisms are differentiated from the groundwork by being stained a different colour.

(a) Simple staining.

A blood film or a scraping of an organ may be stained with any of the dyes described in Chapter VIII.

The stain most generally used is carbol-thionin (p. 138). The technique is as follows :

A. Cover-glasses.—1. Hold the cover-glass in a pair of Cornet's or Debrand's forceps and pour on to the film sufficient stain to cover the surface.

Allow the stain to act for 30–60 seconds.

2. Wash in distilled water.

3. Mount the cover-glass on a slide film downwards in a drop of water. Examine with a $\frac{1}{12}$ objective and No. 2 ocular.

4. If the preparation be satisfactory it may be mounted permanently, by drying it in the air or gently heating it and then mounting in Canada balsam.

To sum up : stain, wash in water, dry, mount in balsam.

B. Slides.—Films made on slides are stained in a similar manner. Hold the slide in the left hand or in a pair of Debrand's forceps ; flood the slide with stain ; wash in water, dry ; place a drop of cedar-wood oil on the film and examine with an immersion lens. The preparation may be mounted by placing a drop of balsam on the film and covering with a cover-glass.

Dilute carbol-fuchsin, the various carbol-violet stains, Kuhne's or Lœffler's or Roux's blue, etc. may any of them be used in suitable cases in place of carbol-thionin. The particular stain which is most useful for the detection and study of the different species will be referred to in the chapters devoted to those species.

The disadvantage of the simple stains is that as they stain the groundwork and the organisms the same colour (fig. 153) ; the latter fail to stand out

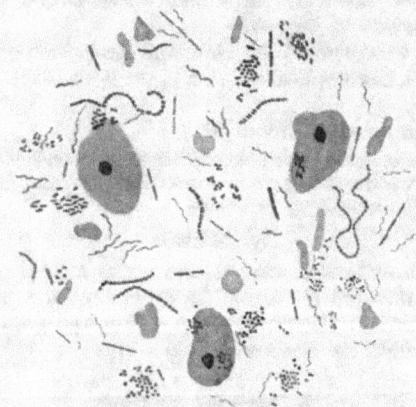

FIG. 153.—Simple staining.

Scraping from gum stained with dilute carbol-fuchsin (oc. 2, obj. $\frac{1}{12}$th, Zeiss).

conspicuously, especially when they are few in number or when the film is thick. The methods of differential staining are adopted to overcome these defects.

Examination of the blood.—In the case of blood-films the necessity for double staining may be avoided by getting rid of the groundwork. Thus if the hæmoglobin (which is the only substance in the red cells which takes the stain) be eliminated there remains after staining a colourless groundwork on which the micro-organisms stand out conspicuously. This result may be effected in one of two ways :

(a) **Gunther's method.—1.** Dry the film by gently heating it and then without passing it through the flame cover it with a 5 per cent. solution of acetic acid, and leave for 30 seconds.

2. Expose to the vapour of ammonia for a few seconds.

3. Wash in water.

4. Stain, wash, dry and mount.

(β) **Vincent's method.**—1. Dry the film by gently heating it, and, without fixing, flood the film with the following solution :

5 per cent. aqueous solution of carbolic acid,	- - -	6 c.c.
Saturated aqueous solution of common salt.	- -	30 ,,
Glycerin (pure),	- - - - - - -	30 ,,

and allow it to act for 1–2 minutes.

2. Wash in water, stain, etc.

(γ) **Direct staining of blood-films.**—Lastly, simple staining with Lœffler's blue gives very good results with blood-films ; the red cells are sharply differentiated from the micro-organisms, the former being stained pale green and the latter deep blue. Carbol-thionin is also useful in that it stains the nuclei of the leucocytes and the organisms but leaves the red cells practically unstained.[1]

(b) Differential staining.

In dealing with micro-organisms which retain the stain by Gram's method it is easy to get a double-stained preparation. But when the organism under investigation does not stain by this method more delicate processes which often give less satisfactory results have to be employed. Finally, in the search for and in the study of certain organisms, such for example as the tubercle and leprosy bacilli, special methods, of which Ehrlich's is a type, have to be adopted. They will be described in the chapter devoted to the tubercle bacillus.

A. Gram's method and its modifications.

The procedure originally described by Gram has undergone various modifications : reference will be made to the more important of these. Meanwhile the beginner must be warned against the danger of practising a large number of methods. The secret of success lies in the thorough understanding of one reliable procedure ; if this advice be neglected the result may be error and failure and consequent discouragement. The method described under (β) is the one recommended.

(a) **Gram's method.**—1. Flood the slide or cover-glass with aniline-gentian-violet (p. 139). Let the stain act for 2–4 minutes.

2. Pour off the stain and, without washing, flood the film with Gram's iodine solution. Let it act for about a minute until the preparation assumes a blackish tint.

3. Wash in distilled water.

4. Decolourize with absolute alcohol (p. 143) until the film assumes a pale grey tint.

5. Wash in distilled water.

6. Flood the film with a solution of eosin :

Water-soluble eosin,	- - - - - - -	1 gram.
Distilled water,	- - - - - - -	200 c.c.

Allow the eosin to remain on for 1–2 minutes.

7. Wash in distilled water and dry.

8. If the preparation has been made on a slide, a drop of cedar-wood oil

[1] In the case of birds' blood, the nuclei of the red cells are deeply stained by these dyes.

may be placed on the film and the preparation examined at once with an oil-immersion lens.

Films made on cover-glasses should first be examined in water, and if satisfactory they can then be mounted in balsam after drying and clearing in clove oil and xylol.

In preparations stained as above the groundwork is pink and the micro-organisms violet. Decolourization must be continued until all traces of violet have disappeared from the groundwork (fig. 154).

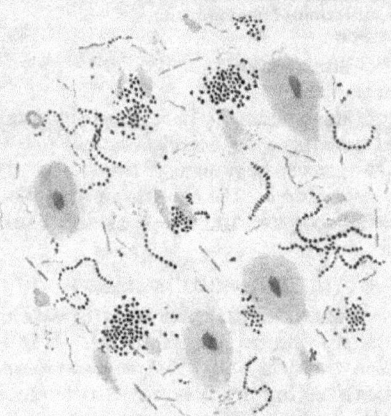

FIG. 154.—Gram's stain.
Scraping from gum stained by Gram's method (oc. 2, obj. ₁₂th, Zeiss).

Blood-films stained by Gram's method give very beautiful preparations. When dealing with the blood of birds the action of the alcohol must be continued until all but the nuclei of the red cells are decolourized and must be stopped short of complete decolourization, so that after counter-staining with eosin the protoplasm of the red cells is stained red while the nuclei of the red cells and the micro-organisms are stained violet.

Note.—Vesuvin can be used as a counter-stain instead of eosin.

Vesuvin,	5 grams.
Distilled water,	100 c.c.

Micro-organisms which retain the stain by Gram's method are then stained deep violet while gram-negative organisms and the nuclei of the leucocytes are dark brown and the protoplasm of the leucocytes light brown.

(β) **Method recommended.**—**1.** Flood the film with carbol-gentian-violet (p. 138). Stain for about a minute.

2. Without washing, replace the violet with Gram's iodine and stain for 1–2 minutes.

3. Wash in distilled water.

4. Decolourize with absolute alcohol.

Instead of using absolute alcohol alone the process may be hastened by washing first with alcohol then with aniline oil and again with alcohol. But it should be pointed out that aniline oil is a very powerful decolourizing agent and should therefore only be allowed to act for a few seconds.

5. Wash in distilled water.

6. Counterstain with an aqueous solution of eosin as before.

7. Wash, dry, mount and examine as above (α).

(γ) **Nicolle's method.**—**1.** Stain with carbol-gentian-violet (p. 138) for 20–30 seconds.

2. Without washing, replace the violet by a modified Gram's iodine solution made as follows :

Iodine,	1 gram.
Potassium iodide,	2 grams.
Distilled water,	200 c.c.

Allow the solution to act for 4–6 seconds, renewing it once or twice during that period.

3. Wash in distilled water.

4. Decolourize with an acetone-alcohol solution :

Absolute alcohol,	5 volumes.
Acetone,	1 volume.

Decolourization does not take place at once and is only fully manifested after washing in distilled water.

5. Wash in distilled water.

6. Stain the ground work rapidly with an alcoholic solution of eosin :

Saturated solution of eosin [1] in 95 per cent. alcohol,	1 volume.
Alcohol (95 per cent.),	2 volumes.

7. Wash, dry, mount, and examine as before.

(δ) **Merieux's method.**—In the author's experience this method has never given results equal to those obtained with the methods already described.

1. Stain with carbol-violet as in (γ).

2. Treat with the following solution for 4–6 seconds, renewing the solution once or twice during that time :

Iodine,	1 gram.
Potassium iodide,	2 grams.
Saturated solution of eosin [2] in 50 per cent. alcohol,	20 c.c.
Distilled water,	200 ,,

3. Wash in distilled water.

4. Decolourize in a 1 in 6 solution of acetone in alcohol (*vide supra*).

5. Wash, dry, mount and examine.

(ε) **Kühne's method.**—1. Stain for several minutes in carbol-blue (p. 138) or in ammoniacal blue (p. 139).

2. Wash in distilled water.

3. Treat with Gram's iodine solution for 2–3 minutes.

4. Wash in distilled water.

5. Decolourize with a saturated solution of fluorescein in absolute alcohol.

6. When the ground-work no longer appears blue, wash in absolute alcohol then in clove oil and xylol, and mount in balsam.

The bacteria appear violet on a background lightly tinted with fluorescein.

B. Claudius' method.

This method as already described on p. 136 can be used for staining smear preparations.

C. Methods available for staining organisms which are gram-negative.

1. Blood-films.

In double staining blood-films containing gram-negative organisms use is made of the property possessed by the red cells of combining with eosin, and also of the marked selective affinity shown by bacteria for the basic aniline dyes.

Note.—The three methods about to be described are the original methods. They have undergone many improvements which, being specially adapted to work on the Hæmatozoa, will be considered in the chapter (LVIII.) dealing with these organisms.

[1] Alcohol-soluble eosin.   [2] Water-soluble eosin.

(α) **Laveran's method. Method recommended.—1.** Flood the film with an aqueous solution of eosin (p. 207). Stain for about a minute.

2. Replace the eosin with a saturated aqueous solution of methylene blue and stain for about 30 seconds.

3. Wash in distilled water.

4. Dry and mount in balsam.

The red cells are stained pink while the bacteria and the nuclei of the white cells are blue. In the blood of birds the nuclei of the red cells are also stained blue.

(β) **Chenzinsky's method.—1.** Lay the cover-glass, film side downwards, in a small ground-glass covered glass dish containing a little of the following solution which must have been recently prepared:

Saturated aqueous solution of methylene blue, - - - 40 c.c.
0·5 per cent. solution of water-soluble eosin in 70 per cent.
    alcohol, - - - - - - - - 20 ,,
Distilled water, - - - - - - - 40 ,,

Leave the film to stain in the glass dish in the warm incubator at 37° C. for 3–6 hours.

2. Then wash the film in distilled water, dry, and mount in balsam.

(γ) **Romanowsky's method.—1.** After drying and fixing in the flame, place the film in a drying oven at 105°–110° C. for about an hour.

2. Then immerse the cover-glass in the following staining solution which must be newly made up and not filtered:

Saturated aqueous solution of Höchst's medicinal methy-
    lene blue, - - - - - - - - 2 parts.
1 per cent. aqueous solution of eosin A.G. (Höchst), - - 5 ,,

Stain for 2–10 hours.

3. Wash in distilled water.

4. Dry, and mount in balsam.

2. Films of pus, etc.

(α) **Kühne's method.—1.** Stain for a few minutes with carbol-blue (p. 138).

2. Wash in water.

3. Wash in dilute hydrochloric acid until the film assumes a pale blue colour (this is rather a delicate proceeding and the time required will vary with the thickness of the film).

Dilute hydrochloric acid.

Pure hydrochloric acid, - - - - - - - 1 c.c.
Distilled water, - - - - - - - - 1000 ,,

4. Remove the excess of acid by washing in an alkaline lithia solution.

Saturated aqueous solution of carbonate of lithia, - - 5 c.c.
Distilled water, - - - - - - - 100 ,,

5. Wash well in water.

6. Dry, clear in clove oil and xylol, and mount in balsam. The groundwork is stained pale blue and the micro-organisms deep blue.

(β) **Nicolle's method. Method recommended.—1.** Stain for a few minutes in carbol-blue.

2. Wash in water.

3. Treat for 2 or 3 seconds with a few drops of the following solution:

Pure tannin, - - - - - - - - 10 grams.
Distilled water, - - - - - - - 100 ,,

4. Wash in water.

5. Treat rapidly with absolute alcohol, clove oil and xylol, and mount in balsam.

The ground-work is stained very pale violet-blue and the organisms deep blue.

SECTION II.—HISTOLOGICAL PREPARATIONS.

For the demonstration of micro-organisms *in situ* in tissues very thin sections (0·05 mm.) must be cut. Hand-cut sections are not sufficiently thin for purposes of bacteriological investigation, so that the tissue must be cut with a microtome, which involves the embedding of the tissue first of all in some suitable material.

The materials ordinarily used in histology for embedding tissues (gum, wax, soap, celloidin and collodion) do not lend themselves to the cutting of very thin sections, so that for bacteriological purposes the tissue is either frozen or embedded in paraffin.

1. Instruments.

Microtomes.—Most of the mechanically-worked microtomes are suitable for cutting the thin sections required in bacteriological work. For paraffin sections, Minot's, Radais' and the Cambridge " rocking " microtome (fig. 155) are among those in most frequent use.

F1G. 155.—Cambridge " rocking " microtome.

It will be unnecessary here to discuss the construction of the different forms of microtome and the method of working them, for a careful examination of the instrument itself will be of far more assistance than any detailed description.

It will suffice to say that microtomes being instruments of precision must be carefully handled ; that they must be cleaned every time after use, and be protected from dust and damp by being kept under a bell jar or in a wooden box.

Microtome razors.—A good razor is indispensable for the cutting of satisfactory sections. One surface of the razor must be flat (the one in contact with the paraffin block). The cutting edge must be sufficiently sharp to sever an hair held between the finger and thumb or one of the fine hairs on the back of the hand.

Always strop the razor before using it, first on the prepared surface of the strop and then on the dry surface, remembering to strop with the back foremost and to pass from heel to tip, stropping each side of the razor alternately.

It is useful also to ensure satisfactory results and to avoid having to send it frequently to the instrument-maker to know how to sharpen a razor on a stone. The razor must be passed with the edge foremost from heel to tip ;

the stone should not be oiled, but simply moistened with a little water or better still with the following solution :

Distilled water,	50 c.c.
Alcohol (95 per cent.),	50 ,,
Glycerin,	50 ,,

After use the razor should be dried on a piece of soft rag, lightly stropped, and returned to its case.

To cut sections embedded in paraffin the blade of the razor should be dry and be placed obliquely to the tissue. The sections, as they are cut, should be picked up from the razor with a pair of fine forceps or a piece of silk paper, never with a needle or scalpel or other similar instrument which might damage the cutting edge of the razor.

2. Freezing methods.

Though frozen tissues cannot be cut so thin as tissues embedded in paraffin, the freezing method has the advantage that sections can be cut in a very short time, and can be stained in a variety of ways; and hence is of particular value for purposes of rapid diagnosis.

Only tissues which have been previously fixed should be cut by the freezing method. Formalin (10 per cent.) is perhaps the best for the purpose (p. 189), as tissues can be frozen without any further treatment. Tissues fixed by other methods should be washed and then put in formalin for a few hours. Tissues for frozen sections should not be more than 5–6 mm. thick.

Microtomes.—The simplest type for frozen sections is a rocking microtome or a Minot. Place the tissue wet with formalin on the carrier of the microtome and direct a jet of methyl chloride on to it until it is firmly frozen to the carrier, then adjust the latter to the microtome and cut the sections.

Of microtomes specially arranged for cutting frozen sections the best are those of Becker and Miller, in which the tissue is frozen by the decompression of liquid carbonic acid. The tissue is placed in an hollow carrier connected by an iron tube to a cylinder of carbonic acid, and when arranged in place on the microtome is frozen by simply turning on the tap of the cylinder. When the tissue is frozen the gas is turned off and the sections cut. If the sections show a tendency to tear, it is because the tissue has been frozen too hard, in which case it must be left for a few seconds.

Transfer the sections to ordinary water in which they will uncurl; when uncurled they are ready for staining.

3. Paraffin embedding methods.

A. Xylol method. Method recommended.—The pieces of tissue after being fixed in the manner described in Chapter XI. are treated as follows:

1. Dehydrate carefully in absolute alcohol or acetone for 24 hours or thereabouts.

2. Transfer to xylol.

Very small pieces (1–3 mm.) for	30–60 minutes.
Small pieces (3–5 mm.),	2 hours.
Medium-sized pieces (5–10 mm.),	3–4 ,,
Thick pieces (10 mm. or more),	4–5 ,,

In the case of the last it is as well to change the xylol once or twice to make quite sure that all traces of alcohol will be removed.

3. After dehydrating, transfer to a mixture of xylol and paraffin melting at 35° C. Such a mixture can be made as follows :

Paraffin [1] (melting point 50° C.),	10–15 grams.
Xylol,	30 c.c.

[1] For embedding, the paraffin sold by Dumaige of Paris is recommended.

The tissue should be placed in the mixture in a well-stoppered bottle and be kept in the warm incubator (37°–38° C.) for from 1–6 hours according to the thickness of the block.

4. After passing through the xylol-paraffin bath transfer to an open flask or tube containing paraffin melting at 50° C. and heated to 52°–53° C. (the

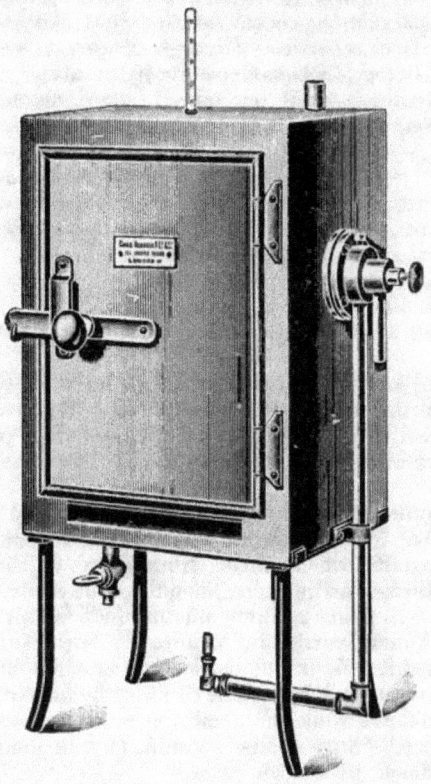

FIG. 156.—Paraffin oven.

temperature must never reach 55° C.) in a paraffin oven (fig. 156) for $\frac{1}{2}$–4 hours according to the thickness of the tissue.

Very thin pieces,	30 minutes.
Thin pieces,	1–2 hours.
Medium-sized pieces,	2–3 ,,
Thick pieces,	3–5 ,,

5. The tissue is now ready to be embedded. Melt some paraffin (melting point 50°, 52° or 55° C.) in a porcelain capsule. (For sections for bacterio-logical examination paraffin melting at 52° C. is, generally speaking, the best, but if the weather is very warm paraffin melting at 55° C. may be preferred.) After the paraffin has been melted allow it to cool until a pellicle forms on the surface.

While the paraffin is melting select a mould and cover the bottom with a thin layer of the melted paraffin, and as soon as it has begun to set (a few seconds is sufficient) place the tissue, which may be conveniently held with a lightly heated needle, on the surface, taking care that it is placed in a good position and suitably orientated ; then fill up the mould with melted paraffin, being careful that the tissue is embedded to a depth of several millimetres to allow for the contraction which will take place during cooling.

As soon as the paraffin has set sufficiently to hold the tissue the needle which was used to retain the latter in position should be taken away. The paraffin should be cooled rapidly by plunging the mould into cold water, being careful first to moisten the bottom and not to immerse the mould completely before the paraffin has cooled sufficiently to allow of the formation of a crust on the surface, otherwise of course the water would penetrate into the paraffin and destroy the homogeneity of the mass.

6. When the paraffin is firmly set, take it out of the mould and the tissue is ready for cutting.

Paraffin moulds.—1. The simplest mould is one made out of paper in the following manner : Select a cork which loosely fits the carrier on the microtome, and roll round it a strip of filter paper—which may be fastened by pinning it to the cork—so as to form an hollow cylinder 2 or 3 cm. deep, the bottom being formed by the upper surface of the cork. This surface may, with advantage, be scored with a few small grooves cut with a scalpel to ensure the paraffin adhering more firmly to it. Oil the inner surface of the paper with a brush avoiding the surface of the cork at the bottom of the cylinder.

Pour the melted paraffin into this cylinder and when it has set take out the pin and unroll the paper ; the paraffin with the tissue embedded in it will remain attached to the cork. Trim up the surface of the paraffin and fix the cork into the carrier of the microtome. The block is then ready for cutting.

2. The lead capsules used for covering the corks of bottles serve the same purpose excellently. When the paraffin has set the capsule is torn off, leaving a solid block of paraffin which can be trimmed up at leisure with a slightly heated scalpel. Blocks cast in such a mould can be easily fitted to the carrier of the microtome. In using a Minot microtome it is only necessary to heat gently the grooved metal carrier and to press the lower surface of the paraffin block lightly against it. To fix the block to the wooden cube or cylinder used with other microtomes apply the blade of a lightly heated scalpel to the lower surface of the block, and while the paraffin is still soft press it on to the wood block ; or if preferred a little melted paraffin may be poured on to the latter and the paraffin block pressed on to it.

In the same way small cardboard or wooden boxes, cover-glass boxes for example, make very good moulds ; these must be painted on the inside with glycerin or oil to prevent the paraffin adhering to the sides.

3. By using Leuckart's moulds blocks of various sizes with perfectly smooth and parallel sides are obtained. These moulds consist of two pieces of brass, which can be placed together in such a way that they form a rectangular box (fig. 157). The surfaces of the two pieces of metal are smeared with

FIG. 157.—Paraffin moulds.

glycerin and laid on a piece of glass which has also been smeared with glycerin and they are then arranged so as to form a box of the size required. The melted paraffin is poured into the box and when it has set the two pieces of metal are pushed apart and the paraffin with the tissue embedded is free.

B. Toluene method.—The technique is exactly the same as when using xylol except that toluene is substituted for xylol.

C. Ether method.—**1.** When the tissue is taken out of absolute alcohol it is transferred to alcohol-ether for from 30 minutes to 6 hours according to the size of the tissue.

2. The tissue is then immersed in pure ether for at least as long as it was in the alcohol-ether mixture.

3. It is then transferred to an hermetically sealed flask containing ether saturated with paraffin melting at 50° C. and placed in the warm incubator at 37°–38° C. (see under xylol for duration of treatment).

4. The block is now immersed in paraffin melting at 50° C. and embedded in the manner described under xylol.

4. Preliminary treatment of sections.

Before sections can be stained the paraffin which has penetrated the interstices of the tissue must be removed.

A. Method recommended.—**1.** As soon as they are cut the sections are placed in a ground-glass stoppered vessel containing ether which rapidly dissolves the paraffin. The length of time required will vary from several minutes to a few hours according to the size and number of the sections treated.

2. When all the paraffin has dissolved the sections are transferred with a platinum or nickel spatula (fig. 158) to a second bath containing absolute alcohol.

3. After being in absolute alcohol for a few minutes the sections are transferred one by one with a spatula to a glass dish full of distilled water. As soon as they come in contact with the water the sections spin round and round very rapidly and at the same time unroll and spread themselves flat.

If the sections are very thin and fragile this gyratory movement may tear them and render them useless, so that it is better to pass such sections from absolute alcohol to 70 per cent. then to 40 per cent. alcohol before placing them in distilled water.

4. To transfer a section to a slide, dip the slide obliquely into the water and beneath one of the sections, then fixing the section with a needle raise the slide and gently draw it out of the water, holding the section with the needle about the centre of the slide on which it will spread out. Blot up the excess of water with a cigarette paper or a piece of silk paper (which should be kept ready cut up into small rectangular pieces, and not torn off as required since the rough edges might pick up the section from the slide) and the section is now ready for staining.

FIG. 158.—
Section lifter.

B. Albumin fixation.—The method just described is the simplest and, in the hands of those used to the work, applicable to the majority of cases. But when the sections are very delicate—sections of lung, for instance, —there is a risk that they may be torn during the various manipulations. In such a case it is invariably necessary to fix the section on the slide immediately it is cut. The fixative generally used in bacteriology is Mayer's albumin.

Mayer's albumin.—Beat up the white of two eggs into a snow, leave them to stand, then filter through filter paper and add an equal volume of glycerin to the clear filtrate. Add a little piece of camphor or thymol as a preservative and keep in a well-stoppered bottle. Before using the solution shake the bottle well to ensure the mixture being homogeneous.

Method of use.—Place a drop of the albumin on the slide and spread it in a very thin layer with the tip of the index finger. Transfer the section with a spatula direct to the prepared slide, carefully spread it out with a fine brush so that there are no folds and press it lightly to make it adhere to the albumin.

Should there be any difficulty in getting the sections to spread, a drop of water may be placed on the slide already smeared with the albumin mixture and the section laid on the drop of water. The slide is then gently heated on the drying stage (fig. 127, p. 141) until the section has spread evenly, the excess of water is then taken up with a piece of silk paper and the process continued as below.

Heat the under side of the slide very lightly over the pilot flame of a Bunsen and in a few seconds the section will have adhered to the surface of the glass. The section is now treated with xylol and then with absolute alcohol to remove the paraffin, after which it is ready for staining.

Note.—The albumin-fixation method has the disadvantage of not being universally applicable: it cannot, for instance, be used with alkaline solutions, Orth's picrocarmine, etc., which dissolve albumin.

5. The staining of sections.

In order to render the detection of micro-organisms as easy as possible and to facilitate their study, it is desirable that they should be stained a different colour from the tissue in which they are contained; hence it is best to use either a double or triple staining method. Unfortunately such methods are of little use when dealing with an organism which is gram-negative and which does not stain either by Ehrlich's or Ziehl-Neelsen's method. In such a case it is sometimes not possible to differentiate further than by staining with a simple stain in such a way that the background (the animal tissue) is only lightly stained while the bacteria (the vegetable tissue) are stained much more deeply. Recently, however, methods of double staining applicable to gram-negative organisms have been devised and two of these will be described.

The description of Ehrlich's and Ziehl-Neelsen's methods will be deferred to the chapter on tuberculosis.

A. Simple staining.
Methods applicable to most organisms.

(α) **Weigert's method.**—1. Cover the section with a few drops of aniline-gentian-violet (p. 139). Allow the stain to act for 30 minutes or so and then blot up the excess.

2. Immerse the section for a few seconds in a vessel containing a 0·5 per cent. aqueous solution of acetic acid.

3. Wash carefully in distilled water and blot up the excess.

4. Dehydrate *very rapidly* in absolute alcohol.

5. Clear in clove oil then in xylol.

6. Mount in Canada balsam.

(β) **Lœffler's method.**—The stains used are Lœffler's alkaline blue (p. 139) (15 minutes) or Ziehl's fuchsin (p. 138) (5 or 6 minutes). The technique is otherwise the same as in the preceding method.

(γ) **Kühne's method A.**—1. Stain for 15 minutes in carbol-blue or ammoniacal blue (pp. 138 and 139).

2. Transfer to distilled water.

3. Treat for a few seconds with dilute hydrochloric acid (1–1000).

4. Transfer rapidly to lithia solution (p. 210).

5. Wash again carefully in distilled water. Blot up the excess of water and leave the section exposed to the air until it is nearly dry.

6. Dehydrate *as rapidly as possible* in absolute alcohol.

7. Clear in clove oil and xylol.

8. Mount in Canada balsam.

(δ) **Kühne's method B.**—This method is not recommended. It is very tedious and only stains a few species of micro-organisms.

1. Stain for about 30 minutes in carbol-blue.

2. Wash in distilled water.

3. Treat with dilute hydrochloric acid (1–1000) until the tissue is pale blue.

4. Wash in lithia solution (p. 210).

5. Wash for several minutes in distilled water and blot up the excess.

6. Dehydrate *very rapidly* in absolute alcohol lightly tinted with methylene blue.

7. Pour off the alcohol and treat with aniline oil similarly tinted with blue for about 2 minutes.

8. Replace the tinted aniline oil with ordinary aniline oil for about 2 minutes.

9. Clear with clove oil and then with two lots of xylol to ensure the removal of all traces of aniline oil.

10. Mount in Canada balsam.

(ε) **Staining with thionin. Method recommended.**—1. Stain with carbol-thionin (p. 138) for several minutes.

2. Wash in distilled water and blot up the excess.

3. Dehydrate *very rapidly* in absolute alcohol.

4. Clear in clove oil and xylol.

5. Mount in Canada balsam.

(ζ) **Gram's method for the typhoid bacillus.**—1. Stain for a few hours in aniline-gentian-violet (p. 139).

2. Wash the section in distilled water.

3. Transfer for 1 minute to a 1 per cent. solution of hydrochloric acid.

4. Wash carefully in distilled water : blot up the excess.

5. Dehydrate *very rapidly* in absolute alcohol.

6. Clear in clove oil and xylol.

7. Mount in Canada balsam.

By this method the bacilli alone are stained.

(η) **Nicolle's tannin method. Method recommended.**—1. Stain the section for 2 or 3 minutes in Lœffler's or Kühne's blue.

2. Wash in distilled water.

3. Treat for a few seconds in a 10 per cent. aqueous solution of tannin.

4. Wash in distilled water and blot up the excess.

5. Dehydrate rapidly in absolute alcohol.

6. Clear in clove oil and xylol.

7. Mount in Canada balsam.

B. Differential staining.

1. Methods applicable to gram-positive organisms.

To demonstrate the presence of gram-positive organisms in a tissue in which they are present the background (the animal tissue) is first stained with an acid dye which has but little affinity for micro-organisms, then by Gram's method. The bacteria being the only structures stained violet stand out sharply from the other tissues.

The background may be stained with one of several dyes.

For double staining, eosin, fluorescein, carmine (Orth's), vesuvin, Bœhmer's hæmatoxylin, aurantia, hæmatein, etc., are used.

For triple staining a selective dye is chosen which will stain the various tissues different colours. This method of staining enables the lesions produced by the organisms to be studied. The stains ordinarily used are Orth's picro-carmine or hæmatoxylin in conjunction with aurantia or eosin. The following are the formulæ most commonly in use:

STAINING SOLUTIONS.

Dilute aqueous solution of eosin.

Water-soluble eosin,	0·50 gram.
Distilled water,	300 c.c.

Filter.

Solutions of fluorescein, aurantia, vesuvin (0·5 per cent.), etc., are prepared in a similar manner.

Bæhmer's hæmatoxylin.

Make up the two following solutions :

(a) Hæmatoxylin crystals,	1 gram.
Absolute alcohol,	10 c.c.

Pour the solution into a well-stoppered bottle.

(b) Potash alum,	20 grams.
Distilled water,	200 c.c.

Dissolve in the warm and filter when cool.

Allow to stand for 24 hours and then mix the two solutions a and b ; leave the mixture exposed to the air for a week, then store in a well-stoppered bottle and filter immediately before use.

Hæmatein.

Prepare the two following solutions :

(a) Hæmatein,	1 gram.
Absolute alcohol,	50 c.c.
(b) Potash alum,	50 grams.
Distilled water,	1000 c.c.

The potash alum is dissolved in the warm and added immediately to the hæmatein solution. Let the mixture cool in the air and then filter.

Orth's carmine.

Saturated aqueous solution of carbonate of lithia,	100 c.c.
Carmine No. 40,	2·50 grams.

Dissolve by trituration in a mortar in the cold.

Orth's alcohol carmine.

Orth's carmine,	5 volumes.
95 per cent. alcohol,	1 volume.

Mix.

The latter solution only should be used for staining sections fixed with Mayer's albumin, as the former—the non-alcoholic solution—dissolves albumin.

Orth's picrocarmine.

Mix.

Orth's carmine,	1 volume.
Saturated aqueous solution of picric acid,	1–2 volumes.

After staining in the picrocarmine solution, the sections should be transferred to the following fixing solution :

Absolute alcohol,	70 c.c.
Saturated aqueous solution of picric acid,	30 ,,
Pure hydrochloric acid,	0·50 gram.

(i) Double staining.

A. Method recommended.—1. Treat the section for about 30 seconds with the dilute solution of eosin (p. 218) until it acquires a pink colour.

2. Wash in distilled water.

3. Stain the section on the slide for about 30 seconds with carbol-gentian-violet or carbol-crystal-violet (p. 138). It now assumes a violet colour.

4. Pour off the violet and treat the section with Gram's iodine for 30 seconds or so, renewing the solution two or three times until the section is black. Wash in distilled water.

5. Wash with absolute alcohol (or absolute alcohol and aniline oil) until the pink colour of the ground-work reappears.

6. Clear with clove oil and xylol.

7. Mount in balsam.

The background is stained pink and those organisms which retain the stain by Gram's method are stained violet.

B. Kühne's method.—1. Stain the section for 5–15 minutes in Kühne's blue or ammoniacal blue.

2. Wash in distilled water.

3. Treat with Gram's solution for 2 or 3 minutes.

4. Wash in distilled water.

5. Decolourize in a saturated solution of fluorescein in absolute alcohol.

6. Treat with pure absolute alcohol, clove oil and xylol.

7. Mount in balsam.

Bacteria are stained blue while the ground-work is faintly stained with fluorescein.

(ii) Triple staining.

A. Method recommended.—1. Stain for about 5 minutes with Orth's picrocarmine.

2. Pour off the stain and fix in the fixing solution for about 30 seconds.

3. Wash in distilled water.

4. Stain with carbol-gentian-violet or carbol-crystal-violet for 30 seconds.

5. Replace the stain with Gram's solution for 30 seconds. Wash in distilled water.

6. Decolourize in absolute alcohol or absolute alcohol and aniline oil.

7. Treat the section in turn with absolute alcohol slightly tinted with picric acid, clove oil and xylol.

8. Mount in balsam.

B. Nicolle's method.—This method is applicable to sections fixed on the slide with Mayer's albumin.

1. Stain with Orth's alcohol-carmine for 15 minutes.

2. Wash in distilled water.

3. Stain in carbol-gentian-violet (p. 138) for 6 seconds.

4. Substitute Gram's strong solution (p. 209) for the gentian-violet and treat for 4 or 6 seconds, renewing the solution twice during the process.

5. Decolourize with alcohol-acetone (1 to 3).

6. Transfer to picric acid in absolute alcohol for a second or two.

7. Clear in clove oil and xylol.

8. Mount in balsam.

C. Claudius' method.—1. Fix the section on the slide with Mayer's albumin.

2. Stain for 10–15 minutes in Orth's alcohol-carmine.

3. Wash in distilled water.

4. Stain for 2 minutes in a 1 per cent. aqueous solution of methyl violet or in carbol-gentian-violet.

5. Treat for 2 minutes with picric acid solution (p. 144).

6. Blot up the picric solution carefully with filter paper and pour a large drop of chloroform over the section. Blot up the chloroform with filter paper and replace it with a drop of clove oil and repeat the process until the section assumes a pink colour.

7. Clear in xylol and mount in balsam.

2. Methods applicable to organisms in general.

A. Foa's method.—This method is particularly useful for the detection of the typhoid bacillus. It depends upon the use of a mixture of methyl-green and pyronin (**Pappenheim's solution**).

When this method of staining is to be used the tissue should not be fixed in alcohol but in the following solution :

Perchloride of mercury, - - - - - - -	2 grams.
Müller's fluid, - - - - - - - - -	100 c.c.

Leave the tissue in this solution for 24–48 hours ; wash in water for 2 hours ; harden in alcohol (p. 188) and embed in paraffin.

1. Stain the sections for 5 minutes in the following mixture :

Saturated aqueous solution of methyl-green (Grübler), -	3–4 volumes.
„ „ „ pyronin, - - - -	1–2 „

2. Wash in running water. Blot up the excess.

3. Pass rapidly through absolute alcohol to xylol and mount in balsam. The bacilli are stained red and the tissues of the section blue or violet.

B. Saathoff's method.—This is a modification of the preceding rendering the latter more convenient and yielding preparations which keep better. Alcohol may be used to fix the tissues.

1. Stain for about 4 minutes in the following solution which must be filtered before use :

Methyl-green, - - - - - - - -	0·15 gram.
Pyronin, - - - - - - - - -	0·5 „
96 per cent. alcohol, - - - - - -	5 grams.
Glycerin, - - - - - - - - -	20 „
2 per cent. aqueous solution of carbolic acid, - - -	Q.S. ad 100 c.c.

2. Wash in running water until the green colour gives place to a bluish-red. Blot up the excess of water.

3. Dehydrate *very rapidly* in absolute alcohol. Wash in xylol. Mount in balsam.

CHAPTER XIV.

IMMUNITY.[1]

THE PROPERTIES OF IMMUNE SERUMS.

IMMUNITY as the word is applied in bacteriology denotes the faculty possessed by a living animal of resisting an infection or intoxication.

Immunity to a particular organism or toxin may be natural or acquired.

Natural immunity is a function of the species and only rarely of the race. In some cases it has a relation to age : thus, adults may be immune while the young of the same species are susceptible to a particular infection or intoxication. Again immunity may be absolute or relative.

Acquired immunity to a specific disease may be a natural condition resulting from an attack of that disease ; for instance, a person rarely suffers from more than one attack of enteric fever, measles or anthrax ; or it may be a condition artificially produced in an individual in response to the inoculation of a virus, a toxin, or the serum of an immunized animal.

Immunity artificially produced may be active or passive.

Active immunity is the result of the inoculation of small doses of vigorous cultures of living organisms, of cultures of living organisms attenuated either by heat or by prolonged artificial cultivation, of dead organisms, or of the toxins which organisms produce. An active reaction takes place in the living tissues in response to the inoculation with the result that the subject has acquired certain new properties and these will have to be studied in detail. Active immunity is only acquired slowly and then at the cost of a real and occasionally serious disease during which the tissues may be highly susceptible to further inoculation of the particular virus ; but on the other hand the

---

[1] It would obviously be beyond the scope of a book such as this to enter into a detailed study of immunity and the theories associated with it. The present chapter is therefore limited to such explanations as are indispensable to the proper understanding of the subsequent chapters and to an account of the principal methods of demonstrating the properties of immune serums.

immunity so acquired is lasting and occasionally absolute.   By increasing the number of successive vaccinating inoculations the animal may in time become so highly immunized that even enormous doses of the specific organism or toxin have no visible effect upon it : this is a special condition of **hyper-immunization** in which the resistance of the animal is raised to its highest limits.

But if a non-immune subject be inoculated with the serum of an immunized or hyper-immunized animal instead of with organisms or toxins a different result ensues.   The former is certainly rendered immune but in this case it is merely a condition of **passive immunity**.   The person or animal passively immunized has taken no active part in the process of immunization but has simply been inoculated with something possessing prophylactic properties. The period during which such immunity lasts, which is always very short (generally a few days only), is dependent upon the time during which the substance inoculated remains in the tissues and as soon as it is eliminated the immunity has gone.

The mechanism of immunity.

If a living animal be immune against a pathogenic organism, the inoculation of that organism into the animal results in an aggregation of leucocytes at the site of inoculation (*chemiotaxis*) which ingest and digest the inoculated organisms.   This is the phenomenon described by Metchnikoff as *phagocytosis*.

Phagocytosis can be easily observed, for instance, with the anthrax bacillus. If a healthy guinea-pig be inoculated with a trace of an anthrax culture the tissues about the site of inoculation soon become the seat of an œdematous infiltration (the œdema consists of a serous fluid containing free organisms but very few leucocytes) : the bacillus quickly generalizes and death rapidly supervenes.   On the other hand, if a guinea-pig previously vaccinated against anthrax be inoculated it can be shown that numbers of leucocytes very rapidly accumulate at the site of inoculation and in a few hours have ingested, killed and digested all the bacilli, the animal suffering no ill-effects from the inoculation.   A similar observation can be made on dogs, animals naturally immune to anthrax.   The inoculation of anthrax bacilli into dogs is followed by a small abscess in which phagocytosis is very active but the infection does not become generalized.[1]

The leucocytes take up the micro-organisms while the latter are still living. Experiments have been devised to show that organisms ingested by leucocytes retain their vitality for a greater or lesser length of time during which they can, in a non-immune animal, set up a fatal infection (Metchnikoff).

On the other hand, in some cases, notably in the case of the cholera vibrio, it has been observed that if the vibrio be inoculated into the peritoneal cavity of an immunized guinea-pig it is killed not after ingestion by the leucocytes— which are present in very small numbers in the exudate—but in the exudate itself : this constitutes **Pfeiffer's phenomenon** (*vide infra*).   Such a phenomenon might be quoted as an objection to the theory of phagocytosis but more extended observation shows bactericidal action of this nature by the body fluids to be exceptional : it may be described as a make-shift in the defence of the individual and only occurs when the leucocytes have undergone changes which prevent them coming in contact with the organisms themselves and is moreover only seen in the case of a few very delicate organisms.

---

[1] Micro-organisms have their own means of defence in their fight with the leucocytes : they secrete soluble substances, agressins, which act on the white cells of the blood and prevent them ingesting and destroying the infecting agents.   In conditions of immunity the leucocytes triumph over these agressins and thus fulfil their function of defence.

According to Metchnikoff the bactericidal substances in the serum are derived from the leucocytes: some (*immune bodies, amboceptors,* or *sensibilisatrices*) are elaborated in the leucocytes and excreted into the plasma as they are formed, whence they pass into the different tissues of the animal; the others (*complement, cytase* or *alexin*) are also of leucocytic origin but are only set free on the death and disintegration of the leucocytes. Petterson and Schneider consider that there are yet other substances in the leucocytes capable of destroying micro-organisms (*endolysins, leukins*).

In the majority of cases the bactericidal substances of the serum of immunized animals intervene to prepare the micro-organisms for the action of the leucocytes and facilitate their ingestion and destruction (*vide* opsonins).

In immunized animals therefore over and above the phagocytic reaction there exist in the fluid part of the blood (serum) certain substances of great importance which play a prominent part in the phenomena of immunity. The properties of these immune serums will be now studied a little more fully.

The serums of immunized animals may exhibit one or more or all of the following properties each quite independently of the other and in different degrees.

1. Prophylactic and therapeutic properties.

2. Antitoxic properties.

3. Agglutinating properties.

4. Bactericidal properties.

5. The property of preparing micro-organisms for ingestion by the leucocytes. This property which is due to the presence of special substances, *opsonins*, would appear to be connected with the bactericidal properties.

SECTION I.—PROPHYLACTIC AND THERAPEUTIC SERUMS.

It has already been pointed out that the serum of an animal vaccinated against a micro-organism if inoculated into a fresh animal confers on the latter an immunity of short duration.

This passive immunization is absolutely *specific* and is only exhibited towards the species of organism with which the first animal was vaccinated.

The serum of an animal vaccinated with toxin if inoculated into a fresh animal confers on the latter an immunity against the same toxin and also against the micro-organism which elaborated the toxin.

Example.—If a normal guinea-pig be inoculated with antidiphtheria serum it is protected against the inoculation of diphtheria toxin and also against inoculation with the diphtheria bacillus.

On the other hand, if an animal be vaccinated with micro-organisms its serum has no protective action against the toxin of the organism though it protects against the organism itself.

Example.—The serum of an animal vaccinated with the cholera vibrio (*vide* Cholera) will protect a normal animal against an inoculation of the vibrio. A trace of the serum, for instance, inoculated into a normal guinea-pig will vaccinate the latter against choleraic peritonitis. On the other hand the serum affords no protection against an inoculation of the toxin and is totally ineffective in intestinal cholera which is an intoxication (Metchnikoff).

In all of the foregoing cases the serum acts as a prophylactic; that is to say, it immunizes the animal to which it is administered provided it be inoculated before or at the same time as the organisms or toxin.

Some serums exhibit therapeutic as well as prophylactic properties. If inoculated after the infection, even though the first symptoms of infection have appeared, they abort the disease and lead to recovery. The curative properties of a serum do not always run parallel with its prophylactic properties. To quote a classical instance : antidiphtheria serum is both prophylactic and curative, but antitetanus serum while exhibiting very marked prophylactic properties has no curative properties. These properties of immune serums will be referred to again in more detail, each serum being dealt with in connexion with its corresponding organism.

SECTION II.—ANTITOXINS.

If an animal be inoculated with progressively increasing doses of a microorganic toxin it will ultimately become immunized against this toxin, and will be able to tolerate without suffering any inconvenience doses infinitely greater than those which if given in the first instance would have proved fatal (Behring and Kitasato).

To this general rule there are however a few exceptions and these have been described by Richet as cases of **anaphylaxis.**

Richet showed that if a dog were inoculated with a small dose of actino-congestine (the poison in the tentacles of sea anemones) it exhibited no ill-effects ; but if 10–20 days after the first inoculation it were re-inoculated with the same or even with a smaller dose than that which before proved harmless the animal quickly died. This result cannot be explained on the theory of an accumulation of toxin because the whole quantity given in the two doses is very much less than that which would be required to produce a fatal result if given in the first instance, and further if the second inoculation be given from 1–6 days after the first, the animal does not die : the phenomena of anaphylaxis do not appear until about the tenth day. The serum of an anaphylactic dog inoculated into a normal dog produces a condition of hypersensibility immediately after inoculation, and hence the serum of anaphylactic animals contains the substance—whatever its nature—causing the phenomena of anaphylaxis (Richet).

Other instances of anaphylaxis may be quoted. If an animal be inoculated once with the serum of another species it is only rarely and then inconstantly that any untoward symptoms develop, but if successive re-inoculations be made the result is quite different, the reaction on the part of the inoculated animal being then very violent and likely to terminate fatally (Arthus). This phenomenon is seen for example when rabbits or, better, guinea-pigs, are repeatedly inoculated with horse serum. According to von Pirquet and Schrick the grave symptoms occasionally observed in the human subject after injections of antidiphtheria serum are of an anaphylactic nature.

Anaphylaxis in connexion with tuberculosis has also been the subject of experimental observation. The reaction to tuberculin is an anaphylactic phenomenon : the inoculation of a trace of tuberculin into man or an animal affected with tuberculosis sets up a severe reaction (*vide* Tuberculosis) and numerous methods of diagnosis are based on this reaction.

Still further examples of anaphylaxis could be given but it must suffice here to have drawn attention to the existence of this phenomenon. To investigate the mechanism of anaphylaxis and to discuss the theories which have been advanced in explanation of it would be altogether beyond the scope of the present work.

The serum of animals which have survived the inoculation of repeated and increasing doses of toxin has acquired **antitoxic properties.**

Antitoxin, like toxin, has its nature altered by being heated, is precipitated by alcohol, and is carried down by precipitates formed in the liquid in which it is in solution. In suitable quantities it saturates toxin both in the tissues and *in vitro*. In mixtures *in vitro* toxin is not destroyed by antitoxin but is simply disguised ; the toxin-antitoxin mixture is nevertheless harmless to animals, though under certain conditions the poisonous nature of the toxin may be made to reappear ; thus, if a neutral mixture of snake venom and antivenomous serum be heated to 70° C. the antitoxin is destroyed but not the toxin so that the mixture is now no longer harmless.

Antitoxic serums are strictly specific. Under the head of each of the pathogenic micro-organisms the antitoxic properties of the corresponding serum will be considered in detail.

SECTION III.—AGGLUTININS.

Durham and Gruber when studying antityphoid serum demonstrated a very remarkable property of the serum. If a small quantity of serum from a typhoid-immunized animal be added to a broth culture of the typhoid bacillus the bacilli distributed through the medium lose their motility, collect together and become agglutinated into masses, retaining however their vitality. This phenomenon is known as *agglutination* and the serum is said to possess *agglutinating properties*.

Previously to Durham and Gruber's experiments, Bordet had demonstrated a similar action of anticholera serum on cholera vibrios, and it has since been shown that in the majority of cases the serum of an animal immunized against a micro-organism agglutinates the organism used for immunization (cholera, dysentery, tuberculosis, mediterranean fever, plague, glanders, etc.).

The property of agglutination however is not limited to the serum of immunized animals. [A. S. Grünbaum showed that] it appears quite early, before a state of immunity has been created, as soon as the tissues have been invaded by a pathogenic organism. The reaction of agglutination is a *reaction of infection*. It remains, moreover, for a long time after recovery has taken place, being found as has already been stated in a marked degree in the serum of immunized individuals.

The agglutination reaction is specific : the serum of an enteric patient agglutinates the typhoid bacillus and (with certain reservations) the typhoid bacillus only. The serum of cholera patients similarly agglutinates only the cholera vibrio.

[A. S. Grünbaum and shortly afterwards] Widal showed that practical use can be made of these facts in the diagnosis of infective diseases and to [the former] is due the method of *serum diagnosis*. Take, for example, the case of a person thought to be suffering from enteric fever : it is only necessary to mix a few drops of his serum with a culture of the typhoid bacillus : then if the patient be suffering from enteric fever the bacilli will be agglutinated ; on the other hand, if he be suffering from some disease other than enteric fever the bacilli will remain separate and motile.

Conversely, suppose it is required to determine whether a bacillus is the typhoid bacillus or not : in this case it is sufficient to prepare a culture of the unknown bacillus and to mix it [in due proportion] with a typhoid-agglutinating serum : if agglutination take place the bacillus may without hesitation be affirmed to be the typhoid bacillus.

To obtain reliable results, there are certain precautions which must be strictly observed in carrying out the reaction. To exemplify : most normal serums—and especially human serums—when used in large quantities agglutinate a considerable number of organisms : if a mixture of serum and organisms be made without knowing the proportions in which they are mixed agglutination might be obtained apart from any specific relation of the ingredients to each other. The following rules should therefore always be followed :

(i) The serum under investigation must be diluted [Grünbaum] and the dilution carried to such a degree that the minimal dose of serum required for agglutination is determined. For purposes of comparison the minimum quantity of normal serum (human or animal) required to produce agglutination must also be determined.

For example, it can be shown that while normal serum frequently agglutinates the typhoid bacillus in a dilution of 1 in 10 a typhoid serum will agglutinate it in dilutions of 1 in 200, 1 in 500, and even in 1 in 5,000.

(ii) It becomes even more imperative to dilute the serum when it is recognized that a specific serum will agglutinate not only its corresponding organism but also, not infrequently, closely related species, provided that the quantity of serum used be sufficient [Grünbaum]. It is obvious therefore that unless a serum be adequately diluted its specific characters will escape recognition.

Take an example : a patient is suffering from a para-typhoid infection. His serum agglutinates both the typhoid and the para-typhoid bacillus in dilutions of 1 in 20 and 1 in 50 : so far there is nothing specific about the serum. Dilute the serum further, say to 1 in 100, 1 in 200, and 1 in 500. In these higher dilutions it has entirely lost all its agglutinating property for the typhoid bacillus but still agglutinates the para-typhoid bacillus. In this case the specific nature of the agglutination is determined by the *titre* of agglutination and not by the mere fact of agglutination itself.

(iii) It is also of the highest importance in studying the phenomena of agglutination that only homogeneous emulsions or cultures be used in which the organisms are as far as possible lying separately, for if they be clumped or massed together the results of the experiments will obviously be misleading. This spontaneous clumping is a source of great difficulty when working at agglutination with organisms which naturally grow in clumps. The difficulty may be overcome either by using very young cultures in broth (typhoid bacillus) or by having resort to one or other of the various methods which have been devised for obtaining homogeneous cultures (of the tubercle bacillus, etc.).

(iv) Finally, in performing agglutination tests with serums care must be taken to add the serum to the culture and never to add the culture to the serum. It can be easily understood that in the latter case the first drops of culture would be mixed with an undiluted serum and that agglutinated masses of organisms might form even though there were no specific relationship between the organism and the serum.

The technique of serum diagnosis will be described in detail in the chapter on the typhoid bacillus, and under the head of each micro-organism data with regard to agglutination will be given.

The mechanism of agglutination.

It would appear that the phenomena of agglutination are not dependent upon any vital activity of the organisms since they can be observed with dead cultures.

The substances in serums producing agglutination are known as *agglutinins*. Agglutinins are distinguishable from bactericidal substances in that unlike the latter they withstand heating at 55° C. for half an hour and are only destroyed at about 60° C. in serum and 70° or 80° C. in milk. They are precipitated by alcohol and do not pass through a Chamberland or Berkefeld bougie. But since they can be demonstrated in the milk, urine, etc., of infected or immunized animals it would appear that they can pass through certain living animal membranes.

The phenomena of agglutination may be explained on the assumption that the *agglutinin* acts on some *agglutinatible substance* present in the bodies of the organisms agglutinated. Organisms which have been separated from the culture medium by filtration, washed and suspended in normal saline solution still retain the property of being agglutinated by a specific serum. But, as Kraus and Ch. Nicolle have shown, if a culture be filtered through porcelain a flocculent precipitate, similar to masses of agglutinated micro-organisms, forms on the addition of a specific serum to the filtrate. It is obvious therefore that the agglutinatible substance is also present in the culture fluid; it may be that as the organisms grow old the agglutinatible substance passes into the culture fluid. The name *precipitins* has been suggested for the substances in serum which cause the precipitate in filtered cultures: there is evidence that precipitins and agglutinins are identical bodies.

Finally, certain chemical substances have the property of agglutinating micro-organisms (Malvoz) but their action is in no way specific and the same substance will agglutinate different micro-organisms (Beco). A mixture of equal parts of commercial formalin, alcohol, hydrogen peroxide, a 1 in 1,000 solution of chrysoidin, vesuvin, safranin, or perchloride of mercury, etc., agglutinates the typhoid bacillus as well as various other organisms.

SECTION IV.—BACTERICIDAL PROPERTIES.

The fact that the serum of immunized animals has the power of destroying bacteria was brought to light by one of Pfeiffer's experiments which has since become classical.

Pfeiffer's experiment.—If a normal guinea-pig be inoculated in the peritoneal cavity with a broth culture of the cholera vibrio the animal rapidly succumbs from peritonitis, and if the peritoneal exudate be examined microscopically in a hanging-drop preparation it is found to contain very large numbers of free motile vibrios, exactly similar to those inoculated.

Let the same experiment be done on a guinea-pig which has been immunized against the cholera vibrio; the animal survives the inoculation and an examination of the peritoneal fluid reveals an entirely different condition.

In a drop of the fluid removed 10–30 minutes after the inoculation it will be found that not only have the vibrios not multiplied but they have also lost their motility, and instead of finding numerous elongated comma-shaped organisms as in the former case, the fluid is seen to contain small granules of no definite shape, which soon disappear altogether being destroyed in the fluid in which they are suspended.

This granular metamorphosis followed by complete destruction of the vibrio may also be demonstrated *in vitro* (Bordet, Metchnikoff).

Bordet's experiment.—Break up a small quantity of an agar culture of the cholera vibrio in a little sterile broth: examine the emulsion under a microscope to see that there are no granular forms and that the vibrios are quite motile: add to the emulsion $\frac{1}{20}-\frac{1}{40}$th of its volume of the serum of an immunized guinea-pig. On examining the mixture a few minutes after the addition of the serum, the vibrios will be seen to have lost their motility and to have become agglutinated and converted into granular dots: the reaction is however not at its maximum until the mixture has been kept at 37° C. for 1 or 2 hours.

From these two experiments it may be concluded that the serum of immunized guinea-pigs, apart from the intervention of any cellular element, contains bactericidal and bacteriolytic substances capable of destroying the cholera vibrio.

These substances are specific so that the serum is only bactericidal for the organism with which the animal has been immunized. The serum of animals immunized with the typhoid bacillus for instance is bactericidal only for the typhoid bacillus and is totally devoid of action on the cholera vibrio, and, *vice versa*, an anticholera serum is not bactericidal for the typhoid bacillus.

The bactericidal action of immunized serums is rapid and at its maximum at 37° C., feeble at the ordinary temperature of the laboratory and altogether paralyzed at 0° C.

The analysis of the phenomena of **bacteriolysis** may now be pushed a step further and an attempt made to investigate the mechanism by which bacteriolysis occurs.

Mechanism of bacteriolysis.

Suppose the serum of a guinea-pig immunized with the cholera vibrio be heated to 55° C. and then mixed with a culture of the vibrio. Bacteriolysis no longer takes place, though the agglutinating properties of the serum remain unaffected (Bordet).

The heated serum, however, has not altogether lost its bactericidal properties; for, if to the mixture of vibrios and heated serum a small quantity of serum from a normal animal be added, bacteriolysis occurs at once—the heated serum is *re-activated*.

It may therefore be concluded that the serum of the immunized animal contains two substances :

(i) One of which is not destroyed by being heated at 55° C. or, in other words is **thermostable** at 55° C. and which is only present in the serum of immunized animals.

(ii) The other of which is destroyed by heating to 55° C. or, in other words, is **thermolabile** at 55° C. and which is present also in the serum of normal animals.

These two substances when present together cause bacteriolysis but either the one or the other acting alone has no action on the vibrio. Let us consider now the part which each of these substances plays.

A. Immune body (Sensibilisatrice).—The thermostable substance is, as has been said, present only in the serum of immunized animals and is a *product of immunization* [hence the term immune body generally used in England]. And, further, it is *specific* and acts only on the organism which was used for immunization.

Experiment.—Treat an emulsion of cholera vibrios with anticholera serum heated to 55° C. and then add a little normal rabbit serum. The vibrios will be bacteriolyzed.

Repeat the experiment using instead of cholera vibrios, typhoid bacilli. Treat the typhoid bacilli with heated (55° C.) anticholera serum and then add the normal rabbit serum. No bacteriolysis takes place.

The immune body in contact with its corresponding micro-organism is fixed by the organism in the same way that a mordant acts on a fabric.

Experiment.—Leave a mixture of cholera vibrios and heated anticholera serum for half an hour at 37° C. Centrifuge to separate the vibrios, wash the latter with normal saline solution and then add a little fresh normal rabbit serum to the vibrios. Bacteriolysis takes place.

This experiment justifies the view held by Bordet that the immune body sensitizes the organisms to the action of the thermolabile substance present in normal serum just as a mordant sensitizes a fabric to the action of a dye. Hence the use in France of the word *Sensibilisatrice* to denote the immune body. The immune body is only destroyed by heating it at 65°–70° C.

Sensitized micro-organisms, that is to say, organisms which have been treated with their specific immune body, continue to grow in the ordinary way and have lost none of their pathogenic properties, but they differ from non-sensitized organisms not only in their susceptibility to the action of the thermolabile substance or complement (*vide infra*) present in all normal serums, but also in that they are more easily ingested and destroyed by leucocytes.

The immune body is known by different names, *substance sensibilisatrice* (Bordet), *fixateur, amboceptor* (Ehrlich). Occasionally it is described as

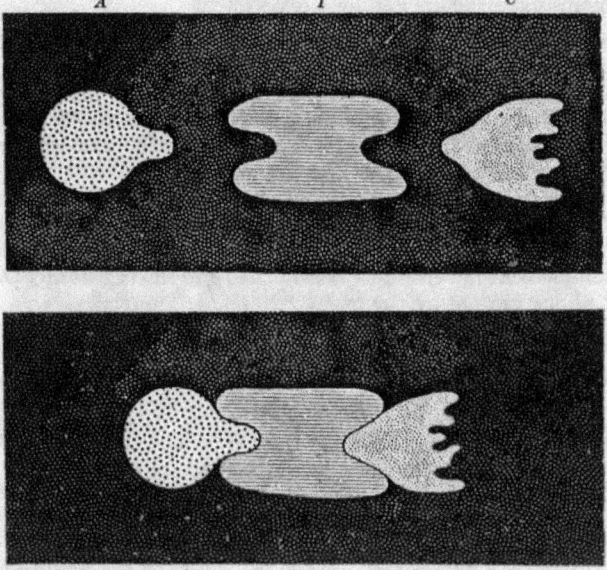

FIG. 159.—Ehrlich's diagram to explain the interaction between the immune body and complement.

A, micro-organism or antigen; *I*, immune body, sensibilisatrice, amboceptor, fixateur or antibody; *C*, complement, alexin or cytase.

the **antibody** because it is antagonistic to the substances inoculated into animals for the purpose of immunizing them. These latter substances are therefore called **antigens**. Thus, in immunizing animals against the typhoid bacillus the antigen is the typhoid bacillus and the antibody the new product appearing in the serum in response to the inoculation of the antigen and which has the property of attaching itself to the typhoid bacillus and so of rendering the bacillus susceptible to bacteriolysis.

B. Complement.—The immune body prepares organisms for the action of the substance contained in the serum of normal animals. This latter substance in conjunction with the immune body produces bacteriolysis, hence the name *complement* by which it is generally described (Ehrlich). By some authors, however, it is occasionally referred to as *alexin* (Bordet) or *cytase* (Metchnikoff and Buchner).

The complement is not a product of immunization and is not specific but is present in all normal serums and is fixed indifferently by all organisms through their specific immune body.

When complement is mixed with micro-organisms it is only taken up by them if they have been sensitized. Complement has no affinity for non-sensitized organisms and remains in the serum. Its action is like that of a dye which only dyes a fabric that has been treated with a mordant.

To sum up : in the serum of immunized animals a specific substance, the immune body, is present which unites on the one hand with its corresponding micro-organism and on the other hand with a substance, alexin or complement, pre-existing in the serum of all normal animals. Upon this interaction of two bodies, in which the immune body plays a part similar to that of a mordant in dyeing, depends the destruction of the micro-organisms in Pfeiffer's experiment.

The interaction of complement with the micro-organism through the immune body is diagrammatically represented in the figure (fig. 159).

Note.—It is important to observe that the phenomena just studied are not seen in the case of all micro-organisms. The combined action of the immune body and complement only leads to destruction in the case of very delicate organisms, *e.g.* the cholera vibrio and the typhoid bacillus.

In the majority of cases the pathogenic micro-organisms are much more resistant to the bactericidal action of the immune serums and no bactericidal action can be seen ; but though not visible, combination of the immune body with the complement nevertheless takes place, and the organisms are rendered more easy of destruction by the leucocytes (*vide* opsonins).

In studying the phenomena of complement fixation (*vide infra*) it will be shown how the action of the immune body on the more resistant organisms may be demonstrated.

Hæmolysins.

Hæmolysis means the destruction of the red cells of the blood with diffusion of the hæmoglobin into the medium in which they are suspended.

If a quantity of red cells be suspended in an hypotonic solution, distilled water for example, they undergo hæmolysis. On the other hand, in an isotonic fluid, normal saline solution for instance, the red cells may remain intact for a very long time. Similarly in the serum of the majority of normal animals the red cells undergo no alteration. To this general rule, however, there are a few exceptions ; dogs' serum, for instance, hæmolyzes guinea-pig red cells ; eel serum hæmolyzes all mammalian red cells, and so on.

The inoculation of large doses of red cells of one species of animal into the peritoneal cavity of another species produces a toxic effect and may kill the animal inoculated.

On the other hand, if small quantities be inoculated on several successive occasions there is a minimal reaction and death does not take place. In the latter case the serum of the inoculated animal is capable of destroying the red cells of the animal species used for inoculation *in vitro* and is therefore said to exhibit hæmolytic properties (Bordet).

For example, if a guinea-pig be inoculated with rabbit red cells, the guinea-pig's serum will become hæmolytic for rabbit red cells.

Preparation of an hæmolytic serum.—Under no conditions must the whole blood be inoculated but only the washed red cells.

1. Collect some blood under aseptic precautions, and after defibrinating it (p. 36) centrifuge. Thus, into each tube of the centrifuge pour equal parts of defibrinated blood and sterile normal saline solution, centrifuge, and when the red cells are all precipitated at the bottom of the tube pipette off the clear supernatant liquid with a bulb pipette. Fill up the tube with fresh saline solution, stir up the deposit and centrifuge again. Repeat the operation three times.

After centrifuging for the third time dilute the red cells with sufficient sterile normal saline solution to bring the total volume up to the volume of blood originally used.

It is perhaps unnecessary to say that these operations should be carried out under aseptic conditions.

2. Inoculate the animal (a guinea-pig if using rabbit cells, a rabbit for sheep cells, etc.) sub-cutaneously, or better into the peritoneal cavity, on five occasions at intervals of 1 week with 5–8 c.c. of a suspension of red cells prepared as described above. Experience has shown that this amount does not produce any toxic symptoms. The serum of the animal is best collected about 1 week after the last immunizing inoculation.

Mechanism of hæmolysis.

The phenomena of hæmolysis are a counterpart of those of bacteriolysis.

The hæmolytic serum is specific and only hæmolyzes red cells of the animal species used for inoculating the animal from which the serum has been drawn.

It contains two substances :

1. A specific thermostable immune body.

2. A complement, non-specific, present in all normal serums and only becoming attached to the red cells through the immune body. The properties of hæmolytic serums can be demonstrated by means of the following experiments.

Experimental illustrations.—Use the serum of a guinea-pig inoculated with rabbit red cells as an hæmolytic serum and prepare an emulsion of red cells by mixing 0·1 c.c. of washed rabbit red cells with 2 c.c. of normal saline solution.

(i) Mix the emulsion of red cells with 0·1 c.c. of the hæmolytic serum and incubate at 37° C. for 1 hour. On taking the tube out of the incubator it will be seen by simply looking at the tube that hæmolysis is complete : the hæmoglobin has been discharged from the red cells and imparts an uniform colour to the solution.

A control tube in which a little normal guinea-pig serum has been added to an emulsion of red cells shows no hæmolysis; the fluid contents are clear.

(ii) To an emulsion of red cells, add 0·1 c.c. of hæmolytic serum previously heated to 55° C. for half an hour and incubate at 37° C.

The red cells are not hæmolyzed but simply agglutinated at the bottom of the tube. In this case hæmolysis has failed because the complement was destroyed by heating the serum to 55° C.

(iii) To the mixture used in the preceding experiment and which is quite clear add 0·1 c.c. of normal guinea-pig serum (complement) and incubate again. The red cells will now undergo hæmolysis.

(iv) Repeat experiment (ii) and after showing that under the conditions of the experiment no hæmolysis occurs, centrifuge the mixture and pipette off the serum from the red cells.

(a) The centrifuged serum has been deprived of its immune body (which has combined with the red cells) and any attempt to re-activate it by the addition of complement (normal guinea-pig serum) fails; it is no longer able to hæmolyze fresh rabbit cells if these be added to it.

(b) The red cells separated from the serum by centrifuging have combined with the immune body ; so that even after being repeatedly washed with normal saline solution and centrifuged, they are rapidly hæmolyzed on the addition of 0·1 c.c. of normal guinea-pig serum (complement) if a mixture of the two be put in the incubator at 37° C.

(**v**) If the foregoing experiment be repeated and instead of rabbit cells, red cells of some other animal, sheep, for instance, be added to the heated hæmolytic serum it can be shown that the sheep cells are not sensitized, since on the addition of complement they are not hæmolyzed.

And further, the serum to which the sheep cells have been added has retained intact its sensitizing properties and is still capable of sensitizing rabbit red cells.

This experiment again demonstrates the specific nature of the reaction.

(**vi**) Hæmolysis does not occur at 0° C. Place a mixture of non-heated hæmolytic serum and a suspension of the corresponding red cells in the ice chest for several hours, then centrifuge the mixture and wash the cells in the cold, add some complement to the cells and incubate at 37° C. : hæmolysis occurs. Add the serum to some sensitized and washed red cells and in this case also hæmolysis occurs. In other words the immune body has been taken out of the serum by the red cells but at the temperature of the experiment the complement remains in solution.

Conclusions.

When a living animal is treated with sublethal doses of micro-organisms or their toxins (antigen) a substance inimical to the antigen (antibody, amboceptor, sensibilisatrice, immune body) appears in the serum which has the property of combining with the antigen, thus rendering the latter susceptible to the action of a third substance (complement, alexin, cytase) already present in the serum of the normal animal and derived probably from the leucocytes.

By the combined action of the immune body and complement, the antigen is either destroyed (in the case of red cells or delicate organisms) or prepared for the destructive action of the leucocytes (as happens with micro-organisms in general).

The fixation of the complement.

(Deviation of the complement.)

Prepare in accordance with the rules elaborated in the preceding paragraphs the following experiment.

Mix in suitable proportions a portion of a culture of the cholera vibrio and some anticholera serum previously heated at 55° C., incubate for 1 hour, and then add a small quantity of non-heated serum (complement) to the mixture. Under these conditions the vibrios sensitized by the specific immune serum are bacteriolyzed by the action of the complement. Now add to the mixture some red cells sensitized with their corresponding inactivated immune serum (*hæmolytic couple*) ; no hæmolysis takes place because there is no complement available, the complement originally present having all been used up in producing bacteriolysis of the cholera vibrios. In other words there has been *fixation*, or *deviation*, *of the complement* by the sensitized vibrios.

Now perform a second experiment. Mix a portion of a culture of the typhoid bacillus with some inactivated anticholera immune serum and after incubating, add a small quantity of guinea-pig complement. In this case, the immune body has not been able to sensitize the bacilli being specific for and combining only with cholera vibrios : consequently, the complement remains unattached, in other words, is not deviated. Now add some sensitized red cells to the mixture and incubate again ; hæmolysis of the red cells occurs because there was free complement in the mixture with which they were able to combine.

From this fundamental experiment Bordet and Gengou deduced a very valuable method of diagnosis for infective diseases which is known as the Bordet-Gengou or *complement-fixation reaction*. The reaction has been applied by Widal and Le Sourd to the diagnosis of enteric fever (fixation reaction, hæmolyso-diagnosis) and is applicable to the majority of micro-organic diseases. Two different cases arise for consideration.

First case.—Given a serum suspected to contain a particular immune body, the serum of an enteric fever patient, for example, a certain diagnosis may be made by the complement-fixation method.

Heat the serum to 55° C. for half an hour, prepare a mixture of typhoid bacilli and the heated serum, add some complement and incubate the mixture at 37° C. for an hour. Then add a mixture of red cells and homologous inactivated hæmolytic serum and incubate again. One of two things may happen.

1. Either the typhoid bacillus is sensitized by the inactivated suspected serum, in which case it fixes the complement so that on the addition of sensitized red cells—there being no free complement—the cells do not undergo hæmolysis. If this takes place it may be affirmed that the suspected serum contains antibodies for the typhoid bacillus and that the patient is suffering from enteric fever.

2. Or the typhoid bacillus is not sensitized by the suspected serum and therefore does not combine with the complement, so that on the addition of sensitized red cells the free complement attaches itself to them and hæmolysis is the result. The suspected serum, therefore, in this case contains no typhoid antibodies.

Second case.—Suppose a given organism is believed to be the typhoid bacillus and it is desired to confirm the diagnosis.

Prepare a mixture containing the suspected bacillus, heated antityphoid serum and complement. Incubate for an hour and then add a mixture of red cells and inactivated hæmolytic serum.

1. The bacillus may be sensitized by the antityphoid serum in which case it will absorb the complement, and on the addition of sensitized red cells—there being no free complement—no hæmolysis takes place ; the complement was deviated or fixed by the suspected bacillus which is therefore the true typhoid bacillus.

2. The bacillus may not be sensitized by the antityphoid serum consequently it cannot fix the complement and this remaining in solution is free to combine with the sensitized red cells. There had been no fixation of the complement so hæmolysis occurs ; the bacillus therefore, not uniting with the antityphoid immune body, is not the typhoid bacillus.

The value of this method of diagnosis to the bacteriologist can be easily appreciated : the results are more constant and more delicate than those obtained by means of agglutination tests but considerable technical skill is required in carrying out the reaction.

Technique of the complement-fixation reaction.

The following materials are required :

Apparatus, etc.—1. A number of narrow test-tubes about 10 cm. long and 5 c.c. capacity.

2. A number of 1 c.c. pipettes graduated in tenths of a cubic centimetre. (Levaditi's pattern is, perhaps, the best (fig. 160).)

The various manipulations should as far as possible be conducted under aseptic conditions, so that the tubes and pipettes must be sterilized in the hot air sterilizer.

3. Sterile normal saline solution. The volume of fluid in each tube used in the test should, if the experiment is to be conclusive, be the same. After the various ingredients have been added sufficient normal saline solution is poured in to bring the volume up to, generally, 2 c.c.

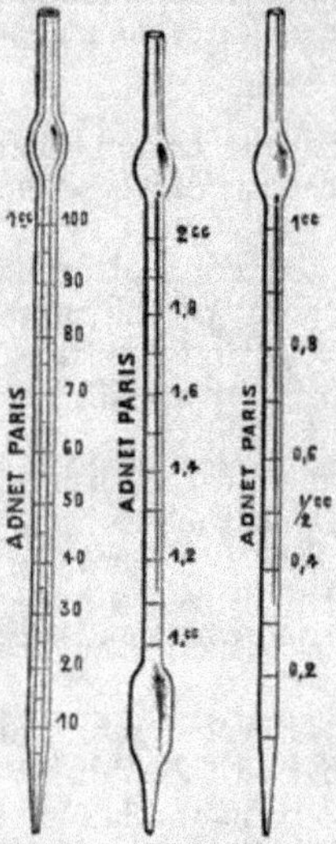

Red cells.—Sheep or rabbit red cells are generally used. They must be separated and washed in the manner already described (p. 230).

After the third washing prepare a 5 per cent. solution of the cells in normal saline solution.

Hæmolytic serum.—The method of preparing hæmolytic serums has been described above (p. 230). If sheep cells are used as the indicator the serum of a rabbit inoculated with sheep cells is employed, and for rabbit cells the serum of a guinea-pig inoculated with rabbit cells.

After collecting the blood in the ordinary way (Chap. XII.) the serum is decanted and then inactivated by heating for half an hour at 55° C. in sealed ampoules. The hæmolytic serum should be stored in an ice chest and it will then retain its hæmolytic properties for several months.

It is absolutely necessary to titrate the hæmolytic serum. This can be done in the manner indicated in the following table. In a series of tubes prepare the mixtures shown in the horizontal lines, incubate at 37° C. for 15–30 minutes and note the results.

FIG. 160.—Levaditi's pipettes.

	5% emulsion of red cells.	Heated hæmolytic serum.	Complement or alexin.	Normal saline solution (to make up to 2 c.c.).	Results.
	c.c.	c.c.	c.c.	c.c.	
Tube No. 1, -	1	0·5	0·1	0·4	Total hæmolysis in 15 minutes.
Tube No. 2, -	1	0·3	0·1	0·6	Total hæmolysis in 15 minutes.
Tube No. 3, -	1	0·1	0·1	0·8	Partial hæmolysis in 15 minutes, complete in 30 minutes.
Tube No. 4, -	1	0·05	0·1	0·85	Incomplete hæmolysis
Tube No. 5, -	1	0·01	0·1	0·9	Very slight hæmolysis.
Control No. 1,	1	—	0·1	0·9	No hæmolysis.
Control No. 2,	1	0·5	—	0·5	No hæmolysis.

From an examination of the table it follows that the amount of hæmolytic serum added to tube No. 3 will, in this particular instance, be the most suitable for subsequent experiments : in this tube hæmolysis is complete in half an hour. Tubes Nos. 1 and 2 contain too much serum, and in Nos. 4 and 5 hæmolysis is not complete and they therefore contain too little serum.

Further, examination of the control tubes shows :

1. That the heated hæmolytic serum only hæmolyzes when complement is added.

2. That complement alone does not hæmolyze the red cells.

In carrying out the reaction of complement fixation then, the quantity of 0·1 c.c. of this particular hæmolytic serum per 1 c.c. of the dilution of red cells will be used.

Complement.—Normal guinea-pig serum collected aseptically will be used as complement. The amount of complement to be added is of the greatest importance : if there be an excess of complement the whole of it will not be absorbed by the antigen-antibody mixture and the excess remaining in solution will hæmolyze the red cell-serum mixture and give an erroneous result. The smallest quantity of guinea-pig serum which will hæmolyze 1 c.c. of the sensitized red cell emulsion must therefore be determined.

If fresh guinea-pig serum be used it will be found to be very rich in complement but the amount rapidly diminishes in the first few hours. It is preferable therefore to use serum collected 8–10 days before [and stored in an ice chest] ; it is not so active but its titre remains constant for several days (Nicolle and Pozerski). The following table illustrates the method of titration. Incubate the tubes for half an hour at 37° C. and then note the results.

	5% emulsion of red cells in normal saline solution.	Heated hæmolytic serum.	Complement (normal guinea-pig serum 8 days old).	Normal saline solution to make up to 2 c.c.	Results.
	c.c.	c.c.	c.c.	c.c.	
Tube No. 1, -	1	0·1	0·2	0·7	⎫
Tube No. 2, -	1	0·1	0·1	0·8	⎬ Complete hæmolysis.
Tube No. 3, -	1	0·1	0·05	0·85	⎭
Tube No. 4, -	1	0·1	0·03	0·9	⎫ Incomplete hæmolysis.
Tube No. 5, -	1	0·1	0·02	0·9	⎭

Tubes Nos. 1, 2 and 3 alone contain enough complement to produce complete hæmolysis. The quantity in tube 3 will be used because it is the smallest quantity which produces complete hæmolysis.

Antigen.—The organisms to be used as antigen should be prepared as follows : Take a young agar culture (in the case of the cholera vibrio or typhoid bacillus, for instance, a 24 or 48-hour culture) and emulsify one loopful in 2 c.c. of sterile normal saline solution. In carrying out the experiment small quantities only of this emulsion are used because large quantities of albuminoid substances may produce, in the absence of a specific reaction, a mechanical deviation of the complement and so give fallacious results.

The bacillary emulsion should be titrated by placing in the incubator at 37° C. a series of tubes containing progressively increasing quantities of the emulsion, an hæmolytic couple (p. 232) and some complement.

	Emulsion of organisms.	Complement.	Emulsion of red cells.	Heated hæmolytic serum.	Normal saline solution.	Results.
	c.c.	c.c.	c.c.	c.c.	c.c.	
Tube No. 1,	0·05	0·05	1	0·1	0·80	
Tube No. 2,	0·1	0·05	1	0·1	0·75	Complete hæmolysis.
Tube No. 3,	0·2	0·05	1	0·1	0·65	
Tube No. 4,	0·3	0·05	1	0·1	0·55	
Tube No. 5,	0·4	0·05	1	0·1	0·45	Incomplete hæmolysis.
Tube No. 6,	0·5	0·05	1	0·1	0·35	

Examination of the table shows that the quantity of bacterial emulsion to be used must be the amount contained in tube 3 or tube 4, namely 0·2 or 0·3 c.c. these being the maximum doses which do not prevent hæmolysis taking place.

In the serum diagnosis of syphilis, as it is not possible to obtain a culture of the treponeme, various other substances are used as the antigen, *e.g.* an extract of the liver of a syphilitic fœtus. This will be referred to later (*vide* Chap. LIV.).

Bacteriolytic serums.—Bacteriolytic serums are obtained either from immunized animals or from man. A small quantity of blood—4 to 5 c.c.—is sufficient and may be obtained in the case of the human subject by puncture of a vein at the bend of the elbow or with the aid of a Bier's cupping glass; in the case of the rabbit by puncturing an ear vein (p. 194). After being collected the blood is put aside for a few hours, and the serum is then pipetted off and heated in sealed tubes at 55° C. for half an hour to destroy the complement. In carrying out the experiment the bacteriolytic serum should be added to the emulsion of bacteria in sufficient quantity to sensitize them but the actual amount required for sensitization should not be greatly exceeded for fear of introducing errors. The serum can be titrated by a method similar to that used for titrating the antigen (*vide ante*).

Experimental details.—The reagents being prepared, assume that it is desired to determine whether a given vibrio is the cholera vibrio or not. The experiment will be carried out as shown in the table on p. 237.

In this experiment no hæmolysis has taken place in the tubes Nos. 2, 3, 4: therefore the vibrio under investigation was sensitized by the cholera immune serum and was able to combine with the complement. The organism, therefore, is the cholera vibrio.

In tube No. 1 the quantity of vibrio emulsion was not quite sufficient, and this is the reason why the fixation has not been complete.

Examination of the control tubes confirms the diagnosis by proving that in the absence of anticholera serum in one case and in the absence of the cholera vibrio in the other no fixation has occurred, and therefore the tubes show hæmolysis.

If, however, the results had been as follows:

Tubes Nos. 1, 2, and 3 = complete hæmolysis.
Tube No. 4 = slight hæmolysis.

it would have been concluded that the vibrio had not been sensitized by the cholera immune serum; that consequently there was no fixation of the complement; and that therefore the vibrio could not have been the cholera vibrio. The assumed occurrence of partial hæmolysis in tube No. 4 is to be explained as due to a slight excess of bacterial emulsion; the micro-organisms alone having absorbed some of the complement in the manner already described (p. 235).

DETAILS OF A COMPLEMENT-FIXATION EXPERIMENT AS ARRANGED FOR THE IDENTIFICATION OF A SUSPECTED CHOLERA VIBRIO (see p. 236).

	(i) Mix and incubate for one hour at 37° C.				(ii) Add at the end of the hour and incubate again for half an hour at 37° C.		Results.
	Emulsion of vibrios.	Heated anti-cholera serum.	Complement.	Normal saline solution.	Emulsion of red cells.	Heated hæmolytic serum.	
	c.c.	c.c.	c.c.	c.c.	c.c.	c.c.	
Tube No. 1,	0·1	0·1	0·05	0·65	1	0·1	{ Slight hæmolysis.
Tube No. 2,	0·2	0·1	0·05	0·55	1	0·1	
Tube No. 3,	0·3	0·1	0·05	0·45	1	0·1	} No hæmolysis.
Tube No. 4,	0·4	0·1	0·05	0·35	1	0·1	
Control, -	0·1	—	0·05	0·75	1	0·1	} Complete hæmolysis.
Control, -	0·2	—	0·05	0·65	1	0·1	
Control, -	—	0·1	0·05	0·75	1	0·1 ·	{ Complete hæmolysis.
Control, -	—	—	0·1	0·80	1	0·1	{ Complete hæmolysis.

To put the result beyond all doubt it is still necessary to show:

1. That the cholera serum in the quantity in which it was used does not fix the complement in presence of any other species of bacterium, and in an actual experiment an additional control tube would have been introduced containing instead of the vibrio emulsion an emulsion of, for instance, the typhoid bacillus. Hæmolysis should, of course, occur under these conditions.

2. That the vibrio under investigation does not fix the complement in presence of another serum. Another control tube would therefore be prepared with the emulsion of the vibrio but substituting, for example, an antityphoid serum for the anticholera serum. Here again hæmolysis should take place.

The above then is the method of applying the complement fixation reaction to the identification of an unknown organism. The data can be reversed and the reaction applied to determine whether a given serum contains anti-bodies for a given organism. In illustration, an example will now be given to show how to determine whether or no the serum of a patient contain typhoid antibodies (hæmolyso-reaction of Widal and Le Sourd).

In this case the suspected serum after heating at 55° C. is mixed with a known typhoid bacillus and complement. The experiment is arranged in the same way as in the preceding experiment.

DETAILS OF A COMPLEMENT-FIXATION EXPERIMENT AS ARRANGED FOR THE IDENTIFICATION OF A SUSPECTED ENTERIC SERUM.

	(i) Mix and incubate for one hour at 37° C.				(ii) Add at the end of the hour and incubate again for half an hour at 37° C.		Results.
	Emulsion of typhoid bacilli.	Heated suspected serum.	Complement.	Normal saline solution.	Emulsion of red cells.	Heated hæmolytic serum.	
	c.c.	c.c.	c.c.	c.c.	c.c.	c.c.	
Tube No. 1,	0·2	0·1	0·05	0·55	1	0·1	} No hæmo-lysis.
Tube No. 2,	0·3	0·3	0·05	0·25	1	0·1	
Control, -	0·3	Normal human serum. 0·3	0·05	0·25	1	0·1	{ Complete hæmolysis.
Control, -	0·3	—	0·05	0·55	1	0·1	{ Complete hæmolysis.
Control, -	—	—	0·05	0·85	1	0·1	{ Complete hæmolysis.

In tubes Nos. 1, and 2 no hæmolysis has occurred ; the typhoid bacillus it is evident has been sensitized by the serum under examination, which must therefore have come from a patient infected with the typhoid bacillus.

Examination of the control tubes shows that the bacillus alone or in presence of normal human serum is unable to fix the complement with the result that hæmolysis has taken place.

If, on the other hand, the serum under investigation had not sensitized the typhoid bacillus hæmolysis would have occurred, and the inference would have been that the serum contained no typhoid antibodies.

Practical applications.

The method of complement fixation has been applied to the diagnosis of a large number of micro-organic diseases (enteric fever, cholera, dysentery, tuberculosis, etc.) and to the identification of most micro-organisms. It is also the basis of **Wassermann's reaction** in syphilis which will be considered later (Chap. LIV.).

SECTION V.—OPSONINS.

In studying the bactericidal properties of serums it has been mentioned that many micro-organisms resist, *in vitro*, the combined action of immune body and complement, but that in the tissues once impregnated with these substances they more readily become the prey of the phagocytes (Metchnikoff).

Wright and Douglas have shown that in the serum of persons convalescent from infectious diseases or vaccinated against these diseases substances are present which prepare micro-organisms for the action of the phagocytes. Without committing themselves to an expression of opinion as to the nature of these substances Wright and Douglas describe them as opsonins (ὀψωνέω I prepare). Neufeld has applied to them the name *Bacteriotropins*.

According to Wright, opsonins play a fundamental part in the phenomena of immunity : he affirms that it is to opsonins that phagocytosis is due and that by means of the opsonic index of the serum it is possible to measure the immunity of the individual and foresee recovery.

Metchnikoff has observed that, as a matter of fact, the ingestion of micro-organisms by phagocytes rendered possible by the intervention of opsonins is only one factor in the problem. Ingestion is only of use in so far as it is followed by destruction and digestion of the organisms. But micro-organisms are not destroyed by leucocytes unless the latter contain bactericidal substances or in other words unless the leucocytes are "living and strong." Resistant micro-organisms may live for a long time in insufficiently active leucocytes without setting up disease but when such leucocytes are destroyed the micro-organisms are set free and exhibit their powers of producing disease. A notable instance of this is seen in the case of the spores of the tetanus bacillus (Chap. XXXVI.). The opsonic content of the serum is not therefore—at any rate in all cases—a sufficient datum upon which to evaluate the degree of resistance of the tissues.

However that may be, opsonins are of sufficient interest in the study of micro-organic diseases and immunizing processes to merit some detailed consideration.

To determine the opsonic content of a given serum for a particular micro-organism, the serum and a culture of the organism are mixed with normal leucocytes and after an interval the average number of micro-organisms ingested by each leucocyte under these conditions calculated. The number of organisms ingested by 50 leucocytes is counted and the total divided by 50 gives the *opsonic power* of the serum.

It is obvious, of course, that the number of organisms phagocyted will depend upon the number of bacteria present in a unit volume of the emulsion. The absolute number obtained—*i.e.* the opsonic power—is therefore of no value in itself, but if this number be compared with the number which represents the opsonic power of a normal serum determined under the same conditions with the same bacterial emulsion then a standard of comparison is obtained ; and the relation of the opsonic power of the serum of an infected individual to that of a normal individual (measured under identical conditions) is known as the *opsonic index*.

The amount of opsonin present in the serum of normal individuals is subject to considerable variation and is dependent upon many factors, *e.g.* the period which has elapsed since food was last taken, pregnancy, etc. (Milhit).

The amount of specific opsonin in the serum of infected persons shows very curious variation. In tuberculosis, for example, if the opsonic index of normal blood for the tubercle bacillus be taken to be about unity that of infected persons is much lower, and a condition of tuberculosis may be diagnosed in every case in which the opsonic index falls below unity (0·3 to 0·8), provided that the experiment be done several times and the same result is obtained on every occasion. The reaction is much more reliable in the "surgical tuberculoses" than in pulmonary tuberculosis.

In suspected cases of enteric fever an opsonic index above 1·7 for the typhoid bacillus affords strong presumptive evidence in favour of enteric fever (Milhit). Similarly in cerebro-spinal meningitis the opsonic index is raised above normal during the course of the disease ; and so on.

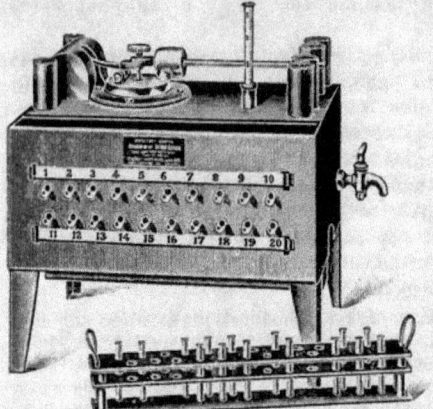

FIG. 161.—Opsonic incubator.

Method of determining the opsonic index.

The following materials and apparatus are necessary :

Apparatus.—Small test-tubes— Pasteur pipettes fitted with india-rubber teats, these should be made from tubing 4–5 mm. in diameter and be drawn out rather long (fig. 162).—Bulb pipettes fitted with a small india-rubber aspirating tube. —A mechanical centrifuge.—An Hearson's opsonic incubator (fig. 161) heated by gas or electricity and regulated at 37° C.

A supply of sterile normal saline solution and of citrated saline solution must be at hand :—

Sodium chloride,	8·5 grams.
Sodium citrate,	15 "
Distilled water,	Q.S. ad 1 litre.

Normal serum.—Collect some blood in a small centrifuge tube either from an animal by puncture of an ear vein or from man by pricking the finger. (Chap. XII.). Centrifuge the blood at once and decant the serum with a Pasteur pipette. The serum is best collected when the individual is fasting (Milhit) : it should be used as soon as possible and always within 2 or 3 hours of collection because the opsonic power rapidly diminishes.

Patient's serum.—This should be collected under the same conditions and at the same time as the normal serum in order that the results may be comparable.

Bacterial emulsion.—[When the rate of growth permits] very young cultures should be used (8–24 hours old according to the organism), and for the same series of experiments cultures of the same age and prepared under identical conditions are necessary.

As a rule, agar cultures are used and one loopful is made into an emulsion with 2 c.c. of normal saline solution. The emulsion must be carefully prepared ; it should be slightly opalescent when held up to the light and must be perfectly homogeneous. If too thick an emulsion be used it will be difficult to enumerate the organisms in the subsequent part of the experiment.

Leucocytes.—**1.** Cleanse the skin of the thumb, tie an india-rubber ligature around its proximal end and then make several little pricks on the dorsal surface of the distal end near the root of the nail.

2. Allow about thirty drops of blood to flow into a small sterile centrifuge tube containing about 10 c.c. of citrated normal saline solution (fig. 147, p. 192).

3. Mix the blood and citrate solution carefully, centrifuge the mixture for 15 minutes, decant the supernatant liquid with a bulb pipette, add an equal volume of normal saline solution to the deposit, mix and centrifuge again. Then decant the saline solution and wash a second time.

4. After washing three times decant the supernatant liquid being careful not to stir up the deposit of cells. Lay the tube as nearly horizontally as possible and leave it for about half an hour. Then collect the upper whitish layer of cells which is composed almost exclusively of leucocytes. The leucocytes ought to be used within 6 hours of the blood being collected (Milhit).

Experimental details.

1. Take a Pasteur pipette ready furnished with an india-rubber teat, cut off the capillary end squarely with a carburundum pencil and make a small mark on the glass about 2 cm. from the point.

FIG. 162.—Preparation of the mixture for the determination of the opsonic index. The figure shows the three equal volumes of fluid aspirated into the pipette and separated by two bubbles of air.

2. Aspirate into the pipette in turn by lightly relaxing the teat:

(*a*) A column of leucocytes up to the mark on the glass, then a small bubble of air.

(*b*) A column of bacterial emulsion up to the same mark, then another bubble of air.

(*c*) A column of the serum to be examined, again up to the mark.

There are now three equal volumes of fluid in the pipette separated by two small bubbles of air (fig. 162).

3. Expel the liquids on to a sterile slide and mix them together, then draw up the mixture into the pipette again, being careful to avoid taking up any air bubbles. Seal the end in the pilot of a Bunsen, and place the pipette horizontally in the opsonic incubator at 37° C. for 15 minutes.

4. Break off the end of the pipette. Place a drop of the mixture on each of several slides, spread rapidly, dry and fix the films by heat or alcohol-ether. Stain with carbol-thionin or in the case of the tubercle bacillus with carbol-fuchsin.

5. With an oil-immersion lens count the number of the organisms phagocyted by 50 leucocytes.

For example, 120 organisms are counted in 50 leucocytes: the opsonic power of the serum examined is therefore $\frac{120}{50} = 2\cdot40$.

6. Repeat the experiment using normal serum. Suppose 90 organisms are counted in 50 leucocytes: the opsonic power of the normal serum is $\frac{90}{50} = 1\cdot80$.

(For clearness of description the two investigations—the opsonic power of the suspected serum and that of the normal serum—have been described successively. In practice, of course, they will be taken in hand together.)

7. The opsonic index being the ratio of the opsonic power of the suspected serum to that of the normal serum is :

$$\frac{2\cdot40}{1\cdot80} = 1\cdot33.$$

PART II.

THE BACTERIA.

CHAPTER XV.

BACILLUS DIPHTHERIÆ.[1]

THE diphtheria bacillus was discovered by Klebs but the first complete description of the organism was contributed by Lœffler, while the specific relationship of the bacillus to the disease was established by Roux and Yersin who experimentally produced symptoms of paralysis in animals.

Distribution of the diphtheria bacillus.
1. In man.

The Klebs-Lœffler bacillus is found in the false membranes of faucial, nasal and cutaneous diphtheria, and in croup. Inflammatory conditions of the throat in which no false membrane is formed are also sometimes due to the diphtheria bacillus, and in these cases the true nature of the disease can only be determined by bacteriological examination.

The bacillus is generally localized in the false membrane or on the infected mucous membrane : it does not as a rule invade the tissues : death is the result of a true

---

[1] The diphtheria bacillus with the pseudo-diphtheria bacillus, the xerosis bacillus and the bacillus of glanders, are by German writers classified together as the Corynebacteria, and known respectively as the *C. diphtheriæ, C. commune, C. conjunctivæ* and *C. mallei.* The group is characterized by the presence of metachromatic granules and club-shaped swellings at the ends of the organisms, and by the appearance of branched forms in old cultures.

intoxication. In a few severe cases of diphtheria however the organism has been found after death in the blood and internal organs (Babès, Spronck, and others): and it is frequently found in the broncho-pneumonic patches which follow an attack of croup (Lœffler, Kutscher).

The bacillus is also found in the mouths and nasal cavities of persons who have suffered from diphtheria, sometimes for many weeks after recovery from the disease. [" In 3 weeks about 30 per cent. of diphtheria patients are free from morphologically typical diphtheria bacilli. In 20 per cent. the bacilli persist for 4 weeks, in 16 per cent. for 5 weeks, and in 11 per cent. for 7 weeks. One per cent. harbour them for 15 weeks and in exceptional cases they remain in the throat for 30 weeks, though even more prolonged periods of persistence are recorded " (Graham-Smith). Fully virulent diphtheria bacilli have been recovered after as long as 335 days (Prip), 230 days (Schäfer), 215 days (Belfanti): these and other observations " conclusively prove that diphtheria bacilli are capable of retaining their virulence during very prolonged persistence in the throats of infected persons " (Graham-Smith).][1]

[Diphtheria bacilli, a very large proportion of which are virulent, are also present in the throats and noses of " contacts "—persons who have recently been in intimate connexion with the disease. It would even appear that less than half the number of individuals in whom the bacillus obtains a lodgment are attacked by the disease. Graham-Smith gives statistics which show that amongst infected families (relatives and attendants) 36·6 per cent. are liable to become infected, while the mean percentage of infection amongst inmates of hospital wards and institutions is 14 per cent. and amongst scholars of infected schools 8·7 per cent.]

Though the fact is denied by several writers there can be no doubt but that the diphtheria bacillus may occasionally be found in the mouths of persons who have not been in contact with diphtheria : [but an investigation in England showed that of 2132 persons who had not so far as could be determined been exposed to infection 0·18 per cent. were found to be harbouring a virulent diphtheria bacillus and 2·62 per cent. non-virulent bacilli, and in the absence of further evidence these figures undoubtedly point to the conclusion " that virulent diphtheria bacilli are seldom if ever present in the throats of healthy persons who have not recently been in contact with cases of diphtheria or infected contacts " (Graham-Smith).]

2. In the lower animals.

[Cows.—According to Klein cows can be experimentally infected with diphtheria, and lesions containing diphtheria bacilli may appear on the teats and udders as a result of the infection : diphtheria bacilli may also be present in the milk after experimental inoculation. From these observations Klein inferred that cows might naturally suffer from diphtheria and that the milk of such cows might be a cause of human infection.

[Klein's experimental results have not been confirmed and most observers hold that there is no evidence that diphtheria is a bovine disease (Graham-Smith). On two occasions however virulent diphtheria bacilli have been recovered from spontaneous lesions of the udder and teats of cows. In one of these cases investigated by Dean and Todd these observers came to the conclusion that though diphtheria bacilli were present, the lesions in the cow were not due to that organism. In the other case Dean and M'Conkey independently isolated the diphtheria bacillus from the lesions in the cow but neither the source of the bacillus nor its relation to the ulcers was determined.

[Horses.—The diphtheria bacillus has only once been isolated from the horse. It was then found by Cobbett in a purulent and slightly sanguineous discharge from the nose of a pony.

[Cats and fowls and other animals.—" Both cats and fowls have frequently been regarded as carriers of the disease, but the bacteriological evidence in support of these statements is unsatisfactory. Instances of natural infection amongst other animals are unknown, though bacilli closely resembling diphtheria bacilli in many of their characters have been found in dogs, guinea-pigs, rats, fowls, turkeys and pigeons " (Graham-Smith).

[These observations on the occurrence of diphtheria in the lower animals may be

---

[1] *The Bacteriology of Diphtheria*, edited by G. H. F. Nuttall and G. S. Graham-Smith; Camb. Univ. Press.

summarized by saying that only once has a true diphtheria bacillus been isolated from an animal spontaneously infected. The disease must therefore be exceedingly uncommon among the lower animals, and statements to the effect that a case has been observed, or an outbreak of diphtheria traced to infection from the lower animals, should not be accepted without rigorous investigation.]

3. In the circumambient media.

The diphtheria bacillus is able to live outside the human body. Park, Wright and Emerson found the organism in the dust of diphtheria wards and on the clothes of the attendants. Abel also found it on the toys with which children infected with diphtheria had been playing.

Bacillus pseudo-diphtheriæ.

The pseudo-diphtheria or Hofmann's bacillus, an organism in some respects closely resembling the diphtheria bacillus but differing from it in being shorter and non-pathogenic to laboratory animals and in other particulars, is fairly frequently met with in the mouths of healthy persons.

Reference is again made to this organism later in the present chapter (p. 273).

SECTION I.—EXPERIMENTAL INOCULATION.

1. Symptoms and lesions produced in animals susceptible to infection.

(a) Guinea-pigs.

The guinea-pig is the most suitable animal for the study of experimental diphtheria. The organism may be introduced either under the skin, into the peritoneal cavity, into the trachea or on to mucous surfaces.

1. Sub-cutaneous inoculation.—0·5 c.c. of a twenty-four-hour broth culture inoculated sub-cutaneously will kill the animal in 24–72 hours according to the virulence of the organism. Following the inoculation there is a slight œdema at the site of inoculation and a rise of temperature ; the animal shows symptoms of illness and finally dies.

If only a slightly virulent culture be used, the animal may recover ; in that case there is some œdema at the site of inoculation, followed by a slough which heals in course of time. [Similar effects result from the inoculation of sub-lethal doses of fully virulent cultures.] In the œdema at the site of inoculation the bacilli multiply for the first 6 or 8 hours following the inoculation, after which multiplication ceases and their numbers decrease so that *post mortem* relatively few organisms are found in the clear œdematous fluid.

The organisms do not pass into the blood and internal organs. *Post mortem* there is a very marked congestion of the internal organs and especially of the supra-renal capsules, and a large pleural effusion. The fluid is occasionally blood-stained : it contains no bacilli.

2. Intra-peritoneal inoculation.—The symptoms following intra-peritoneal inoculation are less severe than after sub-cutaneous inoculation : death is longer delayed and does not occur till between the fourth and the twelfth day. Over and above the ordinary visceral lesions there is an effusion of fluid into the peritoneal cavity and it is only in this situation that bacilli can be found.

3. Intra-tracheal inoculation.—Tracheotomy is first performed ; the tracheal mucous membrane is then abraded and a portion of a culture of diphtheria bacilli applied. A false membrane forms on the abraded surface which sets up a true condition of croup rapidly followed by death. An essential condition of success is that the mucous membrane be traumatized ; the bacilli fail to develop on the uninjured membrane.

4. Infection of mucous membranes.—False membranes may be produced by applying traces of culture to the scarified surfaces of the conjunctiva or vulva of the guinea-pig.

(b) Rabbits.

The rabbit is far less susceptible to the diphtheria bacillus than the guinea-pig and only succumbs to the inoculation of very virulent cultures.

1. Sub-cutaneous inoculation.—The inoculation of 2 c.c. of a very virulent broth culture leads to the death of the animal in about 5 days. There is an œdema at the site of inoculation : the internal organs are congested and dotted with hæmorrhagic points : the inguinal and axillary glands are swollen, the liver jaundiced and friable and shows fatty degeneration : as a rule the lungs are normal : rarely there is some effusion into the pleural cavities. A sub-lethal dose leads to paralysis, affecting chiefly the hind-quarters.

2. Intra-peritoneal inoculation.—The results of intra-peritoneal inoculation are less severe, and death only takes place after some lapse of time. The lesions are similar to those mentioned above.

3. Intra-venous inoculation.—Following the inoculation of 1–2 c.c. of a virulent culture into an ear vein death takes place in from 30–60 hours. At the *post-mortem* examination an acute nephritis in addition to the ordinary lesions is found : the organisms do not multiply in the blood stream being rapidly taken up by the phagocytes (Métin).

4. Inoculation on a cutaneous surface.—Roux and Yersin obtained excellent examples of false membranes by blistering a small area on the internal surface of the ear and then applying a trace of a culture of the diphtheria bacillus to the exposed dermis. It is essential that the infected surface should not be allowed to dry ; the ear may be enclosed in a small rubber bag. care being taken that the vessels are not compressed at the base. To stop the development of the membrane it is only necessary to uncover the ear.

5. Intra-tracheal inoculation.—A typical condition of croup is produced as in the guinea-pig but more easily.

6. Inoculation on mucous surfaces.—The results are the same as in the guinea-pig.

(c) Dogs.

The dog is susceptible to infection with the diphtheria bacillus.

1. Sub-cutaneous inoculation.—Death ensues in 3 or 4 days. Roux and Yersin noted œdema at the site of inoculation, jaundice and finally a progressive paralysis : the fluid of the œdema contained bacilli but the blood was sterile.

2. Intra-tracheal inoculation.—Roux and Yersin produced a swelling of the neck, jaundice, complete paralysis, and death on the fourth day. *Post mortem* no false membrane was found in the trachea.

(d) Cats.

Death follows sub-cutaneous inoculation in 6–13 days. A cat fed on milk from a cow [said to be] suffering from diphtheria with ulcers on the udder also contracted the disease (Klein).

(e) Cows.

Klein [claims to have] shown that the cow can contract diphtheria both by spontaneous infection and as a result of experimental inoculation (*vide ante*).

Cows inoculated with a young and virulent culture of the diphtheria bacillus died with congestion of the internal organs ; but using agar cultures several days old, Klein was unable to set up a fatal disease. This observer on several occasions noted an eruption on the udder [of cows which had not been experimentally infected] which commenced as a papule, became a vesicle, then a true pustule and finally an ulcer : the diphtheria bacillus was found in the vesicles and was traced on many occasions into the milk. Klein observed a similar eruption on two cows which succumbed to the inoculation of a very virulent culture. [Other observers have failed to confirm these experiments.]

(f) Birds.

Pigeons and fowls rapidly succumb to the inoculation of a broth culture of the diphtheria bacillus injected sub-cutaneously or into the pectoral muscles in doses of 1 c.c.: death takes place in less than 60 hours. With doses of less than 0·2 c.c. the animal usually recovers ; sometimes paralysis is observed. *Post mortem* a thin greyish film and a gelatinous œdema is found around the site of inoculation. When the culture has been inoculated into the muscle the latter is swollen and its fibres have an ochre tint : the internal organs are intensely congested.

Following inoculation of the bacillus into the larynx these animals suffer from croup as do rabbits.

Small birds (sparrows, chaffinches, etc.) are highly susceptible and rapidly succumb to sub-cutaneous inoculation.

(g) Rats and mice are immune to diphtheria.

To sum up, *it is characteristic of the diphtheria bacillus that it cannot penetrate the tissues of susceptible animals ; it remains localized at the site of inoculation and even in this situation its development is quickly arrested, so that passage through a series of animals rapidly becomes impossible.*

2. Influence of other organisms on the clinical course of the disease.

Roux and Yersin have shown that certain other organisms may be associated with the diphtheria bacillus and at times play an important part in the clinical manifestations of the disease. The diphtheria bacillus is rarely found in pure culture in the false membranes : occasionally the organisms associated with it are few in number and of no clinical importance, but it often happens that a considerable number of other bacteria are present many of which play an important rôle in the clinical course of the disease and to some extent determine its severity.

Martin has pointed out that the mere presence of a few other organisms with the diphtheria bacillus in cultures sown with the material from a case of diphtheria is not sufficient evidence upon which to base a diagnosis of secondary infection ; such an infection can only be diagnosed when the number of other organisms is very considerable. And it is in this connexion that a preliminary microscopical examination of the material from the throat is of great importance, because it can then be determined whether any associated bacillus has multiplied therein and whether it is or is not the predominant organism present : whereas when the material is sown on culture media one species of organism (perhaps sparsely represented originally) may outgrow all others, and moreover the presence of anaërobic organisms may pass entirely unnoticed. It must be remembered that the surface of the membrane may be contaminated by the different organisms of the mouth, and must therefore be cleansed before material is removed for investigation.

The following are the most important of the organisms which may be found in association with the diphtheria bacillus :—

(i) **Brisou's coccus.**—Roux and Yersin and also Martin drew attention to a small coccus which they frequently found associated with the diphtheria bacillus ; they called it the *Brisou coccus* after the name of the child from whom they first isolated it. The organism occurs either in the form of single cocci or as diplococci or in clusters. It is gram-positive. On coagulated serum the colonies are small, whitish in colour and almost transparent, slightly raised and circular. It is not pathogenic to laboratory animals. Usually, though not always, the association of this organism with the diphtheria bacillus is unimportant from the point of view of prognosis.

(ii) **Staphylococcus pyogenes.**—Staphylococci constitute a more serious complication than the preceding : respiratory complications are frequent. In a case in which the *staphylococcus aureus* was associated with the diphtheria bacillus the author observed a considerable swelling of the neck during convalescence.

(iii) **Streptococci.**—According to Martin a secondary infection with streptococci produces the most severe form of diphtheria ; broncho-pneumonia frequently supervenes in cases where streptococci are found with the diphtheria bacillus.

Métin demonstrated by experiment upon guinea-pigs the unfavourable influence of staphylococci and streptococci upon the course of the infection : he found that the diphtheria bacillus in such cases multiplied in the blood stream and in the internal organs and was present in enormous numbers at the site of inoculation.

(iv) **Bacillus coli.**—The colon bacillus is not infrequently found in the mouths of healthy persons, and it is therefore to be expected that it should be found in the false membranes in some cases of diphtheria. The multiplication of this organism is a serious complication ; three cases recorded by Blasi and Russo-Travalli terminated fatally. These observers grew the diphtheria bacillus and the colon bacillus together and showed that the toxicity of diphtheria cultures was increased considerably.

(v) **Other organisms.**—Association with the *pneumococcus*, with the *pneumobacillus* of Friedländer, with the *proteus vulgaris*, Vincent's *bacillus fusiformis* and the anaërobic organisms of the mouth (Chap. XXXIX.) has also been recorded.

SECTION II.—MORPHOLOGY.
1. Microscopical appearance.

The diphtheria bacillus is a highly pleormorphic organism ; it is non-motile, and does not form spores. [In general terms it may be described as a small, slender, straight or slightly curved, usually irregularly-staining rod with rounded and sometimes swollen ends : curved bacilli with swollen ends, resembling a gherkin in appearance, are very characteristic. In size it is subject to considerable variation.]

Various attempts have been made to classify the different varieties of diphtheria bacilli. Three varieties may be distinguished depending upon the length of the organism, viz.

(a) *Short bacilli* almost like cocci. They measure about $2\mu \times 0.8\mu$ and often occur in pairs arranged parallel to one another (p. 254).

(b) *Bacilli of intermediate size*, measuring $3-4\mu \times 0.8\mu$. These bacilli are arranged parallel to one another or are found in pairs end to end, the latter often forming an acute angle like the letter V or a circumflex accent.

(c) *Long bacilli* $4-5\mu$ or more in length. In culture they are seen to be interlaced and without definite arrangement, very like brushwood. These bacilli generate the most potent samples of toxin (p. 257) and are usually found in severe cases of diphtheria. On the other hand, the short bacilli (group a) are as a rule almost avirulent (Martin).

[Another basis of classification of diphtheria bacilli is that worked out by Cobbett. This observer relied on the staining reactions alone and was thus able to distinguish five groups of diphtheria bacilli in young serum cultures.

(Size and form were as will be seen found to be in close relation to similarity in staining reaction.)

(α) Irregularly beaded bacilli—long and faintly stained—the type most frequently seen.

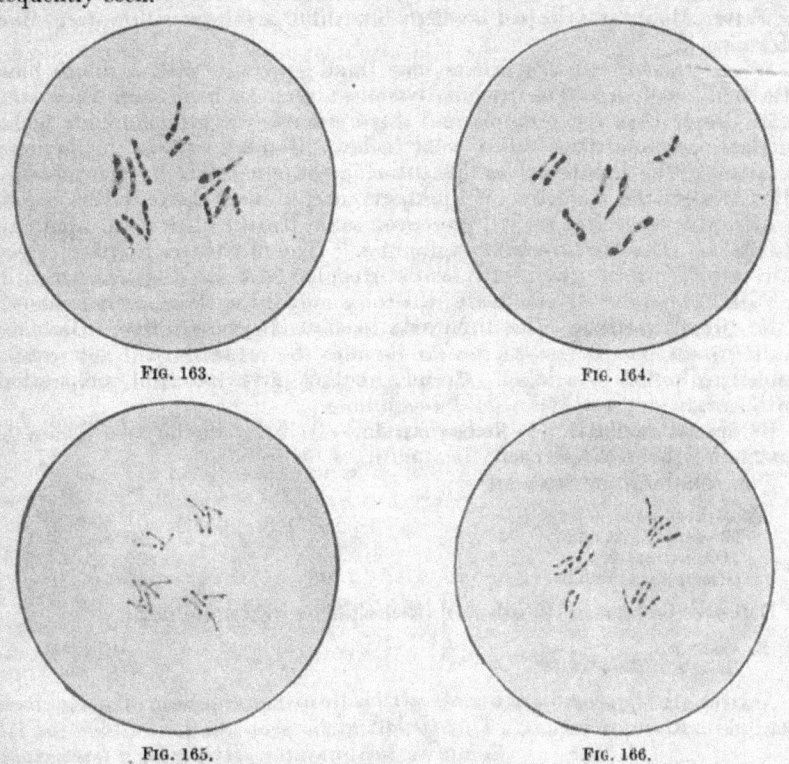

FIG. 163.

FIG. 164.

FIG. 165.

FIG. 166.

FIGS. 163, 164, 165, and 166.—Types of diphtheria bacilli from young serum cultures.
Mounted in dilute Lœffler's blue (1 in 5 with water). Oc. 4; obj. $\frac{1}{12}$th, Zeiss.

(β) Regularly beaded bacilli—streptococcal forms—stain darkly and may be mistaken for streptococci.

(γ) Barred, segmented or banded forms.

(δ) Uniformly stained bacilli.

(ε) Oval bacilli with one unstained septum.

[This last type is found in greatest numbers in very young cultures: they are probably young forms, and it is to be noted that individuals of this type are—morphologically—practically indistinguishable from typical forms of Hofmann's bacillus.]

Branched forms of the diphtheria bacillus have been described by Babès, Escherich, Concetti and others.

Involution forms in which one or both ends are swollen giving the organism a pear-shaped, clubbed, or dumb-bell-like appearance are sometimes met with in old cultures and in smears made from the false membrane.

Staining methods.

(a) The diphtheria bacillus stains readily with the basic aniline dyes. Films from cultures or membranes may be stained with diluted Lœffler's alkaline methylene blue [1 to 4 of water], Roux's blue or with dilute Ziehl's fuchsin. [The first of these is the stain strongly recommended.]

[The method recommended is that devised by Cobbett. Spread films, dry and fix in the ordinary manner. Wash in 10 per cent. acetic acid. Wash in water. Mount in a drop of Lœffler's blue diluted 5 times with water. Blot. Examine.]

When stained with methylene blue [and especially with a dilute blue] the bacilli are found to be irregularly stained, granules being seen which stain more deeply than the protoplasm : these granules or metachromatic bodies of Babès are sometimes called polar bodies. German writers, in the determination of the diphtheria bacillus, attach great importance to their presence. [But though the majority of diphtheria bacilli show these Babès' bodies, some types stain uniformly ; moreover some bacilli other than diphtheria bacilli also show deeply-staining granules.] The diphtheria bacillus, especially in old cultures, frequently shows irregular vacuolated spaces which do not stain, whatever dye be used ; it is to be noted that these are not spores.

(*b*) **Gram's method.**—The diphtheria bacillus is gram-positive. Decolourization must not be pushed too far because the organism will not resist a prolonged action of alcohol. Gram's method gives beautiful preparations with smears and sections of false membranes.

(*c*) **Special methods.** (*a*) **Neisser's stain.**—To bring out the polar bodies the method of Ernst-Neisser may be applied.

Two solutions are necessary—

A. Methylene blue (Grübler),	1 gram.
96 per cent. Alcohol,	20 c.c.
Distilled water,	950 ,,
Glacial acetic acid,	50 ,,

Dissolve the blue in the alcohol, then add the water and acid.

B. Vesuvin,	0·5 gram.
Boiling distilled water,	250 c.c.

A cover-glass preparation is made with a drop of an emulsion of an eighteen-hour-old culture on serum. This is left in the acid solution of blue for 1–3

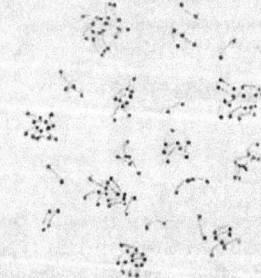

hours, washed in water, stained for a few seconds in the vesuvin solution, washed again and mounted. [The method may be modified by staining for 1 minute in each of the blue and brown solutions, washing in water between the two operations.]

So stained the diphtheria bacillus is brown, with deep blue [or violet] granules situated as a rule at the poles or ends. (The pseudo-diphtheria [Hofmann] bacillus, on the other hand, according to Neisser, never shows polar bodies, but is stained uniformly brown or has some blue grains irregularly distributed through the body of the organism. It is now admitted that this reaction is not characteristic.)

FIG. 167.— Diphtheria bacilli stained by Neisser's method. Oc. 4 Obj. 𝒜th Zeiss.

[(*β*) The following is a modification of Neisser's stain which apparently gives better results than that just described.

Prepare two solutions :

A. Methylene blue (Grübler),	1 gram.
Absolute alcohol.	50 c.c.
Glacial acetic acid,	50 ,,
Distilled water,	1000 ,

B. Crystal violet (Höchst), - - - - - - - 1 gram.
    Absolute alcohol, - - - - - - - - 10 c.c.
    Distilled water, - - - - - - - 300 ,,

1. Stain for 1 second or longer in a mixture made just before use of 2 parts of **A** and 1 part of **B**.

2. Wash rapidly in water.

3. Counterstain for 3 seconds in cresoidin solution.

    Cresoidin, - - - - - - - - - 1 part.
    Warm water, - - - - - . . - - - 300 parts.

When dissolved, filter.

4. Wash in water.]

(γ) For staining the polar bodies Epstein recommends pyronin. Stain for 20 seconds in pyronin, wash, treat with Gram's solution for 10 minutes, dry and mount. The polar bodies are stained bright red, the bacilli pale red.

Note.—The diphtheria bacillus becomes decolourized very quickly when mounted in balsam. For permanent preparations the following method is recommended.

1. Stain in dilute Ziehl's fuchsin for 1–3 minutes. Wash in water.

2. Stain in the same manner with Roux's blue. Wash, dry, and mount.

2. Cultural characteristics.

A. Conditions of growth.—The diphtheria bacillus will grow at all temperatures between 20° C. and 40° C. but not above 42° C., the optimum being 35°–37° C. The bacillus is aërobic : some growth however takes place under anaërobic conditions, but it is poor and the organism rapidly loses its vitality.

B. Media. 1. Broth.—Peptonized-veal-broth gives a better growth than beef-broth.

After incubating at 37° C. for 12–24 hours small white masses of growth will be seen adhering to the sides of the flask and later a film forms on the surface : if examined microscopically this film will be found to consist of tangled masses of bacilli many of the latter being club-shaped. Finally, a deposit forms at the bottom of the tube leaving the supernatant fluid clear. The best way to obtain rapidly a luxuriant growth is to sow the bacillus in a flat flask and pass a current of air over the growth during incubation.

For this purpose select a flask similar to that (Fernbach's) shown in fig. 168. Fill it with broth through the central vertical tubulure until the level of the fluid reaches to just below the openings of the lateral tubulures : plug all three openings with wool, sterilize in the autoclave, sow through the vertical tubulure, replace the wool plug and cover it with an india-rubber cap. Now attach one of the lateral tubes to a water pump and draw a slow current of air through the flask ; the air entering by the other tube sweeps in a continuous stream over the surface of the broth.

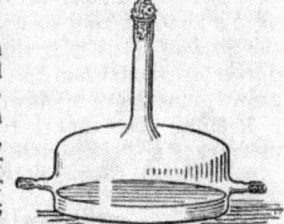

FIG. 168.—Fernbach's flask.

When the diphtheria bacillus is grown in broth the reaction of the latter first becomes acid, but after a few days this initial acidity is converted into an alkaline reaction accompanied by a precipitate of ammonio-magnesium phosphate. In media containing glycerin the acid reaction is very marked and persistent, and the bacilli rapidly die in such media.

Martin's broth (p. 33) is better than ordinary broth and in it growth is very luxuriant, the medium never becomes acid and the virulence of the bacilli is maintained for a long time.

2. Coagulated serum.—On coagulated serum which is the best medium for the diphtheria bacillus growth is very rapid.

(*a*) **Isolated colonies** (grown on the surface).—After 18 hours (at 37° C.) greyish white points are seen which rapidly grow to the size of a pin's head : by transmitted light the colonies are more opaque in their centres than at

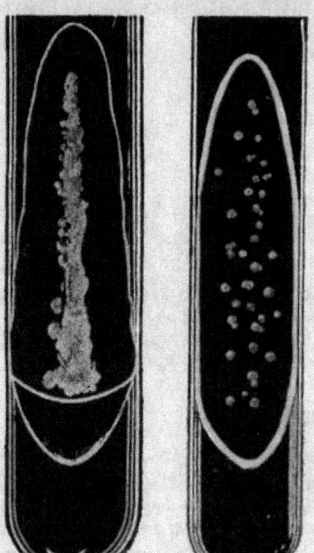

their margins : as growth proceeds they attain a diameter of 5 mm., remain regular and are sometimes pale yellow in colour.

(β) **Stroke cultures.**—Colonies similar to those described above appear along the line of sowing ; these soon become confluent and form a fairly broad greyish band with irregularly serrated margins.

The long, short and medium sized varieties of the diphtheria bacillus cannot be distinguished by the characters of the growth on serum ; colonies of the short variety are, however, sometimes whiter and moister than usual ; these whiter colonies will be found to grow at room temperature, and are in fact colonies of the pseudo-diphtheria or Hofmann's bacillus (*vide infra*).

3. Agar.—Colonies on agar are very similar to those on serum though sometimes larger and whiter. Growth is somewhat slower.

4. Gelatin.—Stab cultures in gelatin (15 per cent.) at 22°–24° C. give rise to a very poor growth consisting of small white punctiform colonies along the line of the stab. The diphtheria bacillus does not liquefy gelatin.

Fig. 169.—Cultures of the diphtheria bacillus.
1. Surface culture on agar (3 days at 37° C.). 2. Isolated colonies on serum (24 hours).

5. Potato.—The diphtheria bacillus does not appear to grow on potato ; some observers have however described a delicate growth consisting of an almost invisible yellowish glaze.

6. Egg albumin.—Sakaroff recommends white of egg coagulated by heat as a medium in place of blood-serum (p. 53, B). When sown on the surface of this medium, and incubated for 24 hours, the diphtheria bacillus grows in the form of small, dull, slightly transparent, hemispherical colonies somewhat darker in colour than the medium : sometimes towards the twelfth day the growth may be of a brown or flesh colour.

7. Milk.—The diphtheria bacillus grows luxuriantly in fresh milk. The medium is not coagulated.

SECTION III.—BIOLOGICAL PROPERTIES.

1. Vitality and virulence.

Vitality.

The diphtheria bacillus remains alive in culture for a considerable length of time : sub-cultures can be grown from a culture 5 or 6 months old.

In moist cultures the bacillus is readily destroyed : exposure to a temperature of 58° C. for a few minutes is sufficient to sterilize a broth culture.

When dried, the bacilli are more resistant to the action of heat and can be subjected to a temperature of 95° C. for several minutes without being killed.

The diphtheria bacillus shows even greater powers of resistance to heat in the false membranes : thus a false membrane may be dried and exposed to a temperature of 95°–100° C. for an hour, and yet the bacilli may be found to have escaped destruction when cultures are sown.

Drying has but little effect on the vitality of the diphtheria bacillus if albuminous matter be present : thus Roux and Yersin dried a piece of membrane and kept it away from the light at room temperature ; portions of it sown 3 months or 5 months later yielded cultures of the diphtheria bacillus. Bacilli from serum cultures are more easily destroyed by drying especially if they be rapidly dried. Light has considerable bactericidal properties : thus if a piece of false membrane be dried and then exposed to air and sunlight the organisms are destroyed in a relatively short space of time. Roux and Yersin, for instance, found no living bacilli in a piece of membrane which had been dried and then exposed to air and sunlight for a period of 6 weeks. A culture on serum which had been dried and spread in a thin layer was found to be sterile after an exposure of 24 hours to diffused light (Ledoux-Lebard).

Antiseptics rapidly sterilize cultures of the diphtheria bacillus : 1 per cent. carbolic acid, 2 per cent. bichromate of potassium etc. kill cultures instantly.

If silk threads be dipped in a culture of the diphtheria bacillus and dried, it will be found that the organisms on the dried threads are more resistant to the action of antiseptics and can withstand the action of 1 per cent. carbolic acid, 5 per cent. salicylic acid in alcohol, etc. for several minutes (Chantemesse and Widal). The resistance of the bacilli in false membranes to antiseptics is even greater.

Virulence.

The virulence of a given diphtheria bacillus must be tested as follows :

Sow the organism in broth, incubate at 37° C. for 24 hours, inoculate 1 c.c. beneath the skin of a guinea-pig weighing 400–500 grams. One of several results may follow :

(α) In the case of a very virulent bacillus death will occur in 24–30 hours.

(β) In the case of a bacillus of intermediate virulence the animal will succumb in 2–6 days.

(γ) In the case of a slightly virulent organism death may not take place for 8–10 days.

(δ) In the case of a bacillus of very low virulence the animal will survive but a local œdema followed by a slough will form at the site of inoculation.

(ε) Finally, should the bacillus be avirulent no lesion whatever will follow the inoculation.

[It will be necessary, as pointed out above, to make certain that the bacillus used for inoculation is growing well in broth.]

The virulence of the organisms isolated from false membranes is very inconstant : in severe cases, virulent bacilli are very numerous : in mild cases in addition to colonies of virulent bacilli, numerous colonies of non-virulent organisms will be found.[1]

[The *pathogenicity* of different strains of the diphtheria bacillus when first isolated, as tested with two-day broth cultures, varies greatly (minimal lethal dose varies as 400 to 1) ; and the *virulence* of washed bacilli from two-day broth cultures of different strains varies at least as much as the *pathogenicity* of whole cultures (Arkwright).]

(a) **Attenuation of virulence.**—In old cultures the diphtheria bacillus loses much of its virulence but the latter can be fully recovered by sowing the

[1] It is not justifiable to assume that because a diphtheria bacillus is non-pathogenic to laboratory animals that it is therefore non-pathogenic to man : on the other hand there is some evidence to show that "non-virulent" diphtheria bacilli are capable of producing diphtheria in a susceptible human subject.]

organism in broth. On the other hand, when the organism is grown at 39° C. on glycerin-agar or on broth with a current of air passing over it, it loses its virulence rapidly, so that on inoculation into a guinea-pig nothing more than a local œdema results (Roux and Yersin).

The same fact may be observed by drying a false membrane from a case of diphtheria and exposing it to the air : the organism remains alive for a long time, but if fragments of the membrane be sown from day to day it will be found that the number of non-virulent colonies increases. The bacilli thus artificially attenuated have all the characteristics of the pseudo-diphtheria—Hofmann's—bacillus (see p. 273). [It would seem that the explanation of this result is to be found in the supposition that both the diphtheria bacillus and Hofmann's bacillus were present in the original membrane and that the former died out. All attempts to convert a diphtheria bacillus into an Hofmann and *vice versa* have invariably failed (p. 274).]

(*b*) **Restoration of virulence.**—It is impossible to restore the virulence of an organism which has become so attenuated as to have entirely lost its virulence for the guinea-pig (Roux and Yersin).

On the other hand, Roux and Yersin succeeded in restoring the virulence of an organism which produced nothing more than a slight œdema in the guinea-pig ; they inoculated the bacillus, and with it a virulent culture of a streptococcus, into a guinea-pig : the animal succumbed with symptoms of diphtheria and the bacillus recovered from the fluid of the local œdema was found to have very distinctly increased in virulence. According to Blasi and Russo-Travalli association with the colon bacillus has a similar effect in restoring the virulence of the diphtheria bacillus. Trumpp, by inoculating a mixture of small doses of diphtheria toxin and an almost non-virulent bacillus, was also able to restore the virulence of the organism.

(*c*) **Exaltation of virulence.**—Roux and Yersin failed to raise the virulence of the diphtheria bacillus by passage through guinea-pigs or rabbits. Bardach after passing the bacillus through twenty-five dogs noted a distinct increase in virulence for the dog but only a slight increase in virulence for the guinea-pig.

Martin grew the bacillus in collodion sacs in the peritoneal cavities of a series of rabbits and succeeded in raising the virulence for that animal, but found that the virulence for the guinea-pig was unaltered.

[Thus the virulence of the diphtheria bacillus, while easily lowered, is difficult to increase by laboratory methods.]

[2. Bio-chemical reactions.]

[(*a*) **Action on carbohydrates.**—" All strains of the diphtheria bacillus produce acid from glucose, galactose, lævulose, and maltose. Most form acid out of dextrine and glycerine. On lactose the action is very variable, and only a few strains act on saccharose. All tests on mannite yielded negative results " (Graham-Smith). No gas is formed in any of the media.

[In testing the action of diphtheria bacilli upon carbohydrates the most suitable medium to which to add the carbohydrate is Hiss's serum-water medium (Chap. XL.). To this solution, 1 per cent. of the carbohydrate is added.

[If sugar-free broth be used the results may not be so uniform, because as has already been pointed out some strains of the diphtheria bacillus do not grow readily on such broth when first isolated from the body.

[(*b*) **Indol.**—The diphtheria bacillus does not appear to form indol (Theobald Smith and others), but some observers are said to have obtained an indol reaction after prolonged cultivation.]

3. Toxin.

Diphtheria, as Roux and Yersin showed, is an intoxication with the highly poisonous products of the diphtheria bacillus. These products, as the same observers proved, are also present in broth cultures of the bacillus.

In their first experiments Roux and Yersin were only able to manufacture a very weak toxin of which 30 c.c. were required to kill a guinea-pig. Martin now prepares a toxin which is fatal to adult guinea-pigs in doses of $\frac{1}{200}$ c.c. and even $\frac{1}{300}$ c.c. [Toxins even more powerful than this have been prepared.]

(a) Preparation of diphtheria toxin.

Conditions under which toxin is elaborated.—Diphtheria toxin is obtained by growing a toxigenic[1] bacillus in presence of air.

A. Selection of the organism.—The strain to be used for the preparation of toxin should first of all be tested on animals. To be suitable for the purpose, 1 c.c. of a broth culture of the organism should when inoculated beneath the skin of a guinea-pig weighing 300–400 grams prove fatal in 24–36 hours. But as it has been found that bacilli isolated from very severe cases of diphtheria are not always powerfully toxigenic [and on the other hand some bacilli are toxigenic and non-virulent] the toxigenic capacity of the organism should always be tested before embarking upon the manufacture of large quantities of toxin. [At the present time a bacillus known as Park and Williams' bacillus No. 8 is extensively used for the preparation of toxin. Park and Williams recovered this organism from a mild tonsillar case of diphtheria.]

To preserve the toxigenic properties of a diphtheria bacillus sow it on Martin's broth, and after incubating for a week at 33°–35° C. remove the tubes from the incubator and keep them in the dark. Martin keeps 48-hour cultures on coagulated serum at 10°–12° C. Old cultures stored in this way on being revived by a couple of sub-cultivations yield a bacillus which has a very considerable power of toxin production.

B. The choice of medium.—The amount of toxin produced depends upon the composition and especially upon the reaction of the medium.

It has already been pointed out that when the diphtheria bacillus is cultivated upon an alkaline broth the medium is first turned acid but after a few days again becomes alkaline; and it is just when the acid reaction begins to diminish that the toxin begins to be formed, after that the toxicity of the broth increases *pari passu* with the alkalinity of the medium, and the more rapidly the alkalinity increases the more rapidly does the amount of toxin increase. If the formation of acid be prevented toxin will be formed both more rapidly and also in larger quantity. As the result of experiment many methods have been devised for diminishing or altogether preventing the initial formation of acid. Roux, Yersin and Martin shortened the period of acid reaction by growing in a current of air, and this method has been used for a long time in the preparation of toxin on a large scale. Many observers have aimed at excluding from the culture medium all those substances (*e.g.* glucose, lævulose, saccharose, galactose, glycerin) from which the diphtheria bacillus forms acid. Nicolle, for instance, obtained a satisfactory toxin by using fresh meat—meat, that is, in which the glycogen had not had time to be converted into glucose (glycogen not being convertible into acid); Spronck on the contrary suggested the use of meat which was

---

[1 The disease-producing power of a diphtheria bacillus may be regarded as made up of two elements, toxicity and virulence. The former represents the rate of accumulation by it of toxin in culture fluids, the latter the behaviour of the bacillus towards living tissue (Theobald Smith).]

rotten and from which the sugars had vanished.   Park and Williams in their investigations used a broth previously made alkaline with soda : Macé employed a broth to which calcium carbonate had been added.   None of these methods however give results as good as those obtainable by the method devised by Martin.   This observer succeeded in finding a medium in which no acid reaction is developed and which yields highly toxic cultures.

1. Method of Roux and Martin.—The bacillus is grown in a current of air, and for this purpose a flask (modified from Fernbach's) similar to that shown in the illustration (fig. 170) is very convenient.

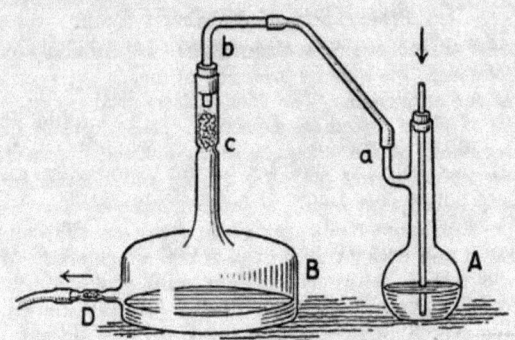

FIG. 170.—Flask arranged for the growth of the diphtheria bacillus for toxin production (Roux and Martin's method).

Pour into each flask 400–500 c.c. of veal broth ; this quantity should not form a layer more than 2 or 3 cm. deep, and the surface of the broth should be below the opening of the lateral tube D.   Plug the lateral tube D and the neck of the flask C with cotton-wool : autoclave the flask and its contents : allow to cool and sow the medium through the neck C.   Incubate at 37° C. and after about 24 hours or so— when the growth is well started and the broth has become cloudy—arrange the flask so that a current of air can be passed over the surface of the broth, thus :—Into the neck of the flask B and above the wool plug fit an india-rubber plug through which a piece of glass tubing, *b*, bent at right angles is passed, and connect this with a second flask A containing a little water through which the air is made to bubble. Attach the tube D by means of a piece of india-rubber tubing to a water pump. When the water is turned on air is bubbled through the water in A—where it is saturated with moisture—and drawn over the surface of the culture in B which is thus aërated.   By means of a clip on the tube connecting the flasks A and B the amount of air can easily be regulated so that a constant but not violent stream of air can be drawn through the flask.

After incubating for 3 or 4 weeks the culture is sufficiently rich in toxin. At the bottom of the flask there is a deposit of micro-organisms and on the surface a thin layer of young bacilli.   The reaction is strongly alkaline.

The culture is now filtered through a Chamberland bougie by one of the methods already described (Chap. I.).   The filtrate kills adult guinea-pigs in doses of 0·1 c.c.

2. Martin's method.   Method recommended.—The need for the current of air is obviated, and hence also the necessity for complicated apparatus : it is quicker than the method just described and yields moreover a more powerful toxin.   A dose of $\frac{1}{300}$th c.c. of this toxin is sufficient to kill a guinea-pig.

Martin uses a peptonized veal broth (p. 33) sterilized by filtration [1] and

---

[1] If sterilized at 120° C. this medium does not give such good results.   It is better to filter or, failing filtration, to sterilize on three successive days at 100° C.

distributed in thin layers (3–4 cm. deep) in large flasks. The organism soon becomes accustomed to the medium, and grows well in it.

Character of growth.—A film forms on the surface during the first 24 hours and increases both in area and in thickness during the next 24 hours (reject all organisms which do not form a film about the end of the third day). If the culture is growing well the film breaks up and falls to the bottom ; a new film then forms and this in turn sinks to the bottom about the sixth day, after which no further film is formed. The medium is never acid to litmus, but on the other hand about the second to the fourth day it is alkaline to phenol-phthalein.

The culture should be filtered about the end of the first week when its toxicity is at its maximum ; after about a fortnight the toxin content begins to diminish.

3. [G. Dean's method. Recommended.—This is a less complicated method than Martin's and is somewhat similar to that described by Park and Williams.

[The broth is prepared with "silverside" of beef and the meat may be used either perfectly fresh or after hanging for 7–12 days at a temperature of 8° C. Free the beef from all fat and fascia and then mince very finely. To each pound (about 500 grams) of beef add 1 litre of fairly alkaline tap water. Put the meat and water into an enamelled saucepan, cover with the lid, and allow to boil quietly for $\frac{1}{2}$–2 hours. Filter through Swedish filter paper, thoroughly squeezing out all the juice from the beef. To the filtrate add 2 per cent. Witte's peptone and 0·5 per cent. sodium chloride. Steam at 100° C. for 1 hour. Filter. Make the filtrate neutral to litmus and then add 7 c.c. of normal soda per litre while still hot. Steam again for 1 hour. Filter. Distribute in Erlenmeyer flasks. Steam at 100° C. for 20 minutes and then allow the temperature to run up to 120° C. before turning out the gas.

[The toxicity in the case of 39 toxins prepared by Dean with this medium and the American bacillus (p. 257) is shown in the following table.

$$
\begin{array}{llll}
1 & \text{killed within 6 days at} & & \frac{1}{700}\ \text{c.c.} \\
1 & \text{ ,, } & \text{ ,, } & \text{ ,, } & \frac{1}{600}\ \text{ ,, } \\
8 & \text{ ,, } & \text{ ,, } & \text{ ,, } & \frac{1}{500}\ \text{ ,, } \\
9 & \text{ ,, } & \text{ ,, } & \text{ ,, } & \frac{1}{400}\ \text{ ,, } \\
7 & \text{ ,, } & \text{ ,, } & \text{ ,, } & \frac{1}{300}\ \text{ ,, } \\
7 & \text{ ,, } & \text{ ,, } & \text{ ,, } & \frac{1}{200}\ \text{ ,, } \\
1 & \text{ ,, } & \text{ ,, } & \text{ ,, } & \frac{1}{100}\ \text{ ,, } \\
5 & \text{did not kill at} & & \frac{1}{200}\ \text{ ,, }]
\end{array}
$$

4. Other methods.—The methods of Spronck, Park and Williams, Nicolle, and Macé which are in use in some laboratories are less reliable than those given above.

Spronck's first method.—This was based upon the absence of glucose in stale meat. Prepare 2 per cent. peptone broth in the ordinary way but with meat which has been hanging for some days until it has acquired a slight smell. Make alkaline and then add 0·5 per cent. salt and a little calcium carbonate. Distribute in quantities of 300–400 c.c. in half litre flasks. Sterilize. Sow when cool. Incubate at 37° C. for 3–4 weeks. Filter.

Spronck's later method.—This was based upon a possibly beneficial action of yeast in promoting the production of toxin.

Revive the bacillus by sowing first on coagulated blood serum and then on peptone yeast water (p. 37). Sow from the latter on to a shallow layer of peptone yeast water contained in a large flat flask. After incubating for 24 hours the growth has formed a continuous pellicle on the surface and at the end of a week the content of toxin is at its maximum : a dose of $\frac{1}{200}$ c.c. suffices to kill a guinea-pig. Spronck does not use a porcelain filter but adds 3 grams of carbolic acid per litre and filters through paper.

Massol's method.—Proceed as in Spronck's method but use the following medium :

High veal, - - - - - - - -	500 grams.
Peptone (Witte), - - - - - - -	20 ,,
Water, - - - - - - - - -	1000 ,,

Neutralize. Add 7 c.c. normal soda solution. Filter through filter paper. Sterilize by filtration through a Chamberland bougie.

Nicolle's method.—Use beef killed the same morning. Mince the meat. Add twice its weight of water and allow to stand for an hour at 10° or 12° C. Add 2 per cent. peptone and 0·5 per cent. common salt. Heat to boiling point. Filter. Make "sufficiently" alkaline. Heat to 120° C. for 10 minutes. Filter. Distribute in sterile plugged vessels. Heat to 115° C. for 15 minutes.

Grown in this medium for 7 days at 37° C. the diphtheria bacillus will yield, without having a current of air passed over the culture, a toxin quite as powerful as that obtainable by Roux and Martin's method.

Macé's method.—Use ordinary peptone broth containing in addition 10 per cent. calcium carbonate. Distribute in litre or two litre flasks and autoclave. Incubate after sowing for 4–6 weeks at 37° C. The product is said to be equally as toxic as the filtrate prepared by Roux and Martin.

Park and Williams' method.—[*vide ante* Dean's method.] These observers grow the bacillus on ordinary peptone broth made alkaline to the extent of 7 c.c. normal soda solution per litre (p. 31).

Protein-free media.—Utchinsky has shown that it is possible to get toxin by growing the organism on media containing no protein and consisting merely of salts and asparagin (p. 39). Hadley prefers to substitute glycocoll (1 gram per litre) for the asparagin in Utchinsky's medium.

These methods always yield a filtrate very weak in toxin. Nicolle, by adding to Utchinsky's medium peptone Chapoteaut and gelatin liquefied by *B. subtilis*, gets a very powerful toxin.

(b) The testing and storing of toxin.

(a) **The testing of toxin.**—The toxin content of the product manufactured with the same bacillus under apparently identical conditions is subject to considerable variation [*vide* Dean's results, p. 259]. It follows therefore that every sample of toxin must be tested.

To be suitable for the immunization of animals (for the purpose of preparing a therapeutic serum) a toxin must kill a guinea-pig weighing 400–500 grams in 48 hours or less when inoculated in quantities of 0·1 c.c. beneath the skin.

The toxins now used are often much stronger than this so that a dose of $\frac{1}{100}$ or $\frac{1}{200}$ c.c. will kill a guinea-pig weighing 500 grams in 36 hours.

For convenience of comparison Ehrlich has suggested the adoption of a unit of toxin. A unit of toxin is the quantity necessary to kill a guinea-pig weighing 300 grams in 96 hours; and a toxin is said to contain 100, 200 etc. units per cubic centimetre.

For measuring small quantities of toxin it is convenient to make dilutions in sterile water; for example, 1 c.c. of toxin added to 9 c.c. of sterile water gives a dilution of which 1 c.c. represents 0·1 c.c. of toxin. [Similarly 1 c.c. of No. 1 dilution added to 9 c.c. of sterile water affords a dilution of which 1 c.c. represents 0·01 c.c. of toxin.]

(β) **To store toxin.**—For the purpose of storing toxin use sterile [amber-coloured] bottles, which must be exactly filled, well plugged and kept in the dark. Even under these conditions the toxin slowly loses its toxic properties.

(c) Action of toxin on animals.

The symptoms which follow the inoculation of diphtheria toxin into susceptible animals are identical with those produced by the inoculation of living cultures of the diphtheria bacillus. It is immaterial whether the toxin be administered by inoculation beneath the skin, into the peritoneal cavity, into the veins or into the brain. But given by the mouth toxin has no effect whatever.

On the guinea-pig.—If a fraction of a c.c. of toxin (0·1–0·25 c.c.—the exact amount depending upon the toxin content of the filtrate) be inoculated beneath the skin of a guinea-pig an œdema rapidly forms at the site of inoculation ; this is soon followed by panting respiration and death super-venes in 20–30 hours.[1]

Post mortem the lesions found are similar to those described as following the inoculation of living bacilli.

Smaller doses (0·005–0·002 c.c.) of a powerful toxin kill guinea-pigs after the lapse of 5–30 days, but if the dose be too small the animal will survive. Paralysis is very rarely seen in guinea-pigs.

On the rabbit.—The administration of 0·25–0·5 c.c. toxin either sub-cutane-ously or into a vein terminates in death accompanied by the usual lesions. If the dose be not large enough to kill the animal very rapidly typical diph-theria paralyses develop.

If toxin be applied to mucous membranes, local lesions and occasionally true false membranes form even though the surface be intact (Roger and Bayeux, Morax and Elmassian).

On the dog.—Dogs are very susceptible to the action of diphtheria toxin. A dose of 1 c.c. sub-cutaneously is sufficient to kill a dog rapidly with symptoms of jaundice and diarrhœa ; lesions will be found in the liver *post mortem*. Smaller doses are followed by paralyses : the animal may recover but if death takes place it does not occur so rapidly as when larger doses are used.

On birds.—Fowls, pigeons and small birds rapidly succumb to very small doses of toxin whether the inoculation be made beneath the skin or into the pectoral muscle.

Ruminants.—Goats are very susceptible to diphtheria toxin ; similarly, cows often succumb to the inoculation of a fraction of a c.c. of toxin. Sheep are somewhat less susceptible.

Horses.—The horse is less affected by diphtheria toxin than ruminants ; but a dose of 0·1 c.c. of a toxin of which the lethal dose for guinea-pigs was $\frac{1}{200}$ c.c. has been known to kill a horse weighing 400 kg. The ass is more susceptible.

Rats and mice are nearly immune to the action of toxin when inoculated sub-cutaneously : the dose of toxin required to kill a mouse would kill as many as 24 to 100 guinea-pigs (Roux and Yersin).

On the other hand intra-cerebral inoculation of 0·1 c.c. of toxin kills rats with symptoms of diphtheria paralysis (Roux and Borrel).

The brain of the rat is therefore sensitive to the action of diphtheria toxin, and the reason why the animal does not die as the result of sub-cutaneous inoculation of large quantities of the poison is because the latter is fixed by certain cells in the tissues (probably by the phagocytes), and so never reaches the cerebrum.

(d) On the nature and properties of diphtheria toxin.

The problem of the nature of diphtheria toxin has been the subject of prolonged and extensive investigations. Brieger and Frænkel as well as Wassermann and Proskauer regarded toxin as a tox-albumin, and Gamalèia considered it to be a nucleo-albumin : but these observers only succeeded in obtaining very impure products containing relatively very little toxin. Roux and Yersin have shown that the active principle in filtered cultures has the chief properties of enzymes. A temperature of 100° C. destroys diphtheria toxin : an exposure to a temperature of 58° C. for 12 hours lowers its toxicity to such an extent that 1 c.c. of the heated toxin fails to kill a guinea-pig ; and the effect of heating to 70° C. is to attenuate the toxin even more. In common with the diastases, diphtheria toxin has the property of being carried

[1 Sub-cutaneous inoculation is always followed by an incubation period before symptoms appear.]

down in the precipitates which can be produced in the solutions in which it is dissolved (*Miahle's reaction*).

By adding a solution of chloride of calcium drop by drop to diphtheria toxin phosphate of lime is precipitated as the result of the combination of the calcium with the phosphates present in the liquid. This precipitate when collected on a filter and washed, is very toxic; the sub-cutaneous inoculation of a mere trace of it rapidly causes death in a guinea-pig accompanied by a swelling at the site of inoculation and the formation of a small greyish false membrane. The precipitate is more toxic in the moist than in dry state. Nevertheless after desiccation it retains most of its toxic properties, and in this condition it is more resistant to the action of heat and can be raised to a temperature of 70° C. without losing any of its toxicity; and further a very small amount of the desiccated precipitate if inserted beneath the skin will kill three guinea-pigs in succession if transferred from one animal to the other as each dies.

After filtering off the first precipitate the clear filtrate is still toxic, and precipitates may be produced time after time; every time the precipitate contains toxin but in a progressively diminishing quantity, until finally the filtrate will no longer produce a precipitate though it is still slightly toxic, as is shown by the fact that when inoculated into guinea-pigs in very large doses, it sets up a chronic intoxication.

Diphtheria toxin is soluble in water but again like the diastases is precipitated by alcohol: but precipitation with alcohol diminishes its toxicity.

To precipitate the toxin it is best to evaporate the filtrate first to one-tenth its volume *in vacuo* at 25° C., and then to add to the liquid extract 4–5 volumes of strong alcohol: the toxin mixed with numerous impurities is carried down in the precipitate. Toxin may also be precipitated by ammonium sulphate.

Toxin obtained by filtration can be dried *in vacuo* to the consistency of a dry extract: this extract is soluble in water and contains the true toxin mixed with a very large proportion of impurities. On dialysis the watery solution quickly loses the mineral salts in solution, but the toxin is only removed with great difficulty. This method may be used for purifying diphtheria toxin.

The toxic content of diphtheria toxin is considerable.

1 c.c. of filtered cultures yields 0·01 gram of dry residue: thus, if 0·005 c.c. of filtered culture suffice to kill a guinea-pig the lethal dose of the dry residue is $\frac{0\cdot01}{200}$ gram (0·00005 gram) and of this small quantity the greater part consists of mineral salts, peptone etc. This will give some idea of how infinitely small is the fatal dose of the real toxic substance.

Many chemical substances alter the toxic nature of diphtheria toxin. For instance, toxin is destroyed by peptic ferments, while alcohol, acids, and antipyrin diminish the toxin content. Oxidizing agents, again, are remarkable for the capacity they exhibit of changing its character: thus, hydrogen peroxide, alkaline hypochlorites and especially iodine and iodine terchloride lower its toxicity considerably. The action of these substances is turned to practical account in attenuating toxin which is to be used for the immunization of animals.

4. Vaccination.

(i) In **laboratory animals** it is difficult to produce immunity by repeated inoculation of very small doses of diphtheria toxin because the toxin accumulates and the animals become cachectic and die.

(a) G. Hoffmann was the first to successfully immunize guinea-pigs. He inoculated them first with cultures attenuated by keeping, and later with fully virulent cultures. Subsequently, Behring and Wernicke employed a similar method.

(b) F, the animals sub-cutaneously with cultures heated for an hour to 65°–70° C.: altogether the animals received from 10–20 c.c. of these cultures at various times. Immunity was acquired at the end of 14 days.

(c) Behring in immunizing guinea-pigs and rabbits used the pleural fluid obtained from guinea-pigs which had succumbed to the inoculation of virulent cultures. Immunity was acquired in about a fortnight after the inoculation of the vaccine, but the results were very inconstant.

(d) Behring immunized guinea-pigs and sheep by inoculating them with cultures 3 weeks old and to which 1 part of iodine terchloride had been added to 500 parts of culture. A few c.c. of this mixture were inoculated into an animal and then 10 days or so later a second inoculation was given of a culture to which a smaller quantity of iodine terchloride had been added. Immunity was acquired in about a fortnight. This method fails in the case of the rabbit.

(e) Brieger, Wassermann and Kitasato conferred immunity on guinea-pigs by inoculating them on several occasions with 2 c.c. of a culture on thymus broth warmed to 70° C. for 15 minutes. But this method is not so effective as the iodine terchloride method of Behring.

(ii) Roux, Nocard and Martin succeeded in immunizing various animals (rabbits, sheep, goats, cows, horses) by inoculating them first with a virulent toxin mixed with Gram's iodine solution, then with gradually increasing doses of pure toxin.

A **rabbit**, for example, was inoculated sub-cutaneously in the first instance with 0·5 c.c. of the following mixture which was prepared immediately before use.

Toxin (Roux and Martin's method),	2 vols.
Gram's iodine solution,	1 vol.

The injection was repeated every few days for some weeks, then the proportion of iodine was gradually diminished and last of all pure toxin was inoculated. The animals were weighed at frequent intervals and if they showed any loss of weight the inoculations were stopped for the time being, otherwise they died of cachexia.

Goats and **cows** may be immunized in a similar manner, but these animals being very susceptible to diphtheria toxin, very small doses of iodized toxin must be used for the initial inoculations, and pure toxin should only be given when the blood shows some content of antitoxin. It should be borne in mind that pregnant animals are more susceptible to diphtheria toxin than non-pregnant animals.

Horses stand toxin well and especial interest attaches to the immunization of these animals because they are the source whence antitoxin for therapeutic purposes is derived.

The horses selected should be young (6–9 years old) well fed and free from disease. After having been tested with mallein to exclude a possible infection with glanders (*vide* Glanders) [and with tuberculin to exclude tuberculosis] the horse is inoculated in the first instance with a small quantity of a virulent toxin to which Gram's iodine has been added: at subsequent inoculations the doses are gradually increased, and after the eighth inoculation pure toxin is used: different animals vary enormously in susceptibility, and care should always be taken that the dose used in the initial experiment shall be so small that no violent reaction results, as this might imperil the steady progress of the immunizing process. The injections should be made sub-cutaneously into the neck or behind the shoulders.

The following table exhibits an actual record of the immunization of an horse by Roux and Nocard.

Horse, 7 years old and weighing about 400 kg.

The iodized toxin contained one-tenth its volume of Gram's solution. The toxin used killed guinea-pigs weighing 500 grams in 48 hours in doses of 0·1 c.c. The injections were made beneath the skin of the neck or behind the shoulder.

1st day of experiment. Injection of 0·25 c.c. of an iodized toxin. No local nor general reaction.

2nd day of experiment. Injection of 0·5 c.c. of an iodized toxin. No local nor general reaction.

4th, 6th, and 8th days of experiment. Injection of 0·5 c.c. of an iodized toxin. No local nor general reaction.

13th and 14th days of experiment. Injection of 1 c.c. of an iodized toxin. No reaction.

17th day of experiment. Injection of 0·25 c.c. of a pure toxin. Slight œdema. No rise of temperature.

22nd day of experiment. Injection of 1 c.c. of a pure toxin. Slight œdema. No rise of temperature.

23rd day of experiment. Injection of 2 c.c. of a pure toxin. Slight œdema. No rise of temperature.

25th day of experiment. Injection of 3 c.c. of a pure toxin. Slight œdema. No rise of temperature.

28th day of experiment. Injection of 5 c.c. of a pure toxin. Slight œdema. No rise of temperature.

30th, 32nd, and 36th days of experiment. Injection of 5 c.c. of a pure toxin. Slight œdema. No rise of temperature.

39th and 41st days of experiment. Injection of 10 c.c. of a pure toxin. Slight œdema. No rise of temperature.

43rd, 46th, 48th and 50th days of experiment. Injection of 30 c.c. of a pure toxin. Fairly well marked œdema which vanished in 24 hours.

53rd day of experiment. Injection of 60 c.c. of a pure toxin. Fairly well marked œdema which vanished in 24 hours.

57th, 63rd, 65th, and 67th days of experiment. Injection of 60 c.c. of a pure toxin. Fairly well marked œdema which vanished in 24 hours.

72nd day of experiment. Injection of 90 c.c. of a pure toxin. Fairly well marked œdema which vanished in 24 hours.

80th day of experiment. Injection of 250 c.c. of a pure toxin. Fairly well marked œdema which vanished in 24 hours.

This horse therefore had received in 2 months and 20 days 800 c.c. of toxin, without showing any symptoms other than a transient local œdema, some loss of appetite and a rise of temperature of about 1° C. on the evenings following the larger injections. The animal was bled on the 87th day and was inoculated into the jugular vein with 200 c.c. of toxin without showing any reaction.

Vaccinated horses withstand equally well enormous doses (many hundred cubic centimetres) of living cultures.

As has been pointed out some horses are more susceptible to diphtheria toxin than others, and in the more susceptible individuals an extensive, firm œdema lasting many days may follow inoculation, and in some cases, the horse may sweat and show a marked rise of temperature.

Occasionally a highly immunized horse will die of paralysis 1–3 weeks after the last inoculation of toxin.

With the very powerful toxins at present in use immunization should be carried out still more carefully. With a toxin containing 200 units of toxin per c.c. horses should be inoculated three times a week for 6 weeks with a mixture of toxin and Gram's solution (commencing with a mixture consisting of 2 parts Gram's solution and 1 part toxin), then with toxin alone in progressively increasing doses: the initial dose being 0·5 c.c. and the final inoculation 200 c.c.

When small doses are inoculated at frequent intervals the antitoxic content of the serum is greater than when large doses are given at longer intervals (Roux).

To maintain horses in a state of immunization it is necessary to inoculate a dose of toxin from time to time : this may be done in different ways.

1. After bleeding a horse 300–500 c.c. of culture may be inoculated at intervals of 20 or 25 days into the jugular vein.

2. At the Pasteur Institute Martin inoculates beneath the skin of the shoulder, 13 days after bleeding the animal, 300 c.c. of toxin ; and 4 days later on the opposite side—also sub-cutaneously—a further 500 c.c. The horse may be bled again a week after the last inoculation.

3. Another method is to inoculate sub-cutaneously every 2 or 3 days for 3 weeks quantities of toxin increasing from 25–150 c.c. until about a litre has been injected. The animal is bled 12 days after the last inoculation of toxin.

(iii) Horses may also be immunized by inoculating them with a mixture of antidiphtheria serum and toxin (Babès, Madsen and Dreyer, Park). The yield of antitoxin is good and the method is more rapid than the iodine terchloride method (Park).

The following table shows the details of an experiment by Martin.

Inoculations twice a week. Toxin killed guinea-pigs weighing 500 grams in 36 hours in doses of 0·1 c.c. Antidiphtheria serum contained 200 units per c.c.

1st inoculation,	-	-	-	-	-	25 c.c. serum + 25 c.c. toxin.		
2nd ,,	-	-	-	-	-	10 ,, ,, + 25 ,, ,,		
3rd ,,	-	-	-	-	-	25 c.c. pure toxin.		
4th ,,	-	-	-	-	-	40 ,, ,,		
5th ,,	-	-	-	-	-	60 ,, ,,		
6th ,,	-	-	-	-	-	80 ,, ,,		
7th ,,	-	-	-	-	-	100 ,, ,,		
8th ,,	-	-	-	-	-	150 ,, ,,		
9th ,,	-	-	-	-	-	200 ,, ,,		
10th ,,	-	-	-	-	-	250 ,, ,,		
11th ,,	-	-	-	-	-	300 ,, ,,		
12th ,,	-	-	-	-	-	350 ,, ,,		
13th ,,	-	-	-	-	-	400 ,, ,,		
14th ,,	-	-	-	-	-	450 ,, ,,		
15th ,,	-	-	-	-	-	500 ,, ,,		

5. Serum therapeutics.
Antitoxin.

Behring and Kitasato in 1890 were the first to demonstrate the antitoxic properties of the blood of animals immunized against diphtheria.

These observers found that the blood of immunized animals had the property of destroying diphtheria toxin both *in vivo* and *in vitro*; that this property was also present in the serum of blood deprived of all cellular elements; and that the serum was both therapeutic and prophylactic when used on rabbits and guinea-pigs intoxicated with diphtheria toxin or inoculated with living diphtheria bacilli.

Having established these facts Behring, Ehrlich, and their collaborators turned their attention to the application of antidiphtheria serum to the treatment of human diphtheria (Behring, Ehrlich, Boer, Wassermann, Rossel). But the serum treatment of diphtheria did not become an accomplished fact in medical practice until after the Congress of Hygiene at Buda-Pesth in 1894 when Roux and Martin communicated a summary of the work they had carried out during the years 1891-4.

(a) Preparation of the serum.

The horse is chosen as the source of antitoxin for these reasons, viz.:— Horse serum, even in large doses, is innocuous to man and to the lower animals; horses withstand the action of diphtheria toxin very much better than other animals; lastly, very large quantities of serum are available (Roux; Nocard and Martin).

The immunization of the horse which is carried out as described above generally occupies about 3 months. In practice toxin is inoculated in gradually increasing doses until some 1000–1500 c.c. have been administered: the final inoculations should consist of quantities of 150–200 c.c.

The animal is bled 8–10 days after the date of the last inoculation: about 6 litres of blood are withdrawn and a further quantity may be taken a few days later. It is best to bleed the horse from the jugular vein according to the directions given on pp. 49 and 50 (Nocard's method and Latapie's apparatus). Six litres of blood yield nearly 4 litres of serum.

The horse is maintained in a state of immunization by the inoculation of toxin from time to time.

When several animals have been immunized it is highly desirable that the serum of the various horses should be mixed ; by doing this a product is obtained of which the antitoxin content is uniform. Moreover, the serum of some horses is liable to provoke erythematous rashes when used in the human subject, which though harmless are nevertheless irritating, and by mixing different serums this inconvenience may be minimized.

For the purpose of storing serum it is distributed with aseptic precautions in small sterile bottles stoppered with sterile india-rubber plugs and kept in the dark.

Serum prepared under strictly aseptic precautions may be kept in these climates many months in a sterile condition without losing any of its antitoxic properties. Occasionally the serum after bottling becomes distinctly cloudy, but this is of no importance with respect either to the purity or efficacy of the serum. A deposit is less likely to occur and the keeping property of the serum is better assured if, immediately after filling, the bottles are heated for an hour at 57° C. in a water bath. This degree of heat has no effect upon the properties of the serum.

Dried serum.—Serum may be dried by evaporation *in vacuo*. Just before use the dried serum is dissolved in eight or ten times its volume of sterile water ; this solution frequently gives rise to a local but transient swelling which is not the case with liquid serum. In these latitudes liquid serum should always be administered in preference to the dry product : the value of the latter is apparent in warm climates where liquid serum quickly loses its properties.

Antitoxic milk.—The milk of immunized females possesses antitoxic properties (Ehrlich). This fact however is merely of theoretical interest because the extreme dilution of the antitoxin in the milk renders the latter incapable of being used in practice.

Still, milk containing antitoxin may be condensed to a sufficiently small volume to allow of laboratory experiments being conducted with it (Wassermann). The milk of cows or goats can be used ; for experimental purposes it is scarcely possible to get a milk with a preventive strength of one-fifth (*vide infra*, p. 268).

(b) Properties of the serum.

The serum of immunized animals is antitoxic, that is to say if the serum be mixed with toxin in suitable quantities the mixture is harmless on inoculation into animals.

This property of the serum is due to a special substance known as *Antitoxin*, the nature of which is as little understood as is the nature of toxin : like toxin, antitoxin is altered by heat, precipitated by alcohol and carried down with the precipitates formed in liquids which contain it in solution. In the living body it is formed in response to the absorption of toxin ; "under the influence of toxin, certain cells of the living body acquire a new and persistent secretory property" (Salomonsen and Madsen).

Antitoxin saturates toxin both *in vivo* and *in vitro* (p. 224) : it has both *prophylactic* and *curative* properties : a guinea-pig inoculated with an adequate dose of serum can withstand the subsequent inoculation of such a quantity of toxin as would be sufficient to kill with certainty a non-inoculated guinea-pig. Even if the toxin be inoculated first and the serum not until several hours later, the animal will be protected. Immunity is rapidly produced but is short-lived : in a few days or weeks it has entirely disappeared. The amount of serum necessary to cure an animal inoculated with toxin depends upon many factors : among others upon the weight of the animal, the amount and toxicity of the toxin used and upon the antitoxic strength of the serum. It is very important to know the antitoxic strength of the serum used, and rules have been devised by which this may be determined (*vide infra*).

Antitoxic serum is *not bactericidal* and contains *no immune body* (*sensibilisatrice*); it has some power of *agglutinating* the bacillus but only in a feeble and inconstant manner. For instance, it may agglutinate in dilutions of 1 in 10 and 1 in 20 and the agglutination is sometimes visible to the naked eye, flocculi falling to the bottom of the tube (Nicolas); on the other hand agglutination is often absent. The serum also of patients suffering from diphtheria sometimes shows some slight property of agglutination (Bruno).

Bacilli recently isolated from the living subject are often unaffected or but slightly affected by the specific agglutinin; prolonged sojourn in artificial culture, however, seems to develop or increase the power of being agglutinated. It is important in testing the agglutinating properties to use a good emulsion of bacilli, and for this purpose a culture in 2 per cent. glucose broth, which is generally cloudy, is suitable. Martin recommends collecting the bacilli from flasks which have been used for toxin preparation and heating them to 100° C. For use, shake up these bacilli well with normal saline solution and let the emulsion stand: a large number of the bacilli will deposit, but the supernatant liquid remains cloudy permanently and serves very well for the agglutination reaction.

Antitoxic serum is both *prophylactic* and *therapeutic* in the case of animals inoculated with a *living culture* of the diphtheria bacillus: the therapeutic properties are exhibited even if the serum be not administered until 12 or 18 hours after the virus.

In the case of a guinea-pig which has been inoculated with a living culture on the mucous membrane of the trachea or vulva, the inoculation of serum, even though it be administered before the virus, will not prevent the formation of the characteristic false membrane, but does entirely prevent symptoms of intoxication or of disturbance of the general health of the animal: and further the false membrane becomes detached on the second day and the infected surface commences to heal. If instead of being inoculated before the living virus, the serum be administered after the false membrane has formed, it leads to the disappearance of the œdema and swelling in a few hours and after two days to the casting off of the false membrane.

The false membranes formed in the trachea of the rabbit as the result of infecting the mucous membrane with a mixture of *streptococci* and *B. diphtheriæ* are not so readily affected by antitoxic serum. In such a case 5, and even 10 c.c. of antitoxin are insufficient to save the life of the animal: but Roux and Martin have been able to effect a cure in parallel cases by repeating the inoculation of antitoxin several times.

Roux and Martin have tested the value in these cases of mixing antistreptococcal and antidiphtheria serums, but with only moderately successful results (*vide* The Streptococci).

(c) The standardization of antitoxin.

1. Behring estimated the antitoxic content of an antiserum in terms of the amount of the serum necessary to immunize 1 gram weight of animal against the minimal fatal dose of toxin inoculated 12 hours after the serum.

Thus, for example, if 1 c.c. of a serum immunized 1 kg. weight of guinea-pig against the inoculation 12 hours later of the smallest dose of toxin which would kill a control animal of the same weight within a given time the serum was said to have a strength of $\frac{1}{1,000}$. This method of titration is not very exact but it has the advantage of being simple.

2. Behring then altered his test inoculation. Instead of toxin he used living bacilli and measured the value of the serum against an infection and not against an intoxication. The unit of serum, now, was the amount necessary to immunize 5,000 grams of guinea-pig (or 10 guinea-pigs of 500 grams each) against the inoculation 24 hours later of ten times the amount of a forty-eight-

hour old culture of the diphtheria bacillus which was certainly fatal to control animals.

[Thus if 0·01 c.c. of a serum would immunize a guinea-pig weighing 500 grams against 10 times the lethal dose of a 48 hour old culture of a virulent diphtheria bacillus, it follows that 0·1 c.c. would be required to immunize 5,000 grams. That amount then was the unit and 1 c.c. of the serum would contain 10 units.]

3. Ehrlich adopted another unit of measure. The **unit of antitoxin (I.E.)** is the amount of antitoxin necessary to neutralize 100 minimal lethal doses of normal toxin (*i.e.* of a toxin which is fatal to guinea-pigs in doses of 0·1 c.c.). Thus if 0·01 c.c. of a given serum neutralizes 100 fatal doses of toxin that serum is said to contain 100 antitoxic units per c.c.

It was difficult to get comparable results by this method because the toxicity of a given toxin diminishes with the lapse of time. On the other hand the antitoxin content of a carefully standardized serum dried *in vacuo* and without heating is known to remain constant almost indefinitely. Ehrlich therefore having determined an antitoxic unit (as above) prepares a glycerin solution of the serum containing 17 units per c.c. This preparation is now used in antitoxin laboratories as the standard for testing serums. One c.c. of Ehrlich's glycerin solution diluted with 16 c.c. of water gives a solution containing 1 unit of antitoxin per c.c.

It is then easy to titrate a toxin against the standard antitoxin and after standardizing the toxin very carefully the latter is used to titrate the antitoxin under investigation.

4. The French method of standardizing antitoxin.—Roux prefers to standardize antitoxin according to its preventive strength (*pouvoir préventif*). **The preventive strength** is the numerical ratio between the weight of a given animal—guinea-pig—in grams and the amount of serum necessary to save its life if it be inoculated 12 hours later with 0·5 c.c. of a young and highly virulent culture. For example, if the guinea-pig weigh 500 grams and 0·1 c.c. of serum has to be inoculated to protect it against the subsequent inoculation of culture the preventive strength is $\frac{0·1}{500}$, and the serum is said to be active in $\frac{1}{5,000}$.

In practice the Ehrlich and Roux methods may usefully be controlled against each other; for instance, a serum is said to contain 100 antitoxic units per c.c. and to have a preventive strength of $\frac{1}{50,000}$. It must be borne in mind however, that the maximum of preventive strength may not coincide with the maximum of antitoxic strength (Martin, Momont and Prévot).

A serum which has a preventive strength of $\frac{1}{50,000}$ is suitable for the treatment of diphtheria in man, but a more powerful serum ($\frac{1}{70,000}$ or even $\frac{1}{100,000}$) is easily prepared. The serum made at the Pasteur Institute with the older toxins contained 100 antitoxic units per c.c. and had a preventive strength of $\frac{1}{50,000}$; but at the present time, using toxins prepared by growing the bacillus in Martin's broth which are ten times stronger than the older toxins, the serum supplied contains at least 200–300 antitoxic units per c.c. and has a preventive strength of $\frac{1}{100,000}$.

(d) Serum therapeutics.

The application of serum therapy to the treatment of diphtheria in the human subject has yielded results which might have been expected from animal experiment. The serum therapy of diphtheria is one of the most striking successes of modern therapeutics.

[1] Animals treated in this way do not survive indefinitely but die generally in from 1–6 months (Roux).

The serum is inoculated in doses of 10–40 c.c., either in one dose or at different times according to the severity of the disease : the technique of administration need not be dealt with here.

The serum has been used as a prophylactic during epidemics of diphtheria ; the immunity so produced is only of short duration and varies from 3–6 weeks. The dose to be used as a prophylactic should be 5–10 c.c. (Roux).

6. Agglutination.

Antitoxic serum obtained by the inoculation of toxin possesses, as has been pointed out, no agglutinating properties and contains no immune body (*substance sensibilisatrice*). By the inoculation of the bacilli themselves a serum containing both agglutinins and immune body can be obtained. Wassermann, Bandi, Martin, thought that such a serum might be of use in clinical practice to facilitate the disappearance of diphtheria bacilli from the pharynx of those cases in which it remained an unduly long time even after the use of antitoxin.

(i) Wassermann inoculated into the veins of a rabbit an extract of bacilli to which antitoxin had been added to neutralize the toxin. After several inoculations the serum of the rabbit precipitated the bacillary extract and agglutinated the diphtheria bacillus.

It would not appear to be true, as Wassermann thought, that this serum can be used to differentiate the pseudo-diphtheria from the diphtheria bacillus. As Lipstein points out, a serum obtained by the inoculation of a given strain of bacilli may agglutinate that strain in high dilution, while having no effect whatever on other strains.

(ii) Lipstein obtained a serum which agglutinated the diphtheria bacillus in a dilution of 1 in 1,000 by inoculating into the peritoneal cavity of a rabbit first a mixture of dead bacilli and antitoxin and later a mixture of living bacilli and antitoxin. This serum has no prophylactic properties.

(iii) Bandi inoculated a dog sub-cutaneously several times during the course of a month with *sensitized* diphtheria bacilli—bacilli, that is, which had been treated with antitoxin and had then been washed and centrifuged to remove any excess of serum. This observer obtained a serum which besides possessing agglutinating and sensitizing properties was also feebly antitoxic (15 units per c.c.). He had good results in seven cases of diphtheria which he treated with the serum.

(iv) Martin also prepared a serum which exhibited agglutinating, sensitizing and immunizing properties. The serum was obtained from a horse by inoculating it sub-cutaneously, intra-peritoneally, or better into the veins with bacilli which had been heated to 100° C. for an hour. Martin treated a number of cases of diphtheria locally with this serum. Repeated application to the false membranes gave very little result—some diminution of pain —but by making it into pastilles with gum and so ensuring a more prolonged contact with the membrane the results were more satisfactory. Under these conditions the false membranes swelled up, became softened and were soon detached. Cultures showed a rapid and marked diminution in the number of bacilli.

SECTION IV.—DETECTION, ISOLATION, AND IDENTIFICATION OF THE DIPHTHERIA BACILLUS.

The diagnosis of diphtheria is often impossible by clinical methods alone ; hence in practice the nature of the infecting agent in all cases of croup and sore throat, especially if any trace of a false membrane be present, has to be determined by bacteriological investigation.

By this means it can be ascertained whether any given case of croup or sore throat be the result of an infection with the diphtheria bacillus either alone or in association with other organisms. If a diphtheria bacillus be found, its virulence should be ascertained ; but this is not a matter of great importance in practice, because as a rule it may be assumed that the long bacillus is the most virulent of all the varieties of the diphtheria bacillus— and so the virulence of the strain under examination may be gauged from the relative number, or entire absence, of such forms—and because from the point of view of treatment the mere fact that the diphtheria bacillus has been found demands the administration of antitoxin.

When dealing with a case of sore throat three investigations are necessary before the bacteriological examination can be said to be scientifically complete. First films from the inflamed surface must be examined, then cultures must be sown with the material from the throat, and lastly the causal organism must be isolated and injected into an animal. In clinical work, however, the two former investigations are all that is required and these occupy 24 hours at the outside.

1. Collection of the material.

Remove the false membrane on a small piece of absorbent cotton-wool fixed in a pair of pressure forceps : in those cases in which the membrane is very adherent it is better to tear it off with the forceps. For the collection of material in the ordinary way it is convenient to have a small cotton-wool plug fixed on the end of a metal rod ; this is placed in a test-tube which is then plugged with wool and the whole sterilized in the hot air sterilizer. If there be no membrane scrape the surface of the tonsils or pharynx with a small platinum or nickel spatula.

When it is desired to send a fragment of false membrane to a laboratory situated at a distance, place it in a small sterilized tube plugged with wool; or wrap it in a piece of thin cloth which has been passed through boiling water and then place it in a new glass tube carefully plugged.

Most laboratories send the necessary apparatus for the collection of material to practitioners. A convenient form is that which contains two sterilized plugs in test-tubes, a sterile tube for false membrane and a small spatula, as well as two tubes of serum.

2. Methods of examination.

A. Microscopical examination of the fresh material.—This part of the investigation is of considerable importance.

Before using the false membrane for bacteriological examination press it lightly between sterile filter-paper to blot up any mucus which may be present on the surface.

1. Prepare films with small portions of the exudate, and stain with Roux's blue, wash and dry. Examine with an oil immersion lens. [Cobbett's method (p. 252) is recommended as giving more characteristic appearances.]

The absence of the diphtheria bacillus must not be assumed if the microscopical examination be negative as it is a well known fact that it often passes unrecognized when mixed with a number of other organisms (Martin).

Should bacilli resembling diphtheria bacilli be found in the preparation, the diagnosis may be advanced a stage by staining other films by Gram's method. The diphtheria bacillus is gram-positive, and a certain number of bacilli frequently found in the mouth and morphologically resembling it, but gram-negative, can by this means be excluded.

Cultures must be sown in every case.

One or two important inferences may be drawn from the result of microscopical examination, when positive. As a rule, the finding of numerous bacilli of the " long " type denotes a severe infection, and a similar inference may be drawn if streptococci are found associated with the specific microorganism. On the other hand the presence of the Brisou coccus generally indicates a mild infection.

2. Sections.—Harden in alcohol, embed in paraffin and cut sections perpendicularly to the surface of the membrane. For staining, Gram's method with double counterstain gives very good preparations.

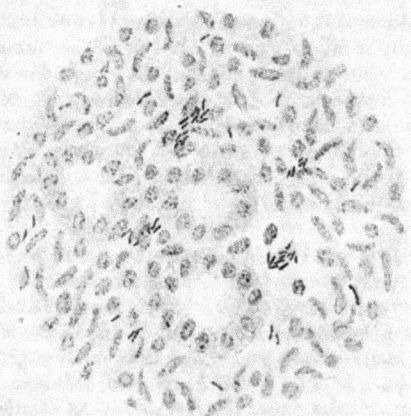

FIG. 171.—Section of tracheal membrane from a case of diphtheria, showing diphtheria bacilli (Eosin and methylene blue), oc. 2 ; obj. $\frac{1}{12}$th, Zeiss.

From an histological study of diphtheria membranes it can be shown that they are made up of three layers : the deepest layer—that next the body surface (mucous membrane or skin)—consists of a network of fibrin enclosing epithelial cells and leucocytes : the middle layer is made up of granular fibrin with but few cellular elements, while the most superficial layer consists almost entirely of microorganisms ; the bacilli, many of which are swollen at the ends, are arranged in masses parallel to one another, and side by side with the diphtheria bacillus are found the organisms associated with it.

B. Cultures.—Cultures should be sown on coagulated blood serum. If however this medium be not available, white of egg may be used instead.

The various serum media of Lœffler, Tochtermann, Joos, etc., give no better results than coagulated blood serum while they have the disadvantage of needlessly complicating the technique. [Many observers, however, state that **Lœffler's serum** (p. 52) is by far the best medium for the cultivation of the diphtheria bacillus.

[**Lorrain Smith** prepared a transparent serum medium by adding 0·1 per cent. to 0·15 per cent. of caustic soda to ox serum and heating the mixture at 120° C.

[**Cobbett** added about 1 c.c. of a 10 per cent. solution of caustic soda to 100 c.c. of ox or horse serum and after thoroughly mixing added 1 per cent. of glucose and sterilized at a temperature of 87° C.]

[**Coplans** has recently introduced the following medium for the routine recognition of the diphtheria bacillus :

Sheep's serum,	75 parts.
Broth,	25 ,,
Glucose,	0·5 per cent.
KCNS.,	1 ,,
CaCl₂,	1 ,,
1 per cent. aqueous solution of neutral red,	0·25 ,,

[The medium is adjusted so that on coagulation the reaction is but faintly alkaline. When throat swabs are sown on surface slopes and incubated at 37° C. for 18 hours, colonies of the diphtheria bacillus appear almost invariably to yield a bluish-pink tint, with diffusion of like tint through the medium; with Hofmann's bacillus the growth is yellowish and alkaline with diffusion of a yellowish tint. Staphylococci usually yield a straw-coloured raised growth with discrete colonies, but certain varieties produce either discrete pink colonies with strictly local diffusion of tint into the medium; and again, other varieties, more especially such as are derived from the throats of adults, yield acid with pink colouration of the medium in varying intensity. The colonies of torulæ are usually raised and straw-coloured but they may be brownish or red.]

Sow a tube of blood serum with a small piece of membrane held in a platinum loop and after rubbing it all over the surface of the serum and without recharging the loop sow two other serum tubes (p. 82 **B** 1). In the absence of membrane the spatula or cotton-wool swab, as the case may be, which has previously been applied to the tonsils or pharynx is rubbed over the surface of the serum. Should the cotton-wool swab be dry on arrival at the laboratory wash it in a little sterile water and then use the latter for sowing cultures. Dried membranes should similarly be softened in sterile water before being sown.

Incubate the cultures at 37° C. and examine about 20 hours later: colonies of the diphtheria bacillus are easily recognized at this stage; some cocci indeed produce a very similar growth but it is moister and more homogeneous than that of the diphtheria bacillus. A mere naked eye examination of the growth is however insufficient, and must always be supplemented by microscopical examination. If examination of the cultures be delayed beyond 24 hours difficulty may arise from the development of micro-organisms which are either associated with the diphtheria bacillus or which are present as an impurity. Select the culture which shows the greatest number of discrete colonies.

[In examining cultures sown with swabs from infected throats Cobbett picks off single colonies one by one with a straight platinum wire, sows a separate tube of broth with each colony, and then smears the wire in a straight line across a cover-glass. The first colony is smeared along one edge of the cover-glass, the others at right angles to it. In this way not only is it possible to make 4 to 9 separate preparations on one cover-glass from the different colonies of a single culture-tube, but pure cultures of each colony also are available for further examination.]

A diagnosis of diphtheria [infection] must be given in all cases in which colonies consisting of organisms having the morphological appearance and giving the staining reactions of the diphtheria bacillus are found on the serum sown with the suspected material. [In cases of faucial diphtheria such colonies will generally be present in large numbers; in laryngeal diphtheria they are sometimes few in number—this is often the case also with convalescents and "contacts."]

C. Inoculations.—When absolute confirmation of the positive microscopical and cultural results is desired resort must be had to animal inoculation. Several of the suspected diphtheria colonies must be inoculated, because bacilli of different degrees of virulence may be present in the same membrane.

Each colony is dealt with as follows: Sow a portion of a colony in broth (taking every care that the needle touches no other colony), incubate for 24 hours, and after verifying the purity of the culture inoculate 1 c.c. subcutaneously into an adult guinea-pig: the animal dies more or less rapidly according to the virulence of the organism (p. 247). Should the organism prove non-virulent for the guinea-pig test it on a small bird in a similar manner.

[3. Summary of diagnostic tests.]

[As some difficulty may be experienced in differentiating the diphtheria bacillus from other organisms—and especially from Hofmann's bacillus—it will be convenient to summarize the purely specific characteristics of the organism : references are given to the pages on which these characteristics are discussed in detail and also to the pages on which the reactions under similar circumstances of Hofmann's bacillus are considered : the reader will thus be in a position readily to form a diagnosis.

The true diphtheria bacillus is characterized by its :

(α) Macroscopic growth on *serum* and morphology (p. 250 diphtheria bacillus: and p. 273 Hofmann's bacillus).

(β) Power of producing acid in glucose broth (pp. 256 and 274).

(γ) Invisible growth on potato (p. 254).

(δ) The lesions produced in inoculated guinea-pigs (pp. 247, 255 and 274).

Bacillus pseudo-diphtheriæ.

(Hofmann's bacillus.)

In the mouths of healthy persons and in some cases of non-diphtheritic sore throat a non-virulent bacillus described by Lœffler as the pseudo-diphtheria bacillus is, as has already been indicated, not infrequently present.

[According to Graham-Smith, Hofmann's bacillus is most commonly found in the throats of the poorer classes, especially the scholars in the public schools (51 per cent. to 56 per cent.). The children attending better-class schools are less commonly found to harbour this bacillus in their throats (8 per cent. and 15 per cent.). In adults the extent of infection is less than in children, but it is greater amongst the poor (20 per cent.) than amongst the well-to-do (9 per cent.).]

This organism is by some observers (Lœffler, Hofmann, [Cobbett] and others) sharply differentiated from the diphtheria bacillus, while others (Roux and Yersin) have brought forward arguments in favour of its identity with the diphtheria bacillus : in the view of the latter school the pseudo-diphtheria bacillus is merely a diphtheria bacillus devoid of virulence (see also pp. 274 and 275).

[**Morphology.**—When taken from young serum or alkalized glucose serum cultures, stained with a weak solution of Lœffler's blue and mounted in the stain, the bacillus of Hofmann exhibits great uniformity of type ; it is oval, stains deeply, has no granules, but shows one unstained septum. The arrangement too is quite different from that of the diphtheria bacillus: Hofmann's bacillus ranges itself in parallel groups resembling a paling. Occasionally the organism departs from this typical form, and numerous many-banded forms occur ; on sub-culture however these many-banded forms revert to the type already described.

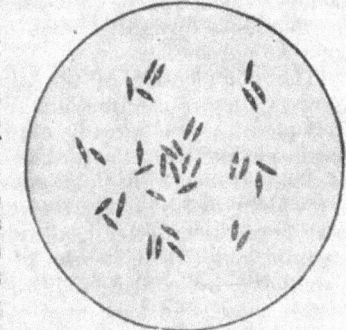

FIG. 172.—Hofmann's bacillus from a young serum culture (18 hours). Mounted in dilute Lœffler's blue (1 in 5 with water). Oc. 4, obj. $\frac{1}{12}$th, Zeiss.

[**Staining reactions.**—Stained with a weak solution of Lœffler's blue and mounted in the stain, the organism will be found to be a deeply and uniformly stained oval bacillus with one unstained septum. On running a drop of acetic acid (5 per cent.) under the cover-glass the whole organism decolourizes. (Occasionally a minute granule may be seen at the poles of some of the bacilli, but these minute specks present a very different appearance to the granules seen in the diphtheria bacillus, and are relatively few in number.) If stained with Neisser's blue, washed in water and counterstained with Bismarck brown (1 minute in each stain) the bacilli will be stained uniformly brown; blue granules will be absent or very indistinct.

[The organism like the diphtheria bacillus retains the violet by Gram's method.

[**Cultural characteristics.**—On serum or alkalized glucose serum the growth is more rapid than that of the diphtheria bacillus, and the colonies are larger and whiter and do not take up the pigment in the serum.

[**Bio-chemical reactions.**—When grown for 48 hours in a nutrient broth neutral to litmus and containing 1 per cent. glucose no acid is formed; on the contrary a slight increase in alkalinity takes place.]

Rothe recommends the following medium.—

Neutral broth,	1 part.
Ox serum,	4 parts.
10 per cent. glucose litmus solution,	½ part.

Solidify.

The diphtheria bacillus on the other hand turns this medium red.

[**Virulence.**—A guinea-pig inoculated with 2 c.c. or more of a 48-hour culture in broth remains perfectly well, not even a local œdema resulting.

[**Immunity.**—Hofmann's bacillus produces no substances toxic to laboratory animals. Petrie experimenting with Hofmann's bacillus finds that "no substances capable of neutralizing diphtheria antitoxin are present in the filtrates of the pseudo-diphtheria bacillus"; and his attempts to immunize horses with this bacillus against diphtheria toxin were negative.]

The relation of Hofmann's bacillus to the diphtheria bacillus.—[Petrie's experiments on immunity accentuate the differences between the two organisms—which have been detailed above—and diminish the probability that they stand in close relation to each other.

[Further support of the latter view is afforded by the fact that no satisfactory evidence of the conversion of the one organism into the other has yet been brought forward, though numerous experiments have been conducted for that purpose.

[The non-identity of the diphtheria bacillus with Hofmann's bacillus receives further confirmation from certain practical observations. In the first place, as has already been shown, the distribution of the diphtheria bacillus is limited to the throats of those who have been in contact with cases of diphtheria or with diphtheria-infected contacts, while Hofmann's bacillus is a widely distributed organism present in the throats of a considerable proportion of the ordinary healthy population. Secondly, Cobbett successfully stamped out two epidemics of diphtheria, one at Colchester and one at Cambridge, by isolating only those in whose throats diphtheria bacilli were found, those who harboured the bacillus of Hofmann and no diphtheria bacilli being treated as non-infected individuals. Other observers have had similar experiences.

[The above then is the case of those who regard the Klebs-Lœffler and Hofmann's bacillus as distinct species.

[The following arguments, which are advanced in favour of the identity of the two organisms, may be prefaced by a quotation from Graham-Smith supporting an earlier statement to the same effect by Cobbett. "The authority of Roux (1890) whose opinion justly carries great weight has often been quoted in support of the idea that Hofmann's bacillus is related to the diphtheria bacillus. But this is not right, for his remarks on the pseudo-diphtheria bacillus were made in comparatively early days, when the importance of acid production had not been generally recognized and before Hofmann's bacillus had been clearly distinguished from the so-called non-virulent diphtheria bacillus."]

A. From the morphological point of view the differences between the two organisms are insignificant. The pseudo-diphtheria bacillus is as a rule shorter than the diphtheria bacillus but Martin has shown that this is not always the case : the length of the diphtheria bacillus is subject to great variation and the variations in length afford in some degree an index of its virulence.

Too great importance must not be attached to certain characteristics said to be possessed by the pseudo-diphtheria bacillus by those who consider it a distinct species, and the constancy of which have not been confirmed. Deguy, for instance, says that the pseudo-diphtheria bacillus is motile (?) ; according to Barbier it shows no granules and stains more deeply than the diphtheria bacillus by Gram's method, etc. Characteristics based upon the staining of the metachromatic granules are of no value whatever for purposes of identification (p. 252).

B. The pseudo-diphtheria bacillus occasionally produces an œdema at the site of inoculation when inoculated into guinea-pigs but never leads to a fatal result. The same organism may kill small birds. Roux and Yersin were able to increase the virulence of an organism which produced a local œdema in guinea-pigs by mixing it with a streptococcus. The virulence is therefore not a fixed quantity.

C. Objection to the identity of the two bacilli may be taken on the ground that a totally non-virulent pseudo-diphtheria bacillus has so far never been made virulent. The objection, however, cannot be upheld in view of an experiment of Roux and Yersin : virulent bacilli were isolated from the throat of a person suffering from diphtheria ; as the patient progressed to convalescence, the bacilli became less virulent and were finally totally avirulent and their virulence could not be restored.

Martin has proved that some short bacilli which are not fatal to guinea-pigs are degenerated diphtheria bacilli.

Martin's experiment.—An eight months old broth culture of a long, very virulent bacillus was subcultivated in broth ; the organism failed to grow, but when sown on agar covered with a film of recently prepared veal broth gave a growth of a short bacillus which remained short in subsequent subcultivations. On inoculation, it did not kill guinea-pigs but gave rise to a local œdema from which the short bacillus was recovered.

This artificially obtained short bacillus was pathogenic for sparrows, and on inoculating a normal sparrow and a sparrow which had been treated with anti-diphtheria serum each with 0·1 c.c. of a broth culture of the organism, the normal sparrow died while the other survived. This short bacillus, non-virulent for guinea-pigs, was therefore undoubtedly a diphtheria bacillus.

D.—In suitable culture media, non-virulent diphtheria bacilli frequently produce toxin neutralizable by antidiphtheria serum (Martin).

[In this case apparently a non-virulent diphtheria bacillus was under observation. Such bacilli admittedly exist. See p. 255.]

E.—According to Spronck, a prophylactic inoculation of antidiphtheria serum prevents the development of a local œdema when a diphtheria bacillus is inoculated but not when a pseudo-diphtheria bacillus is inoculated. This however does not appear to be a constant phenomenon.

CHAPTER XVI.

BACILLUS PYOCYANEUS.

THE cause of blue suppuration was discovered by Gessard.

Blue pus is rarely seen nowadays though it was very common before the introduction of antiseptics. The *bacillus pyocyaneus* is always associated in these conditions with the ordinary micro-organisms of suppuration ; its presence in a wound is simply a complication and that not of a serious nature.

The *bacillus pyocyaneus* may invade the tissues of the body when the resistance of the latter has been broken down by some pre-existing pathological condition. It has, for instance, often been found in the internal organs in cases of enteric fever ; Calmette found it in the blood of persons suffering from chronic dysentery [and Williams and Lartigan in association with diarrhœa]. [Of twenty-three cases of pyocyaneus infection occurring in British Guiana and recorded by Minett and Duncan six were cases of acute filariasis and six others showed intestinal ulceration.] On the other hand it may be the primary cause of disease : *e.g.* of enteritis (Legros), appendicitis (Coyne and Hobbs), otitis, pseudo-membranous sore throat (Calvo Ignacig), etc. : some twenty cases of a generalized infection with the *bacillus pyocyaneus* have moreover been recorded.

[Dogs are liable to infection with the *bacillus pyocyaneus* and in these animals the symptoms may clinically resemble rabies ; and moreover the inoculation of brain tissue from the affected animals into normal rabbits and guinea-pigs produces symptoms similar to those seen in the original animal.]

The organism is sometimes found in the soil, in dust and in water. Besson has recorded its almost constant presence in the waters of the regency of Tunis where blue suppuration is very common and where serious infections are often complicated by the *bacillus pyocyaneus*.

SECTION I.—EXPERIMENTAL INOCULATION.

The *bacillus pyocyaneus* is pathogenic to rabbits, guinea-pigs, rats and mice.

Rabbits.—Sub-cutaneous inoculation in rabbits is rarely fatal: intra-peritoneal inoculation does not give constant results.

The injection of 1 c.c. of a broth culture into the ear vein of a rabbit, which is the most certain method of producing infection, causes an acute disease. The animal suffers from fever, albuminuria and diarrhœa and dies in 24–48 hours. The organisms are very numerous in the liver, spleen and kidneys but only a few are to be found in the blood.

Inoculation of smaller doses of culture leads to a chronic but not necessarily fatal disease in the rabbit which is characterized by wasting, paralysis of the limbs and convulsions. If death take place it is not uncommon to find *post mortem* a true nephritis with small contracted kidney, and even hypertrophy of the left ventricle of the heart. In some cases there is amyloid degeneration of the kidneys and infarcts may be present in the alimentary mucosa (Charrin).

By feeding rabbits with the organism Brau produced fatty degeneration of the liver and ulceration of the intestine followed by generalization of the bacillus in the tissues of the animal.

Guinea-pigs.—In guinea-pigs, sub-cutaneous inoculation produces a local swelling followed by ulceration: the organism then becomes disseminated and the animal dies. Intra-peritoneal inoculation is more severe and death takes place rapidly; the bacillus is found in the blood.

Rats. Mice.—The results are the same as in the guinea-pig: mice are very susceptible.

An organism isolated by Besson from the water of Zaghouan killed white rats in 20–36 hours when inoculated sub-cutaneously in doses of 0·20 c.c. *Post mortem* the abdominal organs were congested, the peritoneum contained a small quantity of an almost clear fluid and the intestines showed an early stage of ulceration in many of the Peyer's patches: in two cases the animals had hæmaturia and the organism was found in the blood, liver, kidneys, peritoneal fluid, intestinal contents, and in the urine in the bladder.

Increase of virulence.—The virulence of the *bacillus pyocyaneus* can be increased by passage through rabbits to such an extent that after a few passages a dose of 0·1 c.c. of a broth culture rapidly kills animals of this species.

SECTION II.—MORPHOLOGY.

1. Microscopical appearance.

The *bacillus pyocyaneus* is a small motile rod-shaped [non-spore bearing] organism with rounded ends, and of variable size. Its average length is about 1·5μ and its breadth 0·5 to 0·6μ. It has one flagellum situated terminally [monotrichous].

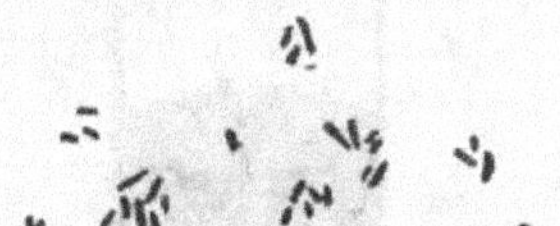

Staining reactions.—The *bacillus pyocyaneus* is easily stained by the basic aniline dyes and is gram-positive.

FIG. 173.—*Bacillus pyocyaneus.* Film from an agar culture (dilute carbol-fuchsin). (Oc. iii, obj. $\frac{1}{12}$th, Reich.)

The bacillus stains rather badly by Gram's method: some strains stain feebly and irregularly, and decolourization takes place easily if the action of the alcohol be prolonged.

Morphological variations.—The morphology of the *bacillus pyocyaneus* undergoes considerable change if sown in media containing small amounts of antiseptics (Guignard and Charrin). For instance, in broth containing 0·02 per cent. of carbolic acid the organism is long and filamentous. The addition of alcohol and bichromate of potassium to the medium have a similar effect. In broth containing boric acid the bacillus assumes a spiral form, and in media containing creosote it looks like a coccus (figs. 174 to 177).

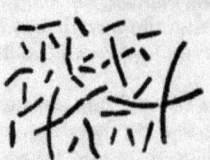

FIG. 174.—Culture in 4 per cent. alcohol broth.

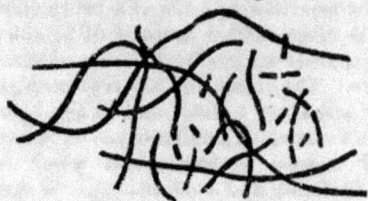

FIG. 175.—Culture in 0·015 per cent. potassium bichromate broth.

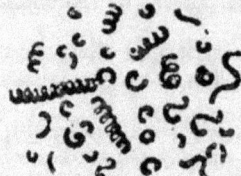

FIG. 176.—Culture in 0·70 per cent. boric acid broth.

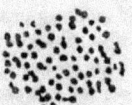

FIG. 177.—Culture in 0·10 per cent. creosote broth.

FIGS. 174–177.—Different morphological appearances presented by the *Bacillus pyocyaneus* when grown in broth containing traces of antiseptics (after Guignard and Charrin).

2. Cultural characteristics.

Conditions of growth.—The *bacillus pyocyaneus* is a facultative aërobe but the pigment is only formed in presence of air. It grows at all temperatures between 15° and 43° C., the optimum temperature being about 35°–37° C.

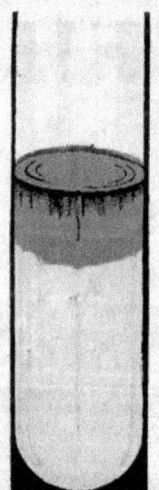

FIG. 178. — *Bacillus pyocyaneus*—broth culture—1st day.

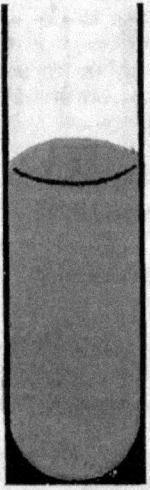

FIG. 179. — *Bacillus pyocyaneus*—broth culture—3rd day.

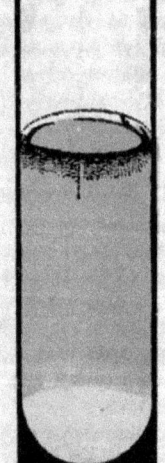

FIG. 180. — *Bacillus pyocyaneus*—broth culture—7th day.

Characters of growth on various media. (i) **Broth.**—After incubating at 37° C. for 8 hours the medium becomes cloudy, and then a greenish fluorescence appears at first limited to the upper part of the medium then extending throughout. During the next few days a white wrinkled pellicle forms on the surface, which as growth proceeds becomes thicker, dry and brown and falls to the bottom of the tube where it forms a dirty white deposit, the broth at the same time becoming deep green in colour and afterwards brownish. The culture is viscous and ropy and has a peculiar odour.

(ii) **Gelatin.** *Stab culture.*—After incubating for 2 days at 20° C. small colonies appear along the line of the stab : these coalesce and form a white streak : liquefaction commences about the third day (champagne glass liquefaction) and rapidly extends to the walls of the tube. The medium is coloured green.

Isolated colonies.—Small, yellowish, granular colonies appear on the plates after incubating for 2 days. Liquefaction occurs round them and gradually extends throughout the plate. The gelatin assumes a green tint.

(iii) **Agar.**—After incubating for 24 hours at 37° C. a greenish streak appears on the agar which rapidly spreads over the surface, the agar taking a fluorescent green colour.

(iv) **Potato.**—Along the line of sowing a thick brown layer is formed, and if this be removed the surface of the potato beneath becomes green on exposure to air.

FIG. 181.—Surface culture on agar—3 days at 37° C.

SECTION III.—BIOLOGICAL PROPERTIES.

1. Pigments (Gessard).

When a broth culture of the *bacillus pyocyaneus* is shaken up with a little chloroform and allowed to stand for a moment the chloroform separates at the bottom of the tube and is coloured pure blue, while a beautiful fluorescent green watery liquid floats to the surface.

The *bacillus pyocyaneus* secretes three pigments, one blue (*pyocyanine*) ; another fluorescent and green and identical with the pigment produced by saprophytic fluorescent bacilli ; the third is greenish and non-fluorescent and of little importance.

In contact with air pyocyanine oxydizes and forms a brown substance, *pyoxanthose.*

Pyocyanine is easily obtained by extracting a broth or agar culture with chloroform. In the case of agar it is only necessary to leave the chloroform on the culture for a few hours without shaking. The chloroform acquires a blue colour, and if evaporated long blue needles of pyocyanine crystallize out. Solutions of pyocyanine are turned red by dilute acids but the blue colour is restored on the addition of an alkali. Cultures in broth or peptone solution retain their colour after filtration through a Chamberland bougie. Pyocyanine is not toxic.

The formation of pyocyanine and of the green pigment may be varied at will and even suppressed by growing the organism on different culture media.

In a solution of peptone, Gessard was able to suppress the formation of the green pigment and the culture then had a very pretty blue colour (this phenomenon cannot be obtained with all peptones). In the same way, on glycerin-peptone-agar (the test medium of Gessard) the amount of pyocyanine produced is considerably increased. Pyocyanine is the only pigment formed when the organism is sown in a 10 per cent. gelatin medium containing a little glycerin and incubated at 35° C.

On the other hand, the green pigment is formed to the exclusion of the others when a medium containing 2 per cent. glucose is used, and the same result is obtained with white of egg.

No pigment at all is formed in broth containing 5–6 per cent. of glucose, or when the bacillus is grown on the serum of immunized animals.

Gessard was successful in producing strains of the bacillus some of which secreted the green pigment, and others pyocyanine. Wasserzug grew the bacillus on slightly acid media and found that it had altogether lost its power of pigment production. Charrin obtained a similar result by sub-culturing in broth and incubating the cultures at 42° C.

Melanogenic variety.—Cassin and Gessard studied a strain of the *bacillus pyocyaneus* which when sown in broth produced in the first instance the ordinary pigment but later a dark brown and finally a black pigment. Cultures on potato formed a deep brown layer which soon turned black. This production of black pigment was found to be possible only when tyrosin was present in the media. In a " mineral " medium such as the following :—

Ammonium succinate,	1 gram
Sodium phosphate,	1 ,,
Magnesium sulphate,	2·5 grams.
Calcium chloride (crystals).	1·25 ,,
Water,	1000 ,,

this bacillus produces no black pigment, the growth having all the characteristics of an ordinary *bacillus pyocyaneus* ; but by adding 0·5 per cent. of tyrosin to the medium a rose colour is at first produced which later becomes a deep brown.

2. Toxins.

Filtered cultures of the *bacillus pyocyaneus* inoculated in sufficient quantity into rabbits either cause death with all the symptoms of the acute experimental disease or lead to cachexia and paralysis which may also terminate fatally.

Wassermann obtained a very toxic product by incubating broth cultures for 40 days and then sterilizing them by leaving them to stand under toluol for a week. These cultures killed guinea-pigs in doses of 0·5 c.c. when inoculated intra-peritoneally.

The toxicity of the cultures is not due to pyocyanine but to certain other substances, some of which are volatile, easily destroyed and have merely a transitory action, while others are non-volatile. The non-volatile products may be divided into two groups ; those of the first group—the most toxic—are precipitated by alcohol, the others are soluble in alcohol (Arnaud and Charrin).

If injected into the veins of a rabbit the products of the *bacillus pyocyaneus* rapidly lead to the death of the animal without an incubation period. This absence of an incubation period is to be referred chiefly to the action of the volatile constituents which are not a part of the true toxins.

Pyocyanolysin.—Cultures in neutral peptone-broth, 7–30 days old, filtered or killed by toluol or heat (15 minutes at 60° C.), have a powerful hæmolytic action on freshly defibrinated ox, sheep and rabbit blood (Bulloch and Hunter).

Cultures 3–4 weeks old are the best for demonstrating these properties. The cultures are strongly alkaline in reaction.

Pyocyanolysin withstands high temperatures. The hæmolytic property is not destroyed by heating cultures at 100° C. for 15 minutes, and it is also said to be unchanged by heating at 120° C. for 30 minutes (Weingeroff, and Breymann).

3. Vaccination—Serum therapy.

If an average dose of a culture be inoculated beneath the skin of a rabbit the animal suffers no harm. Rabbits can be immunized by inoculating them in this manner with doses of 0·5–1 c.c. of a broth culture on five or six occasions at intervals of 3 or 4 days : or by inoculating them with small doses of filtered cultures or cultures heated at 115° C.

The blood and urine of animals treated with filtered cultures will also immunize animals.

Wassermann immunized guinea-pigs by inoculating them in the peritoneal cavity with gradually increasing doses of living culture or toxin. Guinea-pigs vaccinated with living cultures show a permanent immunity to the organism, but an inoculation of toxin is fatal to them. Their serum is prophylactic and has feeble therapeutic properties, but is not antitoxic. Guinea-pigs vaccinated with toxin are immune against both the organism and the toxin, and their serum is prophylactic, distinctly therapeutic and antitoxic.

The *bacillus pyocyaneus* will grow in the serum obtained from immunized animals : it preserves its shape, its vitality and its virulence, but forms *agglutinated colonies* (Charrin and Roger, Gheorghiewsky) and produces no pigment. The serum of immunized animals is therefore not bactericidal ; it is simply agglutinating *in vitro*. The agglutinating property does not run parallel with the prophylactic property.

4. Agglutination.

The agglutinating property of the blood of vaccinated animals has just been referred to. The blood of infected persons when diluted 1 in 40 to 1 in 100 similarly agglutinates the bacillus, and normal human serum has in some cases an agglutinating action in a dilution of 1 in 10. To demonstrate the agglutination it is best to add the serum to an emulsion prepared by diluting in normal saline solution the centrifuged deposit of broth cultures at least 24 hours old.

5. Antagonism.

The *Bacillus pyocyaneus* impedes the growth of anthrax in cultures. In the same way by inoculating a mixture of the anthrax bacillus and the *bacillus pyocyaneus* into animals susceptible to anthrax the animals do not become infected with anthrax. Porcelain-filtered cultures of the *bacillus pyocyaneus* possess the same properties (Blagovetschensky).

Similarly the *bacillus pyocyaneus* will prevent the development of the cholera vibrio (Kitasato).

Rumpf has recorded a parallel antagonism between the *bacillus pyocyaneus* and the typhoid bacillus ; he successfully treated 65 cases of enteric fever by inoculating the patients with sterilized cultures of the *bacillus pyocyaneus*. It does not appear that much faith should be put in these statements. Analogous investigations undertaken on the guinea-pig by Besson led him to the conclusion that these animals when treated with filtered cultures of the *bacillus pyocyaneus* are more than normally susceptible to the action of the typhoid and colon bacilli : moreover, infection with the bacillus pyocyaneus has been recorded coincidently with a fatal attack of enteric fever in man.

SECTION IV.—DETECTION, ISOLATION AND IDENTIFICATION OF THE BACILLUS PYOCYANEUS.

The presence of the *bacillus pyocyaneus* in pus is detected by the blue colour of the dressings and by the characteristic smell of the wound.

Pus should be examined by staining films with gentian-violet or thionin. The bacilli can be easily isolated on gelatin plates on which they produce a characteristic appearance. At the *post mortem* examination cultures should also be sown in broth with blood and scrapings of tissues.   The characteristic colour produced by the bacillus is sometimes only apparent after several sub-cultures in broth or on agar.

Inoculations should be made into the peritoneal cavity of a guinea-pig or into the ear vein of a rabbit.

Besson isolated the organism from water by sowing in Metchnikoff's gelatin-peptone-salt medium (p. 33).   When growth appeared after 12–15 hours a sub-culture was made in the same medium, and a trace of this second culture was sown on gelatin plates from which the organism was easily obtained in pure culture.

CHAPTER XVII.

THE BACILLUS OF SWINE ERYSIPELAS.[1]

SWINE erysipelas (measles) is due to a bacillus discovered by Pasteur and Thuillier, and of which the classical description was given by Lœffler.

A very large number of deaths among swine are attributable to swine erysipelas and the disease becomes therefore of considerable economic importance. The acute form of the disease is nearly always fatal, and in infected herds about 50 per cent. of the animals die.

Swine are liable to the disease between the ages of 6 months and 2 years; under 3 months old they are immune and beasts more than 2 years old are rarely infected.

Highly bred swine, such as the English breeds, are the most susceptible, while wild animals are immune.

Swine become infected by feeding upon the excreta of infected animals; they are almost the only animals susceptible to the spontaneous disease; pigeons, [mice], and rabbits which have frequented infected pig-sties are however sometimes attacked.

The flesh of suspected animals, and even of those dead of the disease, is frequently consumed as food without apparently causing any harmful effects in man: but cases of painful erythema have been noticed following the accidental inoculation of the virus.

Swine erysipelas occurs in two forms, acute and chronic.

The **acute** form of the disease is characterized by the appearance of bright or dark red purpuric spots on the skin, chiefly about the ears, the anus and vulva, the internal surface of the thighs and the groins. The animal suffers from diarrhœa: it grunts dismally and remains lying down hidden in its bedding with its tail uncurled and hanging down: its temperature is raised and death takes place in from 48–72 hours.

Chronic swine erysipelas is the less severe form of the disease; recovery after an attack is not infrequent, though some animals never recover completely. When an animal begins to recover, desquamation occurs about the spots on the skin. The characteristic swelling of the joints is responsible for a peculiar gait noticed in infected animals, and for the disease being sometimes called *Gout*.

Post mortem, in swine dead of the disease, there is frequently in addition to the spots on the skin an intense congestion of the serous membranes and of the intestines; the lymphatic glands especially those of the abdomen are swollen and congested; the spleen is very much enlarged and diffluent and shows bosses on the surface;

[1] [(Fr. Rouget du porc, érysipèle, rougeole. Ger. Schwein Rothlauf.)]

the liver is congested and the blood very dark in colour. More rarely, a thickening of the walls of the intestine and patches of broncho-pneumonia are present.

The specific organism is found in the liquid discharges from the bowel, in the spleen, lymphatic glands, bone marrow, and also but in smaller numbers in the blood, liver and kidneys.

The bacillus of swine erysipelas appears to be frequently present as a saprophyte in the tonsils and intestinal canal of healthy pigs (Olt, Pitt, Overbeck). Pitt found the organism in the intestine in 26 out of 66 and in the tonsil in 28 out of 50 normal animals examined by him.

SECTION I.—EXPERIMENTAL INOCULATION.

1. Animals susceptible to the disease.

Swine, pigeons, mice and rabbits are all susceptible to swine erysipelas but in different degrees. Guinea-pigs are immune.

Inoculation into the pectoral muscle of a pigeon or into the sub-cutaneous tissue of a mouse is fatal in 3 or 4 days. Rabbits are more resistant and to produce a fatal result the virus must be inoculated into a vein.

When passed through a series of rabbits the virulence of the virus is increased for rabbits but diminished for swine. The first rabbit is inoculated intra-venously with a culture from a pig and the spleen of the first rabbit is used for inoculation of the second and so on.

On the other hand, the virulence is increased for all susceptible animals by passage through pigeons.

It is a curious fact that swine are not very susceptible to experimental infection, and even when the virus is inoculated into a vein it seldom leads to a fatal result; so that to produce the disease in these animals a virus experimentally increased in virulence must be used for inoculation. It is however possible to infect *pure bred* swine by feeding them on the organs of animals which have died of the disease.

2. Technique of inoculations.

The general rules applicable to the inoculation of animals must be observed and special attention given to the following points. The material for inoculation may be taken directly from the spleen, lymphatic glands or blood of an animal dead of the disease, though it is better to sow a broth culture with a fragment of the spleen and to inoculate a little of the culture after incubating it for 36–48 hours.

3. Symptoms and lesions.

The *symptoms* have already been detailed.

The most prominent lesion in experimentally-infected animals is the swelling and softening of the spleen. The organism will be most easily detected in the spleen, bone marrow, tonsils, lymphatic glands and blood, but may also be found in the liver and kidneys. Sections of the lymphatic glands, spleen, liver and kidneys should also be cut.

SECTION II.—MORPHOLOGY.

1. Microscopical appearance.

The micro-organism of swine erysipelas is a small, non-motile bacillus, visible only with difficulty in unstained preparations, and measuring $0.5–1.5\mu$ by $0.2–0.3\mu$. In the blood and internal organs it occurs singly, in pairs or

in groups. In broth cultures it forms short chains (fig. 182). The bacilli
are more numerous in the spleen and lymphatic glands than in the blood.
They are frequently found within the leucocytes, and in sections masses
of bacilli will be seen within the capillaries.

The bacillus is not known to form spores.

Staining methods.—The bacillus of swine erysipelas stains readily with

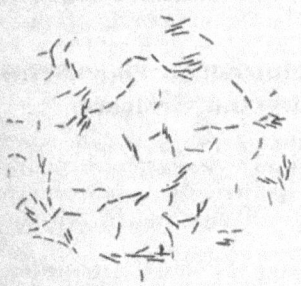 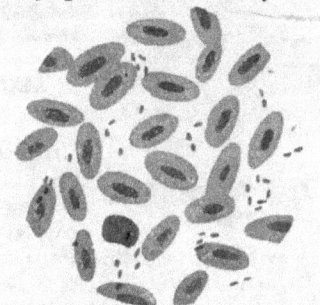

FIG. 182.—Bacillus of swine erysipelas (broth
culture). Carbol-thionin. (Oc. iv, obj. $\frac{1}{12}$th,
Reichert.)

FIG. 183.—Bacillus of swine erysipelas in
pigeon's blood—Gram's stain. (Oc. iii, obj.
$\frac{1}{12}$th, Reichert.)

the basic aniline dyes, is gram-positive, and retains the violet in Claudius'
method. The best methods to use are :

(a) *Cultures.*—Stain with carbol-thionin or dilute carbol-fuchsin.

(b) *Blood-films and smears of tissues.*—Carbol-thionin or carbol-methylene-
blue may be used, but Gram's method is preferable.

(c) *Sections.*—Gram's method should be used with either double or triple
staining (p. 219).

2. Cultural characteristics.

Conditions of growth.—The bacillus of swine erysipelas
is indifferently aërobic. Growth is better under anaërobic
conditions but is always rather scanty.

Cultures, which should be sown with the blood, pulp of
organs, or bone marrow of an animal recently dead of the
disease, are easily obtained at temperatures between 15°
and 40° C. on the ordinary media.

Broth.—The medium soon becomes slightly opalescent
when incubated at 33°–38° C. The growth which is always
scanty ceases about the fourth day, and subsequently forms
a very small white precipitate.

Gelatin. *Stab culture.*—The growth in gelatin stab-culture
is characteristic. Along the stab a thin opaque line develops, from which numerous small very delicate branching
filaments radiate. The growth is more luxuriant in the depth
of the stab. Towards the twentieth day, the characteristic
appearance vanishes and the culture becomes cloudy. There
is never any liquefaction of the gelatin (fig. 184).

FIG. 184.—Bacillus
of swine erysipelas.
Stab culture in gela-
tin (8 days).

Stroke culture.—Cultures on the surface of a sloped
gelatin tube radiate from the line of sowing like the feathers
of a quill.

Single colonies.—Fine downy flocculi giving off delicate radiating filaments are seen embedded in the gelatin, then the appearance becomes

woolly-looking and the centre of the colony forms a small brownish spot.

Agar.—At first the growth is similar to that on gelatin, but it soon assumes an homogeneous appearance and forms a delicate scanty layer.

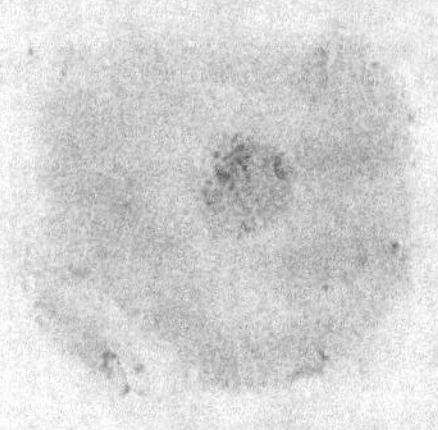

Potato.—On potato the organism only grows under anaërobic conditions and then forms a barely visible streak.

SECTION III.—BIOLOGICAL PROPERTIES.

1. Vitality and virulence.

FIG. 185.—Bacillus of swine erysipelas. Single colony on gelatin × 60 (8 days).

The bacillus of swine erysipelas remains alive for several months in cultures under anaërobic conditions, and shows an equal vitality in deep stab cultures in ordinary gelatin : it will give rise to sub-cultures and even kill pigeons after 6 months.

In aërobic broth cultures kept in the warm (37°–39° C.) incubator the virulence as well as the vitality vanish much more rapidly. The virulence becomes progressively enfeebled and after about 20 days the culture is harmless. As already pointed out the virulence of an attenuated virus can be restored by passage through pigeons.

2. Vaccination.

One attack of the acute disease confers immunity on swine ; moreover an attack of the chronic form (*Gout*) protects an animal from the acute disease. Pasteur and Thuillier considered it possible to immunize animals by inoculating them with attenuated viruses, and at the present time vaccination of swine is very extensively practised, especially in Austria.

For the purposes of vaccination broth cultures which have been attenuated (through the action of the oxygen of the air) by incubation for a longer or shorter time in the warm incubator may be utilized.

The pigs are inoculated first with a very weak virus, and then with a virus which has not been in the incubator so long and which is therefore somewhat more virulent. Pigs should be inoculated before they are 4 months old as they are then less susceptible to the disease. The immunity so conferred, which is complete 12 days after the second inoculation, lasts about a year, and this is a sufficient length of time for fattening purposes. If the animal is to be kept for breeding it is well to repeat the vaccination at the end of a year.

As stated above, the organism may also be attenuated for the pig by passage through rabbits. After several passages the virus becomes very virulent for rabbits, but attenuated for swine, and may then be used for inoculation as a vaccine. For this purpose a tube of broth is sown with the spleen of the last rabbit of the series and incubated, and the pig is vaccinated with the culture.

3. Soluble products. Serum therapy.

(i) Negative results follow the injection of filtered cultures. The amount of growth, as has already been pointed out, always remains very scanty and toxins are not formed in any appreciable quantity. But if rabbits be inoculated sub-cutaneously with small quantities of unfiltered cultures they quickly recover, and it soon becomes possible to inoculate large doses into the veins without producing any morbid symptoms. Emmerich, Leclainche and others killed rabbits which had been treated in this manner, made an emulsion by pounding and extracting the muscles and organs, and obtained a product which after filtration possessed both vaccinating and therapeutic properties.

(ii) Lorenz prepares a serum which has distinct therapeutic properties. A rabbit is inoculated sub-cutaneously first of all with a few cubic centimetres of specific serum (1 c.c. per 1 kg. of body weight); two days later, and again on the twelfth day, the rabbit is inoculated (sub-cutaneously) with a virulent culture, the second and third doses being larger than the first. After a further interval of 10 days a large dose of culture is administered intra-venously.

(iii) Mesnil, adopting Pasteur's method of attenuated viruses, immunizes rabbits as follows : At intervals of 1 week a rabbit receives, first, 0·25 c.c., then 1 c.c. of a very attenuated virus—the first vaccine. This is followed by inoculations of a less attenuated virus—the second vaccine—in doses of 0·25 c.c. first and 1 c.c. later. Finally, at periods varying from a week to a month increasing doses of virulent culture—0·25 c.c., 1 c.c., 3 c.c., 4 c.c., 5 c.c., 10 c.c.—are used. Immunization requires about 6 months' treatment, and in spite of the small doses used some of the animals die. It is only after about 3 months that the animals under experiment can resist the inoculation of a large dose of culture every week or 10 days without showing considerable reaction. The serum of animals prepared in this way if given in doses of 0·05 c.c. protects mice against an inoculation of a virulent virus given the following day. In doses of 0·25 c.c. it exhibits therapeutic properties provided it be administered within 24 hours of infection. The serum is also efficient in the case of pigeons and rabbits. It is not bactericidal, for if the bacillus be sown in the serum it grows in long chains and moreover retains its virulence.

(iv) Leclainche uses horses for the preparation of his prophylactic serum : 200 c.c. of a virulent culture (one which will kill pigeons in doses of 0·25 c.c. when inoculated into the pectoral muscle) are inoculated into the jugular vein and the inoculations repeated at intervals of about 10 days. The resulting serum is prophylactic for rabbits in doses of 0·5-1 c.c. and the immunity conferred lasts 1 or 2 days. Inoculation of 1 c.c. of this serum mixed with an equal volume of a virulent culture confers a lasting immunity (*Sero-vaccination*).

Leclainche has applied this method to the vaccination of swine. He inoculates healthy pigs on the inner side of the thigh first with a mixture of 5–10 c.c. of serum and one-half a virulent culture, and 12 days later with one-half a virulent culture without serum. If he has to deal with an herd already infected he inoculates them with 10–20 c.c. of serum without any culture 2 days before he inoculates the first vaccine.

The serum has no appreciable therapeutic property.

4. Agglutination.

The serums prepared by Mesnil's and by Leclainche's methods agglutinate the bacillus of swine erysipelas to a marked degree. The serums of infected rabbits and pigeons show no agglutinating properties (Overbeck).

SECTION IV.—THE DETECTION, ISOLATION AND IDENTIFICATION OF THE ORGANISM.

The recognition of the disease is of supreme importance from the point of view of vaccination, and the demonstration of the organism is of great help in diagnosis and in differentiating the disease from *hog cholera* [*swine fever*].

The following observations and experiments should be made :

1. Stain blood films and smear preparations of the lymphatic glands, spleen and bone marrow by Gram's method, etc. and examine them for the bacillus (*vide* "Microscopical appearance ").

2. Sow cultures from the spleen in broth and on gelatin.

3. Inoculate pigeons and guinea-pigs with broth cultures.

Guinea-pigs it will be remembered are immune to swine erysipelas, but they are very susceptible to *hog cholera* [or swine fever]. This experiment therefore will differentiate between the two diseases.

The bacillus of mouse septicæmia (Koch).

(*Bacterium murisepticum.*)

The septicæmic disease of the domestic mouse (*Mus musculus*), investigated by Koch, is due to a small bacillus, similar to that of swine erysipelas, but on inoculation somewhat less rapidly fatal than the latter organism. It is not pathogenic for the field mouse (*Arvicola arvalis*) nor for pigeons nor rabbits. But if its virulence be increased by numerous passages through mice a virus is ultimately obtained which, on intra-venous inoculation, proves fatal to pigeons.

The symptoms observed in experimentally-infected mice are drowsiness, blepharitis and a spasmodic form of respiration. The disease finally terminates fatally. Bacilli are to be found in large numbers in the blood and in the internal organs.

The morphological characteristics and staining reactions of this organism are the same as those of the bacillus of swine erysipelas. The cultures of the two organisms are very much alike ; there is, however, this difference that the growth of the *bacterium murisepticum* in gelatin is more cloudy and the radiating filaments are not so well marked.

The serums of animals immunized against the bacillus of swine erysipelas agglutinate the bacillus of mouse septicæmia (Overbeck). This affords definite proof that the latter bacillus is not a distinct species [but merely a variety of the bacillus of swine erysipelas adapted to its special host.]

CHAPTER XVIII.

BACILLUS TUBERCULOSIS.

THE tubercle bacillus, discovered by Koch, is the cause of tuberculosis in man and the lower animals.

In accordance with established practice, the infecting agent in tuberculosis will, in the present chapter and elsewhere in this book, be spoken of as a bacillus although it is now agreed that it should be classed with the genus Discomyces (Streptothrix of Cohn). Metchnikoff has suggested for it the name *Sclerothrix kochi.*

1. Types of tubercle bacilli.

Of the tubercle bacilli recoverable from human tissues and from the tissues of the lower animals four types can be distinguished, differing from one another in various characteristics. It is customary therefore to speak of human, bovine, avian and ichthic tubercle bacilli, meaning thereby the type of bacillus [most commonly] obtained from human, bovine, avian or ichthic sources respectively.

(*a*) **The human and the bovine types of tubercle bacilli.**—Most bacteriologists, Koch's opinion notwithstanding, are agreed in regarding the human and the bovine types of tubercle bacilli as identical, for [in the opinion of these observers] the facts which can be brought in support of this view are numerous and conclusive. Each bacillus though best adapted to the particular species

T

of animal—man and bovine respectively—in which it finds its normal habitat may nevertheless infect either of them, and though the bovine bacillus appears to be more virulent than the human bacillus the latter, according to de Jong, may by passage through goats be made as virulent for bovine animals as the bovine bacillus itself.

[The findings of the English Commission [1] are not altogether in agreement with the statements contained in the preceding paragraph.

[In the opinion of this Commission the human and bovine types are not identical but "varieties of the same bacillus." They point out that since the human and the bovine tubercle bacilli are "morphologically indistinguishable" the question of their identity or non-identity resolves itself into a consideration of their cultural and pathogenic differences or similarities.

[With regard to the former, the human type consistently grows more luxuriantly in culture than the bovine type and this difference in cultural characteristics is quite definite though "the gap which separates the human type from those strains of the bovine type which grow most abundantly is not wide."

[A study of their pathogenic resemblances and differences shows on the one hand that the disease produced in certain species of animals such as guinea-pigs and monkeys by the two types is "histologically and anatomically identical" and on the other hand that in man fatal tuberculosis due to infection with bacilli of the bovine type is identical with that caused by the human type.

[That the bovine bacillus can infect man is certain. Many cases of tuberculosis in children and a few in adults investigated by Cobbett and A. S. Griffith (working for the English Commission) were shown to be caused solely by the bovine tubercle bacillus. An infection of the bovine species by the human tubercle bacillus on the other hand did not occur: the human tubercle bacillus was in fact incapable of producing in cattle anything but a slight and non-progressive tuberculosis, however large the dose.

[Neither did the human type of bacillus cause anything more than a slight non-progressive tuberculosis in goats, pigs and, with rare exceptions, in rabbits, while the bovine bacillus readily caused a fatal tuberculosis in these animals as well as in cattle.

[Certain tubercle bacilli isolated during the investigations of the Commission from cases of lupus and equine tuberculosis had the cultural characteristics of the bovine bacillus but were only slightly virulent for calves and rabbits (the animals usually relied on for differential tests) and were of relatively low virulence also for monkeys and guinea-pigs. These bacilli, it would seem, in no way bridge the gap between the two types; for while they approach the human tubercle bacillus in their low degree of virulence for calves and rabbits, they recede from it in virulence for monkeys and guinea-pigs (A. S. Griffith). At the same time, as the Commissioners point out, "the discovery of these exceptional bacilli makes it impossible to regard differences of virulence for the calf and rabbit as sufficient to establish the non-identity of the human and bovine types." Several of these attenuated bacilli isolated from human (lupus) and equine sources were raised to the full virulence of a typical bovine bacillus by passage through calves and rabbits.

[To establish the complete identity of the two types it would appear to be necessary to demonstrate that both cultural and pathogenic differences were unstable, i.e. that the transmutation of the human type of bacillus into the

[1] The references to the "English Commission" in this chapter are to the Reports and Appendices thereto of the Royal Commission on Tuberculosis appointed in 1901.]

bovine or *vice versa* was possible, and on this point after reviewing the numerous prolonged passage experiments on various species of animals carried out under their direction the Commissioners conclude that " transmutation of bacillary type " is " exceedingly difficult if not impracticable of accomplishment by laboratory procedure."

[Though it has been considered desirable to introduce thus briefly the conclusions arrived at by the English Commission,] it is altogether beyond the scope of the present work to enter upon a discussion of the arguments which have been brought forward in support of their theses by those who hold that the human tubercle bacillus is identical with the bovine and by those who are of contrary opinion. For these arguments the reader is referred to the publications of the authors whose names are mentioned in the text and to those of other writers on the subject.

(*b*) **The avian tubercle bacillus.**—Straus and Gamaléia regard avian tuberculosis as due to a special organism which, though closely allied to the human bacillus, constitutes a separate species. The view long held by Arloing and others that the human and the avian bacillus are identical has been [held to be] proved by certain experiments of Nocard (*vide infra*). [It is largely on the results of these latter experiments that] the bacillus of avian tuberculosis has been regarded merely as a strain or race of the human tubercle bacillus.

Nocard [claims to have] converted an human tubercle bacillus into an avian tubercle bacillus by growing it for a long time in collodion sacs in the peritoneal cavities of fowls. Nocard filled a collodion sac (p. 175) with a thick emulsion of a glycerin-potato culture of a human bacillus. The sac, after remaining at least 4 months in the peritoneal cavity of a fowl, contained a thick mass of bacilli which, when sown on culture media, gave, at first, a scanty growth, and this on sub-culture became more luxuriant and had all the characteristics of a culture of the avian bacillus (a soft, greasy, fatty, easily dissociated and wrinkled layer of growth). These cultures were only slightly virulent for guinea-pigs but highly virulent for rabbits which succumbed to a generalized miliary tuberculosis on inoculation with bacilli from the first passage and when inoculated with bacilli from the second passage the animal died of a tuberculous septicæmia without apparent lesions exactly as though it had been inoculated with an avian bacillus. After three passages of 6–8 months in collodion sacs the human tubercle bacillus killed fowls with symptoms identical with those of the spontaneous disease.

[A. S. and F. Griffith (working for the English Commission) entirely failed to confirm the results obtained by Nocard. No modification of human or bovine tubercle bacilli into avian, or of avian tubercle bacilli into mammalian, was demonstrated.

["With ten mammalian viruses, eight of which were *bovine*, sixteen collodion capsule experiments on fowls and twenty on pigeons were performed, lasting 55–186 days. In certain of the cases cultures, which were obtained from the capsules on removal from the bird's peritoneal cavity, were placed, again in capsules, in the peritoneal cavities of other birds, the total duration of residence being in one series as much as 475 days. In 20 of these experiments cultures were obtained from the capsules and found to be unchanged in character. In the remaining 16 cases the bacilli in the capsules were apparently dead."

["Similar experiments were performed with *human* tubercle bacilli obtained from 12 different sources. These experiments lasted from 59 to 685 days." The results were similar to those obtained with mammalian tubercle bacilli.

["With cultures of five *avian* viruses 25 collodion capsule experiments were performed on guinea-pigs. The duration of residence in individual guinea-pigs ranged up to 253 days and the total periods during which the cultures were in the peritoneal cavities of series of guinea-pigs varied up to 424 days." In two instances the bacilli in the capsules were dead: "from all the other capsules cultures were obtained and the bacilli were found to be unchanged " in cultural characteristics and virulence.]

Lydia Rabinowitsch isolated thirty-four strains of tubercle bacilli from birds. Two of these, isolated from birds of prey, had all the characteristics of the human bacillus. Rabinowitsch concluded from this investigation that the human and avian tubercle bacilli are merely two varieties of the same species adapted to different conditions.

(c) **The ichthic tubercle bacillus.**—The ichthic tubercle bacillus is more sharply differentiated from its congeners (*vide infra*) but some observers, notably Mœller, Sorgo and Suess [report that they] have succeeded in converting an human into an ichthic tubercle bacillus.

It appears to be true that all tubercle bacilli have a common origin, and that acclimatization under parasitic conditions in different animals has led to the creation of the four different types.

2. Human tuberculosis.

Man becomes infected with tuberculosis either by way of the respiratory or digestive tracts, more rarely by the skin and genital passages.

The tubercle bacillus is found in all tuberculous lesions in the human subject.

[As to the ætiology of *human tuberculous phthisis* opinion is somewhat sharply divided. The original theory was that tuberculous phthisis is commonly caused by the inhalation of tubercle bacilli. This doctrine received the support of Koch, Cornet, Pflügge and others. Chauveau however put forward the view many years ago that phthisis was not uncommonly caused by bacilli which had been ingested and absorbed from the intestine, and in recent years this doctrine has been strongly advocated by Behring.]

Behring thinks that infection generally takes place through the alimentary canal and that pulmonary tuberculosis of adults is merely a later stage of an intestinal infection contracted in the early years of life. Calmette and Guérin confirm this opinion [in so far as it relates to the channel of infection in phthisis [1]] and bring forward numerous experiments to show that pulmonary tuberculosis (not inoculated) is always a sequela of a primary intestinal infection of which in the adult no trace of the original lesions in the mesenteric glands or abdominal viscera can be detected. [Calmette bases his opinion upon experiments on goats and bovines and on the researches of his pupils, Van Steenberghe and Grysez, on experimental anthracosis. Ravenel also from his own observations is led to believe that infection of the tonsils is the most frequent cause of apical tuberculosis but that infection may take place from any part of the alimentary canal and that the bacilli may pass through the wall of the intestine without leaving any indication of the site of infection in the form of a local lesion.

[The view that phthisis is commonly caused by the inhalation of tubercle bacilli is, however, supported by many recent investigations. Cobbett, for instance, believes " that phthisis is commonly caused by the inhalation of tubercle bacilli " and from the results of an elaborate series of experiments devised to ascertain the ætiology of pulmonary tuberculosis in which he repeats many of Calmette's experiments this observer concludes that the intestine is not a common portal of entry for the tubercle bacilli which cause phthisis. The experiments of the English Commission again though they demonstrate " that a considerable amount of the tuberculosis of childhood is to be ascribed to infection with bacilli of the bovine type transmitted to children in meals consisting largely of milk of the cow " nevertheless do not entirely support the theory that pulmonary tuberculosis is a sequela of a primary intestinal infection as may be seen from the widely different propor-

[1 Calmette does not state that the infection in pulmonary tuberculosis is necessarily an infantile infection but merely that at whatever age infection of the lungs occurs the channel of infection is invariably the alimentary canal.]

tion of bovine and human tubercle bacilli found respectively in alimentary and in pulmonary lesions in man. Thus of nine cases of cervical gland tuberculosis in children three were found to be caused by the bovine tubercle bacillus and six by the human tubercle bacillus; and of twenty-seven cases of primary abdominal tuberculosis in children, fourteen were caused by the bovine tubercle bacillus and thirteen by the human tubercle bacillus (Cobbett and A. S. Griffith). "In these cases," the Commission remarks, "the tubercle bacillus had unquestionably been swallowed." The examination of tissues from fourteen fatal cases of primary pulmonary tuberculosis (A. S. Griffith and Cobbett) showed that in all of the cases the human tubercle bacillus alone was responsible for the disease. A. S. Griffith subsequently examined the sputum from twenty-eight cases of pulmonary tuberculosis: in twenty-six the human tubercle bacillus was the infective agent and in the remaining two the bovine tubercle bacillus (confirmed by repeated examination of the sputum).

[Baumgarten holds that tuberculous phthisis is due to infection during intra-uterine life, but this view receives very little support at the present day.]

Attempts have been made to draw a distinction between the disease as it affects the internal organs, pleuræ and peritoneum on the one hand and that form of it which affects the skin, glands, joints, etc. According to Arloing, the latter, the so-called surgical tuberculoses, are due to an attenuated bacillus which must be regarded as a separate variety. But seeing that in these localized lesions the bacillus is fully virulent, it is more likely that the slight tendency to dissemination which it exhibits is to be explained on other grounds, such as the personal resistance of the infected individual, the influence of the particular tissues in which it is growing, and the small number of the invading organisms which grow but feebly in a soil relatively unfavourable to their multiplication (Krompecher and Zimmermann).

[Arloing's theory, in so far as it relates to tubercle bacilli which infect the skin, is in part supported, and greatly amplified, by A. S. Griffith (working for the English Commission). Twenty cases of lupus were examined. The tubercle bacilli isolated from nine of them showed the cultural characteristics of the bovine tubercle bacillus, but only one had the pathogenicity ordinarily associated with that type, while the rest showed varying degrees of lesser virulence: the least virulent being no more virulent for calves and rabbits than a human tubercle bacillus but differing from the latter in that they were also of relatively slight virulence for guinea-pigs and monkeys.

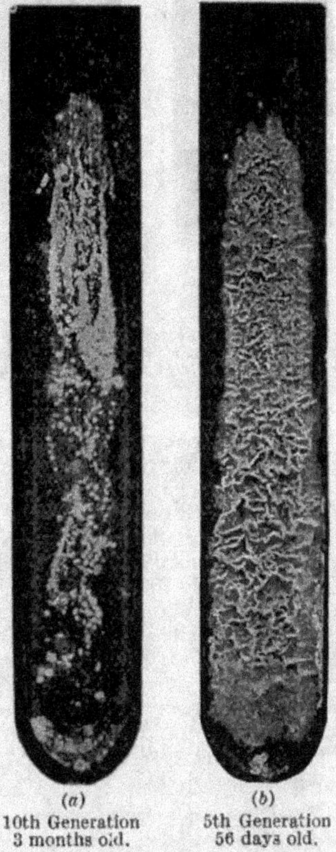

[From the remaining eleven cases tubercle bacilli were isolated which had the cultural characteristics of the human tubercle bacillus; two had the full virulence of the human tubercle bacillus, the others being of lower virulence.

[It was found possible in two of the cases in which a degraded bovine bacillus was the infective agent to "increase the virulence of the culture from the original

(a)   (b)
10th Generation   5th Generation
3 months old.   56 days old.

FIG. 186.—Tubercle bacilli from cases of Lupus growing on glycerin-agar. (a) The dysgonic or bovine type: (b) the eugonic or human type. (This and the succeeding figure (187) are from the Final Report of the Royal Commission on Tuberculosis (Human and Bovine)—Part II. Appendix, Vol. II; Dr. A. Stanley Griffith—by permission of the Controller of H.M. Stationery Office.)

material by residence in the tissues of calf and rabbit so as to bring it up to the high virulence of the bovine tubercle bacillus ": and one of the strains of degraded human tubercle bacilli attained the full virulence of the human tubercle bacillus after residence in the body of a monkey. No correspondence suggesting any relation between the duration or extent of the disease in the human patient and the degree of attenuation of the bacillus isolated was demonstrable in these cases.

[From thirteen cases of joint and bone tuberculosis the human tubercle bacillus with the full virulence of the type alone was isolated; in a fourteenth case both human and bovine tubercle bacilli appear to have been present (Cobbett and A. S. Griffith). These investigations therefore afford no confirmation of Arloing's theory so far as it applies to joint or gland tuberculosis.]

3. Tuberculosis in the lower animals.

The majority of the lower animals are susceptible to infection with tuberculosis; [The infecting agent however is not always of the same type.]

Bovine animals.—Adult animals are frequently tuberculous (3–60 per cent. varying according to the locality), [young] calves very rarely so (1 in 10,000 at the most).

(a)
8th Generation
4 months old.

(b)
5th Generation
28 days old.

FIG. 187.—Tubercle bacilli from cases of Lupus growing on glycerin-potato. (a) The dysgonic or bovine type: (b) the eugonic or human type. (A. S. Griffith.) (See fig. 186.)

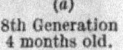

Generally speaking the disease runs a chronic course. Cattle may suffer from the disease for a long time without showing any loss of weight.

The respiratory organs are most frequently affected: large, occasionally calcified, tuberculous masses (*Grapes* [1]) are found in the lungs: the pleuræ and especially the bronchial glands are affected at the same time: occasionally the abdomen (mesenteric glands, liver and more rarely the spleen and kidneys) is invaded. Sometimes, especially in young cattle, the disease is mainly confined to the alimentary tract: the lymphoid structures of the intestine, the mesenteric glands, peritoneum, liver and spleen being infected. Other local manifestations of the disease are sometimes found in cattle; for instance, mammary tuberculosis (in about 1 per cent. of tuberculous animals), tuberculosis of bone, etc.

Finally, bovine tuberculosis may occur as a rapidly-spreading generalized infection resembling the miliary tuberculosis of man.

[The bovine type of tubercle bacillus has been shown to be the sole cause of bovine tuberculosis.]

Monkeys.—In these climates, monkeys frequently develop tuberculosis, and in them the disease runs a course similar to that of human tuberculosis, a characteristic feature being its tendency to become generalized. In these animals the commonest form of the disease is tuberculosis of the lung [and appears to be due mainly to the human type of tubercle bacillus (Rabinowitsch). Thus of twenty-seven cases of tuberculosis in monkeys the human type of bacillus was found in nineteen and the bovine type in three: the avian type, or modified organisms or mixtures of different types were found in the remaining five].

Dogs.—Tuberculosis is not uncommon among dogs (Cadiot), though the fact has for a long time remained unrecognized. In dogs the lesions often

[1 Fr. Pommelière; Ger. Perlsucht.]

simulate malignant growths and they have been mistaken for neoplasms. Sometimes however they resemble the lesions found in man and this is especially true in cases where cavitation of the lungs has been produced.

Pigs.—Of pigs killed in public slaughter houses [in France] one to ten per thousand are infected with tuberculosis.

As a general rule, the alimentary tract is the part affected. Tuberculous otitis has been recorded in pigs : when it occurs it is probably secondary to some pharyngeal lesion which has spread up the Eustachian tube. Tuberculosis of the respiratory passages and localized tuberculous foci are not often seen. The disease is sometimes of a miliary type and runs a rapid course.

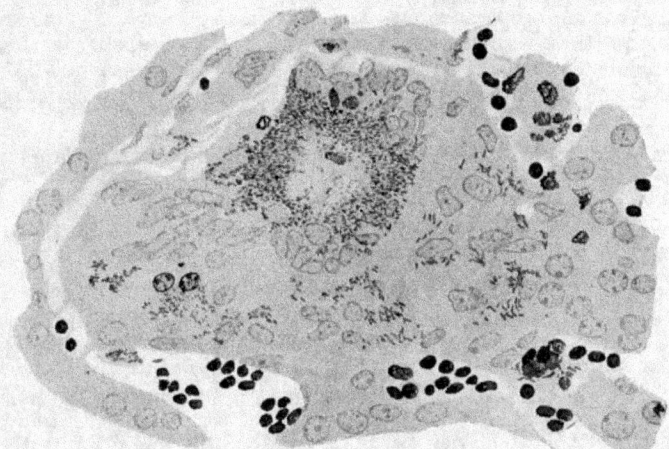

FIG. 188.—Section of the liver of a pig which died 47 days after intra-venous inoculation with 50 mg. of culture of avian tubercle bacilli. This area is typical of the condition found in the liver of this animal. *Note*—(1) the profuse growth of bacilli, with tendency to rosette formation ; (2) the huge " giant cell " showing multiplication of nuclei by irregular longitudinal splitting ; (3) the absence of wandering cells, with the exception of a few small lymphocytes ; (4) the presence of numerous bacilli in the blood stream. × 600. (Eastwood.) [1]

[The nature of the tubercle bacilli occurring in fifty-nine cases of natural tuberculosis in swine was investigated by A. S. Griffith and F. Griffith (for the English Commission). Of these, fifty (including thirty-three cases of generalized tuberculosis) were shown to be due to the bovine tubercle bacillus ; three (in which the disease was localized in the sub-maxillary glands) were caused by the human tubercle bacillus ; five (in which the disease was similarly localized) by the avian tubercle bacillus ; and from one (localized tuberculosis) a mixed culture of avian and bovine tubercle bacilli was obtained.

[Severe and generalized tuberculosis in the pig therefore was by this investigation shown to be due to the bovine tubercle bacillus only.]

Rabbits.—There is no foundation in fact for the popular belief that rabbits are very commonly tuberculous. Spontaneous tuberculosis in the rabbit is, on the contrary, a comparatively rare disease. When it occurs it assumes the pulmonary form.

Goats and sheep.—Both goats and sheep are liable to infection with tuberculosis but the disease in these animals is uncommon.

[[1] This figure as well as figures 191, 192, 193, 198, 199, 200, 201, 202, 203, 205 and 206 are from the Final Report of the Royal Commission on Tuberculosis (Human and Bovine)—Part II. Appendix, Vol. V. ; Dr. Arthur Eastwood—by permission of the Controller of H.M. Stationery Office.]

Horses.—Tuberculosis is rarely seen in horses. When it occurs it is generally of the abdominal type. A pulmonary infection is occasionally seen which may assume the character of a miliary tuberculosis or of diffuse infiltration of the lung, and large sarcoma-like masses may also occur.

[F. Griffith (for the English Commission) investigated five cases of equine tuberculosis. From three of these bovine tubercle bacilli of standard virulence were isolated; the bacilli obtained from the remaining two had the cultural characteristics of the bovine tubercle bacillus associated with a low degree of virulence for all the test animals—calf, rabbit, monkey, guinea-pig, etc. By prolonged passage experiments the virulence of the latter bacilli was increased to that of the bovine tubercle bacillus.]

Cats.—Cats are rarely tuberculous but when the disease occurs the lesions are similar to those seen in dogs. The commonest form is a localized infection of the alimentary canal. [Investigations by A. S. Griffith and F. Griffith show that the bovine tubercle bacillus is the cause of natural tuberculosis in the cat.]

Birds.—Tuberculosis is a very common disease among birds: fowls, pheasants, guinea-fowl, partridges, peacocks, parrots, birds of prey, etc. are, all of them, very frequently infected. [The disease sometimes appears as a rapidly fatal epizootic among farm-yard fowls.]

Tuberculosis in birds is usually primary in the alimentary tract developed [it is affirmed] as a result of swallowing the excreta of tuberculous animals or infected human sputum.

[The investigations of the English Commission do not support this view of the ætiology of avian tuberculosis. Their experiments would tend to show that birds (excluding the parrot) are not susceptible to mammalian tubercle bacilli.]

Tuberculosis in parrots is often associated with a bacillus of the human type and is due to infection from the human subject (Eberlein, Cadiot, Straus). From the experimental point of view parrots are most easily infected with the human tubercle bacillus, next with the bacillus of the bovine type, they appear to be least susceptible to the avian type.

In birds the liver and spleen are the organs most commonly affected: pulmonary lesions are rare though the lungs may become infected in the last stages of the disease. Except in parrots, tuberculosis of the skin, mucous membranes or joints is rarely seen. The disease may be congenital in origin the egg becoming infected in the oviduct (Baumgarten, L. Rabinowitsch, Weber and Bofinger).

The histological appearances of tuberculous lesions in birds are unlike those in mammals: and moreover present different features in the various species. Not uncommonly the viscera will be found to be infiltrated with bacilli while there are no visible tubercles.

Cold-blooded vertebrata.—Tuberculous lesions have been found in the boa-constrictor, the python, the ringed snake [Coluber natrix—Linn.] and the frog. Dubard investigated a tuberculous condition in the carp caused by a bacillus apparently very closely related to the human bacillus.

The *bacillus of ichthic tuberculosis* is very similar to the human tubercle bacillus except that it grows badly at temperatures above 25° C. and in this respect resembles the *para-tubercle* or *acid-fast bacilli* (*vide infra*).

Cultures obtained from the carp are pathogenic to frogs, toads, lizards, tortoises, adders, the common grass snakes, carp and other fish of the same genus, etc. The bacilli are non-pathogenic to guinea-pigs and birds; but by passage through guinea-pigs the organism becomes virulent for that rodent. Ichthic tubercle bacilli when inoculated into rabbits or guinea-pigs behave in the same way as human tubercle bacilli which have become avirulent by prolonged culture on artificial media (Krompechen). Tuberculin prepared from a culture of an ichthic bacillus, which Ramond and Ravaud believe to be the same as the tuberculin obtained from a culture of the human bacillus, does not, when administered in ordinary doses, give the same reaction as Koch's tuberculin but behaves more like that produced from a culture of an avirulent human bacillus (Krompecher).

Friedmann has recovered from two cases of spontaneous pulmonary tuberculosis in tortoises a bacillus which, in many of its characteristics, differs from the ichthic bacillus and which appears to be intermediate between the ichthic and human types of tubercle bacilli. Friedmann's bacillus grows both at ordinary room-temperature and at 37° C.: it is not pathogenic to mammals but in the guinea-pig sets up a local lesion which undergoes spontaneous resolution.

4. Organisms associated with the tubercle bacillus.

In tuberculous lesions in man the tubercle bacillus is found frequently associated with various other organisms, the latter being generally of a pyogenic nature. In cavities in the lungs a rich microbial flora is encountered; the following among other organisms may for instance be found in the lungs in conditions of human pulmonary tuberculous phthisis: *staphylococci, streptococci*, the *pneumobacillus* of Friedlander, *pneumococci, bacillus pyocyaneus, micrococcus tetragenus*, and the bacteria of putrefaction. The hectic fever of patients suffering from tuberculosis is due to the absorption of toxins secreted by these micro-organisms of secondary infection. In glandular and meningeal tuberculosis, etc. it frequently happens that *pneumococci, streptococci*, and *staphylococci* are found together with the tubercle bacillus.

SECTION I. EXPERIMENTAL INOCULATION.

Guinea-pigs or rabbits are generally inoculated with a pure culture of the bacillus emulsified in a little sterile water [or, better, sterile normal saline solution] or with tuberculous tissues pounded in a mortar with a few drops of water [or saline solution]; or the material (sputum, pus, small pieces of tissue, etc.) may be introduced directly either beneath the skin or, in the case of tissues, into the peritoneal cavity.

A. Guinea-pig.

Guinea-pigs inoculated with material containing even a few tubercle bacilli of mammalian or avian origin invariably become infected with tuberculosis.

[The high degree of virulence of the avian tubercle bacillus here suggested was not confirmed by the English Commission.]

Generally speaking, the guinea-pig is less susceptible to avian than to human or bovine tubercle bacilli. According to Weber and Bofinger, the [sub-cutaneous] inoculation of an avian tubercle bacillus leads to a localized infection in guinea-pigs, never to the typical disease. [This opinion is confirmed by the English Commission which finds that " the avian bacillus never produces a progressive tuberculosis in the guinea-pig."] This conclusion however is not supported by the work of numerous other observers.

For purposes of description the infection set up by the inoculation of a mammalian tubercle bacillus will be taken as a type.

1. Sub-cutaneous inoculation.—After 10 days or so there appears, at the site of inoculation, a small indurated nodule which later softens and then forms an abscess; this abscess opens externally leaving an ulcer, the so-called tuberculous chancre. At the same time, the adjacent glands become enlarged, the animal wastes, becomes cachectic and dies in from 1–3 months. *Post mortem* the most conspicuous lesions are those in the spleen and liver: the spleen is much enlarged, ochre-coloured, speckled with caseous tubercles as well as with more recent yellowish granulations; the caseous points may have become confluent giving rise to irregular whitish-yellow mammillated masses: the liver shows similar, though, as a rule, less extensive lesions. The surface of the lungs and of the kidneys and the serous membranes will be found covered with a fine sprinkling of miliary granulations. The lymphatic glands in the neighbourhood of the site of inoculation are caseous.

If the animal be killed within a fortnight to three weeks after inoculation the lesions, especially the tubercles on the spleen and liver, will be found to have attained their characteristic appearances.   At this period of the disease the

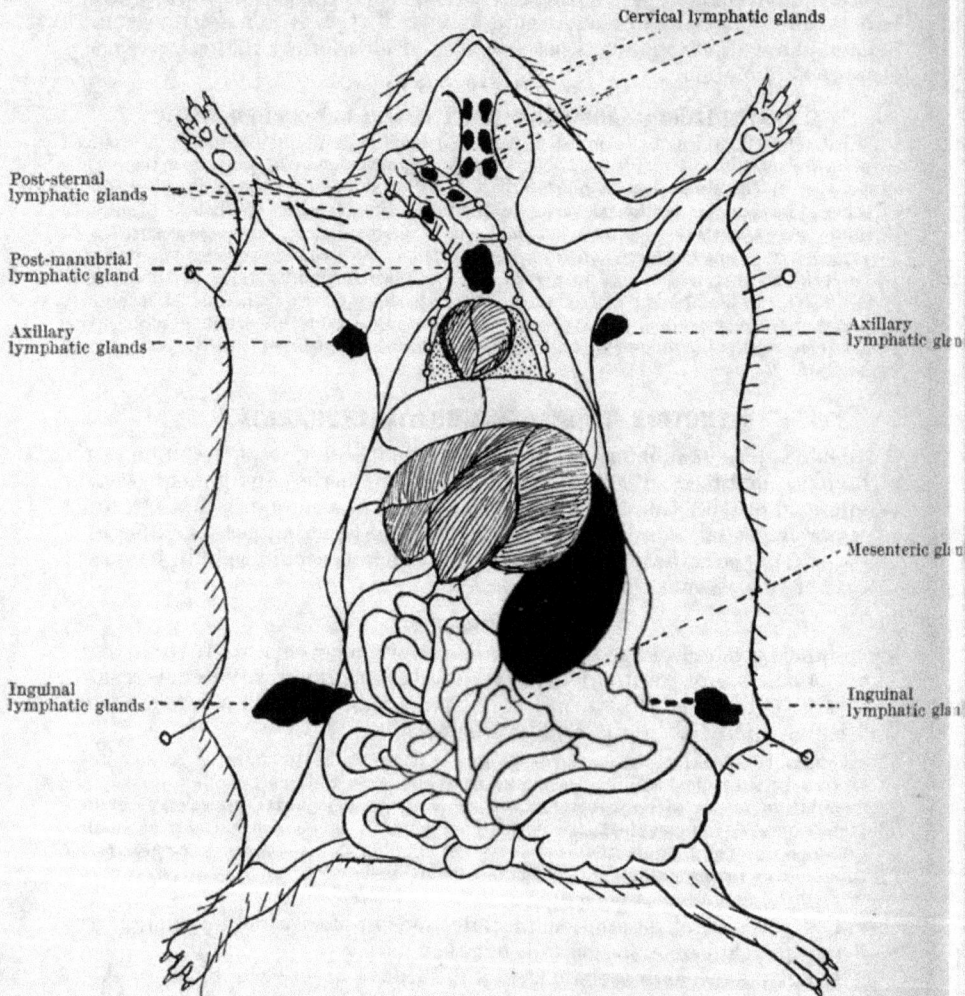

Fig. 189.—Tuberculous guinea-pig (sub-cutaneous inoculation) (3½ weeks).
The areas marked black show the structures mainly affected, viz. the inguinal, axillary, post-manubrial, post-sternal and cervical lymphatic glands and the spleen.

lesions are most marked in the glands on the same side as and adjacent to the site of inoculation.   It is only towards the second month that tubercles appear in the bronchial glands and lungs.

These appearances were first described by Villemin hence this type of generalization of the disease is sometimes known as the *Villemin type.*

[The course of the disease in guinea-pigs following the sub-cutaneous inoculation of bovine tubercle bacilli was worked out by A. S. and F. Griffith.   The material was inoculated in the inguinal region. The guinea-pig killed five days later showed a local thickening only. The ten-day guinea-pig showed in addition to a local lesion, lesions in the superficial inguinal glands and in the axillary and sternal glands. The twenty-day animal showed extension of the disease to the deep inguinal, iliac and manubrial glands and to the spleen, liver and portal glands : one tubercle was found in the lung. In the thirty-day guinea-pig the disease had reached the lungs and bronchial glands, the intestines and mesenteric glands as well as the cervical, lumbar and coeliac glands. The thirty-eight day guinea-pig showed tubercles in the kidneys.

[The duration of life of the guinea-pig will depend upon the dose of tubercle bacilli administered : but the extent of the disease is not found to vary much, since an extremely small number of either bovine or human tubercle bacilli is able to set up general progressive tuberculosis in the guinea-pig (English Commission).]

2. Cutaneous inoculation.—If the inguinal region of a guinea-pig be shaved and rubbed with a piece of absorbent wool soaked in sputum containing tubercle bacilli, the corresponding glands will become enlarged a week or fortnight later and the animal will die of tuberculosis in about two months. *Post mortem*, lesions typical of the disease will be found (Osman Nouri). This method of inoculation is very useful for diagnosis, because it involves no risk of death from septicæmia, an accident very likely to happen if the material be inoculated sub-cutaneously.

3. Intra-peritoneal inoculation.—The course of the disease is similar to that just described but is more rapid. Death occurs in 2–6 weeks being preceded by an increasing degree of cachexia. Lesions similar to those already described are found in the tissues : the peritoneum is infiltrated with tubercles and the omentum forms a compact, caseous mass, while the mesenteric and inguinal glands are also caseous. The indurated nodule at the site of inoculation (chancre) is, of course, non-existent.

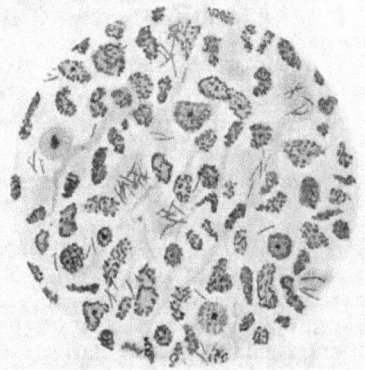

A large dose of an human or avian culture is fatal to guinea-pigs in a few days when inoculated intra-peritoneally. *Post mortem* there is an excess of fluid in the pleuræ but no tubercles are visible in the internal organs (Koch, Straus and Gamaléia).

[Following intra-peritoneal inoculation the course of the disease is as follows : in guinea-pigs which die in under 14 days, there is a local lesion

FIG. 190.—Scraping from the spleen of a tuberculous guinea-pig (carbol-fuchsin and methylene blue). (Oc. 2, obj. $\frac{1}{12}$th, Zeiss.)

in the wall of the abdomen : the omentum is thickened, the mesentery and peritoneum are inflamed and covered with a thin membrane, the mesentery is also thickened : the spleen is enlarged and speckled with minute points, the liver shows minute foci, the kidneys are normal : the pleural cavities are filled with fluid, the lungs are collapsed and often consolidated and show minute grey points in the dark red areas : the pyloric, lumbar and ventral mediastinal (or sternal) glands are severely affected, the portal and other abdominal lymphatic glands less affected while the bronchial glands are usually only slightly affected. In guinea-pigs which survive for 3 weeks to a month there is severe tuberculosis of the peritoneum, omentum and mesentery : the spleen is enlarged : tubercles are visible in the spleen, liver and lungs and sometimes in the kidneys : the pleural cavities sometimes contain an excess of fluid and the pleurae are covered with small grey tubercles. " With smaller and smaller doses of tubercle bacilli the lesions in the organs begin to resemble more and more those produced by sub-cutaneous inoculation " (English Commission).]

4. Intra-pulmonary inoculation.—There is a caseous focus at the point of entry of the inoculation needle and the lungs are spotted with grey tubercles in the adjacent area. In the abdominal viscera the lesions are similar to those following sub-cutaneous inoculation.

5. Inhalation.—Guinea-pigs are readily infected with tuberculosis by causing them to inhale dried and finely powdered tuberculous sputum or dust mixed with tubercle bacilli from cultures. The animal dies with well-marked caseous broncho-pneumonia.

6. Ingestion.—Guinea-pigs have been infected by feeding them with the milk of a cow suffering from tuberculous phthisis (Villemin and Parrot, Klebs). In animals infected in this way it occasionally happens that there are no lesions in the intestines. [When feeding guinea-pigs with tubercle bacilli Cobbett sometimes observed a generalized infection involving the lungs but pulmonary tuberculosis apart from a generalized infection did not occur.]

B. Rabbits.

Rabbits are not so susceptible to tuberculosis as guinea-pigs. A fatal result does not always follow the inoculation of a small amount of tuberculous material. Occasionally the local lesion is of long standing before the disease becomes generalized. Inoculation of bovine or avian tubercle bacilli is followed by a more severe infection than inoculation with human tubercle bacilli.

1. Sub-cutaneous inoculation.—According to the amount of virus inoculated the animal will live for from one to several months. A local induration (tuberculous chancre) is formed but the glands are often not affected while the lungs are, as a rule, the first to show tubercles.

The human tubercle bacillus often fails to kill rabbits when inoculated sub-cutaneously. On an average, four out of five strains of human tubercle bacilli produce only a local lesion and this undergoes spontaneous resolution. The disease resulting from inoculation of bovine bacilli is always more severe.

2. Intra-peritoneal inoculation.—The course of the disease is more rapid. Tubercles are found on the peritoneum, spleen, liver etc. Death often occurs before the disease has reached the thorax.

3. Intra-pulmonary inoculation. Inhalation.—The disease runs the same course and presents the same lesions as in the guinea-pig. Fränkel and Troje have produced caseous pneumonia in rabbits as a result of intra-tracheal inoculation.

4. Inoculation into the anterior chamber of the eye.—By inoculating the bacillus into the anterior chamber of the eye the progress of the lesions can be easily followed. During the third week, the iris becomes covered with tuberculous granulations, the eye swells, the aqueous humour becomes cloudy and occasionally the whole eye is transformed into a purulent abscess : the glands of the neck hypertrophy and infection becomes disseminated giving rise to a generalized tuberculosis of the Villemin type (p. 298).

5. Intra-venous inoculation.—The disease produced by intra-venous inoculation may be of one of two types :

(a) *Miliary tuberculosis.*—According to the amount of material inoculated death may take place in 2 or 3 weeks. The viscera and serous membranes are covered with fine miliary tubercles.

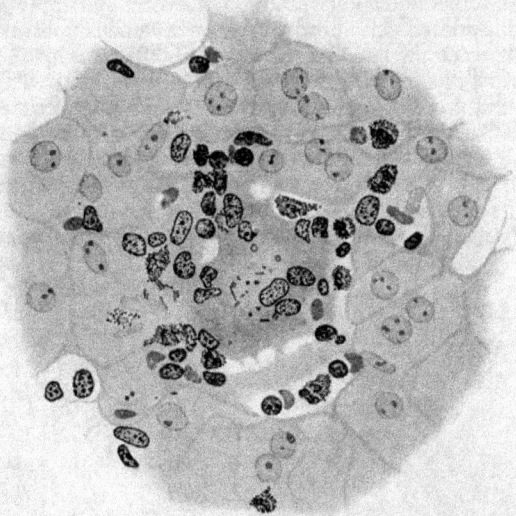

FIG. 191.—The figure represents the lesion produced in a liver of a rabbit 3 days after inoculation in the vein of an ear with 1 mg. of finely emulsified culture of mammalian tubercle bacilli and illustrates a typical tubercle, with a peripheral infiltration of small lymphocytes and finely granular oxyphil leucocytes. × 600. (Eastwood.) (See footnote p. 295.)

(b) *The Yersin type.*—Death takes place in 12–25 days. The animal loses flesh and rapidly becomes cachectic. The temperature is very much raised. *Post mortem* the only visible lesion is a marked hypertrophy of the liver and spleen. There are no visible tubercles. The liver, spleen and bone marrow contain enormous numbers of bacilli.

Strauss and Gamaléia held that this second (Yersin) type of infection is the characteristic change following intra-venous inoculation of the bacillus of avian origin, but numerous facts have been brought forward to prove that intra-venous inoculation of the bacillus of human origin may result in this type of infection (Yersin, Nocard, and others). Granchez and Ledoux-Lebard have produced either of these two types of infection at will by regulating the amount of material inoculated.

Generally speaking, however, intra-venous inoculation of the rabbit with the avian bacillus produces a simple tuberculous infiltration of the organs without any visible lesions.

[The experiments of the English Commission showed that when death occurred about a fortnight after the inoculation of the *avian* tubercle bacillus into the veins, disease of the Yersin type is seen *post mortem*. But if death be postponed to a later period tubercles are visible in the internal organs.

[When rabbits were inoculated intra-venously with *bovine* tubercle bacilli they sometimes died within a fortnight of a very acute disease which did not altogether correspond to the Yersin type. In these cases the lungs were solid with masses of grey tubercles, the bronchial glands were œdematous and the spleen enlarged, there were indefinite tuberculous foci in the liver and kidneys, and on microscopical examination tubercle bacilli were found to be numerous in all the tissues of the body.

[The intra-venous inoculation of *human* tubercle bacilli will occasionally lead to an acute condition similar to that just described as sometimes following intra-venous inoculation of bovine tubercle bacilli.]

[The results of the inoculation experiments carried out on rabbits by Cobbett, A. S. Griffith and F. Griffith on behalf of the English Commission may be summarized here.

[The bovine tubercle bacillus produces a severe and fatal general tuberculosis whether inoculated sub-cutaneously, intra-venously or intra-peritoneally.

[The human tubercle bacillus very occasionally produces a fatal general tuberculosis when inoculated intra-venously or intra-peritoneally but as a rule the lesions found are those of a slight and retrogressive tuberculosis. Sub-cutaneous inoculation never leads to a fatal result : for example, 125 rabbits inoculated sub-cutaneously with doses varying from 1–100 mg. of culture of the human tubercle bacillus and killed after 3–24 months all showed retrogressive tuberculosis.

[The avian tubercle bacillus usually produces a fatal general tuberculosis by whichever of the three methods it be inoculated. This type is less virulent for rabbits than the bovine tubercle bacillus and it causes generally less disease of the internal organs.]

6. Infection by feeding.—The ingestion of tuberculous material mixed with food does not always lead to infection of the rabbit: some animals entirely escape the disease, others show lesions of the alimentary and respiratory tracts while the majority contract an infection strictly limited to the respiratory passages (Weleminsky). The sub-maxillary glands are first infected then the cervical glands, followed by the bronchial; finally the pulmonary parenchyma is attacked. The rabbit is more susceptible to the ingestion of bacilli of bovine or avian origin than of bacilli of human origin.

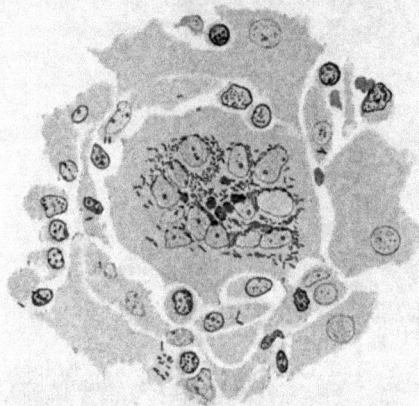

FIG. 192.—An early lesion produced in the liver of a rabbit 14 days after inoculation in the vein of an ear with 1 mg. of culture of avian tubercle bacilli. An example of a "giant cell" produced by the avian tubercle bacillus. The bacilli have been growing abundantly within the "cell" and are very small. × 600. (Eastwood.) (See footnote p. 295.)

[Cobbett, A. S. Griffith and F. Griffith found that feeding with bovine tubercle bacilli was always followed by a progressive tuberculosis in rabbits while when fed with the avian tubercle bacillus only one rabbit out of seventeen fed showed severe generalized tuberculosis. Progressive tuberculosis cannot be produced in rabbits by feeding them on human tubercle bacilli.]

C. Dogs.

Dogs may be infected by inoculating them with large doses of the human tubercle bacillus but they are much more resistant to the avian bacillus though not absolutely immune to it (Grancher and Héricourt). [The dog is one of the few species of animals in which the effects produced by the bovine and human tubercle bacilli are identical. Dogs "have shown themselves insusceptible to avian tubercle bacilli inoculated by the most severe method and in relatively large doses" (work of the English Commission).]

1. Sub-cutaneous inoculation.—The disease following sub-cutaneous inoculation is not necessarily fatal : it may remain localized or become a generalized infection. [The English Commission found that the dog is resistant to the sub-cutaneous inoculation of either the bovine or the human tubercle bacillus.]

2. Intra-peritoneal inoculation.—Death occurs 2–3 months after inoculation of a pure culture of tubercle bacilli into the peritoneal cavity. Inoculation is followed by peritonitis with excess of fluid, the formation of false membranes, adhesion of the coils of the intestine, and infection of the glands. The disease, ultimately, becomes generalized. [Intraperitoneal inoculation with moderate doses of cultures of either the human or the bovine tubercle bacillus is usually but not invariably fatal (work of the English Commission).]

3. Intra-venous inoculation.—Death takes place 1–2 months after inoculation into a vein of 0·25 c.c. of a thick emulsion of bacilli from a glycerin-agar culture. The pulmonary lesions are the most marked while the liver, spleen, etc. may also show tubercles.

4. Inhalation.—Tappeiner infected dogs by causing them to breathe an atmosphere charged with dried and powdered tuberculous sputum. Lesions were found, *post mortem*, in the lungs, spleen and kidneys.

5. Infection by feeding.—Arloing fed dogs with cultures of the tubercle bacillus and in three out of seven cases found lesions in the alimentary canal; in two other cases the disease was generalized (in the spleen and lungs). [Dogs are very resistant to infection with tuberculosis by feeding, especially adult animals (A. S. Griffith and F. Griffith, for the English Commission).]

D. Cattle.

(*a*) *Cattle are very susceptible to infection with the bacillus of the bovine type.*

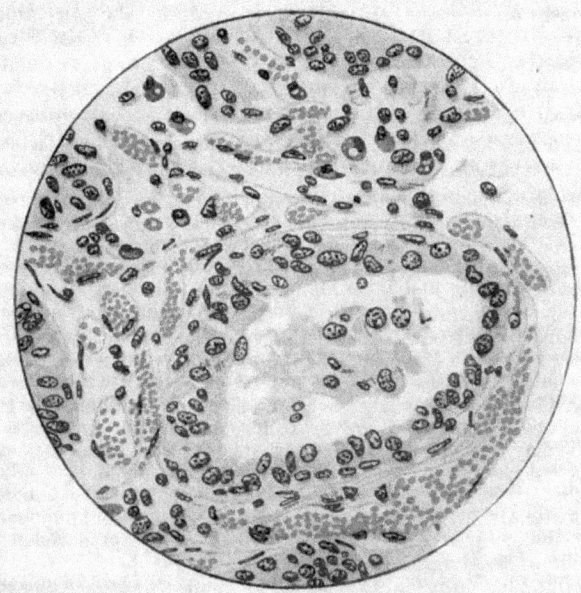

FIG. 193.—Section of the udder of a cow which died 299 days after intra-mammary inoculation of bacilli derived from a typical bovine virus. The specimen is taken from an affected portion of the mammary tissue showing early infection. *Note*—(1) The high vascularity, (2) the interstitial infiltration with bacilli, (3) the entrance of bacilli into the glandular epithelium, and (4) the excretion of bacilli into a mammary tubule. It is of importance to note that lesions such as this, which are obviously unrecognizable clinically, excrete numerous bacilli. Carbol-fuchsin and methylene blue. × 400. (Eastwood.) (See footnote p. 295.)

In calves, infection by way of the alimentary canal leads to very severe symptoms (Vallée). ["Feeding with the bovine tubercle bacillus does not

so readily set up general progressive tuberculosis in the calf as does inoculation " (English Commission).] Feeding is the most certain method of
infecting the tracheal and bronchial glands (Vallée, Calmette and Guérin).
The bacilli may pass through the intestinal wall without producing any
apparent lesion either of the mucous membrane or of the mesenteric glands
provided that very small doses of bacilli and young animals be used, conditions, that is, similar to those obtaining in the spontaneously contracted
disease (Vallée).

Calmette, Guérin and Delearde fed calves with 0·1 gram of bovine bacilli and
found that they reacted to tuberculin 45 days later. The tracheal and bronchial
glands were swollen and hard but not caseous, the mesenteric glands were normal in
appearance, but, on inoculation, both sets infected guinea-pigs.

(b) *Cattle can also be infected with some strains of bacilli of human origin*
(Chauveau, Ravenel, Arloing, M. Wolff, Schottelius, Spronck and others).
[These strains were no doubt strains of bovine tubercle bacilli infecting
human tissues (*vide ante*). The English Commission has demonstrated that
the human tubercle bacillus is incapable of causing progressive tuberculosis
in bovine animals.]

Schottelius fed bovine animals on several occasions with tuberculous sputum.
In cows he found a tuberculous enteritis with caseous glands ; and, in calves,
caseation of the sub-maxillary and mesenteric glands. [This should be read
in conjunction with the comment above. The English Commission investigated two cases of pulmonary tuberculosis in which the sole infecting agent
was the bovine tubercle bacillus.] In calves, bacilli from human lesions
whether inoculated beneath the skin or into the lungs or veins produced
general tuberculosis (De Jong, Sturmann). Inoculation (sub-cutaneous and
intra-peritoneal) of human tuberculous material into calves may lead to a
rapid and generalized infection (Fibiger and Jensen, Eber) [if the bacilli of
human origin are of the bovine type not otherwise (English Commission)].

In two cases in which Eber obtained a very severe infection in calves the bacilli
were derived from children in whom only intestinal lesions were present. It may
be admitted that the children were infected by swallowing bovine bacilli but it is no
less true that human tuberculosis can infect calves and bovine tuberculosis children.
[It seems to be a perfectly justifiable inference from the work of the English Commission that the bacilli used by Eber in which he produced a severe infection in
calves must have been derived from children suffering from an infection with bovine
bacilli. Tubercle bacilli of the human type merely give rise to a slight and retrogressive type of tuberculosis in calves and in the sense that human tuberculosis
due to the bovine tubercle bacillus can infect calves the statement in the preceding
paragraph is true. With regard to the reciprocal infection of children by bovine
tuberculosis it may be pointed out that fourteen out of the twenty-seven cases of alimentary tuberculosis investigated by the English Commission were due to bovine
tubercle bacilli.] Moreover, Eber produced an acute miliary tuberculosis with
tuberculous material from an adult human being suffering from pulmonary tuberculosis and tuberculous meningitis. [Probably an infection produced by the bovine
tubercle bacillus. See English Commission results, *ante*.]

Such facts [may be considered to] constitute a sufficient basis for rejecting Koch's
hypothesis of the existence of two separate and distinct species of tubercle bacilli.

E. Birds.

(a) *Birds are easily infected with the avian tubercle bacillus.* Fowls may
be infected by any method of inoculation (sub-cutaneous, intra-venous,
feeding etc.), and the ingestion of cultures, infected tissues or other pathological tuberculous products readily produces the disease. *Post mortem*
tubercles are found on the abdominal viscera but chiefly in the [spleen and]
liver.

Intra-venous inoculation of the avian bacillus leads to the death of the fowl in a fortnight to three weeks with a disease of the Yersin type (*vide* also p. 301). [In the experience of the English Commission it was generally longer —5 to 6 weeks.]

(*b*) Those who believe that the human and avian tubercle bacilli belong to different species hold that the fowl cannot be infected with the human tubercle bacillus : but this conclusion [in the opinion of many] can no longer be maintained in view of the experiments carried out by Koch, Nocard and Cadiot, and Gilbert and Roger. These observers [appear to] have shown that the fowl becomes infected with tuberculosis as the result of the ingestion of human tuberculous material and of pure cultures of human tubercle bacilli. It is [by some considered as] certain that the fowl may become infected by the ingestion of the sputum of phthisical persons.

It has to be remembered that fowls often resist infection with bacilli of human origin and that when infection does occur the disease is chronic and leads to the formation of tubercles in the internal organs. [Fowls are resistant to mammalian tubercle bacilli of whatever source when inoculated intra-peritoneally, sub-cutaneously and by feeding, but frequently succumb to intra-venous inoculations. Tubercle bacilli killed by exposure to steam at 100° C., whether avian or mammalian, may produce however·similar effects when inoculated intra-venously into the fowl; these effects are therefore not a true tuberculosis but are to be attributed to the toxic action of the bacilli (F. Griffith, for the English Commission).]

F. Cold-blooded animals.

Frogs and fish do not appear capable of infection with bacilli of human and avian origin. But there is an observation to the effect that true tubercles have been produced by inoculating bacilli of human origin into the peritoneal cavity of frogs and carp (Moret).

Bertarelli [is stated to have] succeeded in infecting snakes (*Varanus varius*) by inoculating them under the skin with human tuberculous sputum but failed with cultures of bacilli of avian origin.

Moeller infected the blind worm [*Anguis fragilis*] with bacilli of human origin (p. 334).

Sorgo and Suess produced tuberculous lesions (caseating masses [at the site of inoculation] and occasionally generalization) in two blind worms and four snakes with bacilli of human origin, though many of their experiments were negative. In blind worms the bacilli retain all the characteristics associated with the human tubercle bacillus but in snakes they [are said to] undergo a partial change and to develop some of the characteristics of bacilli of ichthic origin.

SECTION II.—MORPHOLOGY.

1. Microscopical appearance.

Human, avian and ichthic tubercle bacilli all have, in the main, the same characteristics. In cultures they are small, very slender, generally non-motile, rods.

Ferran says that the tubercle bacillus is motile, but the conclusions arrived at in his paper cannot all be accepted unreservedly. Arloing confirms Ferran's opinion. By sub-cultivating a glycerin-potato culture on to glycerin-broth this observer obtained motile bacilli. Schumowsky, in a similar experiment, also found motile bacilli. Auclair [is said to have] succeeded in converting the tubercle bacillus into a motile saprophyte, etc.

In cultures on solid media the bacilli are arranged in long wavy coils some-

U

thing like a moustache due to the regular interlocking of the bacilli with a common orientation.

This arrangement of the bacilli can be readily shown by lightly pressing a cover-glass on the surface of a glycerin-agar culture and lifting it off without friction. The film should be fixed by heat and stained by one or other of the methods described below. On examining with an oil-immersion lens the appearances reproduced in fig. 194 will be seen.

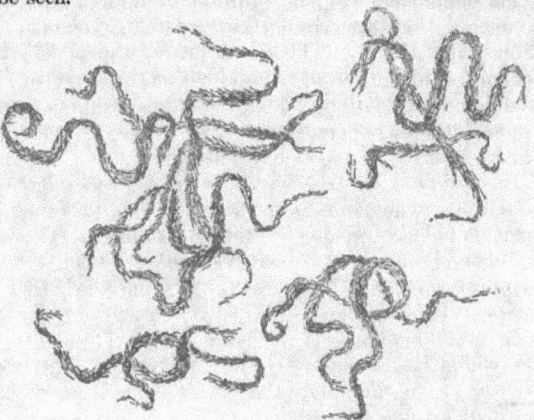

FIG. 194.—Tubercle bacillus: impression preparation. (After Koch.) $\frac{7\,0\,0}{1}$.

The bacillus must be stained before it can be found in fluids and tissues and it will be necessary to describe the various methods of staining before embarking upon a description of its characteristics.

Staining methods.

Special methods have to be adopted in order to stain the tubercle bacillus.

The tubercle bacillus is difficult to stain with the basic aniline dyes but once stained it resists the decolourizing action of such powerful agents as dilute mineral acids. Only two other pathogenic bacilli share this characteristic with the tubercle bacillus viz. the leprosy bacillus from which it is easily distinguished and the bacillus of Verruga peruana.[1]

This acid-fast property of the tubercle bacillus serves as a means of diagnosing the organism in fluids and tissues in which it is present. The property of resisting the decolourizing action of acids seems to be due to the presence of a fatty or waxy substance insoluble in alcohol and ether (Koch, Tavel, Viquerat). By treating the bacilli with warm xylol, Borrel extracted a waxy substance which was acid-fast while the bacilli had lost this property.[2]

[1] Besides these two there are a few other bacilli, like the tubercle bacillus, capable when deeply stained of resisting the decolourizing action of dilute acids. Such, for instance, are the smegma bacillus and the bacillus of Tavel (the so-called syphilis bacillus of Lustgarten)—but these, unlike the tubercle bacillus, are decolourized by absolute alcohol or ether—likewise the various acid-fast bacilli of Bienstock, Gottstein, Möller, Rabinowitsch, etc. (*Vide infra, The acid-fast bacilli.*)

[2] In opposition to the opinion expressed by H. Aronson, Sabrazès has shown that by treating tissues for sections with ether, xylol and chloroform the characteristic staining properties of the tubercle bacillus are in no way interfered with. And the same is true of picric acid, carbolic acid and perchloride of mercury none of which prevent subsequent staining by the Ziehl-Neelsen method. On the other hand, undiluted mineral acids, 2 per cent. chromic acid, formalin, bichromates, osmic acid, etc. either interfere with or entirely prevent subsequent staining by the carbol-fuchsin method. [Eastwood, however, working for the English Commission hardened tissues in formalin (p. 338).]

Numerous methods of staining the tubercle bacillus have been suggested, but they all depend upon the principle enunciated above.

The various methods in most frequent use will now be described, but the necessity for beginners to limit themselves to *one* method which they thoroughly understand and upon the results of which they can rely cannot be too strongly emphasized. The Ziehl-Neelsen method is by far the best.

The tubercle bacillus stains by Gram's method but with difficulty and the stain, carbol- or anilin-gentian-violet (Nicolle), must be allowed to act for, at least, 10 minutes. The bacilli are always granular by this method.

Much has shown that in cattle inoculated with tuberculosis, tuberculous nodules are often seen, *post mortem*, in the lungs in which no bacilli can be demonstrated by Ziehl's method, though the presence of bacilli in the lesions is proved by the result of inoculations (the same is true of "cold abscesses" in man). If, however, Gram's method of staining be adopted, leaving the preparations in the violet for 48 hours and in the iodine solution for 24 hours, large numbers of bacilli can be seen in the lesions. From this observation Much concludes that in addition to the acid-fast tubercle bacillus there is a virulent form which is non-acid-fast.

A. The staining of films.

1. Ziehl-Neelsen method.

Method recommended.

The principle of the method.—If a film stained with carbol-fuchsin be treated with a diluted mineral acid, the background and all the organisms, with the exception of the tubercle bacillus (and also those of leprosy, verruga and the "acid-fast" bacilli, pp. 350 and 345), will be decolourized. The tubercle bacillus retains its red colour. If, now, the preparation be stained with an aqueous solution of methylene blue, the background and the decolourized organisms take up the blue while the tubercle bacillus remains red.

Technique.—**1.** Spread, dry and fix a film on a cover-glass in the ordinary way. Hold it in a pair of Cornet's forceps and flood it with a large drop of

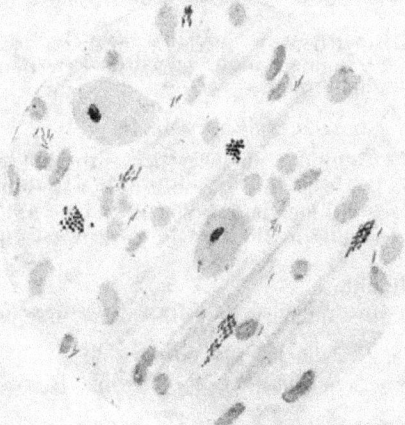

FIG. 195.—Tubercle bacilli in sputum. Carbol-fuchsin and methylene blue.
(Oc. 2, obj. ₁₂th, Zeiss.)

Ziehl's carbol-fuchsin (p. 138). Hold the cover-glass over a small flame (the pilot light of a Bunsen, for example) and heat very gently until steam just begins to rise ; continue the heating for two minutes, being careful not to boil the stain and to see that the staining solution does not dry up.

2. Pour off the stain and treat for a few seconds with a 33 per cent. solution of nitric acid (distilled water 2 volumes, pure nitric acid 1 volume) or a 25 per cent. solution of sulphuric acid (distilled water, 3 volumes, pure sulphuric acid, 1 volume), [or 25 per cent. hydrochloric acid (Eastwood)]. The preparation now assumes a yellowish tint.

[This method of decolourization appears to be perfectly satisfactory in the case of bacilli from cultures and was moreover the method adopted by Eastwood in his work for the English Commission. But it is undoubtedly true that tubercle bacilli direct from human and animal tissues—in sputum, for example—will sometimes lose the stain in these strong acids. In searching for the tubercle bacillus therefore in fresh material in which its presence is suspected it is recommended that a 2·5 per cent. solution of sulphuric or hydrochloric acid be used and that the film be not exposed to the acid for a longer time than is absolutely necessary.]

3. Wash freely in water: the preparation should now be pale pink, and if the pale pink colour does not appear, the decolourization has been insufficient and the film must be treated with acid again.

4. Pour a few drops of absolute alcohol on the film: when decolourization is complete the film should be a very faint pink colour.

By using alcohol decolourization can be pushed much further than would be possible with acid since the latter would ultimately decolourize the tubercle bacillus. A further advantage in using alcohol is that it decolourizes the smegma bacillus and thus a possible source of error is eliminated.

5. Wash well in water. Stain for a few moments with an aqueous solution of methylene blue.

6. Wash in water. Dry. Mount in balsam.

Note.—When it is merely a question of searching for the tubercle bacillus it is a great advantage not to counterstain the background after decolourizing with alcohol; the tubercle bacilli are much more easily seen when they appear stained deep red on an unstained or faintly pink background.

For this purpose the above procedure is stopped at the end of Stage 4 and, after washing, the preparation is examined in water. If after examination it be thought desirable to keep the film it may be counterstained with blue and treated as described above.

This simpler method is particularly applicable when the bacilli are likely to be present in small numbers; in any case it renders the detection of the bacilli more rapid, and beginners will find it of great use.

2. Ehrlich's method.

1. Stain the film for 5 minutes in the warm with aniline-violet.

2. Decolourize in 33 per cent. nitric acid for a few seconds.

3. Wash in water: continue the decolourization with absolute alcohol.

4. Stain for a few seconds in the cold in a saturated aqueous solution of vesuvin.

5. Wash. Dry. Mount.

The tubercle bacilli are stained violet, other structures brown.

3. Gabbé's method.

This method is merely a modification of Ziehl's but is less reliable and more difficult.

1. Stain with carbol-fuchsin as above.

2. Decolourize and counterstain at the same time by dipping the film for a minute in the following solution.

| Methylene blue, | · | | · | · | · | · | · | 2 grams. |
| 25 per cent. sulphuric acid, | | · | · | · | · | · | · | 100 c.c. |

3. Wash. Dry. Mount.

The methods described by Stocquart, by Pithion and Roux (of Lyon) are modifications of the above but are of no interest.

4. Spengler's method.

1. Stain the films by gently warming them in Ziehl's solution.

2. Treat for a few seconds with picric-alcohol.

Saturated aqueous solution of picric acid,	60 c.c.
95 per cent. alcohol,	40 ,,

3. Wash three times in 60 per cent. alcohol.

4. Decolourize rapidly (about 20 seconds) in 1 in 6 nitric acid, then in 60 per cent. alcohol.

5. Treat again with picric-alcohol. Wash. Dry. Mount.

5. Fränkel's method.

1. Stain the films in the warm for 5 minutes with aniline-fuchsin (prepared in the same way as aniline-violet using an alcoholic solution of fuchsin instead of alcoholic gentian-violet).

2. Transfer the films direct to the following solution for 1 minute.

90 per cent. alcohol,	50 c.c.
Aniline water,	30 ,,
Pure nitric acid,	20 ,,
Saturated alcoholic solution of methylene blue,	Q.S. to obtain a deep blue colour.

3. Wash in distilled water. Dry. Mount.

6. Herman's method.

Prepare the following solutions :

A. Krystal violet,	1 gram.
95 per cent. alcohol,	30 c.c.
B. Ammonium carbonate,	1 gram.
Distilled water,	100 c.c.

Immediately before use, pour three parts of solution B into a watch-glass, add one part of solution A and mix intimately.

1. Heat the staining bath until it just begins to boil and leave the films in it for a minute.

2. Transfer the films to a 10 per cent. solution of nitric acid for 4 or 5 seconds.

3. Wash in absolute alcohol to complete the decolourization.

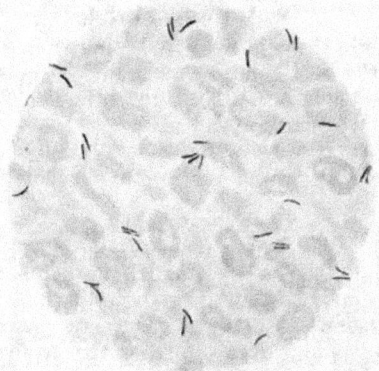

FIG. 196.—Tubercle bacilli in sputum : Herman's method.
(Oc. 2, obj. ₁⁄₁₂th, Zeiss.)

4. Transfer the films for 30 seconds to the following solution :

Eosin,	1 gram.
60 per cent. alcohol,	100 c.c.

5. Wash very rapidly in alcohol. Dry. Mount. The tubercle bacilli are stained violet and the background bright pink.

7. Lustgarten's method (modified).

Sabouraud, by slightly modifying the method devised by Lustgarten for staining his so-called bacillus of syphilis, has perfected a method of staining the tubercle bacillus which he affirms to be very delicate and precise. The method is as follows :

1. Stain the film in Ziehl's solution in the cold for 1 or 2 hours or at 50° C. for 15 minutes.

2. Treat the film for 1–3 seconds with a 1·5 per cent. solution of potassium permanganate.

3. Dip at once into a freshly prepared, saturated, aqueous solution of sulphurous acid for a few seconds until the film is decolourized.

The sulphurous acid solution can be conveniently prepared by bubbling the gas from a cylinder of liquefied sulphurous acid through distilled water.

4. Wash in water and counterstain with an aqueous solution of methylene blue for 1–3 minutes.

5. Wash in water. Dry. Mount in balsam.

8. Koch's method.

This, the earliest method used for the detection of the tubercle bacillus, is chiefly of historical interest.

1. Place the films for 1 day at room temperature or for a few hours at 45°–50° C. in the following bath :

Saturated alcoholic solution of methylene blue, - - -	1 c.c.
10 per cent. aqueous solution of potash, - - - -	2 ,,
Distilled water, - - - - - - - -	200 ,,

2. Transfer the films to a saturated aqueous solution of vesuvin; in about a quarter of an hour a brown tint takes the place of the original blue colour save in the tubercle bacilli which still retain the blue stain.

B. The staining of sections.

The methods just described are applicable with slight modification to the staining of sections : but in this case it is essential that the *staining should always be done in the cold*.

1. Ziehl-Neelsen's method.
Method recommended.

1. Stain the section in the cold for 15–30 minutes in Ziehl's fuchsin.

2. Decolourize in the acid solution for a few seconds. Wash in water.

3. Continue the decolourization with absolute alcohol until the section is a pale pink colour. Wash in water.

4. Stain the groundwork with an aqueous solution of methylene blue.

5. Wash. Pass rapidly through absolute alcohol, clove oil, and xylol. Mount in balsam.

2. Kühne's method.
Method recommended.

The following unpublished method of Kühne has been quoted by Borrel. It is particularly useful for staining sections of lung. The action of the hydrochloride of aniline, which is the decolourizing agent used, is not so rough as that of mineral acids and does not alter the arrangement and shape of the cells.

1. Stain the section for 2 minutes in Bœhmer's hæmatoxylin or hæmatein (p. 218) to stain the nuclei of the cells. Wash in distilled water.

2. Stain in the cold with Ziehl's fuchsin for a quarter of an hour.

3. Decolourize for 30–60 seconds in a 2 per cent. aqueous solution of aniline hydrochloride.

4. Continue the decolourization with absolute alcohol.

The cells of the groundwork are now unstained with the exception of the nuclei.

The section may be treated with orange-yellow which stains particularly the red cells of the blood.   After staining with orange, dehydrate in absolute alcohol.

5. Clear in clove oil and xylol.   Mount in balsam.

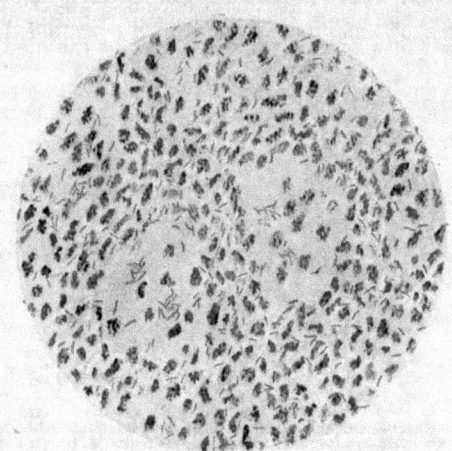

FIG. 197.—Section of human tuberculous lung. Carbol-fuchsin and methylene blue. (Oc. 2, obj. $\frac{1}{12}$th, Zeiss.)

3. Ehrlich's method.

1. Stain in aniline-violet for 12 hours in the cold.

2. Decolourize in 33 per cent. nitric acid for a few seconds.   Wash.

3. Complete the decolourization in absolute alcohol.

4. Counterstain with a saturated aqueous solution of vesuvin.

5. Dehydrate rapidly in absolute alcohol.   Clear with clove oil and xylol. Mount in balsam.

4. Letulle's method.

1. Stain the nuclei with hæmatoxylin as in Kühne's method. Wash in distilled water.

2. Stain with Ziehl's fuchsin in the cold for a quarter of an hour.   Wash rapidly in distilled water.

3. Wash in absolute alcohol for 30 seconds.

4. Treat with the following solution for 5 minutes :

Iodgrün, - - - - - - - - - -	1 gram.
2 per cent. solution of carbolic acid, - - - - -	100 c.c.

5. Decolourize in absolute alcohol.

6. Clear in clove oil and xylol.   Mount in balsam.

The groundwork is stained very pale grey-lilac ; the nuclei, violet ; the bacilli, deep red.   This method can be used for tissues hardened in Müller's fluid.

5. Herman's method.

Herman's method (p. 309) can, according to its author, be applied to the staining of frozen sections of tissues fixed in acetic perchloride solution.   The technique is the same as for films, the stain being allowed to act for 1 minute with steam rising.

6. Lustgarten's method (modified).

1. Stain for some hours in the cold in carbol-fuchsin.

2, 3, 4, 5. As for films (p. 310).

6. Wash in water.   Dehydrate rapidly in absolute alcohol.

7. Clear in clove oil and xylol.   Mount in balsam.

This method is useful when searching for bacilli in sections of the liver, where they are often difficult to find.   It is also available for tissues hardened in Müller's fluid.

Appearance of the bacilli in stained preparations.

In stained preparations, tubercle bacilli vary in length from 2–5μ and in breadth from 0·3–0·5μ. They are generally of the same thickness throughout. In some preparations the bacilli are homogeneous, while in others they appear as though composed of a number of small oval or rounded grains separated

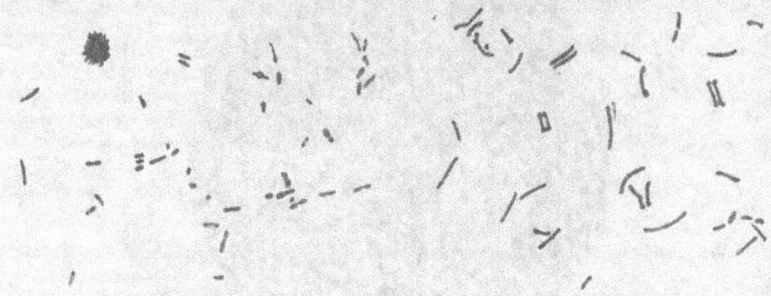

FIG. 198.   FIG. 199.

FIG. 198.—From a culture, 20 days old, on inspissated horse serum, of a mammalian virus of low experimental pathogenicity to bovines and rabbits and of vigorous cultural growth. The virus was isolated from a human being. The bacilli are very short; some of them show a central constriction. The clump at the bottom of the figure illustrates the tendency of mammalian bacilli to stick together and the difficulty of separating them by emulsification. The bacilli from serum cultures of this virus proved shorter than the average vigorously growing bacillus of human origin. This figure illustrates the impossibility of distinguishing with certainty, under the microscope, "human" from "bovine" bacilli. Carbol-fuchsin. ×2150. (Eastwood.) (See footnote p. 295.)

FIG. 199.—From a culture, 22 days old, on glycerinated broth, of a typical mammalian virus of low pathogenicity to experimental animals and of vigorous cultural growth. The virus was isolated from a human being. The bacilli, obtained from a copious growth, are for the most part long and curved, and with a tendency to beading. Carbol-fuchsin. ×2150. (Eastwood.) (See footnote p. 295.)

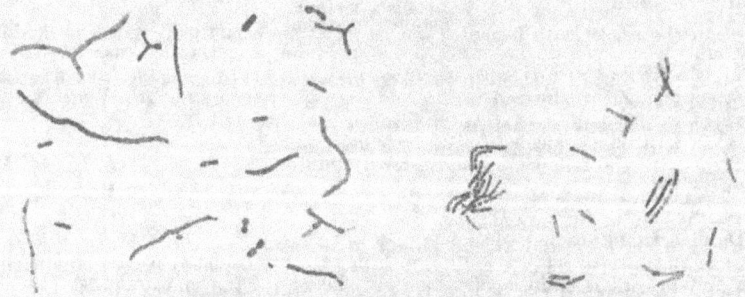

FIG. 200.   FIG. 201.

FIG. 200.—From a culture, 44 days old, on glycerin-agar, of a mammalian virus which was isolated from the bronchial gland of a human being. The occurrence of branched forms of the mammalian tubercle bacillus in cultures obtained from ordinary media, such as glycerinated agar, broth, or potato, is, in Eastwood's experience, extremely rare. The figure also shows other forms of bacilli, some very long, some very short, and many with globular or oval, darkly stained bodies variously situated. Bacilli with such appearances as these are common in agar cultures. Carbol-fuchsin. ×2150. (Eastwood.) (See footnote p. 295.)

FIG. 201.—Tubercle bacilli from milk obtained from the udder of a cow, which had received an intra-mammary inoculation with a typical bovine virus. The bacilli are of various lengths; many of them are curved and regularly beaded. The bacilli here shown are such as are commonly met with in cow's milk; it would obviously be impossible for anyone to decide, on morphological grounds, that they were of bovine rather than of human origin. Carbol-fuchsin. ×2150. (Eastwood.) (See footnote p. 295.)

by clear unstained intervals. They are sometimes straight but more often somewhat S-shaped or bent on themselves.

[According to Eastwood (working for the English Commission) mammalian tubercle bacilli when grown on serum are of a very uniform character; straight, uniformly stained and about 1μ long but ranging from $0.75–2.5\mu$. On media containing glycerin the average length is greater: in the same film, short (1μ) bacilli, longer ($2–4\mu$) forms and very long ($5–7$ or 8μ) forms

FIG. 202.   FIG. 203.

FIG. 202.—Tubercle bacilli from a culture, 2 months old, on glycerin-agar, of a typical avian virus. In addition to short forms, long branching forms are found. The branching frequently emanates from a darkly stained spherical point, which in some instances is of a greater diameter than the breadth of the bacillus. Branching is much commoner with the avian than with the mammalian bacillus. Carbol-fuchsin. ×2150. (Eastwood.) (See footnote p. 295.)

FIG. 203.—Tubercle bacilli from a culture, 2 months old, on glycerinated potato, of a typical avian virus (same virus as fig. 202). The bacilli show a tendency to grow in long, parallel threads. Carbol-fuchsin. ×2150. (Eastwood.) (See footnote p. 295.)

may all be encountered and on these media the bacilli are frequently curved and many are stained irregularly (beaded and globular forms). Branched forms are very rarely indeed seen in cultures.

[Avian tubercle bacilli grown on glycerin serum are generally very short ($0.5–1\mu$) and rather thick. On other media they are as a rule longer and more irregular. Among these irregular forms can be found examples of all the forms assumed by the mammalian bacilli; and large club-shaped swellings are not uncommonly seen while branching occurs more frequently than with the mammalian bacillus.]

Koch regarded as spores the unstained intervals which are sometimes seen in the bacilli. There is now a tendency to regard the deeply stained granules seen at the ends or in the length of some bacilli as spores (Babès, Ehrlich). Gavina thinks that he has stained terminal spores in bacilli grown in presence of antiseptics.

In cultures extraordinarily short bacilli are occasionally seen. In other cases, particularly in old cultures, large, branched bacilli ending in a club-shaped swelling are found (fig. 204). These giant forms afford

FIG. 204.—Involution forms of the tubercle bacillus. (After Metchnikoff.)

ground for grouping the tubercle bacillus with the streptothrices. Coppen-Jones [and Eastwood] have described ray structures with club-shaped ends in tuberculous lesions exactly similar to the structures seen in the grains of actinomycosis (figs. 205 and 206).

In sputum and in tuberculous tissues the bacilli are found singly or arranged in groups and in the latter case may lie parallel to one another. Occasionally two bacilli are seen crossing one another at a more or less acute angle or arranged like a V.

[Bacilli obtained from the living tissues are longer and not so uniform in appearance as bacilli cultivated on serum (Eastwood).

[Both mammalian and avian bacilli when growing freely in the tissues of their host are usually shorter and more uniformly stained than those which are growing under adverse conditions.]

2. Cultural characteristics.

A. Conditions of growth.

The tubercle bacillus only grows in artificial culture provided that the medium contains serum (Koch), glycerin (Nocard and Roux), yolk of egg (Dorset), or fragments of tissues (Lumière).

It is an aërobic organism and only grows at temperatures above 30° C. In the case of human tubercle bacilli growth ceases at 41° C. and in the case of bovine bacilli at 44°–45° C. The optimum temperature is 38° C.

Certain precautions must be observed in *sowing* the tubercle bacillus. For preference, the material will be taken from a lesion in the guinea-pig or rabbit (bacilli taken directly from human tissues grow badly on artificial media), rubbed up in a sterile mortar and portions of it transferred with a stout wire to tubes of coagulated serum. It is better [when sowing cultures of the human tubercle bacillus] to use serum to which 4 per cent. of glycerin has been added before coagulation or blood agar. Never sow tuberculous tissue directly on to glycerin agar: the cultures are more than likely to fail. It is immaterial if the surface of the medium be slightly torn. Sow a large number of tubes as many of them will remain sterile [and others are likely to be contaminated with other organisms]. Incubate at 37°–38° C. Growth only becomes visible to the naked eye after an interval of 12 days or so but continues to increase for about 4 weeks. As soon as colonies appear in any tube cover the mouth with an india-rubber cap to prevent evaporation and the consequent drying up of the medium. [It is perhaps even better to seal the tube with paraffin or sealing-wax.]

When a growth has been obtained, sub-cultures can be sown on various media; it is always advisable to sow a good deal of material and [until a fair amount of experience has been acquired] to sow several tubes.

[The human tubercle bacillus grows more luxuriantly than the bovine tubercle bacillus in artificial culture so that the former is sometimes described as the *eugonic* and the latter as the *dysgonic* tubercle bacillus.]

Cultures of the tubercle bacillus have a characteristic but rather pleasant odour.

B. Characters of growth on various media.

1. Coagulated serum. (a) *The bacillus of the human type.*—After the culture has been incubating at 37°–38° C. for 12 days or so a number of small, white, round, scaly, dry-looking colonies are seen scattered over the surface of the medium. On further incubation the colonies become raised but retain their scaly appearance, and the margins are irregular in outline. Generally speaking, and especially when recently isolated, the colonies do

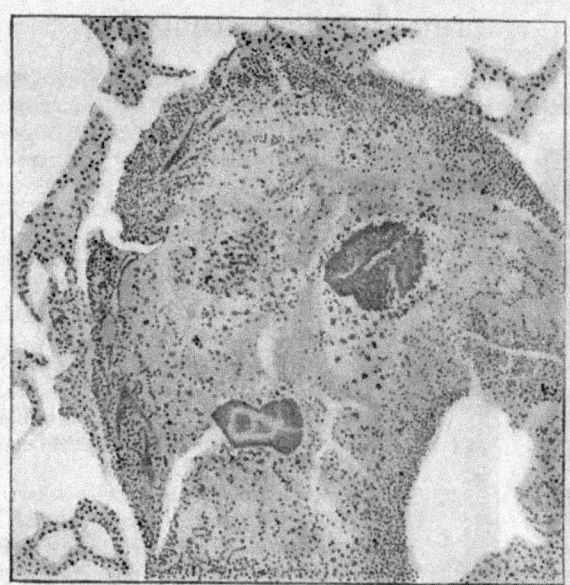

FIG. 205.—Lung of a rabbit, killed 63 days after inoculation, partly intra-venously and partly intra-peritoneally, with a total of 140 mg. of killed culture of bacilli of bovine origin. This figure is a representative example of numerous lesions found in the lungs. The material which stains strongly with eosin bears a curious resemblance to actinomyces. The lesion as a whole is abundantly infiltrated with finely granular oxyphil leucocytes. As carbol-fuchsin has not been applied, no bacilli are stained. Eosin and methylene blue. × 112. (Eastwood.) (See footnote p. 295.)

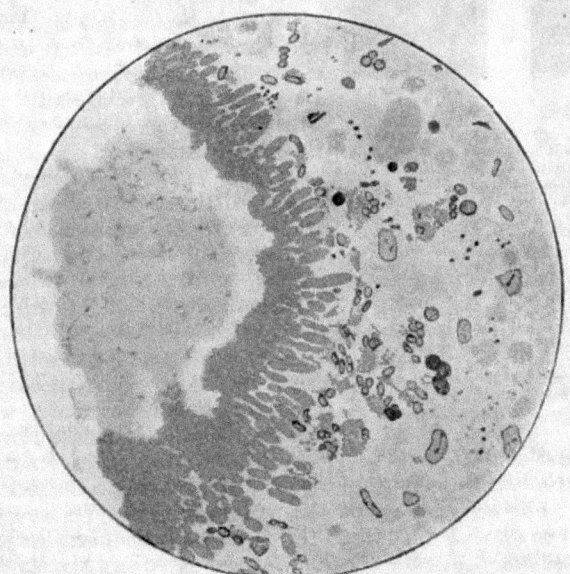

FIG. 206.—From the same lung as fig. 205, the specimen having been stained with carbol-fuchsin before counterstaining with eosin and methylene blue. A portion of one of the actinomyces-like nodules. The club formation may perhaps be attributable to dissolved constituents of the large number of dead bacilli inoculated. Some bacilli not yet disintegrated are seen in the lower part of the field. The abundance of multinuclear leucocytes, which are shedding their granules, suggests that the disintegration of the dead bacilli is due to the digestive action of these cells. Carbol-fuchsin, eosin and methylene blue. × 865. (Eastwood.) (See footnote p. 295.)

not become confluent. After sub-cultivating three or four times, however, they may coalesce to form a dried wrinkled layer.

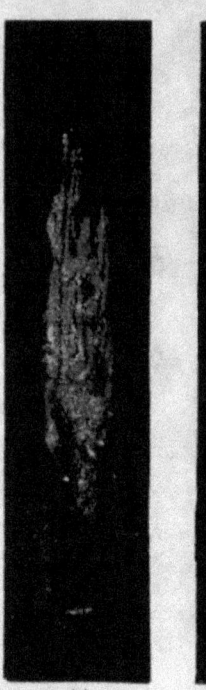

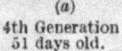

(a)   (b)

4th Generation   4th Generation
51 days old.   3 weeks old.

FIG. 207.—Tubercle bacilli of human origin cultivated on glycerin-agar. (a) A culture from the tuberculous mesenteric glands of a child aged 8½ years, who died from multiple stricture of the gut (due to tuberculous ulceration). The culture grew moderately well on artificial media, and resembled the more easy-growing cultures of bovine origin : it had high virulence for the calf, rabbit and guinea-pig. (b) A culture from a mesenteric gland of a case of general tuberculosis originating in the alimentary tract in a child aged 10 months. The strain grew luxuriantly on media containing glycerin and was only slightly virulent for the rabbit (A. S. Griffith).[1]

[(β) *The bovine type.*—On pure serum the growth of the bovine tubercle bacillus presents no marked differences from that of the human tubercle bacillus.]

(γ) *The avian type.*—The bacillus of avian origin yields a more abundant growth on serum than the human bacillus. The culture is thick and generally has a moist, greasy lustre.

2. Glycerin agar.—Except for primary cultures, this is the best medium upon which to grow the tubercle bacillus. A little glucose may with advantage be added to the medium (p. 44).

(a) *The human type.*—Growth begins as on serum but the colonies are both larger and more numerous. They rapidly become confluent and form a thick, whitish, dry, rough, scaly, mammilated layer even in recently isolated specimens and after being sub-cultivated a few times on glycerin-agar the growth becomes very abundant, moist, greasy and folded. When old the growth has a reddish tint.

[(β) *The bovine type.*—The bovine tubercle bacillus grows much less luxuriantly on media containing glycerin—such as glycerin-serum, glycerin-agar, or glycerin-potato—than the human tubercle bacillus (English Commission).]

(γ) *The avian type.*—Some authors have contrasted the growth of the avian bacillus on glycerin-agar with that of the human bacillus. The latter, they say, gives rise to a dried, wrinkled layer, while the former produces a moist, fatty growth. But as has been seen, the human type frequently gives a copious growth of a moist, fatty appearance : and it is equally true that the avian type occasionally produces a dry, scaly growth (Nocard, Grancher, Fischel).

["The most characteristic point of difference between the mammalian and avian tubercle bacilli is that the cultures of avian bacilli are moist and easily emulsified, while on most media the mammalian cultures are dry and can only be broken up with difficulty" (F. Griffith, for the English Commission).]

3. Egg medium.—A useful medium for the growth of the tubercle bacillus consists of the white and yolk of eggs coagulated and sterilized by discontinuous heating at 72°–74° C. (Dorset, Capaldi, A. S. Griffith).

Bezançon and Griffon mix one part of uncooked yolk of egg with two parts

[1 Fig. 207 and also figs. 208, 209, 210, and 211 are from the Final Report of the Royal Commission on Tuberculosis (Human and Bovine)—Part II. Appendix, Vol. I.; Dr. A. Stanley Griffith—by permission of the Controller of H.M. Stationery Office.]

of 6 per cent. glycerin-agar melted in a water bath and kept at 50° C. The ingredients are mixed as thoroughly as possible and the tubes allowed to set in a sloped position. From human lesions, moist, greasy-looking colonies can be obtained in a week on this medium.

Phisalix prepares a medium by mixing yolk of egg with a potato mash containing a little glycerin. The medium is sterilized in the autoclave. Lubenau gives a method which has already been described (p. 54).

According to Park, yolk of egg media are particularly useful for the isolation of tubercle bacilli from tuberculous material and for the differentiation of the human from the bovine type. On media made with yolk of egg but containing no glycerin bacilli of the bovine type grow easily : and on the same media but containing glycerin bacilli of the human type grow poorly at first while the bovine type does not grow at all.

[According to A. S. Griffith (English Commission) the egg medium is invaluable for obtaining the tubercle bacillus in pure culture from tissues or other material. It is of great value also for sub-cultures; on it the tubercle bacillus retains its vitality for a longer period than on any other medium, and sub-cultures can often be obtained on this medium from old cultures which fail to grow when sown on other media. He gives the following method of preparation of the medium—"The yolk and the white of fresh eggs are thoroughly mixed by shaking in a flask; salt solution is added in the proportion of one to three of egg; the mixture is filtered through muslin, distributed into tubes, sloped and coagulated in a serum inspissator at 80° C."]

4. Blood-agar.—Bezançon and Griffon recommend the addition of rabbit-blood to agar for starting cultures from human or animal tissues. Growth appears early and soon becomes very copious, the colonies absorb the hæmoglobin and become chocolate-coloured.

(a)   (b)

Fig. 208.—Primary cultures of tubercle bacilli on the egg medium. (a) Bovine tubercle bacilli of human origin obtained direct from sputum by means of antiformin : (b) Human tubercle bacilli (A. S. Griffith). (See footnote p. 316.)

5. Tochtermann's agar.—Dissolve 10 grams of peptone, 5 grams of sodium chloride, 5 grams of glucose and 20 grams of agar in a litre of water. Add half-a-litre of calf serum, mix, boil for 15 or 30 minutes, filter in the warm, distribute into tubes and sterilize at 100° C. for 50 minutes.

6. Hesse's agar.—For obtaining cultures of the tubercle bacillus from human sputum Hesse recommends sowing the material on the surface of a special agar prepared as follows :

Dissolve 5 grams of salt, 30 grams of glycerin, and 20 grams of agar in a litre of water. Add 5 c.c. of a normal solution of carbonate of sodium and 5 grams of Heyden's albumose (*Nährstoff Heyden*) dissolved in 50 c.c. of water.

Boil for 15 minutes. Filter in the warm. Sterilize at 100° C. and pour into Petri dishes.

7. Fragments of tissue as culture media.—A. and L. Lumière obtain very copious growths commencing in about 36 hours on fragments of liver or spleen.

Wash the liver and spleen of an adult bovine animal or calf in distilled water, heat in the autoclave for three-quarters of an hour to shrink them and then cut into rectangular pieces; after soaking in a 6 per cent. aqueous solution of glycerin for an hour the pieces are placed in potato tubes and sterilized in the autoclave for 15 minutes. It is best to sow from a potato culture.

Gioelli uses pieces of human placenta immersed in broth, or, better, placenta-broth containing 0·5 per cent. of sodium chloride and 6 per cent. glycerin in place of liver.

8. Bile.—Calmette recommends pieces of potato soaked in a mixture of 95 parts fresh bile and 5 parts sterilized glycerin. The bile-glycerin mixture ought to be kept for 3 weeks at the temperature of the laboratory before being used. [On an ox-bile-glycerin-potato medium tubercle bacilli of the bovine type grow more rapidly and more luxuriantly than on the usual media while bacilli of the human type grow with difficulty on this medium and the avian type not at all. On the other hand bacilli of the human type will grow rapidly on an human-bile medium as will bacilli of the avian type on a fowl-bile medium ; the cultivations of these two types on these media respectively are similar in appearance to cultivations of the bovine type on ox-bile (Calmette)].

9. Glycerin-broth.—Glycerin-broth, or better, glucose-glycerin-broth is a very useful medium for the growth of the tubercle bacillus. To sow a broth culture of the tubercle bacillus, raise as large a piece of growth as possible from the surface of an agar tube or other solid medium (it is even better to lift the film of growth from the water of condensation) with a fairly large platinum loop and float it very carefully on to the surface of the broth. It is advisable to transfer three or four loopsful if a large flask is to be sown.

Growth generally takes place in the form of a pellicle. After incubating for about a fortnight a whitish area appears around the piece of growth which was sown, this gradually extends and ultimately forms a thin delicate film covering the whole of the surface of the medium. The film is at first dry and fragile but becomes thicker in course of time : sometimes it remains dry and scaly and sometimes becomes greasy, moist and wrinkled. Not infrequently the film creeps up the sides of the vessel, sometimes to a height of 1 cm. Rarely no film at all is formed and in this case the growth consists of a flocculent deposit. Whatever the form of growth the broth remains quite clear.

[To obtain a successful growth on glycerin-broth requires considerable attention to details. The material used for sowing the medium must be young and actively growing—perhaps the film growing on the water at the bottom of a glycerin-potato culture gives the best results. If the culture be sown at the right moment the growth will, in the case of the human type, spread and cover the surface of the glycerin-broth in a Roux's bottle laid on its flat side in a fortnight. At other times no change whatever is visible in the material sown for weeks, then little white nodules appear which are the precursors of a growth which once it starts to spread covers the surface very rapidly. The flask must be carefully sealed.]

The tubercle bacillus can also be grown on ordinary broth containing no glycerin (Gioelli [and others]).

Pour a layer of vaseline oil about 1 mm. deep on the surface of the broth. Sterilize and sow the medium from an agar culture or with pieces of tuberculous tissue which have been carefully crushed. The material used for sowing should float on the surface of the vaseline or between the vaseline and the broth. Should it fall to the bottom it is only necessary to shake the flask carefully and the drops of oil will float the material to the surface again. To make microscopical preparations blot up the oil from the slide with blotting paper.

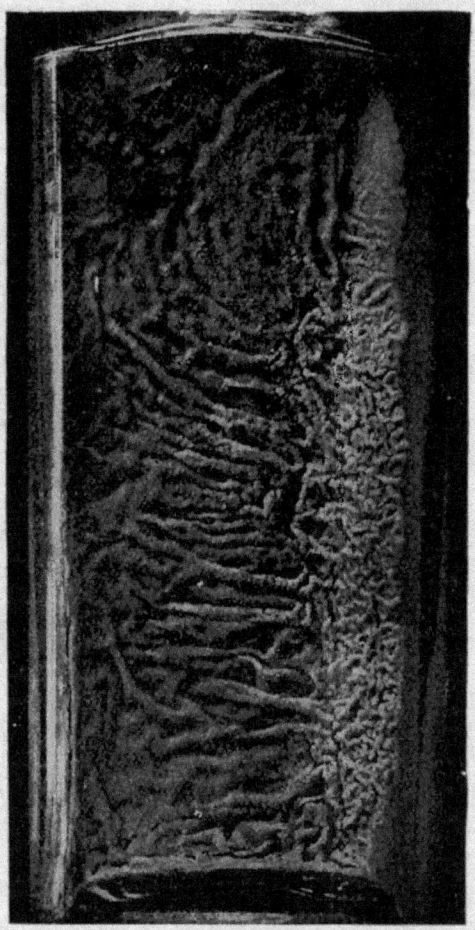

7th Generation
5 weeks old.

Fig. 209.—A culture of the human tubercle bacillus on glycerin broth. The culture was isolated from the lung of a man aged 33, who died of phthisis; it grew luxuriantly on all media containing glycerin, was virulent for chimpanzees, monkeys and guinea-pigs but had only slight virulence for calves, rabbits and horses (A. S. Griffith). (See footnote p. 316.)

10. Glycerin-fish-broth.—This medium has been recommended by Martin. The cultural characteristics are the same as on ordinary glycerin-broth.

Mince the flesh of an herring and add to it one and half times its weight of water. Heat slowly and keep it boiling for three-quarters of an hour. Filter through Chardin paper several times while warm and when clear add 6 per cent. of glycerin. Neutralize if necessary. Distribute in tubes and sterilize in the autoclave.

11. Potato.—The tubercle bacillus will grow on ordinary potato but to obtain the best results it is necessary to add glycerin (Nocard).

Cut the potatoes into suitably-shaped pieces and leave them to soak in a large quantity of a 15 per cent. solution of glycerin in water for two days in the ice chest, then transfer them to a number of Roux's tubes and sterilize in the ordinary way.

On this medium growth appears about the twelfth day generally taking the form of a thick, folded, soft layer, very occasionally it is dry and wrinkled. The growth often extends in the form of a film over the liquid which has drained away into the lower part of the tube. This film will be found very useful for sowing liquid cultures.

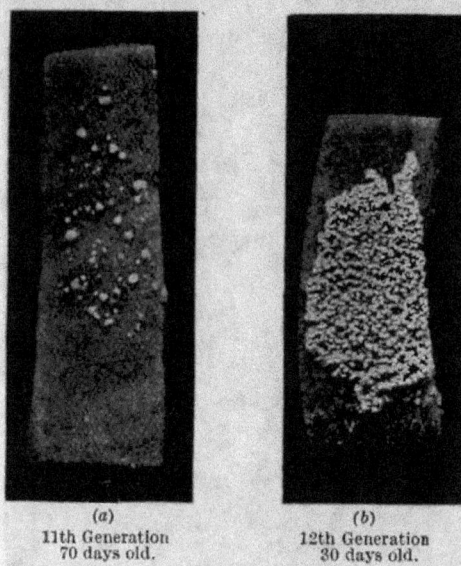

(a)   (b)
11th Generation   12th Generation
70 days old.   30 days old.

FIG. 210.—Cultures of tubercle bacilli of human origin on glycerin-potato. (a) A bovine tubercle bacillus isolated from the meninges of a child aged 2 years who died of tuberculous meningitis. The culture proved highly virulent for the calf, rabbit, guinea-pig and cat. (b) An human tubercle bacillus isolated from the sputum of a youth aged 16 years suffering from pulmonary tuberculosis. The culture grew luxuriantly on artificial media and was slightly virulent for rabbits (A. S. Griffith). (See footnote p. 316.)

Jurewitch recommends a potato broth. Leave a potato mash to macerate in its own weight of water for a day, then filter through muslin and to the filtrate add an equal volume of meat extract, 0·5 per cent. peptone (Chapoteaut), 0·5 per cent. salt and 3 per cent. glycerin: make distinctly alkaline and complete the preparation as in making ordinary broth (p. 30).

C. Differentiation of the various types of tubercle bacilli by cultural methods.

[A. S. Griffith (for the English Commission) investigating the cultural characters of mammalian bacilli, proceeds as follows. From the primary culture on egg the bacillus is transferred to pure serum until it is growing vigorously: it is then tested on the differential media. On these media human tubercle bacilli produce luxuriant growths while bovine tubercle bacilli grow much less luxuriantly. It is possible thus to determine the type of bacillus by cultural characteristics alone. The differential media are serum, agar, potato and broth, all containing glycerin. The serum, agar and broth contain 5 per cent. glycerin and are prepared in the customary way. " The potato is cut into slices put in water and soaked for 24 hours in a 5 per cent. glycerin blue-litmus solution, which is poured off before the first sterilization ; at the bottom of each tube is some absorbent cotton-wool soaked with glycerin solution which helps to keep the potato moist."

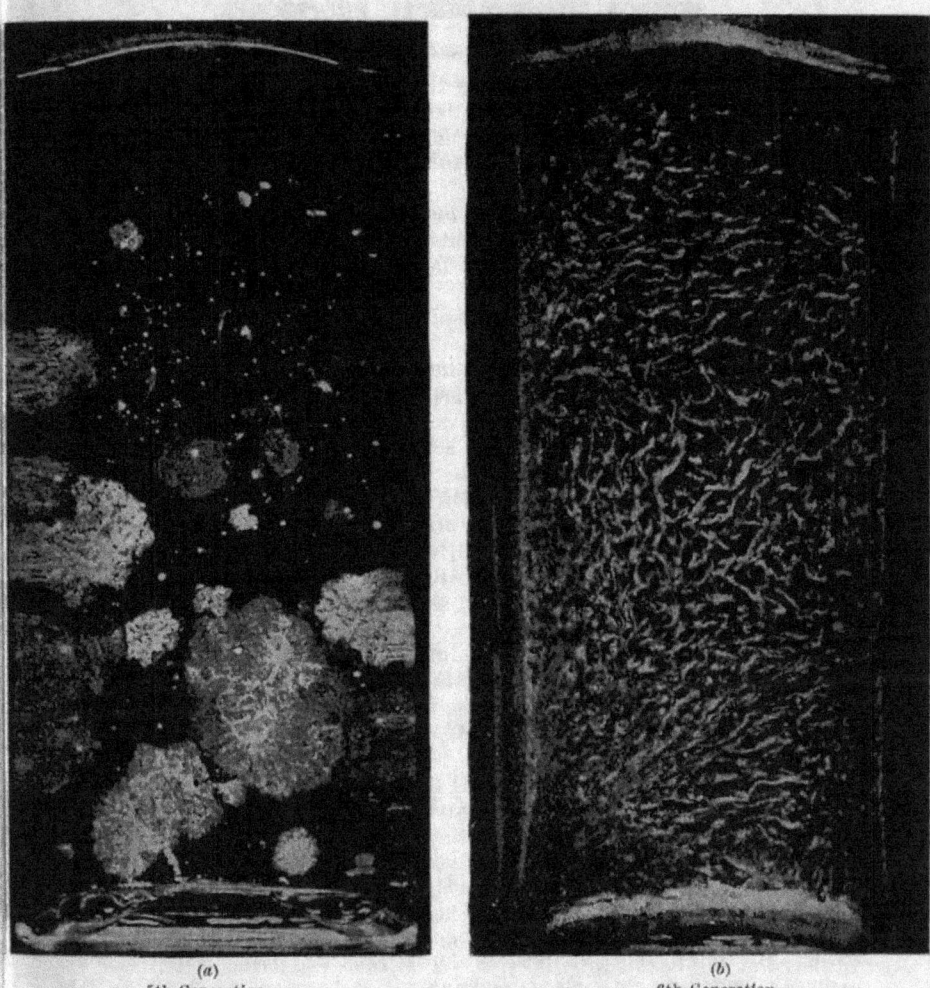

(a)   (b)
5th Generation   6th Generation
over 3 months old.   55 days old.

FIG. 211.—Cultures of tubercle bacilli of human origin on glycerin-gelatin. (a) A bovine tubercle bacillus of intermediate virulence for calves and rabbits ; (b) an human tubercle bacillus slightly virulent for rabbits (A. S. Griffith). (See footnote p. 316.)

[F. Griffith states that " it is possible to differentiate between the avian and the mammalian types of tubercle bacilli from cultural characters alone, but it is necessary in order to avoid error, that a sufficient variety of media should be used. In primary cultures on serum or serum agar the transparent colonies of avian tubercle bacilli are easily recognized. On glycerin-agar

x

and glycerin-potato the avian bacillus frequently forms a wrinkled or warty growth resembling a culture of human tubercle bacilli; but the characteristic difference is evident when the growth is touched with the spatula."]

SECTION III.—BIOLOGICAL PROPERTIES.

1. Viability and virulence.

In determining the viability of the tubercle bacillus, animal inoculation, not artificial cultivation, must be the test [because sub-cultures often fail—though the culture used for sowing them may be alive—especially if the operator has not had considerable experience in dealing with cultures of the tubercle bacillus].

In cultures, the tubercle bacillus is only slightly resistant to adverse conditions and this is one of the arguments in favour of the view that it does not form spores. Exposure to 70° or 75° C. for 10 minutes kills cultures of the bacillus in liquid media.

[F. Griffith (for the English Commission) made numerous tests on bacilli grown on various media and concludes that "cultures" (of mammalian or avian tubercle bacilli) "will maintain their vitality for long periods whether kept in the incubator or at room temperature." Avian bacilli in one culture-tube were found alive after 1067 days, and bovine bacilli similarly after 990 days : "but no distinction could be drawn between the types of bacilli tested."]

Agar cultures [are said to] lose their virulence after a few months [but F. Griffith found that "little if any attenuation" (of the bovine bacillus) was caused by residence on artificial media" (serum and glycerin-serum) for periods up to 1487 days. Human tubercle bacilli tested after from 2–3 years' artificial cultivation (on serum) showed no appreciable diminution in virulence (A. S. Griffith).]

[Calmette and Guérin state that cultures on glycerin-bile-potato (p. 318) are at first increased in virulence but that repeated sub-cultivation on that medium diminishes the virulence of the bacilli for certain animal species.]

In moist sputum the bacillus is not destroyed at 75° C. but is killed in 5 minutes at a temperature of 100° C.

Cultures of the tubercle bacillus, or tubercle bacilli in sputum, retain their vitality for a long period when dried at the ordinary temperature of the air. Under these conditions, the bacilli may retain their virulence for several months (Galtier) ; they are not destroyed by exposure to a dry temperature of 100° C. for 2 or 3 hours and are capable of resisting a temperature of 70° C. for more than 7 hours (Welch, Grancher, Ledoux-Lebard).

The combined action of desiccation and sunlight [is said to] attenuate the virulence of the bacillus (Candler, Koch, Migneco and Ransome).

Zilgen mixed some dust with dried tuberculous sputum and exposed the mixture to the action of sunlight : under these conditions, the tubercle bacillus retained its virulence for about 140 days. According to de Thoma the virulence of the bacilli in sputum left in the patient's room disappears after two months and a half but is retained indefinitely when the sputum is kept in the dark under the same conditions.

The virulence of tubercle bacilli in sputum exposed to the alternate action of moisture and desiccation is retained for several months (Malassez and Vignal).

The tubercle bacillus appears to maintain its vitality in water for a long time : it has been recovered from sterile water after 70 days (Chantemesse and Widal) and after exposure to running water for 150 days (Cadéac and Malet).

Putrefaction has little effect on the tubercle bacillus. Tuberculous tissues left to decompose in water for 20 and 40 days did not lose their virulence (Galtier). Tuberculous lungs buried for 167 days have been found to be virulent (Cadéac and Malet). Schottelius found that the bacillus was virulent in tissues which had been buried for 2 years. Gærtner made the same observation after burying the tuberculous tissues for a winter.

Action of antiseptics.—In cultures, the tubercle bacillus is somewhat sensitive to the action of antiseptics. According to Yersin it is killed in 30 seconds in a 5 per cent. solution of phenol; in 5 minutes in absolute alcohol, and in 1 per cent. iodoformed ether; in 10 minutes in a 1 in 1000 solution of perchloride of mercury, and after several hours in 0·3 per cent. thymol or 0·25 per cent. salicylic acid. It resists the action of 4 per cent. boric acid for more than 12 hours.

According to Koch, the following substances readily hinder the growth of the bacillus in cultures, viz. :—the essential oils, naphthol, fuchsin, methylene blue, gentian-violet, and especially the cyanides of gold and silver. Cyanide of gold in the proportion of 1 part in 2 millions stops the growth of the bacillus.

But in tuberculous fluids and tissues the resistance of the bacillus is much greater : thus 1 in 500 salicylic acid, 1 in 1000 bromine, creosote, quinine, 1 in 1000 perchloride of mercury, and formalin vapour have no effect on the bacillus. Six per cent. carbolic acid has a doubtful influence, a 1 in 4000 solution of hydrofluoric acid (a very caustic liquid) has hardly any action (H. Martin). The numerous experiments of Vallin, Mairet, Cavalier, Coze and Siamon and others have given contradictory results.

2. Toxins.

A. Toxic properties of dead tubercle bacilli.

Koch and many other observers have found that agar cultures sterilized by heat at 100° C. are injurious to animals, and that in sufficiently large doses they lead to suppuration, cachexia and death in guinea-pigs. Dead tubercle bacilli inoculated into the blood or peritoneal cavity [of the guinea-pig] lead to the formation of true tubercles in which the presence of the dead bacilli can be demonstrated, but these lesions do not become generalized nor are they capable of being passed on to another animal. When only a small dose of dead bacilli is inoculated the lesions disappear spontaneously after a few months and the animal recovers (Cantacuzène). The more virulent the living bacilli the more toxic the dead bacilli.

[F. Griffith (English Commission) observing that the intra-venous inoculation of living mammalian tubercle bacilli in doses of 10–50 mg. caused death in about 50 per cent. of fowls, inoculated 3 fowls each with 10 mg. of bovine tubercle bacilli killed by exposure to steam at 100° C. and found in the lungs of two of them after 38 days numerous minute caseating tubercles. He concluded that the fatal infections were due to the direct toxic action of the bacilli and were not true tuberculosis.

[Similar results were obtained in fowls inoculated with dead human and avian tubercle bacilli; the latter were not more toxic for the fowl than the mammalian types.]

Hammerschlag treated dried tubercle bacilli with alcohol and ether. He obtained a product which was toxic to guinea-pigs and rabbits. The inoculated animals had convulsions and died.

Auclair extracted tubercle bacilli with ether and with chloroform. On inoculation into the trachea of a guinea-pig, these extracts produced lesions of tuberculous pneumonia.

Borrel also extracted a toxic substance from tubercle bacilli with xylol.

B. Koch's old tuberculin.

The tuberculin prepared by Koch in 1890 by a method at first kept secret is prepared in the following manner at the Pasteur Institute in Paris.

A culture of the tubercle bacillus on glycerin-broth is grown in a flask. (Bacilli of mammalian and avian origin yield the same tuberculin.) [Certain experiments recorded by the English Commission however show that tuberculin of avian origin cannot be relied on to produce a reaction in animals suffering from mammalian tuberculosis.] The growth must be on the surface of the broth. The film appears after incubating for 15–20 days at 38° C. and is complete about the 40th day.

The whole is sterilized at 110° C. for 15 minutes, then evaporated on a water bath to one-tenth its volume, and filtered through filter paper. The filtrate constitutes *crude tuberculin*.

Tuberculin is a brownish syrupy liquid with a faint, pleasant, characteristic smell. It has no definite composition but is a simple extract prepared from sterilized cultures on glycerin-broth, and contains in addition to the products secreted by the bacilli the substances originally present in the broth. The active principle has not yet been extracted. Tuberculin is very resistant to heat but is destroyed at 150° C.

Attempts have been made to purify the crude product :

(*a*) On the addition of 20 volumes of strong alcohol a brown precipitate is thrown down which contains the active principle and a number of other extraneous substances. Tannin, picric acid, metallic salts, ferrocyanide of potassium and acetic acid also form an albuminoid precipitate which carries down the active principle. Koch, Hunter and Klebs have failed in their attempts to purify this precipitate.

(*b*) Koch precipitated crude tuberculin with three volumes of 66 per cent. alcohol and obtained a flocculent precipitate which on drying formed a white powder. This constitutes *purified tuberculin* : it contains numerous extraneous substances but is a very toxic product and kills guinea-pigs when inoculated in doses of 1 mg. This is a very expensive method of preparation, as nine-tenths of the tuberculin remain in solution and are lost. Calmette precipitates crude tuberculin with 95 per cent. alcohol.

Except for Calmette's ophthalmo-reaction no advantage is to be gained by using purified tuberculin—it has the same properties as crude tuberculin.

1. The effect of Koch's old tuberculin on man and animals.

1. Healthy (non-tuberculous) subjects.—Crude tuberculin inoculated into healthy *animals* in small doses has no untoward effect except, possibly, a very slight rise of temperature. Guinea-pigs can be inoculated with 2 c.c. of tuberculin without harm. Rabbits stand an injection of 5 c.c. of crude tuberculin very well, there is a slight rise of temperature and a transitory loss of weight but the animal quickly recovers. Cattle and dogs do not react to doses of 10 c.c.

In *man* an injection of 0·25 c.c. into a healthy adult leads to somewhat severe symptoms : rigors, diarrhœa and vomiting with a rise of temperature to perhaps 39° C. (Koch). As small a dose as 0·01 c.c. may produce a slight rise of temperature. Man is therefore about 1000 to 1500 times as sensitive to tuberculin as the guinea-pig.

The toxicity of tuberculin can be considerably diminished by adding to it a calculated amount of anti-tuberculous serum (*vide infra*), the toxicity of the mixture is then due to *toxones*. Tuberculin neutralized in this way gives no better results in the treatment of tuberculosis than tuberculin or anti-tuberculous serum given alone. It [is said to] assist the production of the disease experimentally (Arloing and Descos).

2. In persons infected with tuberculosis and in tuberculous animals, the inoculation of small doses of tuberculin gives rise to a marked reaction and severe symptoms which may terminate fatally.

A dose of 0·5 c.c. of tuberculin rapidly kills a guinea-pig infected with tuberculosis 5 or 6 weeks before ; there is a sudden rise of temperature followed by a gradual fall and the animal dies in a state of coma. *Post mortem* there is intense congestion around the tuberculous foci and the internal organs are red, congested and ecchymosed.

In tuberculous cattle the inoculation of quantities of 0·30–0·40 c.c. causes a rise of temperature about 6 hours afterwards from 38°–39° C. (the normal bovine temperature) to 40°–41° C. The animal recovers its normal condition in a few days. Large doses of tuberculin are liable to kill the animal.

Persons suffering from tuberculosis react very sharply indeed to the inoculation of tuberculin : 0·25 c.c. invariably leads to a fatal result.

The so-called curative doses employed by Koch were 0·003–0·004 c.c. Following the injection the patient had rigors, and a rise of temperature to 41° C. The inoculation was frequently followed by coughing, nausea, vomiting, jaundice, etc. Around cutaneous tuberculous lesions there was an intense inflammatory reaction. According to Koch these symptoms ought to last 12–15 hours and then give way to a progressive improvement of the pre-existing lesions. Nothing would be gained here by recalling the disasters which followed the use of tuberculin. The treatment of tuberculosis with tuberculin has recently been revived especially in Germany as the result of the work of Denys, Sahli, and Beraneck. The doses used are much smaller than those used by Koch. The results obtained by this method of treatment have not shown it to be of any great value.

Intra-cerebral inoculation.—A guinea-pig weighing 500 grams which will stand the inoculation of 1 c.c. sub-cutaneously dies when inoculated in the brain with 3–4 mg. of the same tuberculin (von Lingelsheim, Borrel).

A guinea-pig inoculated 12 days previously with tuberculous material succumbs to the intra-cerebral inoculation of 0·1 mg. of tuberculin. The inoculation of 0·001 mg. of tuberculin into the brain of a guinea-pig which has been infected with tuberculosis 6 weeks previously produces symptoms of hiccough, convulsions, muscular twitchings, etc. and the animal very soon dies. These facts afford an explanation of the symptoms of tuberculous meningitis—the only form of tuberculosis in which the action of the poison on the nerve cells can be demonstrated (Borrel).

The toxins of tetanus, plague, etc., are no more toxic in the brain of tuberculous guinea-pigs than in the brain of healthy animals. *Mallein* alone acts like tuberculin : unconcentrated mallein which is harmless to tuberculous guinea-pigs when inoculated sub-cutaneously in doses of 3–4 c.c. leads to a fatal result when inoculated intra-cerebrally in doses of 0·01 or 0·001 c.c. (Borrel).

2. Koch's old tuberculin in the diagnosis of tuberculosis.

A. In Cattle.—Nocard showed that tuberculin is a valuable reagent in the diagnosis of tuberculosis in cattle. In bovine animals the early diagnosis of tuberculosis is clinically impossible in the majority of cases, but it is very important from the point of view of prophylaxis that the disease should be recognized in these animals in its very earliest stages.

Tuberculous cattle, however small the lesions may be, react to the inoculation of 0·30–0·40 c.c. of crude tuberculin. The temperature rises 1·5°–3° C. Animals free from tuberculosis do not react under similar conditions.

The method of diagnosis is as follows.

1. The animal to be tested is kept quiet and its temperature taken in the rectum the day before as well as on the day on which it is to be tested.

2. Dilute the tuberculin.

Crude tuberculin, - - - - - - - 1 c.c.
Boiled water containing 0·5 per cent. carbolic acid, - - 9 „

This solution will not keep and should be newly prepared for each experiment.

3. With the usual aseptic precautions inoculate the animal sub-cutaneously in the neck with 3–4 c.c., according to its size, of the diluted tuberculin.

4. Take the temperature 12 hours after inoculation and again 16 hours, 20 hours and 24 hours after. [Twelve hours is too late. The temperature should be taken 6, 9, 12, 15, 18, 24 and finally 36 hours after inoculation.]

Any animal which during that period shows a rise of temperature of 1·4° C. ought to be regarded as tuberculous. Animals which suffer a minimal rise of temperature (0·5°–0·8° C.) are healthy. When the rise of temperature is between 0·8° and 1·4° C. there is a suspicion of tuberculosis and the animal should be tested again a month later.

Note.—The tuberculin reaction though of great diagnostic value, is not absolutely reliable. In severely infected animals there may be no reaction. On the other hand, a rise of temperature of 1° C. is not sufficient upon which to base a diagnosis. A few cattle which showed a rise of 2° C. were subsequently found to be free from tuberculosis (Arloing, Rodet and Courmont). Infection of the lung with echinococcus is particularly likely to give the temperature reaction in cattle.

B. In man.—Tuberculin has been used in the diagnosis of doubtful cases of the disease in man.

The inoculation of tuberculin is always attended with a certain amount of danger and very great care must be exercised in its use. In any case the dose injected must never exceed 0·002 gram, and the following solution should be used.

Crude tuberculin, - - - - - - - 4 c.c.
Boiled water containing 0·5 per cent. carbolic acid, - - 996 „

One c.c. of this solution contains 0·004 gram of crude tuberculin and not more than a fraction of 1 c.c. should be inoculated. The temperature is taken for 2 or 3 days before the inoculation and every 8 hours for 36 hours afterwards. The part suspected to be infected must be carefully watched, the local reaction being of the greatest importance from the point of view of diagnosis. In patients who have suffered from tuberculosis for a very long time, small doses of tuberculin often produce no temperature reaction (Freymuth).

As to the amount of tuberculin to be inoculated to obtain a reaction observers differ. There are three methods of using tuberculin for purposes of diagnosis.

1. Single inoculation.—A single dose of 0·5 c.c. of the above solution (0·002 gram of crude tuberculin) is inoculated. This method is not free from danger.

2. Inoculation of increasing doses.—German observers, among whom tuberculin is largely used for the diagnosis of tuberculosis in man, do not hesitate gradually to increase the amount of tuberculin until the dose inoculated is very considerable. Generally a dose of 0·5 mg. of crude tuberculin is inoculated deeply into a muscle in the first instance and then gradually increasing doses of 1, 3, 6 and even 10 or 20 mg. every 3 or 4 days until the specific reaction is obtained.

In soldiers in apparently good health, Franz found that tuberculin in doses of 1–3 mg. set up a reaction in about 64 per cent. of the men and when the dose was increased to 10 mg. 96 per cent. reacted.

By gradually increasing the dose of tuberculin a non-specific reaction can [it is said] be produced in healthy persons. [But] in any case it would be most unwise to inoculate such dangerous doses.

3. Repeated inoculation of small doses.—Many observers advise that four or five small doses, 0·1–0·2 mg., should be given at intervals of 3 or 4 days. This is the least dangerous method and the one to be adopted to avoid the untoward results which so frequently follow the inoculation of tuberculin. Repeated small doses lead to a state of hypersensibility [1] and the diagnosis can be made without running the risks attendant upon the use of increasing doses. About 95 out of every 100 tuberculous subjects react to the third or fourth inoculation.

When several successive inoculations of tuberculin are made signs of inflammation appear after each fresh injection around the sites of the former inoculations. The lesions produced are similar to those set up [in the early stages of an infection] by the tubercle bacillus (Klingmueller).

4. Cuti-reaction.—Von Pirquet has shown that when tuberculin is inoculated through scratches on the skin of a tuberculous child, in nearly every case (save in acute miliary tuberculosis and tuberculous meningitis) a small papule—occasionally a vesicle—appears lasting about 8 days; at first it is bright red but subsequently becomes darker in colour. This reaction is of great diagnostic value in the early years of life. Older children often, and adults nearly always, react even when tuberculosis cannot be demonstrated clinically (which calls to mind the fact that *post mortem* nearly all adults show lesions of tuberculosis). In cachectic individuals the reaction often fails.

To effect the reaction make four small scarifications, not deep enough to draw blood, on the skin of the outer and upper part of the arm and cover the lower three with a small drop of diluted tuberculin (p. 326). (Tubes containing tuberculin for the cuti-reaction can be purchased at the shops.) The vaccination marks should be about 2 cm. apart. The upper mark which has not been treated with tuberculin serves as a control. When the reaction is positive, a swelling begins to appear about 48 hours after inoculation.

H. Vallée has shown that von Pirquet's cuti-reaction is of value as a diagnostic test for tuberculosis in the lower animals (cattle, horses and guinea-pigs). While healthy animals do not react, tuberculous animals show, about 36–48 hours after the inoculation, an œdematous infiltration of the vaccination mark with a painful grey-red swelling. The reaction occurs in healthy animals which have been previously treated with a subcutaneous inoculation of tuberculin (this confirms Klingmueller's phenomenon described above).

5. Intra-dermo reaction.—This method of diagnosis is recommended by Martoux but does not seem to offer any advantages over the cuti-reaction. The reaction consists in inoculating a drop of tuberculin into the skin with a fine needle. Calmette advises a 1 in 5000 solution of dry tuberculin precipitated with alcohol. In persons affected with tuberculosis a red, or bright pink, œdematous infiltration surrounded by a more or less extensive area of erythema is seen about 24 hours after the inoculation.

6. Ophthalmo-reaction.—Calmette and Wolff-Eisner have shown that the instillation of a small amount of tuberculin into the eye produces in persons affected with tuberculosis a very marked congestion of the palpebral conjunctiva. The test is easily performed and is very delicate and, provided it is not used when any lesion of the eye is present nor in old people, is quite free from danger.

---

[1] Repeated injections of large doses sometimes diminish the reacting power of the tissues in animals.]

Calmette says that crude tuberculin should never be used for the ophthalmo-reaction because the glycerin present in solution acts as an irritant. He uses a recently made 1 per cent. solution of dried tuberculin (twice precipitated with 95 per cent. alcohol) in distilled water. One drop of the tuberculin is instilled into one eye. Three to five hours later, in all tuberculous persons, there is congestion of the inner end of the palpebral conjunctiva, and more or less œdema and swelling of the caruncle which is covered with a slight fibrinous exudate. Lachrymation occurs and there is a little discomfort but no pain. The reaction reaches its maximum in 6–10 hours, and after 18 hours in children and 24 hours in the adult the signs subside and disappear.

In persons not infected with tuberculosis instillation occasionally produces a little redness but there is never lachrymation nor any fibrinous exudate.

The advantage of the ophthalmo-reaction is, according to Calmette, that it indicates the presence of active lesions only and as a means of diagnosis is of more value than the cuti-reaction. A negative reaction is sufficient to exclude a tuberculous infection except in tuberculous persons suffering from cachexia who have lost the power of reacting. Persons who have recovered from a tuberculous infection give a negative reaction. [Other observers confirm Calmette, finding that] with rare exceptions (some cases of enteric fever, for instance,) a positive reaction indicates an active tuberculous infection.

[**7. Percutaneous reaction.**—Moro rubs a lanolin ointment containing 50 per cent. of Koch's old tuberculin into the skin of the chest or abdomen over an area of about three square inches. From 1–2 days after the application numerous small red papules appear on the anointed surface in tuberculous individuals. The eruption is transitory and the reaction quite painless. The results are very comparable with those given by von Pirquet's method.]

C. Tuberculins TA, TO, and TR.

Koch investigated a number of complex products derived in various ways from cultures of the tubercle bacillus. These are known as tuberculin TA, tuberculin TO, and tuberculin TR.

1. Tuberculin TA (*alkaline tuberculin*).—Tubercle bacilli separated by filtration from a virulent culture are treated with a 10 per cent. solution of soda and after 3 days' exposure at room temperature the bacilli are killed and the liquid can be filtered through paper. After neutralization the filtrate is clear, slightly yellowish and contains numerous dead bacilli. Inoculation of the filtrate produces results similar to, but rather more persistent than, ordinary tuberculin, and often leads to the formation of an abscess containing sterile pus. After filtering through a bougie TA gives results identical with ordinary tuberculin.

2. Tuberculin TO and tuberculin TR.—Bacilli from a young, virulent culture are dried *in vacuo* in the dark and then rubbed up in an agate mortar for a long time. This is a dangerous proceeding and should be carried out with the utmost care. The powder so obtained is mixed with distilled water and the emulsion centrifuged for 40 or 45 minutes (4000 revolutions a minute). Two layers are formed; the upper (*obere*), fluid and opalescent and containing no bacilli, is decanted off and forms tuberculin TO.

The muddy lower layer is dried, rubbed up again, mixed with distilled water and centrifuged. The residue from the centrifuging is treated in the same way and the operation is repeated several times. Finally the liquids poured off after each centrifuging are mixed together and form tuberculin TR [*Rückstand*].

Tuberculin TO is very different from tuberculin TR.

Tuberculin TO is not altered by the addition of 50 per cent. of glycerin: its properties are almost identical with those of ordinary tuberculin and its immunizing properties are nil or very little marked.

Tuberculin TR gives a flocculent, white precipitate on the addition of 50 per cent. of glycerin. Its characters are unaltered by the addition of 20 per cent. of glycerin which, on the other hand, preserves it.

According to Koch tuberculin TR has distinct immunizing properties. Repeated inoculation of small doses into the human subject confers an immunity against ordinary tuberculin and TO, as well as against itself. These statements have not been confirmed (Bounhiol); tuberculin TR, whatever Koch may have said, appears to have no power of arresting tuberculosis. In tuberculous persons, a reaction similar to that produced by ordinary tuberculin occurs but very inconstantly. It appears to be less dangerous than ordinary tuberculin, but, on account of the irregularity of its effects, it cannot be used for purposes of diagnosis.

D. Maragliano's tuberculin.

This is a watery tuberculin obtained by maceration of tubercle bacilli. Recover the bacilli from a glycerin-broth culture, add to them a volume of distilled water equal to the volume of the culture fluid and heat to a temperature of 95°–100° C. for 50 hours or so. Then evaporate on a water bath to one-tenth its original volume and filter through filter paper.

The filtrate has the same properties as Koch's old tuberculin and is said to possess vaccinating properties. Doses of 5 c.c. are fatal to healthy guinea-pigs weighing 500 grams. Tuberculous guinea-pigs succumb to the inoculation of 0·10–0·20 c.c. of this tuberculin.

When precipitated with alcohol, Maragliano's tuberculin yields a powder which kills guinea-pigs in doses equivalent to $\frac{1}{25.000}$th of their weight and rabbits in amounts corresponding to $\frac{1}{33.000}$th of their weight.

E. Toxalbumin.

The different toxic products which have just been studied are contained within the bacilli—endotoxins—and to extract them the bacilli have to be killed by heat or destroyed by trituration.

Maragliano, Bezançon and Gouger have shown that the tubercle bacillus produces a diffusible toxin which passes into the medium in which the bacilli are growing. This toxin is of the nature of a toxalbumin: it is destroyed by heating it to 100° C. and by prolonged exposure to light. It is prepared by filtering a glycerin-broth culture through porcelain and concentrating the filtrate to one-tenth its original volume *in vacuo* at 55° C.

The product differs absolutely from the tuberculins: it is more toxic than the latter and inoculated into animals it never produces a rise of temperature even in non-fatal doses: in fatal doses the animals die with a sub-normal temperature.

Denys uses a porcelain-filtered glycerin-broth culture of the tubercle bacillus in the treatment of tuberculosis.

Beraneck prepares a *toxin-broth* by growing the tubercle bacillus in a maceration of veal made in the cold to which 0·5 per cent. salt and 5 per cent. glycerin are added before sterilization. By the time that the surface is covered with a pellicle, the medium, originally alkaline, has become acid. It is then made alkaline with lime water, filtered through a Chamberland bougie, and the filtrate evaporated *in vacuo* to one-tenth its original volume.

Beraneck recommends for the treatment of tuberculosis the use of this toxin broth diluted with an equal volume of an acid extract of the bodies of bacilli containing endotoxin. This extract, or *acido-toxin*, is made by macerating tubercle bacilli in a 10 per cent. aqueous solution of ortho-phosphoric acid for 2 hours at 60° C., neutralizing the product, filtering and diluting nineteen times with water (1 in 20 solution of endotoxin).

3. Vaccination.

A. Laboratory animals.

(i) Grancher and Martin.—Rabbits are inoculated with avian tubercle bacilli which have been grown on artificial media for prolonged periods. By using successively less and less attenuated cultures they have succeeded in a few cases in producing a certain degree of immunity in the animals.

(ii) Héricourt and Richet sterilize cultures of avian bacilli by heating them several times to 80° C. and then inoculate them into rabbits in doses of 10–20 c.c. This method has enabled them to confer immunity on a few animals.

(iii) Courmont and Dor filter glycerin-broth cultures and inoculate the filtrate into rabbits at the same time that or before they inoculate a tuberculous virus. With bacilli of *avian* origin they succeeded in producing immunity twice in four experiments but they failed with the bacillus of human origin.

(iv) In the foregoing experiments, attempts to vaccinate guinea-pigs had failed. E. Levy immunizes guinea-pigs by inoculating them with a tubercle bacillus [which he states to have been] attenuated by being kept in glycerin (p. 322).

Two guinea-pigs are inoculated, one into the peritoneal cavity, the other subcutaneously, with a slightly opalescent emulsion of tubercle bacilli which has been kept in 80 per cent. glycerin for 6 days in the incubator at 37° C. When they have recovered from the first inoculation they are again inoculated on successive occasions with bacilli which have been kept in glycerin for 4, 3 and 2 days. When they have completely recovered they, as well as two control animals, are inoculated with a tubercle bacillus of standard virulence. All the animals develop an abscess at the site of inoculation but at the end of about 4 weeks the lesion in the case of the vaccinated guinea-pigs has healed while in the controls the disease has spread to the glands. Towards the end of the third month the controls are suffering from a generalized tuberculosis (liver, spleen and lungs) while the most minute search fails to reveal any trace of tuberculosis in the vaccinated animals killed at the same time.

B. Cattle.

(i) Von Behring succeeded in vaccinating young bovine animals (healthy calves under one year old) by inoculating them intra-venously with living attenuated bacilli of human origin, non-virulent for cattle (bovo-vaccin).

The bacillus used by von Behring in his experiments was a human tubercle bacillus which had been in artificial culture for 8 years and had lost much of its original virulence (*vide* p. 322, attenuation of tubercle bacilli). A five-week-old culture of this bacillus on glycerin-serum could be inoculated with impunity into the veins of a calf in doses of 0·005 gram. A first injection of 0·001 gram provoked no reaction and was followed by several other inoculations with increasing doses at intervals of several weeks. Animals treated in this way were finally able to resist doses of bacilli of bovine type which were fatal to the control animals.

Behring has now modified his original procedure. He first inoculates, intra-venously, 0·004 gram of a glycerin-agar culture of a bacillus of human origin completely dried *in vacuo* at the ordinary temperature (the bacilli are ground up in a mortar and emulsified in 4 c.c. of a 1 per cent. saline solution). A month later the animal is similarly inoculated with 0·01–0·02 gram of the same culture.

Animals treated in this way and exposed to contagion or inoculated with a virulent bacillus of bovine origin never develop any tuberculous lesion. And, further, when tested with tuberculin a year after immunization they always fail to react.

With other observers this method of vaccination has always given favourable results, but complete immunity has not been attained.

Vallée and Rossignol undertook a series of control experiments on twenty calves each about 5 months old which had not reacted to tuberculin. These animals were first inoculated in the jugular vein with 4 mg. of dried tubercle bacilli (bovo-vaccin of von Behring) and 3 months later with 20 mg. of similar bacilli. They were tested by intra-venous and sub-cutaneous inoculation and by being kept in contact with animals suffering from open tuberculous lesions. From these experiments, Vallée and Rossignol conclude that von Behring's method of vaccination is harmless to animals protected from sources of accidental infection during the period of immunization and for 6 weeks afterwards: that the vaccination confers considerable but not absolute powers of resistance to the most severe methods of infection and that the vaccinated animals are able to resist for some months spontaneous infection which might have been expected to arise from prolonged contact with infected animals.

Von Behring's method of vaccination should be performed on animals less than 3 months old. In calves more than a year old the vaccinating process occasionally sets up a violent reaction and is dangerous to life. The severity of this reaction seems to depend upon an anterior infection with tuberculosis and it is well known that in cattle the younger the animal the more rare the disease. The resulting immunity is especially noticeable when the animals are tested by intra-venous inoculation but is only effective for a few months when tested by feeding the animals with infected food stuffs or keeping them in contact with cattle suffering from open tuberculous lesions.

(ii) With a view to getting more constant results and a more efficient method of vaccination some observers have experimented with living and virulent bacilli of the human type. Experience has shown that the immunity produced in cattle by the inoculation of living cultures runs parallel with the virulence of the tubercle bacillus used in the experiments, the higher the virulence of the bacillus the greater the degree of immunity produced.

[A. S. Griffith and F. Griffith (English Commission) investigated the production of immunity in calves by the inoculation of living tubercle bacilli. Twelve calves were inoculated sub-cutaneously or intra-venously, ten with human and two with bovine tubercle bacilli and were subsequently tested as to their resistance by the sub-cutaneous inoculation of a large dose (50 mg.) of bovine tubercle bacilli. Nine of the calves " had their resistance so far increased that 50 mg. of bovine tubercle bacilli were unable to set up in them progressive tuberculosis." Of the remaining three two died of acute tuberculosis and the third when killed showed general tuberculosis but in a less severe degree than in any of the four control animals. Thus " by the inoculation of large doses of human or small doses of bovine tubercle bacilli, the resistance of a calf can be raised sufficiently to protect it against the inoculation of a dose of bovine tubercle bacilli capable of setting up a severe and fatal tuberculosis in a calf not so protected "; but " this degree of resistance is not always produced." There was no evidence to show that vaccination with bovine tubercle bacilli produced a higher degree of immunity than vaccination with human tubercle bacilli.]

Bacilli killed by heat are devoid of vaccinating properties.

But vaccination with an active virus is not free from danger; inoculation of living bacilli of the human type may [it is said] set up latent lesions which can be re-awakened and constitute a permanent menace of re-infection; and finally, it is a danger to the consumer of the meat [and milk].

[A. S. Griffith (English Commission) found that if milking cows are " inoculated sub-cutaneously or intra-venously with tubercle bacilli of relatively slight virulence such bacilli quickly appear in their milk and may continue to be eliminated therein for long periods." It is obvious from these experiments that the vaccination of milch cows with living bacilli is undesirable. Other experiments by this observer also seem to indicate that this method of vaccination is not free from danger. In "seven out of eleven heifers (ages 6–10 months) tubercle bacilli of various types which had been inoculated in large doses into the sub-cutaneous tissues had found their way into the milk sinuses of the undeveloped mamma" and in four cases at least appeared to have undergone multiplication since their arrival there. There is therefore reasonable probability that the first milk of some of these animals would have contained living tubercle bacilli.

[Immunization with living tubercle bacilli, whether human or bovine, is therefore not free from risk.]

Thomassen injected 2–3 cg. of fresh tubercle bacilli from an human source grown on glycerin-potato into the jugular vein of calves; as a rule, the animals were resistent to a subsequent inoculation of bacilli of bovine origin but the animals might succumb to the vaccinating inoculation and in those that recovered it was possible that the disease might subsequently recur. Thomassen preferred to give increasing doses (1 mg. at the first inoculation, 10 mg. a month later and finally 20 mg.). The results were good but not constant. It was not uncommon to find that in vaccinated animals which had failed to react to the test inoculation, the bronchial glands, though normal in appearance when the animal was killed a few weeks later, produced tuberculosis on inoculation into guinea-pigs. [Cobbett (for the English Commission) found that human tubercle bacilli inoculated into calves remained alive for prolonged periods, in lesions so minute as to be hardly visible.]

Hutyra had good results with a similar method (inoculation of a recent potato culture of tubercle bacilli of human origin).

Baumgarten obtained very favourable results by simple *sub-cutaneous* inoculation of human tubercle bacilli of standard virulence. According to the author there was a non-specific, local, inflammatory lesion.

(iii) Koch and others immunized calves by inoculating them with an attenuated bacillus of bovine origin. The inoculations were made into the jugular vein, 10 mg. on the first occasion and 25 mg. 3 weeks later.

It did not appear to be a true vaccination but a more or less brief increase of resistance to the action of the bovine type of bacillus. An heifer which had been vaccinated did not react to the test inoculation and remained in apparent good health but died 14 months later while suckling a calf. Lactation would appear to re-awaken a latent infection (Pepere).

[Pregnancy appears to render the sex-organs peculiarly susceptible to tuberculosis. Cobbett (for the English Commission) records that three out of six heifers inoculated when pregnant with bovine tubercle bacilli developed tuberculosis of the uterus. In none of the three was the generalized disease very acute, and in one (in which the mammary gland also was affected) there was very little tuberculosis elsewhere. The non-pregnant uterus, he continues, has never been found affected in calves suffering from general tuberculosis however severe.]

Klemperer endeavoured to treat persons suffering from tuberculosis by inoculating them with tubercle bacilli of bovine type. He is said to have noticed improvement in the condition of the patients but his experiments are not conclusive. Similarly the inoculation of bacilli of the human type into tuberculous cows appeared to have no curative action.

(iv) Vallée succeeded in vaccinating cattle against the effects of the injection of tuberculin and in conferring some degree of immunity against living cultures by inoculating them intra-venously with dead tubercle bacilli from which the fat had been extracted.

The bacilli were rapidly washed in distilled water, drained and dried for several days *in vacuo* over sulphuric acid, then placed in a flat-bottomed flask with pure petroleum ether and glass balls. The flask was placed in a shaking apparatus and the contents shaken for 60 hours.

The bacilli had then lost their acid-fast properties. The emulsion was dried in a desiccator to remove all traces of the ether from the bodies of the bacilli.

Tubercle bacilli treated in this manner behave like a very powerful tuberculin. They kill guinea-pigs in doses of 70 mg. When inoculated several times in doses of from 25–100 mg. into the jugular vein they [are said to] render young calves to a certain degree immune to the bacillus of bovine type. Horses and cattle rendered immune to the inoculation of these bacilli cease to react to the intra-venous inoculation of the various tuberculins.

(v) Animals vaccinated by inoculating them intra-venously show themselves very slightly immune to intestinal infection—a mode of infection particularly common in nature. Behring, Calmette and Guérin, Roux and Vallée have tried to immunize animals by feeding them with tubercle bacilli and have met with some success. The method is only practicable in the case of young cattle and the resulting immunity is particularly efficient against intestinal infection. Although the immunity is only relative the method gives better results than intra-venous inoculation: the animals remain uninfected for a year when kept with cattle suffering from open lesions.

Two young calves were fed at intervals of 45 days with 5 and 25 grams of tubercle bacilli of the human type. Four months later failing to react to tuberculin they were fed with 0·05 gram of freshly isolated bacilli of the bovine type: 32 days later they failed to react to tuberculin while two controls reacted in the ordinary way.

Vallée also immunized a young calf by feeding it, through an œsophageal sound, on two occasions, when it was 2 days and 90 days old respectively, with 20 cg. of a well-made emulsion of a tubercle bacillus of equine origin which was only slightly virulent for guinea-pigs.

Roux and Vallée, Calmette and Guérin have shown that small doses of virulent bacilli of the bovine type when introduced into the alimentary canal of the calf are absorbed into the mesenteric glands and give rise to a substantial immunity in the animal. A certain amount of risk attaches to this method of immunization (Vallée).

(vi) For the purpose of producing immunity Arloing uses homogeneous cultures of bacilli of the human type (*vide infra*) which have lost much of their capacity of producing tubercles; by submitting them to gradually increasing temperatures Arloing was able to grow them at 43°–44° C. These cultures vaccinate calves and appear to act as a true pasteurian vaccine.

C. Vaccination with a virus of chelonian origin.

Friedmann showed that a bacillus which he had recovered from a tortoise (p. 297) produced when inoculated beneath the skin of a guinea-pig a typical localized tuberculous focus which soon completely healed while the animal never showed any sign of generalized tuberculosis. Further, guinea-pigs treated in this way resisted the inoculation of a dose of bacilli of the human type which killed control animals in 4–6 weeks.

In vaccinated guinea-pigs the inoculation of a virus of human origin gave rise to a transitory swelling of the glands and to a caseo-purulent tuberculous focus which healed and left no trace of the injury: when the animals were killed about 3 months later no lesion was found: it is true that small whitish points were seen in the internal organs but these were in no way suggestive of true tubercles and similar lesions are found in animals immune to tuberculosis or vaccinated in various ways against the disease (Koch, von Behring, Neufeld, Thomassen, and others).

The method is available for the immunization of bovine animals against tubercle bacilli of the bovine type. Intra-venous inoculation of calves with a bacillus of chelonian origin [is said to] produce a lasting immunity against bacilli of the bovine type. The method might also be applicable to the treatment of animals suffering from tuberculosis. Libbertz and Ruppel have not been able to confirm these results.

Conversely, Friedmann succeeded in vaccinating tortoises against bacilli of chelonian origin by inoculating them with a virus of the human type.

Mœller, carrying out similar researches, succeeded in infecting blind-worms with human tubercle bacilli and for the purpose used as a vaccine a bacillus of the human type which had been passed through a series of blind-worms. Mœller did not hesitate to practise his method of vaccination on himself. On three separate occasions he inoculated himself intra-venously with a culture of the bacillus from the lesions in the blind-worm and a month after his last vaccination he was inoculated intra-venously with an emulsion of virulent bacilli of the human type which was rapidly fatal to guinea-pigs. This inoculation produced merely a transitory loss of weight without any disturbance of health a year afterwards.

4. Serum therapy.

Attempts at serum therapy in tuberculosis have up till the present given no conclusive results.

(i) Richet and Héricourt by inoculating rabbits with dog-serum before infecting them with tubercle bacilli have been able to delay the course of infection in some of the inoculated animals. Unfortunately, the success of the experiments was very relative and inconstant. Bertin and Picq, experimenting with inoculations of goat serum, obtained similar results.

(ii) Von Behring and Niemann also failed with the serum of animals treated with tuberculin.

(iii) Bernheim tried the blood of animals inoculated with filtered but unheated cultures of tubercle bacilli. His experiments were unsuccessful. The results obtained by Babès and Broca were no more encouraging.

(iv) Maragliano obtained a serum of obvious *antitoxic* properties. He injected animals with increasing doses of a mixture of three parts of ordinary tuberculin and one part of an extract of porcelain-filtered unheated cultures (p. 329). The treatment was continued for 6 months and when 3 weeks had elapsed since the last inoculation the animals were bled. *In vitro*, the serum destroyed the toxic properties of tuberculin and protected guinea-pigs against this poison : 1 c.c. of the serum protected a healthy guinea-pig against a fatal dose of tuberculin : 2–4 c.c. rendered a tuberculous guinea-pig capable of standing without harm a dose of tuberculin which in the ordinary way would have killed it in a few hours. Experiments to determine whether the serum protected healthy animals against infection with the tubercle bacillus were not conclusive.

(v) Marmorek for the treatment of tuberculosis suggests the use of a serum obtained by inoculating horses with a special toxin (a filtered culture of bacilli grown on a leucotoxic calf serum containing glycerin-liver-broth).

(vi) Lannelongue, and Achard and Gaillard found that the serum of asses possessed therapeutic properties after the animals had been treated by inoculation with a toxin extracted from the bacilli by heating in water at 120° C., precipitating with acetic acid and redissolving the precipitate in sodium carbonate.

(vii) Baumgarten and Hegler after vaccinating a bovine animal with bacilli of the human type were able to inoculate bacilli of the bovine type on five successive occasions without producing any appreciable reaction. The tuberculin test was negative and the serum of this animal inoculated prophylactically into a calf (82 c.c. sub-cutaneously in a fortnight) protected the latter against a sub-cutaneous inoculation of bacilli of the bovine type while two controls similarly inoculated with the test culture became infected. The passively immunized calf when killed 6 months afterwards showed no tuberculous lesion. As a curative agent this serum is without effect.

(viii) Vallée treats horses by inoculating them first with progressively increasing doses of bacilli of equine origin then with bacilli of human origin. The serum of such animals exhibits no agglutinating properties but contains immune bodies (*sensibilisatrices*) and has distinct therapeutic properties in the treatment of bovine tuberculosis.

(ix) The serum of bovine animals, guinea-pigs and pigs vaccinated by Friedmann with the chelonian bacillus has proved itself in the case of the guinea-pig to possess undoubted prophylactic properties. While control guinea-pigs died of generalized tuberculosis, the treated guinea-pigs killed at the same time only showed insignificant lesions.

5. Agglutination.

1. The serum diagnosis of tuberculosis with ordinary cultures, consisting as they do of bacilli massed together to form films or scales, is quite impossible. Arloing and Courmont have described a method of obtaining *homogeneous cultures* which are quite suitable for the demonstration of the phenomena of agglutination.

Arloing's homogeneous cultures are obtained from luxuriant, greasy-looking, growths on glycerin-potato. After being sub-cultivated a few times on glycerin-potato the tubercle bacillus is sown in cylindrical flasks with a flat base half filled with a 1 per cent. peptone beef broth containing 6 per cent. glycerin. The cultures are incubated at 38°–39° C. and shaken daily. It is necessary to sub-cultivate several times in order to get a copious growth.

The cultures should be used when about 10 days old: later they are thick and too rich in bacilli and are only partially agglutinated by the specific serum.

Cultures sown in this way are distinctly cloudy after a few days, and when shaken have a watered silk appearance; the cloudiness subsequently increases and becomes uniform, and after 2 or 3 weeks the growth is more or less milky. A drop of the culture examined unstained shows small, isolated rods occasionally slightly motile. The bacilli stain by Ziehl's method (except in rare cases recorded by Arloing and by Ferran). On inoculation into animals they behave like tubercle bacilli of a very low degree of virulence.

Ten-day-old cultures suitable for agglutination can be kept for about a fortnight in the ice-chest or by adding a little formalin (3–4 per cent.) to them; the agglutinability diminishes after a few weeks.

To obtain and to keep an agglutinable culture more easily Arloing and Courmont now advise the use of old cultures which must be diluted as required. The culture is left in the incubator for 30 or 40 days and can be used at once or kept in the ice-chest for about 1 month. When required for use the culture is diluted with sterile normal saline solution until it gives a milky, slightly opaque fluid like a solution of glycogen (about 1 in 50).

When a culture is used for purposes of serum diagnosis a control should always be done with a standard serum of which the agglutinating property is known beforehand (the serum of an experimentally infected animal, or a tuberculous pleural exudate, for example).

Homogeneous cultures are agglutinated by the serum of animals which have been inoculated with tuberculin or living bacilli and by the serum of most tuberculous human subjects (66–90 per cent.) when the serum is diluted from 1 in 5 to 1 in 50. The degree of agglutinating power in serums from the same animal species appears generally speaking to bear no relation to the severity of the lesions or the virulence of the infecting organism. Serums from cases of miliary tuberculosis often fail to produce the reaction. Tuberculous serous exudates (peritonitis, pleurisy) in most cases have the property of agglutinating the tubercle bacillus.

The technique of agglutination is as follows :

Take four small sterile test-tubes 5–6 cm. long. A little of the pure culture is put into the first and acts as a control : into the other three a mixture of serum and culture is introduced in different proportions : thus, into the second 1 drop of serum and 5 drops of culture, into the third 1 of serum and 10 of culture and into the fourth 1 of serum and 15 of culture. After shaking, the tubes are placed at an angle of 45° and kept at laboratory temperature. Agglutination only appears after the mixtures have been put up a few hours. The tubes should be examined several times during the next 24 hours and against a black ground. When the reaction is complete the fluid is clear and there is a slight deposit at the bottom of the tube. A complete reaction giving this appearance is alone of diagnostic value. The diagnosis should always be completed by microscopical examination of the clumps (between a slide and cover-glass and unstained).

The power of agglutination possessed by the serum of persons suffering from tuberculosis is of very limited diagnostic value for the following reasons :

1. Homogeneous cultures are often agglutinated by the serum of healthy persons—26 times out of 100 examined (Arloing and Courmont) : 50 per cent. of cases according to German statistics.

2. The serum of persons suffering from febrile diseases other than tuberculosis (enteric fever, puerperal fever, pneumonia, etc.) nearly always agglutinates homogeneous cultures of the tubercle bacillus.

3. The serum of tuberculous subjects does not always agglutinate homogeneous cultures of the infecting organism. A negative reaction is especially frequent in cases of advanced phthisis, miliary tuberculosis, and " galloping " consumption. The reaction is often absent in the very early stages of the disease, and is practically never obtained in cases of lupus. Bezançon, Griffon and Philibert obtained a positive reaction in about 83 per cent. of cases of undoubted tuberculosis : Ivanova observed agglutination in 66 per cent. of cases of tuberculosis and in 60 per cent. of non-tuberculous persons.

4. The method is very delicate in practice and the most minute precautions must be taken in the preparation of the homogeneous cultures.

(ii) Vasilescu describes a medium which he says gives homogeneous cultures more easily than Arloing's method.

The medium has the following composition :

Clear sterile calf serum,	25 c.c.
Distilled water,	75 ,,
Neutral glycerin,	3 ,,

Heat the mixture to 100° C. in a water bath, distribute into tubes of 2 cm. diameter in quantities sufficient to give a column 3 cm. high in each tube. Sterilize at 120° C. for 15 minutes (the medium does not coagulate). Sow with a bacillus of the human type. Incubate at 37° C. and shake the tubes every day. After two sub-cultures in this medium the growth is homogeneous.

(iii) Hawthorn suggests a simplification of Arloing's technique. He subcultures an homogeneous culture in glycerin-broth in 2 per cent. peptone saline solution. By this means an homogeneous culture consisting of motile bacilli is obtained in 48 hours without shaking the flask. These cultures are more sensitive for the agglutination reaction than Arloing's.

Vincent repeated Hawthorn's experiments and always found that the cultures in the peptone saline medium were agglutinated even when the serum of non-tuberculous persons was used (in dilutions of 1 in 30 and 1 in

40). The method therefore is not available for the agglutination reaction in tuberculosis.

(iv) For the agglutination-reaction German authors prefer an emulsion of tubercle bacilli ground up in a mortar. Von Behring uses an emulsion of bacilli triturated in an agate mortar and afterwards desiccated. Koch recommends the use of the powdered bodies of bacilli obtained in the preparation of tuberculin TR, dried and emulsified in a 1 per cent. solution of sodium chloride. In bacillary extracts, obtained by triturating tubercle bacilli as described in the preparation of tuberculin TR, and freed from bacilli by centrifugation, agglutinating serums produce agglutination in dilutions of 1 in 10 to 1 in 50. This method is recommended by Koch and Romberg.

Kœppen uses an emulsion obtained by saponification which has the advantage of keeping well and of being of such a degree of concentration that, after agglutination, the fluid becomes quite clear.

To prepare the emulsion described by Kœppen filter a glycerin-broth culture through filter paper : wash the bacilli with normal saline solution and dry, *in vacuo*, at the ordinary temperature. To 1 gram of dried bacilli add 3 c.c. of a warm aqueous solution of potash (33 per cent.) : leave to stand for a few hours, then rub up the mixture and place the emulsion in the warm incubator (37° C.) for several hours ; finally heat in a water bath for 15 minutes at 100° C. The emulsion is now of a thick consistency and is again heated, the water which evaporates being replaced by alcohol drop by drop. Soaps are formed which are dissolved in 100 c.c. of warm distilled water. For the purposes of the agglutination reaction 1·5 c.c. of the milky emulsion is mixed with 50–100 c.c. of normal saline solution.

(v) Wright and Douglas heat a glycerin-broth culture of the tubercle bacillus to 60° C. for 1 hour, filter it through paper, grind it up in a mortar, and emulsify in water containing 0·1 per cent. of sodium chloride and 0·5 per cent. carbolic acid. They then centrifuge the emulsion to remove any bacterial masses which have not been resolved into their elements. This liquid is only agglutinated by normal human serum in dilutions of 1 in 2 to 1 in 4 while tuberculous serums agglutinate it in dilutions of 1 in 10 to 1 in 50.

6. Immune bodies (Sensibilisatrices).

The presence of immune bodies is very inconstant in the serum of persons suffering from tuberculosis. The method of complement fixation is, therefore, not applicable to the diagnosis of tuberculosis (Widal and Le Sourd, Camus and Pagniez, Wassermann and others).

SECTION IV.—THE DETECTION OF THE TUBERCLE BACILLUS.

The methods employed for detecting the tubercle bacillus vary in detail according to the nature of the material to be examined but in every case three methods of investigation are available.

1. Microscopical examination.—Fluids, tissues or other material suspected to contain the bacillus must be stained by Ziehl's or Ehrlich's method. The tubercle bacillus is the only organism which will resist the decolourization used in these methods. As a matter of fact two other [parasitic human] organisms might be mistaken for the tubercle bacillus, namely, the bacillus of leprosy and the smegma bacillus.[1]

---

[1] The bacillus of verruga can be left out of account. The disease has so far been very little studied and is unknown in these climates.

The methods of differentiating the former are considered elsewhere (Chap. XIX.). In the case of the smegma bacillus there is hardly any fear of making a mistake if it be remembered (i) where this organism is found and (ii) that though it resembles the tubercle bacillus in resisting the decolourizing action of mineral acids, it differs from it in being rapidly decolourized by alcohol (probably because alcohol dissolves the fatty matter which impregnates it). In short, if Ziehl-Neelsen's method be carried out in the manner described above no confusion is likely to arise (p. 307). To avoid all possible chance of mistake in difficult cases, Bézançon and Philibert advise staining in the warm for 10 minutes, decolourizing in 33 per cent. nitric acid for 2 minutes and in absolute alcohol for 5 minutes.

It must not be forgotten that saprophytic acid-fast bacilli are found in the ambient media, and in milk, butter, etc. (p. 347).

These organisms might conceivably be a serious source of error in the detection of the tubercle bacillus. But they are often only feebly resistant to acid and are frequently decolourized by absolute alcohol alone. If there be any doubt the difficulty can be cleared up by inoculation.

Films are prepared in the ordinary way. Tissues should be hardened in alcohol or acid perchloride. [Eastwood (for the English Commission) hardened tissues in 10 per cent. formalin for a few days then washed well in water and transferred to Muller's fluid.]

2. Inoculations.—In those cases, which, in practice, are far from infrequent, in which microscopical examination fails to reveal the presence of the tubercle bacillus resort must be had to animal inoculation. Guinea-pigs being the most susceptible animals are always used for the purpose. When the material is free from other organisms, as, for instance, in the case of caseo-pus, pleural fluid, etc. it may be inoculated into the peritoneal cavity but in other cases (sputum, pus from a fistula, etc.) it is best to inoculate it beneath the skin otherwise there is the risk of setting up a septic peritonitis which will kill the animal before the tubercle bacillus has had time to produce its characteristic lesions. [Our experience would lead us invariably to inoculate intra-peritoneally when the material contains other organisms. It would seem that the peritoneal fluid possesses a quality lacking in the sub-cutaneous tissues of destroying putrefactive and other organisms.]

For purposes of rapid diagnosis, Nattan-Larrier and Griffon advise the inoculation of the suspected material into the mammary gland of a guinea-pig during the period of lactation. The bacilli multiply in the gland and after a week or a fortnight they can be detected in the milk by staining the latter by Ziehl-Neelsen's method.

Osman Nouri has drawn attention to the advantages of inoculating an animal by rubbing the material into the skin after shaving it (p. 299).

Bloch advises that the suspected material should be inoculated sub-cutaneously into the inguinal region of a guinea-pig and that the inguinal glands should then be squeezed and manipulated between the fingers to bruise them and so render them more susceptible to infection. Under these conditions, if the material (urine, etc.) contained the tubercle bacillus the glands will be found enlarged, inflamed and even suppurating when the animal is killed 9 days after inoculation. The method is not reliable; tubercle bacilli cannot be found on microscopical examination when the animal is killed 9 days after inoculation; moreover under the conditions of the experiment, acid-fast bacilli, Staphylococci, etc. may equally with the tubercle bacillus cause swelling and suppuration of the glands.

3. Cultures.—Cultures are very rarely used as a means of detecting the tubercle bacillus in a pathological product. To obtain cultures not only must the material be rich in tubercle bacilli [and free from contaminating organisms] but there must be a good deal of it. Cultures, however, are particularly successful in the case of sputum.

A. Sputum.

1. Microscopical examination.—The search for tubercle bacilli in sputum is easy when the latter is purulent and the bacilli are present in large numbers. It is much more difficult and often unsuccessful when the sputum is scanty and mucous in character and derived from a recent lesion or again when the sputum consists almost entirely of blood. As a rule, the sputum coughed up by the patient in the early morning gives the most satisfactory results.

In the case of nummular sputum it is only necessary to pick up a small fragment from the centre of a purulent mass with a loop and spread it on a cover-glass [or slide] in the ordinary way. The yellowish lumps found in tuberculous sputum are very rich in bacilli and should therefore be selected; similarly, in mucous sputum, the solid particles suspended in the more fluid portion should, as far as possible, provide the material for examination.

It is very difficult to find the tubercle bacillus in the blood coughed up during an attack of hæmoptysis; it can be more readily found in the consolidated sputum streaked with blood which is expectorated in the days following the hæmorrhage.

Bacilli can seldom be detected in the sputum of persons suffering from miliary tuberculosis; [it would seem that] the bacilli only pass into the sputum when purulent lesions are breaking down.

The tubercle bacillus is present also in the expectoration of tuberculous cattle (Riddoch) and a good method of diagnosing the disease in cattle is to collect some of the muco-purulent material from the partitions in the sheds and to examine it for the bacillus after staining in the ordinary way.

Homogenization.—When the bacilli are likely to be few in number the sputum must be specially treated before they can be detected: the sputum must be liquefied and made homogeneous, and then be left to deposit or else be centrifuged. Under these conditions the deposit contains in a small volume all the bacilli which were in the viscous mass. It is then a simple matter to stain and detect them.

1. Biedert's method.—To 15–20 c.c. of sputum add 30–40 c.c. of water and a few drops (6–15) of soda—the thicker and more viscous the sputum the larger must be the quantity of soda added. Boil the mixture in a porcelain dish until it is quite homogeneous, then dilute with one or two volumes of water and boil again for a minute or two. Put the mixture aside for 48 hours, then pour off the supernatant fluid and prepare films with the deposit. Or, after boiling, the mixture may be centrifuged and the deposit used for making films.

2. Ilkewitsch's method.—Mix 0·5 c.c. of sputum with 20 c.c. of distilled water and about 10 drops of a 3 per cent. solution of caustic potash. Heat the mixture in a porcelain dish but without letting it boil, stirring all the time until the mixture is homogeneous. Add a little casein and a drop or two of a 3 per cent. solution of caustic potash and continue the heating until the mixture has a milky appearance, then pour into centrifuge tubes, add a few drops of acetic acid until coagulation is just beginning, centrifuge for a few minutes and use the deposit for making films.

3. Ellermann and Erlandsen's method.—Mix 10–15 c.c. of sputum with one-half its volume of a 0·6 per cent. solution of sodium carbonate in a small flask and place in the incubator at 37° C. for 24 hours. Then decant the supernatant fluid and centrifuge the remainder. To the deposit add 4 times its volume of 0·25 per cent. soda solution, mix, centrifuge again and use the deposit for preparing films.

According to Ellermann and Erlandsen, and Bertarelli, this method is better than any other which has been described.

4. Abbé's method.—Place 5–10 c.c. of sputum in a cylindrical vessel, add 15–30 c.c. of a perchloride solution (perchloride of mercury, 2 grams; salt, 10 grams; distilled water, 1000 c.c.); mix thoroughly, centrifuge 15 c.c. of the mixture and make films with the deposit.

5. Sprengler's method.—Mix 10 c.c. of sputum with 10 c.c. of warm water and a drop of normal soda solution, add 0·25–0·50 gram of pancreatin and incubate at 37° C. for 2 or 3 hours. Then centrifuge or pour into a conical glass vessel with a crystal of thymol added to hinder putrefaction and leave to stand for 12 or 14 hours : decant the supernatant liquid and prepare films with the deposit.

6. Jousset's method.—Digest the sputum for 2 or 3 hours in the warm incubator (38° C.) with 10–30 c.c. of artificial gastric juice made as follows :

Pepsin, - - - - - - - - - -		2 grams.
Pure glycerin, ⎫		
Hydrochloric acid, ⎬ - - - - - - - - -		āā. 10 c.c.
Sodium fluoride, - - - - - - - -		3 grams.
Distilled water, - - - - - - - -		Q.S. to 1000 c.c.

When completely peptonized, centrifuge and examine the deposit for tubercle bacilli (p. 342).

Films made from sputum or from the deposit after solution of the sputum are stained by one or other of the methods already described, Ziehl-Neelsen's method being the most satisfactory.

2. Inoculations.—In a doubtful case of pulmonary phthisis when no bacilli can be found in the sputum on microscopical examination, animal inoculation must be resorted to for purposes of diagnosis. The sputum must be collected as free from other organisms as possible (p. 192) and an emulsion of it, prepared by rubbing up in sterile water, inoculated beneath the skin of a guinea-pig. Should the sputum contain tubercle bacilli the animal soon shows signs of infection (p. 297). On account of the risk of setting up a fatal septic peritonitis sputum should never be inoculated into the peritoneal cavity [but see p. 338, 2. Inoculations].

3. Cultures.—It is very rarely that cultures are sown with sputum for purposes of diagnosis; indeed for a long time it was thought impossible to obtain a pure culture of the bacillus from sputum. Kitasato, Pastor and others have, however, described methods by which it may be effected.

(a) **Kitasato's method.**—After the mouth has been washed out with sterile water, induce the patient to cough, and collect the sputum in a sterile glass vessel. Wash the sputum in a number (10) of glass vessels containing sterile water (p. 192) then, adopting the ordinary precautions to prevent contamination, lift a small fragment from the centre of the purulent mass and spread it on glycerin-serum. Sow a large number of tubes. Growth appears after the tubes have been incubated at 38° C. for about 10 or 12 days.

(b) **Pastor's method.**—Collect the sputum with the same precautions as in Kitasato's method. Emulsify in a little sterile water and filter through a piece of sterile gauze. Sow a tube of liquefied gelatin with a few drops of the filtrate, pour into a Petri dish and allow it to set. After incubating for 3 or 4 days at 20° C. any contaminating organisms that may have been present will have grown giving rise to numerous colonies. With a sterile scalpel cut out those portions of the gelatin on which there is no growth and transfer them to tubes of glycerin-serum ; if a number of tubes be sown a pure culture of the tubercle bacillus will be obtained on some of them.

(c) **Hesse's method.**—Hesse's medium (p. 317) is most useful for isolating the tubercle bacillus from sputum. The sputum is washed by Kitasato's method and sown on the agar in Petri dishes. After incubating at 37° C. for about 10 hours the number of tubercle bacilli in the fragments of sputum will be found to have considerably increased and colonies will be visible to the naked eye at about the sixth day.

(d) **Jockmann's method.**—Jockmann uses the following medium which is similar to Hesse's :

Heyden's albumose (*Nührstoff Heyden*), - - -		5 grams.
Salt, - - - - - - - -		5 ,,
Neutral glycerin, - - - - - - -		30 ,,
Normal soda solution, - - - - - -		5 c.c.
Water, - - - - - - - -		1000 ,,

Sterilize. To 20 c.c. of this broth add 10 c.c. of sputum, mix thoroughly and incubate at 37° C. After incubating for 24 hours the bacilli will have increased in number and films can be prepared with the deposit and stained with carbol-fuchsin. This method facilitates the detection of tubercle bacilli when they are present in the sputum in small numbers only.

(e) **Spengler's method.**—Tubercle bacilli are more resistant to the action of formalin vapour than most other bacteria and it is upon this fact that the following method is based. Dworetzky states that he has not had good results.

Cover the bottom of a Petri dish with a filter paper, spread 3 c.c. of sputum in a layer not more than 2·5 mm. thick and sprinkle with pancreatin to facilitate digestion of the mucus. Line the lid of the dish with filter paper soaked in a few drops of commercial formalin. Incubate the dish and its contents at 20°-25° C. After incubating for 2 hours it is said that all micro-organisms other than the tubercle bacillus are killed so that if tubes of glycerin-agar be now sown with the sputum a pure culture of the tubercle bacillus will be obtained.

[(f) **Griffith's method.**—A. S. Griffith (for the English Commission) obtained cultures direct from sputum by using antiformin.[1] The method he employed at first was that recommended by Uhlenhuth; diluted sputum and antiformin were mixed together to form a 15 per cent. antiformin-sputum mixture and allowed to stand for 3 hours; the solution was then centrifuged, the deposit washed and recentrifuged, and the second deposit sown on to culture media. In subsequent experiments however he found that a 10 per cent. dilution of antiformin allowed to act for 20–30 minutes gave better results (see fig. 208 (a), p. 317).]

(g) Spengler recommends another method based on the resistance of the tubercle bacillus to heat and only applicable to nummular sputum.

Take up a large fragment of the nummular sputum in a platinum loop and hold it near a flame so that the sputum is roasted but not detached from the loop. Repeat the process three or four times then sow the flamed sputum on 2 per cent. glycerin-serum or on glycerin-agar. Growth appears in a week to 10 days. Spengler acknowledges that this requires a certain amount of skill.

In a case of advanced pulmonary tuberculosis recorded by Bertarelli the sputum [appeared to] consist solely of tubercle bacilli which when sown direct on to glycerin-serum and blood-agar readily gave a pure culture of the bacillus. Such cases are very exceptional; and undoubtedly the most certain method of obtaining a pure culture from sputum is to inoculate a guinea-pig and sow cultures from the lesions which develop (p. 314).

B. Blood.

The tubercle bacillus rarely passes into the blood of persons affected with tuberculosis. Lustig and others have, however, succeeded in staining the bacillus in films prepared with blood obtained by pricking the finger or the spleen (in miliary tuberculosis). The bacillus is more easily demonstrated in the clots formed *post mortem* in the heart and blood vessels. The carbol-fuchsin method should be employed for staining the preparations.

Bezançon and Griffon, and Jousset have described methods designed to facilitate the detection of the organism in the blood. The results of the methods are, however, vitiated by the occurrence in the surrounding air of acid-fast bacilli and of saprophytic bacilli which acquire acid-fast properties in organic products. To secure the best results from these methods they ought to be carried out under strictly aseptic conditions, and this in practice is difficult

[1 Antiformin is a mixture of sodium hydroxide and Eau de Javelle. It has the power of dissolving albuminous substances and of killing and dissolving all bacteria except those which like the tubercle bacillus possess a waxy envelope.]

of accomplishment. Under strictly aseptic conditions, Bergeron has repeatedly failed to detect the presence of tubercle bacilli in blood by Jousset's method.

(a) **Bezançon and Griffon's method.**—To 5 c.c. of blood add 5 c.c. of distilled water and 5 drops of soda and triturate the mixture in a mortar until completely dissolved. Then add 20 c.c. of water and boil in a porcelain dish for 5 minutes. Centrifuge the product for 10 minutes, prepare films and stain with carbol-fuchsin.

(b) **Jousset's method** (Inoscopy).—To 30 c.c. of blood add 100 c.c. of distilled water. Digest the clot with artificial gastric juice (p. 340) for 2 or 3 hours in the warm incubator (38° C.). Centrifuge the product, stain the deposit with carbol-fuchsin, and examine it for tubercle bacilli.

(c) **Nattan-Larrier and Bergeron's method.**—In this method the blood is received direct from the vein into twenty times its volume of sterile distilled water; the blood hæmolyses; the hæmolyzed mixture is centrifuged and the deposit examined for bacilli.

(d) Blood may be collected in sodium citrate solution to prevent it clotting, centrifuged and the deposit examined. Lesieur utilizes the anti-coagulating property of the digestive juices of leeches. A leech is put on the patient and when gorged with blood it is pressed and the product centrifuged.

C. Pus.

In pus from a tuberculous lesion the bacilli are present in small numbers only so that search for the organism in films often has a negative result. One or other of the methods of homogenization described above when dealing with sputum may with advantage be adopted, though it is always preferable to inoculate a guinea-pig. The tubercle bacillus in the majority of cases occurs in pure culture in tuberculous pus but in other cases it may be associated with the ordinary pyogenic organisms and particularly with staphylococci.

D. Exudates.

In the sero-fibrinous exudates which occur in pleurisy, peritonitis, pericarditis, etc., direct examination for the tubercle bacillus by microscopical examination always gives negative results.[1]

Jousset appears to have obtained remarkable results by applying the method of inoscopy to the detection of the tubercle bacillus; he found the bacillus in all sero-fibrinous exudates. Unfortunately this method involves risk of error by reason of the presence of acid-fast bacilli in the surrounding air (p. 341) and the bacilli stained by Jousset were, at least in most cases, evidently not the tubercle bacillus. Jousset himself noted their abnormal forms and the ease with which they were decolourized by too long immersion in acid (p. 345). Moreover, the method of inoscopy usually gives negative results when it is applied under strictly aseptic conditions (Bergeron).

Jousset's technique.—If the liquid be spontaneously coagulable it is allowed to clot and the clot treated as described above in the case of blood. When the liquid does not coagulate spontaneously (cerebro-spinal fluid, for example) some horse-blood plasma is added to form a clot and this is then treated in the ordinary way. Horse plasma is obtained by mixing equal volumes of horse-blood and 10 per cent. solution of sodium chloride, centrifuging and collecting the supernatant fluid.

Satisfactory results may be obtained by sowing the exudate on blood-

---

[1 Though this statement is true in the majority of cases its application is not so universal as the author's experience would lead him to think. Microscopical examination of fluid from cases of tuberculous pleurisy may show the presence of tubercle bacilli and occasionally in extraordinarily large numbers.]

agar. Bezançon and Griffon obtained cultures in 12–15 days from ten cases of tuberculous meningitis by sowing the cerebro-spinal fluid. And the same authors obtained cultures of the tubercle bacillus in two cases of sero-fibrinous pleurisy.

The classical method is to inoculate a guinea-pig with the suspected fluid. But in this connexion it must be borne in mind that inoculation of tuberculous pleural fluid gives negative results in three-quarters of the cases. [This is probably because the fluid is actually free from tubercle bacilli since " an extremely small number of bacilli is able to induce a progressive tuberculosis in the guinea-pig " (A. S. Griffith, for the English Commission).] Inoculation is best done into the peritoneum with a large quantity (10–15 c.c.) of the fluid, which must, of course, be collected aseptically. To ascertain the degree of virulence of the bacillus a rabbit should be inoculated at the same time, for a bacillus which will infect a guinea-pig often produces no lesion in a rabbit (Arloing). [Rabbits inoculated with very small doses of the human tubercle bacillus frequently show no tuberculous lesions when killed (English Commission).]

Debove and Renault's method.—Debove and Renault have devised a very ingenious method for deciding the nature of a suspected tuberculous exudate. They showed that tuberculous exudates contain tuberculin. The inoculation of a small quantity of a pleural or pericardial exudate into a tuberculous guinea-pig gives the characteristic tuberculin reaction (p. 324).

E. Granulomata.

In the majority of cases microscopical examination fails to reveal the presence of the tubercle bacillus. A small piece of the growth should in this event be inoculated beneath the skin of a guinea-pig.

F. Nasal cavities.

Strauss has shown that tubercle bacilli are frequently found (once out of three times) in the nasal fossæ of healthy subjects living in close contact with persons suffering from phthisis. The following paragraph describes Strauss' technique.

Prepare a number of small swabs by rolling a little piece of absorbent wool round the end of a small stick of wood (10–15 cm. long) [or stout iron wire] and sterilize them in wool-plugged test tubes in the hot air sterilizer. Pass one of these sterile swabs into the nasal cavity and by rubbing it gently over the mucous membrane collect the dust and mucus adhering to it. Wash the swab in a little sterile water. Repeat the operation six or eight times in each case and wash the different swabs in the same water, then inoculate the emulsion into the peritoneal cavity of a guinea-pig.

G. Urine.

Microscopical examination for tubercle bacilli of the urine of patients affected with tuberculosis of the urinary passages often gives negative results even when the urine has been centrifuged and the deposit used for examination. It must not be expected that large numbers of bacilli will be found even in the most favourable cases : should, however, a large number of acid-fast bacilli be found on microscopical examination of a urine the result should be regarded with suspicion and the examination done again, decolourizing with alcohol for a long time. This would be a typical case for inoculation.

In cases of acute tuberculosis even when there is no lesion of the urinary passages the tubercle bacillus may pass into the urine (Benda, Weichselbaum, L. Fournier and Beaufumé).

Pour the urine into a conical glass vessel and add a small crystal of thymol or camphor. If there is an abundant deposit of pus homogenize the deposit and centrifuge. If only a small deposit is formed, decant the supernatant liquid and prepare films with the deposit. If after 24 hours there is only a minimal deposit, decant the upper part of the liquid, add an equal volume of 95 per cent. alcohol to the few cubic centimetres of liquid remaining in the vessel, mix and centrifuge.

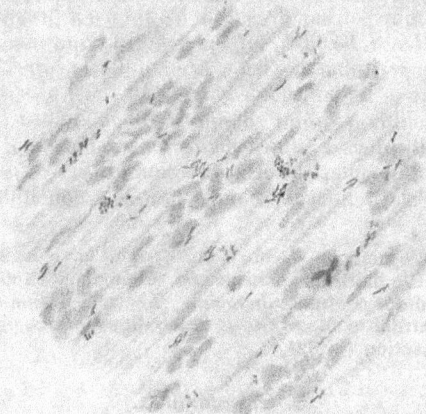

FIG. 212.—Tubercle bacilli in urine. (Carbol-fuchsin and methylene blue.) (Oc. 2, obj. 1/12th, Zeiss.)

When a urine contains only a few cells the deposit adheres badly to the cover-glasses and is liable to be washed off in the staining process. This difficulty is especially encountered when the urine yields a large bulky precipitate of crystals of urates on centrifuging. To overcome this, Trevithic recommends washing the deposit several times in distilled water but the method does not seem altogether reliable. The author prefers to heat the urine to 40–45° C., centrifuge and wash the deposit once with distilled water at 45° C. When the deposit is very small, it may be mixed with 2 or 3 drops of a mixture of fresh egg albumin and distilled water (1–3) which fixes the deposit on the slide better.

Jousset's method has been utilized for the detection of the tubercle bacillus in urine. Add some blood plasma to the urine, digest the clot which forms with artificial gastric juice (p. 340) and centrifuge. Examine the deposit for tubercle bacilli. It must not be forgotten that this method more than any other is liable to lead to error on account of the presence of other acid-fast bacilli—particularly of the smegma bacillus which is easily mistaken for the tubercle bacillus.

The only certain method of detecting the tubercle bacillus in urine is to inoculate a guinea-pig. When the urine can be collected aseptically and is not contaminated either with the colon bacillus or other pyogenic organisms a few cubic centimetres may be inoculated into the peritoneal cavity of a guinea-pig. In the contrary case the urine should be inoculated sub-cutaneously. It has also been recommended that the deposit obtained on centrifuging the digested clot in Jousset's method should be inoculated.

H. Excreta.

[Acid-fast bacilli having the morphological properties and staining reactions of the tubercle bacillus can often be seen in films made with the excreta of tuberculous subjects and sometimes in very large numbers. In order to determine whether these bacilli are tubercle bacilli or no resort must always be had to guinea-pig inoculation.]

I. Milk.

Tubercle bacilli occur only in small numbers in milk and the chances of finding them by microscopical examination are far from great. [Moreover non-pathogenic acid-fast bacilli (*infra*) are of frequent occurrence in cow's milk and no reliance can be placed upon microscopical examination for the detection of tubercle bacilli in milk.] Several methods have been described for detecting the bacillus in milk.

(*a*) Leave the fresh milk to stand for 24 hours and examine the deposit.

(*b*) Centrifuge and use the precipitate for making films.

(*c*) Coagulate 200 c.c. of milk with a little powdered citric acid, filter, dissolve the precipitate on the filter in a solution of sodium phosphate, pour the liquid into a large test-tube, add a few cubic centimetres of ether, shake for 10 minutes or so, decant the ether with the fat in suspension, centrifuge the aqueous fluid and examine the deposit.

With milk as with urine the only certain method of ascertaining whether a given specimen contains the tubercle bacillus is to inoculate a few cubic centimetres collected as aseptically as possible into the peritoneum of a guinea-pig. [Stand the milk in the ice chest for 12 hours. Pipette off some of the cream into one sterile centrifuge tube and the deposit into another. Centrifuge and inoculate 3 c.c. of the cream from the first tube into one guinea-pig and 3 c.c. of the deposit from the second into another guinea-pig.]

[A. S. Griffith and others have shown that tubercle bacilli are found in the milk of cows suffering from tuberculosis quite independently of whether there is disease of the udder or not. The bacilli cannot however be detected on every occasion on which the milk is tested. F. Griffith suggests in explanation of this fact " that the quantity of milk " (50 c.c.) " inoculated was not sufficient to be representative rather than that the elimination of tubercle bacilli was irregular since in several of the animals " (inoculated with the milk) " the slight amount of disease produced showed that only a few bacilli had been inoculated." And this explanation is supported by the fact that if a number of guinea-pigs say eight be each inoculated with the same quantity of the same milk often not more than one animal develops tuberculosis.]

THE ACID-FAST, OR PARA-TUBERCLE, BACILLI.

Besides the leprosy bacillus, the bacillus of verruga and the smegma bacillus there is a number of bacilli which share with the tubercle bacillus the property of resisting the decolourizing action of acids. These organisms, variously described as *acido-phile*, *acid-fast*, or *para-tubercle* bacilli, have been described by Petri, Rabinowitsch, Rübner, Beck, Obermüller, Coggi, and Mœller as occurring in milk, butter, manure, grass, air, and so on. Mœller, in particular, has described several species of these organisms (the manure bacillus or *Mistbazillus*, the grass bacillus or *Grasbazillus*, and the Timothy-grass bacillus or *Timothee bazillus*). Similar bacilli have been found in various pathological conditions in man *e.g.* in gangrene of the lung (Pappenheim, Meyer, Lydia Rabinowitsch, and others) in conditions of the eye simulating tuberculosis (Guisberg), in various pulmonary diseases (Mœller, Flexner, Ohlmacher, and others) and in diseases of the alimentary canal (Mironescu) etc.—The ichthic bacillus described by Dubard also belongs to this group.

None of these bacilli which morphologically resemble the tubercle bacillus can be distinguished from the latter in microscopical preparations : when stained they are not decolourized by acid and even sometimes not by alcohol. According to

Borrel their morphological characteristic as in the case of the tubercle bacillus is to remain a bright red colour when treated in the following manner

1. Stain with carbol-fuchsin in the warm for 5 or 10 minutes.
2. Treat with 2 per cent. aniline hydrochloride for 1 or 2 minutes.
3. Decolourize in absolute alcohol.
4. Differentiate with a dilute aqueous solution of methylene blue.

They are distinguished from the tubercle bacillus, (1) by the ease with which they can be grown on various media containing no glycerin at the ordinary temperature of the laboratory (2) by their cultural characteristics (luxuriant and generally greasy and creamy) and (3) finally and especially, *by the fact that they do not produce tuberculin*. Ramond and Ravaut, Bataillon and Terre have described an ichthic tuberculin similar to human tuberculin but their results have not been confirmed (p. 296).

Some of the members of this group are pathogenic to animals, particularly guinea-pigs, and may set up either local lesions distinctly tuberculous in appearance or pseudo-miliary-tuberculoses (*Timothee bazillus*) which have a tendency to suppurate.

Finally, several authors have directed attention to the existence of *pseudo-acid-fast* bacilli (Bezançon and Philibert, Bienstock, and others). A large number of saprophytic bacilli as a result of living in contact with fatty substances acquire, accidentally, as it were, acid-fast properties (e.g. *B. smegmæ*) which are lost as soon as they are grown on ordinary culture media. Other saprophytic bacilli become acid-fast when grown in blood or sero-fibrinous exudates (Bezançon and Philibert). The *Bacillus anthracis*, *Bacillus subtilis* (Bienstock), *Bacillus entericæ febris* (Ramond and Ravaut), and *Bacillus diphtheriæ* (Bezançon and Philibert) become acid-fast when grown in media containing butter. But all these pseudo-acid-fast bacilli are decolourized by prolonged treatment with acid and especially by alcohol (pp. 306 and 342) ; and moreover, unlike the tubercle bacillus, they can be stained with Unna's blue (10 minutes).

There should therefore be no reason for confusing the tubercle bacillus with the para-tubercle bacilli.

The smegma bacillus.

Tavel, Alvarez, Matterstock have isolated from normal smegma a bacillus which resists decolourization by acids. This is the bacillus which Lustgarten described as the cause of syphilis. It has not been grown outside the body and is decolourized by acid-alcohol ; it should not therefore be difficult to differentiate it from the tubercle bacillus. Houssell's method is the best for purposes of micro-chemical diagnosis. The technique is as follows.

Stain the film in the warm for 2 minutes with carbol-fuchsin. Wash. Treat for 10 minutes with acid-alcohol.

Absolute alcohol,	100 c.c.
Pure hydrochloric acid,	3 „

Wash again. Stain with an aqueous solution of methylene blue.

Saturated aqueous solution of methylene blue,	50 c.c.
Distilled water,	50 „

Wash. Dry. Mount. The smegma bacillus will be stained blue, the tubercle bacillus, red.

The bacillus of verruga peruana.

Verruga is a disease found in certain valleys of the Andes. It affects man and some of the domestic animals. It occurs both as an acute and chronic disease and leads to the formation of granulomata on mucous surfaces, the skin and the viscera (Odriozola). Carrion inoculated himself with the blood of a person suffering from the chronic form of the disease and died of an acute infection. A dog, inoculated by Tamayo with 1 c.c. of the blood of a person suffering from verruga in an acute form, became infected with a typical attack of the disease and recovered. According to Ch. Nicolle and Letulle, the cause of verruga is a bacillus morphologically identical with the tubercle bacillus and staining by the Ziehl-Neelsen method. The organism has not been cultivated.

The Pseudo-tuberculoses.

In addition to the tubercle bacillus there is a certain number of other pathogenic organisms capable of producing tubercles in the tissues.

The bacillus of leprosy and of glanders both lead to the formation of true tubercles. In connexion with the *Discomyces* it will be seen that some of the members of that group produce pseudo-tuberculous conditions in man and the lower animals.

Finally some bacteria give rise to lesions so closely simulating those produced by the tubercle bacillus that they may be mistaken for tuberculous lesions. These pseudo-tuberculous lesions may be classified into two groups, namely the *zoogleic pseudo-tuberculoses* of Malassez and Vignal, Chantemesse, and others: and the *bacillary pseudo-tuberculoses* of Charrin and Roger, Dor, Courmont, and others.

The descriptions given by different authors do not at all coincide: perhaps they relate to a number of varieties of the same organism. It must suffice to have recorded the existence of this group of organisms: a description of them is beyond the scope of this work.

[The disease commonly known as pseudo-tuberculosis in guinea-pigs and rabbits is briefly described at p. 160.]

CHAPTER XIX.

BACILLUS LEPRÆ.

Introduction.

LEPROSY, the cause of which is a bacillus discovered by Hansen, is a contagious disease peculiar to man: the lower animals are never infected. A disease apparently very similar to human leprosy has however been described as occurring in the rat (Rabinowitsch and others); it takes the form of ulcers on the skin and swelling of the glands, and the lesions contain very large numbers of bacilli similar to the leprosy bacillus.[1]

[Though there is no definite and absolutely conclusive proof of the ætiological rôle of the organism hitherto commonly known as the *bacillus lepræ* there can be no reasonable doubt but that the parasite is the cause of leprosy. Recent investigations however would seem to afford amply sufficient ground for believing that the organism is not a true bacterium but rather an hypomycete belonging to the genus Discomyces (Streptothrix). This being so it would be more exact to supersede its present designation by the name proposed by Deycke: *Streptothrix leproides*. These researches are also of interest in that they afford botanical evidence of the close relationship which has on other evidence been known for long enough to exist between the parasites of leprosy and tuberculosis.]

SECTION I.—ATTEMPTS TO REPRODUCE THE DISEASE EXPERIMENTALLY.

Most of the attempts made to reproduce the disease experimentally by inoculating the bacillus have failed.

In man, Arning is said to have succeeded in inoculating with leprosy the condemned criminal Keanu, but in this case the hypothesis of a spontaneously contracted infection may be pleaded: the same objection may be raised against two or three other cases in which leprosy is said to have been successfully transmitted by inoculation. Against these experiments of doubtful validity must be placed the very numerous unsuccessful attempts made by a number of different observers.

[1 This disease is endemic in England, on the Continent, in Australia, in America, and in Japan.]

In the lower animals, with the probable exception of the monkey, the inoculation of leprous tissues or cultures of the organism produces no result [but *vide infra*].

In the lesions found by Melcher and Orthmann, and Tedeschi, after the inoculation of a rabbit with pieces of leprous tissue, the presence of what was probably the tubercle bacillus as well as other organisms unconnected with leprosy was demonstrated. Thiroux, in Madagascar, inoculated four rabbits with leprous nodules: all four animals developed typical and fatal tuberculosis; the tissues were sown and yielded cultures of the tubercle bacillus. Numerous inoculation experiments carried out on monkeys (Babès), pigs (Hilairet and Gaucher, Widal), dogs (Neisser, Danisch), rabbits (Wesener), and cold-blooded vertebrata (Kobner), and more recently the experiments of Besnier and Leloir, have all failed to give rise to leprosy in the inoculated animal.

A piece of leprous tissue inoculated beneath the skin of an animal retains its normal appearance for a long time, and the bacilli contained in it will stain even after the lapse of several months, but they never undergo any multiplication. The tissue is gradually absorbed by leucocytes which congregate around it.

C. Nicolle has succeeded in inoculating macacus monkeys with leprosy.

A non-ulcerated leprous nodule was pounded in a mortar and inoculated into the ear of a bonnet monkey (*Macacus sinensis*): sixty-two days later leprous nodules containing quite typical leprosy bacilli were found to have developed. Six other macacus monkeys (*Macacus sinensis* and *Macacus rhesus*) are said to have been successfully inoculated beneath the skin: by repeating the inoculations the susceptibility of the monkey is increased and the period of incubation can be reduced from two months to a fortnight (fourth inoculation). The lesions resolved spontaneously in 29–160 days. For purposes of inoculation, material rich in bacilli from untreated lesions should be used.

[Rost also successfully infected a monkey by repeatedly inoculating it with an organism which he had cultivated from a case of leprosy. The monkey exhibited all the clinical features of tuberculous leprosy and in the nodules acid-fast bacilli were found situated as in leprosy.

[Rost failed in his attempts to infect guinea-pigs, white rats and rabbits either by inoculation (sub-cutaneous and intra-peritoneal) or by feeding.

[Bayon inoculated rats and mice with an acid-resisting diphtheroid bacillus which he had cultivated from cases of human leprosy and found that the organism when recovered from the organs of these animals had acid-fast properties. This acid-fast organism when inoculated into a further series of rats and mice caused " the identical changes of genuine spontaneous rat leprosy and very striking analoga of the glands and organs of human cases."

[Williams with his pleomorphic streptothrix (*vide infra*) produced in guinea-pigs by sub-cutaneous inoculation a lesion somewhat resembling leprosy with large numbers of cocco-bacilli in the cells of the connective tissue.

[Duval used cultures 2–3 days old on glycerin-blood-agar. By inoculating monkeys (*Macacus rhesus*) sub-cutaneously two or three times at intervals he was able to produce lesions resembling those of leprosy. The monkeys lost all sensation to pain and the skin for a radius of 2–3 cm. about the nodules was distinctly hypersensitive. About 6–8 weeks after the first inoculation the animals exhibited typical signs of disseminated infection and presented the clinical picture of human leprosy of the tuberculous type.]

Sugai inoculated an emulsion made by grinding up young lepromata in normal saline solution into Japanese dancing mice, with the result that small granulomata similar to those of miliary tuberculosis appeared on the peritoneum while the mesenteric and bronchial glands became enlarged. The leprosy bacillus was present in the lesions.

SECTION II.—MORPHOLOGY.

1. Microscopical appearance.

A. In human lesions.—The leprosy bacillus is a slender rod-shaped organism with rounded ends, and of the same size as the tubercle bacillus (5–$6\mu \times 0\cdot5\mu$).

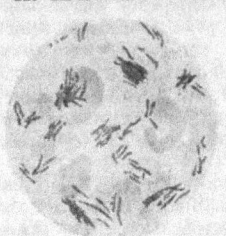

FIG. 213.—Bacillus of leprosy in a film from nasal mucus. Ziehl - Neelsen's method. × 1000.

Though it may be very slightly curved, it is generally speaking straighter than the tubercle bacillus; occasionally the ends are slightly swollen.

Staining reactions.—The leprosy bacillus, like the tubercle bacillus, stains by both Ehrlich's and Ziehl-Neelsen's methods, but is more acid-fast and therefore more difficult to decolourize than the tubercle bacillus. The two bacilli may therefore be differentiated by this characteristic. The leprosy bacillus retains the violet in Gram's method.

The following table gives the differential characteristics of the two bacilli.

LEPROSY BACILLUS.	TUBERCLE BACILLUS.
Stains with aqueous solutions of the basic aniline dyes.	Does not stain with aqueous dyes containing no mordant.
Stains readily by Gram's method.	Stains with difficulty by Gram's method (p. 307).
Stains with Ziehl-Neelsen's and Ehrlich's solutions and resists decolourization for a long time.	Stains with Ziehl-Neelsen's and Ehrlich's solutions but is much more readily decolourized than the leprosy bacillus.
Stains by Baumgarten's method (*vide infra*).	Does not stain by Baumgarten's method.
Bacilli present in very large numbers within the cells of the leprous nodule.	The tubercle cells contain only a few bacilli.

Baumgarten's method of staining.—Stain for 5 minutes in the cold with aniline-violet, decolourize with the following solution :

    Absolute alcohol, - - 10 c.c.
    Nitric acid, - - - 1 ,,

Wash in distilled water. Dry. Mount.

The leprosy bacillus is stained violet : the tubercle bacillus is decolourized.

Weil has shown that the leprosy bacillus only stains with Ziehl-Neelsen's and Baumgarten's methods when taken from young nodules. In lesions undergoing resolution these methods as well as Gram's fail to stain the bacillus.

When stained, the leprosy bacillus is often granular and its protoplasm contains irregular vacuoles. The ends are often swollen and stain easily : by some authors these swellings are regarded as spores.

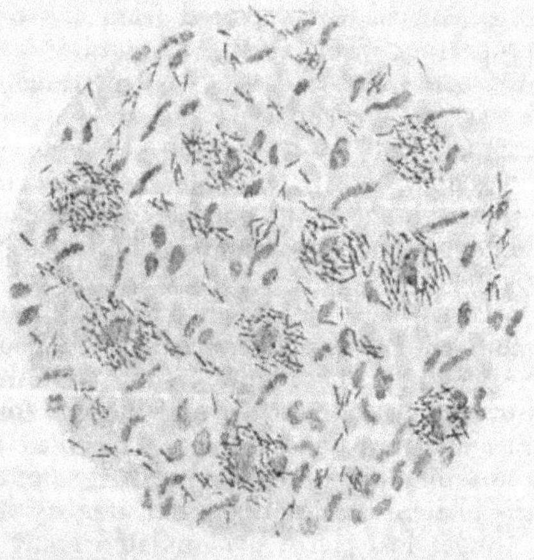

FIG. 214.—Section through a leprous nodule in the larynx. Carbol-fuchsin and methylene blue. (Oc. 2, obj. ¹⁄₁₂th, Zeiss.)

Jamamoto's stain.—By staining in the manner now to be described Jamamoto claims that the leprosy bacillus can be differentiated from the tubercle bacillus in films.

Fix the film in the flame, treat for 10 minutes in a bath of 5 per cent. solution of silver nitrate at 55°–60° C. and then transfer to the reducing solution :

Pyrogallol,	2 grams.
Tannin,	1 gram.
Distilled water,	100 c.c.

Tubercle bacilli are stained black : leprosy bacilli are unstained and may be counter-stained by carbol-fuchsin.

[**B. In cultures.**—Rost found acid-fast bacteria massed together in parallel arrangement in his cultures from leprous nodules on milk-fish broth. When sub-cultivated on agar and broth a feebly acid-fast bacillus developed and it was found that the acid-fastness could be increased by growing the organism on milk. The organism is highly pleomorphic. In sub-cultures after 48 hours' incubation the appearance is the same as in the nodules of a leper. In older cultures or when grown under unfavourable conditions " degenerate forms are found which double or treble their usual length with a moniliform arrangement and lose their acid-fastness. These break down after a few days into clumps of small acid-fast coccoid forms."

[Bayon cultivated from cases of human leprosy an acid-resisting diphtheroid organism which acquired acid-fast properties on being inoculated into rats and mice.

[Williams grew a very pleomorphic streptothrix from the lesions of human leprosy " which in addition to changes in form exhibited marked changes in its staining reactions in regard to the quality known as acid-fastness." Williams describes the following forms :

(α) On broth media and on potato-broth a non-acid-fast streptothrix in the mycelial stage which produced acid-fast rods.

(β) On milk and lemco-broth a non-acid-fast diphtheroid bacillus which also produced acid-fast rods.

(γ) On Rost's medium an acid-fast bacillus which is but the broken-down stage of a streptothrix, and

(δ) On Dorset's egg medium an acid-fast mycelium. This streptothrix which was cultivated from a leper passed through respectively all the stages described above.]

2. Cultures.

Very little is yet known about the cultivation of the leprosy bacillus. Roux, Cornil, and Chantemesse failed in their attempts to grow the bacillus : numerous observers have obtained cultures after sowing pieces of leprous tissues, but in the great majority of cases these were cultures of organisms of secondary infection and not cultures of the leprosy bacillus (*vide infra*).

Bordoni-Uffreduzzi for example described certain growths as cultures of the leprosy bacillus which were obviously cultures of the tubercle bacillus : similarly Neisser's cultures were not cultures of the leprosy bacillus. Babès' cultures were cultures of a bacillus which did not stain either with Ehrlich's or Ziehl-Neelsen's stain : apparently also Ducrey's anaërobic organism may be dismissed without consideration.

Czaplewski, Spronck, Teich, Levy, Rost and others seem to have grown cultures of the leprosy bacillus, though it is to be noted that the descriptions of their cultures do not at all coincide.

Spronck sowed leprous bone marrow and non-ulcerated leprous nodules on neutral glycerin-potato : the growth was said to have been sub-cultivated on coagulated serum, glucose-glycerin-agar and glycerin-fish-broth (p. 319) but not on glycerin-potato. Growth took place at 25° C. and was copious at 37° C.

On glycerin-potato (primary culture).—After incubation for 10 days at 37° C. very small, yellowish, hardly visible colonies appeared.

On glucose-glycerin-agar.—Small, colourless, irregularly circular colonies.

On coagulated serum.—Small, greyish-yellow, irregularly circular colonies.

On fish broth.—A viscous precipitate adhering to the sides of the vessel.

Bacilli from these cultures were agglutinated by the blood of lepers in dilutions of 1 in 70 to 1 in 1000 and only in dilutions of 1–30 or 1–40 by the blood of other persons.

[Rost employed a medium of the following composition :

Distilled volatile alkaloid of rotten fish,	250 c.c.
Weak Lemco broth (without peptone or salt),	250 c.c.
Milk,	50 c.c.

which he sowed with material from cases of nodular leprosy and obtained in 3 days a slight stringy growth at the bottom of the tube which on microscopical examination proved to be masses of acid-fast bacteria. Sub-cultures were sown on agar and in broth (no salt or peptone) and a growth was obtained in 48 hours.

[Clegg grew the parasite of leprosy symbiotically with amœbæ and their symbiotic bacteria on an agar medium. After destroying the amœbæ and bacteria by heat at 60° C. for half an hour pure cultures of the organism were obtained by sub-cultivating on ordinary media—agar, potato, milk, etc.

[Duval used egg-albumin or human blood serum poured into sterile Petri dishes and inspissated for 3 hours at 70° C.

[The excised leprous nodule is cut into thin slices (·5–1 mm.) and distributed over the surface of the albumin. After sowing, the medium is bathed in a 1 per cent. solution of trypsin added with a pipette but the tissue must not be submerged. Incubate at 20° C.[1] for a week to 10 days trypsin being added from time to time as evaporation necessitates.

[Sub-cultures may be sown on the albumin-trypsin medium but after subcultivating three or four times growth can be obtained on a glycerin-agar.

Agar,	20 grams.
Salt,	3 ,,
Glycerin,	30 c.c.
Distilled water,	500 c.c.

Mix, clear and sterilize in the usual manner.

To 10 c.c. of the agar at 42° C. add 5 c.c. of unheated turtle muscle infusion :

Turtle muscle cut into fine pieces,	500 grams.
Water,	500 c.c.

Keep in the ice chest for 48 hours : filter through gauze and then through a Berkefeld filter.

[Duval claims that the organism he cultivated was the true leprosy parasite because with its aid he was able to produce the lesions of leprosy in a monkey.

[Twort has introduced a method of cultivation based upon the addition of sterilized tubercle bacilli to an egg medium. Growth is not visible to the naked eye for about 6 weeks.

[The material used was the nasal discharge and scrapings from a typical leper.

[The nasal discharge was first placed in a 2 per cent. solution of ericolin—a glucoside—at 38° C. for 1 hour to destroy contaminating organisms and the sediment was then used for sowing the culture medium.

[*The culture medium.*—Cultivations of the tubercle bacillus on Dorset's egg medium were steamed and the growth scraped off (care being taken to avoid the medium containing the waste products). The tubercle bacilli were ground up into an emulsion with glycerin and normal saline solution, steamed for 30 minutes and added to the yolk and white of new-laid eggs in the following proportions :

Eggs,	75 parts.
8 per cent. saline solution,	25 parts.

Mix well and add tubercle bacilli 1 per cent. and glycerin 5 per cent. or less.

[The medium was tubed, heated to 60° C. for 1 hour and on the following morning

[1] In his original experiments Duval incubated at 37° C. but now finds room temperature more suitable.]

incubated at 38° C. for 6 hours after which it was again heated in a water bath at 60° C. for 1 hour and then sloped at 85° C.

[The ericolinized nasal discharge was sown and the tubes capped and incubated at 38° C. After 24 hours the medium had absorbed a quantity of the ericolin so the material was transferred to other tubes.

[The bacilli grew and were sub-cultivated in pure culture. In sub-cultures the bacilli were long, thin, beaded rods, well formed and quite acid-fast. Growth appears as a thin colourless film visible to the naked eye in 6 weeks.]

[Bayon states that the most favourable media appear to be either placental-extract-glycerin-agar, or horse-serum-nutrose-agar containing 2 per cent. ground-up smegma bacilli (Twort's method). From the nodules of a case of leprosy this observer isolated an organism which " grew rapidly as a white viscid growth on placental-extract-glycerin-agar." Morphologically it assumes one of three forms : (i) a non-acid-fast and non-acid-resisting streptothrix, (ii) a pleomorphic, acid-resisting diphtheroid bacillus and (iii) a definitely acid-fast bacillus indistinguishable from the bacillus in the tissues.

[The organism appears to be identical with that cultivated from leprous lesions by Kedrowsky so long ago as 1901 and this latter Bayon regards as the true parasite of leprosy for the following reasons : it has been cultivated repeatedly from lepers, it causes leprous lesions in rats and mice, it reacts specifically with the serum of lepers in a way that neither human nor avian tubercle bacilli react and lastly it is not identical with any other known organism.]

According to Ch. Nicolle and Weil, primary cultures of the leprosy bacillus can be obtained by sowing non-ulcerated leprous tissue rich in bacilli in considerable amount in the water of condensation of glucose-glycerin-agar tubes to which serum may or may not have been added ; the bacilli appear to grow solely at the expense of the cells of the material sown and cannot be sub-cultivated. Weil has also succeeded in growing cultures by sowing the material in the yolk of the whole egg and on yolk of egg-agar (p. 53, A) ; but here again sub-cultures could not be obtained.

SECTION III.—SERUM THERAPY.

Carasquilla was the first to attempt the preparation of an antiserum by inoculating large animals with the blood and serum of leper patients. Later, Laverde injected asses, lambs and horses with blood and serum from leprosy patients, with the juice of lepromata and even with the pulp of an epithelioma of the cervix uteri : the serum of the treated animals had a favourable influence on the course of the disease (10–20 c.c. were used for inoculation). Laverde's results were confirmed by Buzzi, Abraham, and Arning ; Hallopeau, Neisser, and Brieger, however, failed to obtain similarly favourable results with the antiserum.

Metchnikoff and his pupils showed that the serum prepared by Laverde contained neither leprous products nor toxin, but that the inoculation of serum, blood, or cellular elements of one animal into an animal of another species leads to the formation in the latter of substances (*cytotoxins*) which have the property of destroying the cells of the animal from which the material for inoculation was taken : and they demonstrated that the improvement noted in the lepers treated with Laverde's serum should be attributed to these cytotoxins. The inconstant results are explicable on the ground of the delicate nature of the cytotoxins, since these are destroyed by transport, the addition of carbolic acid, etc.

Metchnikoff and Besredka inoculated a goat over a period of 36 days with 34 c.c. of defibrinated blood from a healthy man. The goat's serum acquired powerful agglutinating and hæmolytic properties for human blood : a given volume of the serum agglutinated at once and dissolved in 7 minutes all the red cells in an equal volume of human blood.

When injected into lepers in doses of 1, 3, and 7 c.c. this serum to some extent relieved pain, and caused congestion and suppuration of some of the lepromata with the result that sloughs formed which afterwards became detached : in a few cases an insignificant febrile attack was noticed. In short, the results of using Metchnikoff's serum, although not so good, were very similar to those obtained with the serum prepared by Carasquilla, Laverde, and others.

In Metchnikoff's opinion the favourable results following the use of such serums should be attributed to the leucotoxin developed in response to the inoculation into the tissues of an animal of human leucocytic products ; this leucotoxin should in suitable doses lead to stimulation of the leucocytic system : the hæmotoxin is of no therapeutic value, and indeed prevents the employment of sufficiently large doses of serum. The obvious conclusion from this argument is that in the treatment of leprosy an attempt should be made to prepare an antiserum by inoculating an animal with blood serum alone or better still with human lymphatic glands.

SECTION IV.—DETECTION AND IDENTIFICATION OF THE LEPROSY BACILLUS.

Microscopical examination is at present the only means of detecting the leprosy bacillus. Sections should be cut and films made of the suspected tissues and fluids. [Cultures should however be attempted.]

The bacillus of leprosy is found in the leprous nodules, bone marrow, and spleen. The bacillus can also be found in the glands, in the swellings along the nerves, in the discharge from ulcerating lesions, in the saliva when the buccal mucous membrane is affected, in the stools when the disease infects the large intestine, in the secretion of the testicle when that organ is involved, in the milk (Babès), etc. Sticker has drawn attention to the presence of the bacillus in the nasal mucus. Nasal lesions are commonly present from the early stages of the disease and were found in 128 out of 153 lepers examined by Sticker ; examination of films of the nasal mucus will therefore often afford confirmation of the diagnosis.[1]

Leprous nodules consist of large cells, similar to epithelioid cells, having as a rule a single nucleus and crammed full of bacilli : these constitute the lepra cells. The leprosy bacillus is therefore intra-cellular.

During life portions of a leproma can be easily excised, since it is known that in the majority of these lesions there is an absence of all sensation. Manson recommends isolating a succulent leproma in a pile clamp, slowly screwing up the jaws of the instrument so as to drive out the blood, pricking the now pallid leproma and then collecting on a cover-glass the droplet of " leper juice " which exudes from the puncture.

Arning has never found the bacillus in the blood. According to Cornil, Babès and Goujerot the bacillus enters the blood stream a few days before death and especially during the febrile attacks.

The following technique is recommended for the detection of the bacillus :

(a) Stain films by Ziehl-Neelsen's method. The leprosy bacillus is differentiated from the tubercle bacillus by three tests—

　　1. Simple staining with a watery alcoholic solution of fuchsin ;
　　2. Gram's stain ;
　　3. Baumgarten's method of staining.

[1] In examining films of the nasal mucus care must be taken to distinguish the leprosy bacillus from the bacillus of Karlinski. The latter bacillus is found in the nasal mucus of man quite apart from leprosy or tuberculosis : it gives rise to no symptoms, morphologically resembles the bacillus of leprosy, and is acid-fast. In cultures, it grows easily on ordinary media, and is pathogenic to guinea-pigs when inoculated intra-peritoneally.

(*b*) Stain sections of tissues hardened in alcohol and embedded in paraffin by Ziehl-Neelsen's method : if necessary, the differential tests given above can be applied.

A good diagnostic point is afforded, as has been said, by the enormous numbers of bacilli to be seen in the lepra cells : tubercle cells never contain more than a few bacilli. Finally, the inoculation of a guinea-pig [will exclude tuberculosis].

Sub-cutaneous inoculation of tuberculin produces a reaction in persons suffering from leprosy while the ophthalmo-reaction is negative (Nicolle and Uriarti, Gaucher and Abrami).

Associated micro-organisms.

The lesions of leprosy being so frequently situated in the skin and mucous membranes or in the lungs are particularly liable to secondary infection.

Lesions of the skin and mucous membranes are very soon invaded by the ordinary organisms of suppuration (*staphylococci, bacillus pyocyaneus,* etc.). In the case of a leper in Tunis the author was able to demonstrate in the discharge from the specific lesions in addition to a few leprosy bacilli, *staphylococcus aureus, bacillus pyocyaneus* and *bacillus coli.* These organisms of secondary infection may invade the tissues in cases of leprosy and give rise to a rapidly fatal pyæmia (Babès).

Babès has frequently found the tubercle bacillus in persons suffering from leprosy, and the two organisms are frequently found in association especially in the lung : and in pulmonary lesions the pneumococcus may also be present.

In three cases of leprosy, Babès found as a secondary infection in the bone marrow, spleen, and kidneys, a bacillus which was easily cultivated outside the body and did not stain by Ehrlich's or Ziehl's methods.

CHAPTER XX.

BACILLUS DYSENTERIÆ EPIDEMICÆ.

THE bacillus of epidemic dysentery was discovered by Chantemesse and Widal in 1888 : Shiga amplified their observations and adduced additional evidence in proof of the specific relationship of the bacillus to the disease.

The term dysentery is applied to a clinical syndrome indicating certain lesions of the large intestine and ætiologically includes two distinct diseases, one caused by an amœba and the other by a bacillus. The former is an endemic disease of warm climates and is frequently complicated by abscess of the liver : the latter is an epidemic disease not complicated by abscess of the liver and prevalent both in warm and—especially—in temperate climates.

It is possible that in rare cases symptoms of dysentery may be due to certain other parasites which up till now have received little attention, such for example as *Balantidium coli*, *Spirilla*, or *Trichomonas*.

[Dysentery bacilli have been isolated from a number of cases of infantile diarrhœa (quite unrelated to any epidemic of dysentery) by Bassett and Duval in the United States ; and in South Africa, Birt found dysentery bacilli in 7 out of 10 cases of this disease. In London, however, Morgan failed to find any bacilli of the dysentery group in cases of infantile diarrhœa.

[Asylum dysentery has been shown to be a bacillary dysentery and bacilli of the dysentery group have been isolated from cases of the disease in England, Germany and America (Eyre ; Aveline, Boycott and W. F. Macdonald ; Kruse ; Vedder and Duval).

[Sporadic cases of dysentery are said to occur in England though rarely (Marshall ; Bainbridge and Dudfield). Ledingham's investigations would appear to show that dysentery bacilli are occasionally found in the stools of healthy persons.]

In patients suffering from bacillary dysentery the organism is found in large numbers in the intestinal mucous membrane and in the stools especially in the mucous flakes. It does not become generalized, and with the exception of one case recorded by Rosenthal the organism has never been found in the blood stream : according to some observers it is occasionally present in the mesenteric glands (Shiga, Duval and Bassett, [Aveline, Boycott and Macdonald]).

[Aveline, Boycott and Macdonald isolated the organism from the spleen in one out of three fatal cases of asylum dysentery.]

Several varieties of dysentery bacilli have been described, differing from one another in one or more particulars and especially in their action upon sugars. For practical purposes these varieties may be divided into two types: the Shiga-Krüse or non-mannite fermenting type and the Flexner or mannite fermenting type (see also p. 360).

Some authors regard the differences between these two types as sufficient to justify their classification as separate species. But it is held that these differences are not marked enough to warrant so complete a separation, and the view put forward by Gay and Duval which is perhaps of the nature of a compromise, commands general acceptance. These authors consider that the bacilli causing bacillary dysentery are to be regarded as belonging to a group of organisms exhibiting certain variations among themselves rather than as a single sharply-defined species. There is a similar multiplicity of varieties of the causal organism of cholera, as will be shown later.

The Shiga type of bacillus is the common cause of dysentery, and it will be therefore described at length in the following pages, the points of difference between it and the Flexner type being noted in the proper places. The general statement may here be made that all strains of Shiga's bacillus agree in their cultural and other characteristics, while under the title of Flexner's bacillus a number of very closely related though not absolutely identical organisms are included.

SECTION I.—EXPERIMENTAL INOCULATION.

Shiga bacillus.

Speaking generally, the Shiga type of bacillus is much more highly pathogenic to laboratory animals than are bacilli of the Flexner type.

Infection by the alimentary canal. *In man.*—Strong and Musgrave after administering some bi-carbonate of soda to a condemned criminal gave him a two-day-old broth culture of the dysentery bacillus. After an incubation period of 36 hours the man suffered from a typical attack of dysentery with blood-stained stools from which he made a rapid recovery. The bacillus was isolated from the stools.

In animals.—Infection of the alimentary canal whether by feeding or inoculation generally gives negative results (Rosenthal, Shiga, Conradi, and others).

After feeding guinea-pigs with dysentery bacilli, however, Chantemesse found lesions in the intestines similar to those seen in human dysentery ; and Shiga, after introducing a culture into the stomach of a cat, noticed that it suffered from mucous diarrhœa and found the bacilli in increased numbers in the stools. Kazarinow introduced very large quantities of culture into the intestines of rabbits by means of an œsophageal sound, and *post mortem* found characteristic lesions in the intestine.

Intra-peritoneal inoculation.—Inoculation of dysentery bacilli into the peritoneal cavity is rapidly fatal to most animals. *Post mortem*, the peritoneal cavity contains a blood-stained serous exudate and the intestine is very markedly hyperæmic but presents none of the lesions characteristically seen in the human disease.

Intra-venous inoculation.—In the majority of animals death from septi-cæmia rapidly follows the inoculation of bacilli into the veins. *Post mortem,* the intestine is found to be slightly hyperæmic.

Sub-cutaneous inoculation.—The most interesting experimental results are obtained by inoculating animals sub-cutaneously : in rabbits, dogs, cats and young pigs such inoculation is followed by lesions similar to those found in the human subject. Guinea-pigs are less susceptible than other laboratory animals to this method of infection.

Rabbits.—Sub-cutaneous inoculation of 3–4 c.c. of a broth culture is fatal in 4–6 days.

Inoculation leads first of all to the formation of a large inflammatory œdema at the site of inoculation and this is soon followed by a rise of temperature and the onset of diarrhœa, then paralysis of the hind limbs appears and finally the tem-perature begins to fall and continues to decline steadily until death occurs. *Post mortem,* lesions are present throughout the alimentary canal being especially marked in the colon. In that portion of the intestine the mucous membrane is thickened, swollen, intensely hyperæmic and covered with blood-stained mucus ; small foci of superficial necrosis and hæmorrhagic patches are also seen, the former occasionally ending in ulceration. The bacillus multiplies both in the mucus and in the mucous membrane where it is found in pure culture.

Dogs.—A "true representation of human dysentery with its painful and frequent strainings, characteristic stools and lesions" is seen in young dogs as a result of sub-cutaneous inoculation (Vaillard and Dopter).

Following the inoculation of one or two agar cultures, the temperature rises and an œdematous infiltration appears at the site of inoculation : the animal lies down, seems ill and is apparently in pain, and the motions become frequent and in character similar to those of human dysentery. These symptoms are followed by wasting and a fall of temperature to below normal, and death takes place between the third and sixth day. *Post mortem,* lesions similar to those described in the case of the rabbit are found in the intestine which is also extensively ulcerated, and the mes-enteric glands are swollen. The bacillus is present in pure culture in the affected parts of the mucous membrane.

Young pigs.—Sub-cutaneous inoculation in these animals generally leads to a fatal attack of dysentery. *Post mortem,* lesions are found resembling those in the human disease.

Note.—After sub-cutaneous inoculation the bacillus can always be found in the local lesion at the site of inoculation and frequently also in the mesenteric glands, but only exceptionally in the spleen and liver and never in the blood of the heart.

Bacilli of the Flexner type.

Bacilli of the Flexner type are far less pathogenic than the Shiga bacillus. Intra-venous inoculation is not followed by severe symptoms and rarely leads to death in the case of dogs and rabbits : intra-peritoneal inoculation is more dangerous and produces a fatal peritonitis in guinea-pigs : feeding experiments give negative results. Sub-cutaneous inoculation is not followed by the symptoms and lesions of experimental dysentery and does not lead to a fatal result ; after a marked local reaction the animal recovers.

SECTION II.—MORPHOLOGY.
1. Microscopical appearance.

The dysentery bacillus is a small rod-shaped organism morphologically like the colon bacillus; it measures 1–3μ long. In cultures the organism is pleomorphic, and long almost filamentous forms are found side by side with very short bacilli. It does not form spores, and is non-flagellated and non-motile, though it exhibits oscillatory movements which have been compared to those of a compass needle.

Staining reactions.—The dysentery bacillus stains with the ordinary basic aniline dyes, and with weak dyes tends to stain more deeply at the ends than in the centre. It is gram-negative.

Films should be stained with carbol-thionin or carbol-methylene-blue : sections with thionin or by Nicolle's tannin method (p. 217).

2. Cultural characteristics.

Conditions of growth.—The dysentery bacillus grows on all the ordinary alkaline media, and equally well under aërobic or anaërobic conditions. The optimum temperature is 37° C., though growth takes place within wide limits (10°–40° C.).

Characteristics of growth. Broth.—Growth is visible after incubating for 6 hours, and after 24 hours the medium is uniformly cloudy and has a watered-silk appearance. On further incubation a small glutinous deposit forms which continues to increase until towards the end of the second day, when the upper part of the broth is clear : no pellicle is formed on the surface. The cultures have a peculiar spermatic odour.

Gelatin.—The growth is like that of the typhoid bacillus. Isolated colonies are small, delicate and translucent with edges like the edges of a vine leaf. In stroke culture the growth consists of a thin narrow opalescent band. The medium is not liquefied.

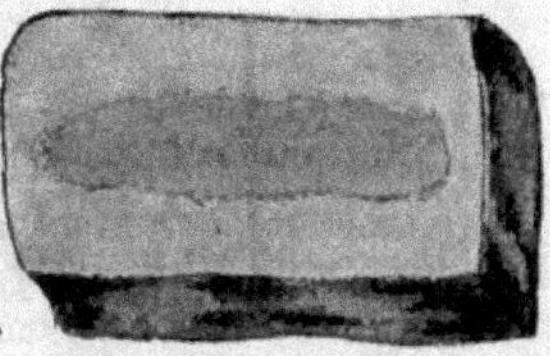

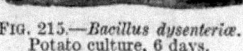

FIG. 215.—*Bacillus dysenteriæ.*
Potato culture, 6 days.

Agar.—On agar the growth resembles that of the typhoid bacillus.

Potato.—The growth on potato consists of a moist, shiny glaze which is at first so scanty as to be hardly visible. Later it acquires a greyish or yellowish tint.

Milk.—Milk is not coagulated.

Bacilli of the Flexner type.

Cultures of bacilli of the Flexner type have the same characteristics as but are more luxuriant than those of the Shiga bacillus. In broth, after incubating for about 3 days a ring of growth adherent to the sides of the tube is formed on the surface of the medium ; this falls to the bottom a few days later.

SECTION III.—BIOLOGICAL PROPERTIES.

1. Biochemical reactions.

Action on carbohydrates.—The Shiga bacillus does not ferment carbohydrates. The blue colour of litmus milk is unchanged, but litmus whey is slightly reddened at first becoming blue again about the second or third day. No gas is formed in litmus-lactose-broth. Neutral-red in glucose media is not reduced.

Bacilli of the Flexner type.—These bacilli ferment mannite and maltose but no other sugar [but see below] : litmus media containing mannite and maltose are turned red, but no gas is formed. In lactose media they behave like the Shiga bacillus.

[All strains of the Flexner or mannite-fermenting type apparently produce acid in mannite, glucose, galactose, arabinose, and raffinose : many ferment (acid, no gas) maltose and dextrin though not necessarily both, while a few form acid in sorbite, inulin, salicin, and isodulcite. One strain of the Flexner type—apparently identical with the dysentery bacillus isolated by Strong—forms acid in dulcite and cane sugar (Morgan).

[Hiss divided dysentery bacilli into four groups according to their fermentation reactions.

[The first or Shiga group ferments the monosaccharides and sometimes, after an interval of several days, maltose.

[The second group represented by Hiss' Y bacillus (isolated from dysenteric diarrhœa in children due to milk and identical with the bacillus found in cases of asylum dysentery by Kruse and others) ferments the monosaccharides and mannite. Maltose and saccharose are sometimes also but with difficulty decomposed with the formation of acid.

[The third group consisting of Strong's Philippine bacillus ferments the monosaccharides, mannite and saccharose, occasionally also maltose.

[The fourth or Flexner (Manilla) group decomposes the monosaccharides, mannite maltose, saccharose and dextrin.

[To these Shiga subsequently added a fifth group; the characteristic of the organisms comprising it being that they give first of all an acid reaction in mannite, which subsequently changes to an alkaline reaction.

[Aveline, Boycott and Macdonald find that the fermentative reactions of the Flexner group towards maltose and cane sugar are variable. Thus, 24 cultures were tested with the result shown—

	CANE SUGAR.				MALTOSE.			
Days incubated, - - -	1	7	14	28	1	7	14	28
Number of cultures acid, -	0	0	1	4	2	3	13	24]

[In litmus milk the mannite-fermenting bacilli first produce a slight acidity (1–3 days) and ultimately become alkaline (15 days). Strong's bacillus is the only strain which forms acid and clot (Morgan).]

Indol production.—The Shiga bacillus produces no indol in culture. Bacilli of the Flexner type vary, some strains produce indol, others do not. [According to Morgan the vast majority of the mannite-fermenting group form indol, some more freely than others.]

Growth in arsenical and carbolic broth.—All dysentery bacilli grow in broth containing carbolic acid (0·075 per cent.) or arsenious acid (0·1 per cent.).

Growth on vaccinated media.—On agar which has already served for the growth of the Shiga bacillus neither the Shiga bacillus nor the bacilli of the Flexner type will grow: the typhoid bacillus gives a very poor growth, the colon bacillus on the other hand grows abundantly.

Similar results are obtained with agar which has served for the growth of bacilli of the Flexner type.

On media which have served for the growth of the typhoid or colon bacillus neither the Shiga bacillus nor bacilli of the Flexner type grow.

2. Vitality.

The dysentery bacillus is a somewhat delicate organism. In culture it does not live for more than 3 or 4 weeks: in infected stools it appears to be quickly destroyed by the other micro-organisms present, and especially by the action of the colon bacillus, so that it cannot be isolated after 48 hours. Direct sunlight and desiccation rapidly destroy the bacillus: it is killed in less than an hour at 58° C. : and in sterile spring water at 20° C. it does not live more than 8–10 days. In water containing saprophytic organisms the larger the number of such organisms the more quickly does the dysentery bacillus disappear, and at the ordinary temperature it cannot be recovered after 2–10 days (Vincent); the higher the temperature also the more rapidly does the organism vanish: this may [partly] explain the frequency of epidemic dysentery in cold and temperate climates (Vincent).

3. Toxin.

(i) Filtered cultures of the dysentery bacillus are generally only slightly toxic, and even in large doses merely cause a temporary loss of weight; but the blood of the inoculated animal acquires the property of agglutinating the bacillus. Todd and Rosenthal however obtained a strong toxin, which was fatal to rabbits in doses of 0·2 c.c. sub-cutaneously, by growing the bacillus in Martin's broth at 37° C. for 3 weeks and then filtering. Their results have been confirmed by Ludke and Doerr.

(ii) Unfiltered cultures of the dysentery bacillus killed by heat (58° C. for 1 hour or 85° C. for 30 minutes) or chloroform, when inoculated into rabbits intra-peritoneally, intra-venously or sub-cutaneously, lead to a fatal result similar to that produced by the living organism and accompanied by diarrhœa and hyperæmia of the mucous membranes of the colon (Drigalski and Conradi).

(iii) **Endotoxin.**—Conradi, Neisser and Shiga, Vaillard and Dopter, and Besredka have extracted an intra-cellular toxin from the bodies of the bacilli.

Conradi's method.—Scrape the growth from an agar culture and make into an emulsion with normal saline solution. Place the emulsion in the incubator at 37° C. and leave for about 30 hours, then decant the clear liquid, filter through a Berkefeld bougie and finally evaporate *in vacuo* to one-tenth its original volume.

2. Neisser and Shiga's method.—Heat an emulsion of bacilli in normal saline solution to 60° C., allow to stand for 48 hours at 37° C. then filter through a Reichel filter.

3. Vaillard and Dopter's method.—Make a thick emulsion of bacilli from an agar culture with sterile water, heat to 58° C. for an hour, distribute in tubes, seal in the blow-pipe and leave in the warm (37° C.) incubator for a month. The clear supernatant liquid is used without filtration.

4. Besredka's method.—Besredka has applied his method of extracting endotoxin (p. 379) to the dysentery group. The endotoxin is very toxic and kills rabbits in doses of 0·05 c.c.

The Shiga bacillus alone produces dysentery toxin, bacilli of the Flexner type being almost invariably atoxic.

Properties of dysentery toxin.—The toxin kills dogs, rabbits and mice with all the symptoms of a dysentery infection. The fatal dose for rabbits varies according to the method of preparation and the method of inoculation (intra-peritoneal. intra-venous or sub-cutaneous) from 0·05 c.c. (with Besredka's endotoxin) to 2–5 c.c. Administered by the mouth it gives rise to no symptoms.

Dysentery toxin is less affected by heat than are many other toxins: it is, for instance, unaltered by being exposed to 70° C. for an hour, but a temperature of 75° C. weakens it and at 80° C. its properties are rapidly destroyed. It is now admitted that dysentery toxin is not a soluble or diffusible toxin but an endotoxin retained within the bodies of the bacilli.

4. Vaccination. Serum therapy.

(i) Shiga has shown that animals can be immunized by inoculating them sub-cutaneously with small doses first of dead bacilli then of living bacilli. The serum of immunized animals agglutinates the bacillus and has both prophylactic and curative properties.

Small animals are very difficult to immunize and it is therefore better to use a goat or an ass or an horse; horses must be treated very carefully.

The serums (ass and horse) obtained by Shiga and by Kruse protect guinea-pigs against the inoculation of a fatal dose of bacilli and have powerful agglutinating properties (1–10,000). Good results have also been obtained in the treatment of human dysentery ; with Shiga's serum the mortality was reduced two-thirds. The results with Martini and Lentz's goat serum and with Krauss and Doerr's serum (prepared by sub-cutaneous inoculation of living cultures) have not been so satisfactory.

(ii) Gay, who repeated and confirmed Shiga and Krüse's experiments, found that the agglutinating and prophylactic properties of dysentery-immunized horse serum were more marked with the strain used for immunization than with other strains. Independently of this Krauss and Doerr, thinking that their experiments showed that a Shiga immune serum had no action on bacilli of the Flexner type and conversely, recommended the preparation of a mixed Shiga-Flexner serum which could be given indifferently in all cases of dysentery whatever the infecting organism : acting on this suggestion, Coyne and Auché prepared a polyvalent serum which has yielded satisfactory results in the treatment of dysentery in man. Vaillard and Dopter however affirm that a Shiga-serum gives as good results in a Flexner as in a Shiga infection, and in consequence consider that polyvalent serums are unnecessary.

(iii) To obtain a serum which was both anti-bacterial and antitoxic Rosenthal immunized horses by inoculating them subcutaneously with toxin (p. 361) and living cultures simultaneously. The serum is both prophylactic and curative : it protects guinea-pigs in doses of 0·5 c.c., and in the treatment of human dysentery very good results have been obtained, especially by Korentchewsky in Manchuria where the mortality from dysentery fell by more than one-half.

(iv) Vaillard and Dopter, relying upon an observation by Besredka to the effect that in the case of endotoxic organisms the most active antiserums are obtained when the bacilli are inoculated directly into the blood-stream, immunized horses by inoculating living cultures and toxin directly into the veins.

The process of immunization must be carried out very slowly. Virulent cultures of the Shiga bacillus were used and the toxin was prepared by Rosenthal's method (p. 361). Increasing doses of cultures were inoculated alternately with toxin, commencing with 0·25 c.c., then 0·5. 1, 2, 3 c.c., rising to 50 c.c., an amount which was never exceeded. The animals reacted violently (fever, prostration, temporary paralysis of the hind limbs, loss of weight). The serum was collected a fortnight to three weeks after the last inoculation.

Vaillard and Dopter's serum is both anti-bacterial and antitoxic. It protects rabbits when given in doses of 0·25–0·5 c.c., and leads to recovery (doses of 1–2 c.c.) even when administered 24 hours after the experimental infection. In doses of 20–100 c.c. it is very efficient in the treatment of human dysentery : the symptoms are alleviated almost at once, recovery is rapid and the mortality lowered more than three-fourths.

Vaccination in man.—Shiga tried to vaccinate men by inoculating them with a mixture of killed bacilli (80 parts) and serum (20 parts). The results were encouraging, but the immunity is only of short duration.

Immunity can be quickly produced by the inoculation of serum alone but does not last long (10–12 days).

In future experiments it would seem to be better to work with sensitized bacilli according to the technique of Besredka : emulsions of bacilli are killed by heat and agglutinated by a non-heated specific serum the excess of serum being removed by repeated washing and centrifuging (p. 382). Such a vaccine is only slightly toxic : in mice an immunity lasting 3–4 months is acquired in 4 days, and the susceptibility of the animals is not increased during the process of immunization.

5. Agglutination.

Shiga was the first to show that the serum of persons suffering from dysentery agglutinates the bacillus.

1. Agglutinating properties of experimental serums.—The serum of an immunized animal has the property of agglutinating the bacillus used for its immunization. The most highly agglutinating serums ($\frac{1}{2,000} - \frac{1}{10,000}$) are obtained by inoculating animals intra-venously.

(α) The serum of normal animals has no action on the dysentery bacillus.

(β) Antidysentery serum is specific and has no agglutinating action on the typhoid, colon, or paratyphoid bacilli.

(γ) The serum of animals immunized with the Shiga bacillus agglutinates that organism but has no action on bacilli of the Flexner type.

(δ) The serum of animals immunized with bacilli of the Flexner type agglutinates these bacilli but not the Shiga bacillus.

[(ϵ) Hiss' Y bacillus is agglutinated by a Flexner serum but not by a Shiga serum. Strong's bacillus is agglutinated neither by a Shiga nor by a Flexner serum.]

2. Agglutinating properties of the serum of persons suffering from dysentery.—In testing the agglutinating properties of the serum of a dysentery patient it is important to recognize that the Shiga bacillus is only agglutinated in low dilutions (rarely in dilutions higher than $\frac{1}{50} - \frac{1}{100}$) and that bacilli of the Flexner type are agglutinated in much higher dilutions ($\frac{1}{500}$) and may even be agglutinated by normal serums in low dilution.

(α) The serums of healthy persons [*exceptis excipiendis*] and of patients suffering from diseases other than dysentery do not agglutinate the dysentery bacilli.

(β) The serum of patients suffering from amœbic dysentery does not agglutinate the dysentery bacilli.

(γ) The serum of patients suffering from bacillary dysentery agglutinates the bacillus causing the infection and with rare exceptions has no action on the other type: the serum of a patient infected with the Shiga bacillus agglutinates the Shiga bacillus while having no action on bacilli of the Flexner type, and *vice versa*.

Whatever the type of bacillus causing the infection agglutination-capacity is present in the serum only in severe or averagely severe cases; it seldom appears before the end of the first week and occasionally not until later, but remains for several weeks after recovery.

(δ) The serums of dysentery patients never agglutinate the typhoid bacillus; but such a serum may agglutinate some varieties of the colon bacillus, the explanation being, possibly, that a colon bacillus infection has been superadded upon the original dysentery infection.

6. Precipitins.

If one drop of Shiga serum be added to ten drops of a filtered culture of the Shiga bacillus a precipitate is formed: a similar but less marked precipitate is also formed if instead of the Shiga culture, a culture of one of the bacilli of the Flexner type be used.

Conversely, Flexner serum precipitates filtered cultures of bacilli of the Flexner type and also but less markedly filtered cultures of the Shiga bacillus.

7. Immune body.

In the serum of persons suffering from dysentery and also in the serum of immunized animals a specific immune body is present which is fixed both by the bacillus causing the infection and by all other types of dysentery bacilli. In human subjects the immune body makes its appearance about the fifth to the seventh day of the disease. It is quite distinct from the agglutinin and may be present in the serum before the latter.

SECTION IV.—DETECTION, ISOLATION AND IDENTIFICATION OF THE DYSENTERY BACILLUS.

To ensure the detection of the bacillus in a case of dysentery it is necessary to examine a recently evacuated stool. The bacilli are most numerous in the latter during the first week of the disease, but subsequently diminish in number, and finally disappear altogether as soon as the stools resume their normal consistency. Bacilli cannot be found in the stools after the twenty-first day of the disease (Rosenthal).

Dysentery bacilli may occasionally be found in the mesenteric glands and very exceptionally in other organs.

The bacilli cannot be differentiated by microscopical examination alone and cultures must be sown in every case.

Select a flake of sero-sanguinolent matter and after washing it thoroughly in sterile water emulsify in a little broth ; use the emulsion for sowing gelatin or agar plates. Plates may also be sown by smearing the surface of the medium with one of the flakes after washing it. Should there be no mucous flakes dilute a trace of the stool in broth.

(i) Gelatin plates should be sown by the dilution method (p. 78) and incubated at 22° C. After 2 or 3 days the surface colonies are examined and any which resemble colonies of the dysentery bacillus picked off for further tests.

(ii) The best method is to use a lactose-agar medium—Chantemesse's (p. 407), Conradi-Drigalski's (p. 407), Endo's (p. 408) [or M'Conkey's (p. 412)].

Dip a fine sterile camel-hair brush in the broth emulsion and smear the surface of a number of plates of this medium without recharging the brush. It is perhaps even better to smear the surface of the medium with a washed mucous flake and spread the material with a Drigalski's spatula (p. 407). Incubate at 37° C. When examined after 20–24 hours colonies of the colon bacillus will appear as red spots, while those of the dysentery bacillus and of some other organisms will not have altered the colour of the medium. From among the latter pick off those which have a translucent iridescent appearance with irregular margins and the centres of which are rather more opaque than the edges, and sow them in broth and other media. Dysentery bacilli will be recognized by the absence of motility, by the cultural characteristics mentioned above and by their being agglutinated by a specific serum. [For purposes of identifying bacilli of the Flexner type the serum of an animal immunized with the Y bacillus is the most generally useful (Morgan).]

Serum diagnosis.

Since specific agglutinins are present in the serum of patients suffering from dysentery it is possible to make a diagnosis by the serum reaction.

Knowing that the serum of patients only agglutinates the bacillus which is causing the infection the serum must be tested both with a Shiga bacillus and with a Flexner bacillus. The reaction towards the Shiga bacillus should be tested in the first instance with a dilution of 1 in 20 or 1 in 30 ; a positive result under these conditions will be conclusive. The reaction towards bacilli of the Flexner type should be tested in a dilution of 1 in 80 ; agglutina tion in a lower dilution cannot be accepted as evidence of a Flexner infection because, as has already been pointed out, normal serum in low dilution often has an agglutinating action on this type of the bacillus.

In dysentery, as in enteric fever, a positive reaction confirms the diagnosis : on the other hand if no reaction be obtained bacillary dysentery cannot be definitely excluded, because the blood may have been collected before agglutinins had developed.

[Bacillus dysentericus El Tor No. 1.[1]]

This organism which was described by Armand Ruffer in 1909, was found to be the cause of the largest percentage of the cases of dysentery among the Mussulman pilgrims passing through the lazaret at El Tor : it has the characteristics of the dysentery group, but appears to differ from all the known sub-groups at present described. Ruffer, however, remarks that the name is merely provisional and that the bacillus may prove to be identical with one of the bacilli already described.

Morphology.—The bacillus Tor No. 1 is similar to, but plumper than, the Shiga bacillus ; filaments were rarely seen ; no spores were found nor could cilia be demonstrated. It showed movement of spiral rotation when freshly isolated but no movement of progression : in sub-cultures it was quite motionless.

Cultures.—Broth.—Uniform turbidity : no pellicle.

Gelatin.—Not liquefied. The colonies have no vine-leaf appearance like those of the Shiga bacillus.

Agar.—Similar to typhoid.

On Endo's medium.—Colourless.

On Conradi-Drigalski's agar.—Like typhoid.

Bio-chemical reactions.—B. dysentericus Tor No. 1 formed acid out of mannite and thus resembled Flexner's bacillus. In saccharose, maltose, salicin, sorbite, dulcite and dextrin reactions were very inconstant and differed with different strains of the bacillus.

A small amount of indol was formed sometimes but the reaction was not constant.

In milk, no clot was formed.

In litmus milk most strains produced first an acid reaction followed by a stage of increased alkalinity. One of the strains however produced a permanent acidity.

Pathogenicity.—The B. dysentericus Tor No. 1 was highly pathogenic to rabbits on intra-peritoneal or intra-venous inoculation, giving rise to fever, diarrhœa and paralysis and causing death in 24 hours to 5 days according to the dose inoculated. The pathogenicity for rabbits rapidly disappeared in sub-cultures.

Guinea-pigs were even more susceptible than rabbits. The symptoms and lesions were the same as in the rabbit.

In horses inoculation of sterilized cultures produced a wide-spread œdema about the site of inoculation, rise of temperature and a general condition of ill-health.

In man the clinical signs and pathological appearances were indistinguishable from those produced by infection with the Shiga bacillus.

Toxin.—None of the cultures of this organism gave a soluble toxin.

Agglutination.—The serum of patients suffering from a Tor No. 1 infection did not agglutinate the Shiga bacillus but agglutinated bacillus Tor No. 1 constantly except in early acute cases or in very feeble old people. The dilution in which agglutination could be obtained varied from 1 in 25 to 1 in 300, the index rising during convalescence.

With the serum of animals specifically immunized it was found that the Tor bacillus was agglutinated in dilutions of 1–1000 to 1–2000 with a Tor serum, while with a Shiga serum agglutination was inconstant in a dilution of 1–100, and with a Flexner serum agglutination was effected with a dilution of 1–200 but not with a dilution of 1–500.

Serum therapy.—Ruffer found that patients suffering from a Tor infection were not benefited by treatment with a Shiga serum while severe cases were quickly cured by inoculation with the serum of an horse immunized with B. dysentericus Tor No. 1.

[1 This section has been added.]

CHAPTER XXI.

BACILLUS FEBRIS ENTERICÆ.

THE causal organism of enteric fever was originally discovered by Eberth in the spleen, lymphatic glands and Peyer's patches of persons suffering from the disease. Its morphological characteristics were more fully described by Gaffky.

The bacillus of enteric fever [1] is always present in the spleen, liver, mesenteric glands, glandular follicles of the intestine and bone marrow and less frequently in the lungs, meninges, testicles, tonsils, etc. A certain number of cases of enteric fever have been recorded in which there was no intestinal localization.

For a long time it was thought that the bacillus did not pass into the blood stream (Chantemesse and Widal, and others). It is, however, now recognized that the failure to find the bacillus in the blood was due to the defective technique then employed and that, as a matter of fact, in enteric fever the bacillus does pass into the blood stream and that the disease is in reality a true septicæmia. In all cases of moderate and severe infection the organism can be isolated from the blood from the fifth day until the end of the third week of the disease (Courmont).

The bacillus can often be isolated from blood taken from the rose spots (Thiemisch and Neuhaus, Besson, etc.).

Rémy's experiments have proved that, contrary to the opinion formerly held, the bacillus is present in the stools of enteric fever patients as early as the third day of the disease and before ulceration of the intestine has begun. The number of organisms present in the stools increases until the end of the first week and then gradually diminishes until at the end of the fourth week they can as a rule no longer be found. The bacilli have, however, been isolated from the stools of persons who have recovered from the disease for more than a month (Rémy, Chantemesse and Decobert) and it will be shown later that in some cases they persist for a still longer period.

[1] In the remainder of the chapter and elsewhere the organism is termed for the sake of convenience the typhoid bacillus.

The bacillus also sometimes passes into the urine of enteric fever patients. Besson from an examination of thirty-three cases came to the conclusion that the bacillus was only present when there was albumin in the urine and found it in 40 per cent. of such cases : the bacillus disappears synchronously with the disappearance of the albumin. Vincent found the bacillus in the urine in about 1 case in 5 of enteric fever ; he noticed that occasionally the bacillus remained in the urine after the patient had recovered, and considered that under those conditions the organism multiplied in the bladder. Horton Smith showed that the bacillus may set up slight cystitis with pyuria.

The bacillus is also the cause of many of the complications of enteric fever, such for instance as inflammation of the fauces, naso-pharynx and larynx, broncho-pneumonia and various suppurative affections : deep seated abscesses, osteitis, adenitis, pleurisy, pericarditis, etc. It may also become localized in lesions existing before the onset of the infection. Widal observed instances of this in a case of ovarian cyst and in a case of tuberculous adenitis.

Chantemesse was the first to put forward the opinion afterwards supported by Remlinger and Schneider that the typhoid bacillus might live a sapro-phytic existence in the intestines of healthy persons. The investigations of Rémy and others, however, seem to prove that the bacillus is only found as a saprophyte in the intestines of those who have recently been either in contact with cases of the disease or in some other way exposed to infection. But further, in a certain percentage (according to Schneider, 3 per cent.) of patients who have recovered from the disease, and especially in women, the bacillus may remain for several months and even years : it is said to take up its abode principally in the gall bladder from whence it is discharged into the intestine. It is easy to appreciate the prominent part which such "carriers," to use Drigalski and Conradi's expression, may take in the dis-semination of enteric fever.

The bacillus has been frequently found in drinking water and in ice destined for human consumption. Wherever enteric fever is epidemic the drinking water should be examined for the presence of the typhoid bacillus.

The organism has also been isolated from soil and from the dust of wards in which cases of enteric fever have been nursed, etc.

The attention of observers has been drawn to the part which flies may possibly play in the propagation of the disease. During an epidemic of enteric fever at Chicago, Mrs. Hamilton on several occasions obtained cultures of the typhoid bacillus by sowing flies which had been caught in water-closets, enteric wards, etc. Ficker has shown that flies which have been in contact with cultures of the typhoid bacillus may specifically contaminate objects on which they settle even as long as 23 days afterwards.

The typhoid bacillus and the colon bacillus are in many ways very like one another and both have their usual habitat in the intestines of man and the lower animals. The analogies which undoubtedly exist between these two organisms have led some observers to express the opinion that they are identical. This view however has not met with general acceptance and it is now clear that the colon bacillus and the bacillus of enteric fever have each their own characteristic properties and are in fact two distinct though closely related species

SECTION I.—EXPERIMENTAL INOCULATION.

The lower animals are not naturally susceptible to enteric fever. The inoculation of laboratory cultures is in most cases without result ; some observers indeed have noticed symptoms of intoxication in guinea-pigs, rabbits and mice, but they have not been able to produce a generalization of the bacillus. If, however, a virus of increased virulence be inoculated these animals die with the lesions of septicæmia. In monkeys and rabbits typical attacks of enteric fever have been induced by feeding with typhoid bacilli.

A. Inoculation of viruses of ordinary virulence.—Even cultures which have been sown with material direct from a case of enteric fever do not as a rule lead to a generalized infection in the lower animals, though occasionally guinea-pigs and mice can be infected by inoculating them in the peritoneal cavity. Sub-cutaneous inoculation generally results in the formation of a small abscess at the site of inoculation from which the animal rapidly recovers.

In rabbits, guinea-pigs and dogs, intra-cranial inoculation of a small amount (0·05–0·1 c.c.) of a fifteen- or twenty-day old culture gives rise, by reason of the toxin it contains, to severe symptoms which terminate fatally. The inoculation of young cultures produces nothing more than a transitory illness (Vincent).

B. Inoculation of viruses of exalted virulence.—Sanarelli, Chantemesse and Widal and others have succeeded in increasing the virulence of typhoid bacilli and with these exalted viruses they can always be certain of producing a typhoid septicæmia in laboratory animals.

Methods of increasing virulence.—(*a*) Sanarelli inoculated 5 c.c. of a twenty-four-hour old broth culture of a typhoid bacillus of ordinary virulence into the cellular tissue of a guinea-pig and at the same time into the peritoneal cavity 10 c.c. of an old sterilized broth culture of the colon bacillus : death supervened in about 20 hours and *post mortem* the typhoid bacillus was found in the peritoneal cavity and occasionally also in the spleen and blood.

A little of the peritoneal exudate from this animal was then sown on broth and it was found that 5 c.c. of the broth culture sub-cutaneously inoculated into a second guinea-pig would kill the animal if, at the same time, 7–8 c.c. of a sterilized culture of the colon bacillus were inoculated intra-peritoneally. By thus passing the bacillus through a series of animals diminishing at each inoculation the dose of colon bacillus culture, it happened that after a short time a strain of the typhoid bacillus was recovered which could, unaided by the simultaneous inoculation of the colon bacillus, lead to an enteric infection in rabbits and guinea-pigs when inoculated sub-cutaneously in doses of 5 c.c.

Similar results were obtained by Sanarelli if instead of the colon bacillus he inoculated sterilized cultures of *Proteus vulgaris*, sterilized cultures of stools, or an infusion of meat a month old sterilized at 120° C. By simply feeding guinea-pigs with small quantities of this infusion he was able to secure the generalization of a typhoid bacillus which before had no pathogenicity for the guinea-pig.

In the case of a virus which is fatal to guinea-pigs in large doses the virulence may be raised by passage intra-peritoneally through guinea-pigs. For this purpose 2 or 3 c.c. of a peritoneal exudate rich in bacilli are inoculated in the first instance, then, as the virulence increases, as evidenced by the fact that the animals die in a shorter space of time and by the diminished quantity of exudate found *post mortem*, the quantity injected is gradually reduced to 0·5 and 0·1 c.c. After fifteen to twenty such passages a single drop is sufficient to kill an adult guinea-pig in 12 hours. After the thirtieth passage the virulence is fixed and cannot be further increased. A few drops of a twenty-four-hour old broth culture of the " fixed virus " is sufficient to kill susceptible animals on intra-peritoneal inoculation. If inoculated sub-cutaneously much larger doses must, however, be employed : thus, for instance, in the case of rabbits and guinea-pigs 1–4 c.c. and for mice 0·5 c.c. are necessary.

Note.—In attempting to raise the virulence of an organism by passage through the peritoneal cavities of guinea-pigs it is important to utilize the peritoneal exudate itself for the successive inoculations and not cultures sown from the exudates. To maintain the virulence after exaltation the organism should be grown on a broth which before sterilization turns phenol-phthalein pink and to which a few drops of guinea-pig blood have been added just before sowing it (Rodet and Lagriffoul).

(b) Chantemesse and Widal also raised the virulence of bacilli moderately virulent [for experimental animals] by passage through guinea-pigs, utilizing to that end a discovery of Vincent relative to the exaltation of the typhoid bacillus when associated with sterile cultures of streptococcus pyogenes. They inoculated into the cellular tissues of a guinea-pig 4 c.c. of a culture of a typhoid bacillus and at the same time into the peritoneal cavity 8–10 c.c. of a culture of a pyogenic streptococcus which had been sterilized at 100° C. for 1 hour. The animal died in less than 24 hours and the typhoid bacillus was found to have become generalized. The organism was passed through a series of guinea-pigs and the dose of sterilized streptococcus emulsion gradually diminished, with the result that the typhoid bacillus soon became so virulent that a few drops introduced into the peritoneal cavity caused the death of the animal.

(c) According to Chantemesse and Balthazard the most efficient method of raising the virulence of a typhoid bacillus to a maximum is to sow a culture in a collodion sac, and after leaving it in the peritoneal cavity of a guinea-pig for 24–36 hours to sow the contents in broth : the growth is very abundant so that in 12 hours the surface is covered with a thick pellicle. This culture is fully virulent.

Infection with viruses of exalted virulence.—Guinea-pigs are the best animals for the study of typhoid infections. A few drops of an exalted virus inoculated into the peritoneal cavity gives rise to a typical attack of the disease.

Two to four hours after inoculation the temperature rises and may reach 41° C. but it soon (6–12 hours) begins to fall to 36° C. and perhaps 32° C. ; synchronously with the fall of temperature collapse sets in and the animal dies 15–30 hours after the inoculation.

During the febrile period the animal is dull and refuses its food. When the temperature has become subnormal it huddles itself up in a corner of its cage, the abdomen is painful and the animal rapidly wastes.

Post mortem examination. The peritoneal cavity is found to contain a variable amount of an opalescent serous fluid very rich in bacilli (the greater the virulence of the organism the less the effusion) : the spleen, liver, kidneys, intestines and notably the Peyer's patches are swollen and congested : the mesenteric glands are swollen and in some cases there is a little pleural effusion : the intestine contains a serous fluid rich in bacilli. According to Chantemesse and Widal these latter are typhoid bacilli but according to Sanarelli they are very virulent colon bacilli.

The organism is found in pure culture in the peritoneal exudate and also in the internal organs, blood etc.

C. Infection by the alimentary canal. 1. Monkeys.—Chantemesse and Ramond fed a *Macacus rhesus* for a fortnight on an exclusively milk diet and then gave it a virulent agar culture of the typhoid bacillus mixed with jam. As early as the third day the animal experienced a rise of temperature, anorexia and diarrhœa, and was dead at the end of a week. *Post mortem* examination revealed lesions characteristic of human enteric fever especially in the neighbourhood of Peyer's patches.

2. Rabbits.—Remlinger succeeded in infecting rabbits by starving them for 2 or 3 days and then feeding them for 5–10 days on vegetables contaminated with cultures of the typhoid bacillus. Many of the animals remained

2 A

unaffected but a few of them towards the end of the first week had a rise of temperature, became emaciated, suffered from diarrhœa and eventually died. *Post mortem* examination showed ulceration of Peyer's patches, enlargement of the spleen, etc. The typhoid bacillus was recovered in pure culture from the spleen.

Chantemesse and Ramond lowered the resistance of rabbits by injecting into the peritoneal cavity some sterile broth containing 50 drops of laudanum and then a quarter-of-an-hour later introduced into the stomach by means of a tube 5 c.c. of a young broth culture of the typhoid bacillus. Animals so treated became infected with a true enteric fever; they developed the characteristic lesions, and their serum agglutinated the bacillus.

By daily inoculation with human blood serum or urine for a period of 3 weeks animals can be rendered more susceptible to infection with the typhoid bacillus.

SECTION II.—MORPHOLOGY.

1. Microscopical appearance.

The typhoid bacillus occurs in the tissues as a short rod measuring about 2–3μ long and 0.6–0.7μ broad.

In cultures its length and breadth vary within wide limits. In broth, for example, the bacillus is shorter and more slender; in old gelatin cultures, it is elongated and shows filamentous forms; on agar and potato it is broader and shorter and has a squat appearance.

The bacilli both in tissues and cultures occur singly or joined together in pairs, and in young cultures they not infrequently look like diplococci.

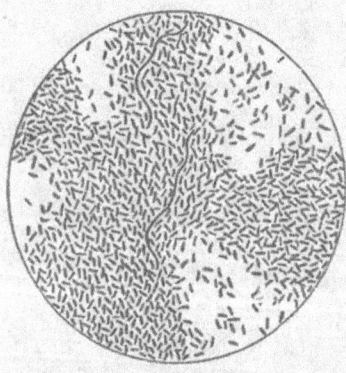

The ends of the bacilli are rounded. The protoplasm stains uniformly, but occasionally, in old cultures, the bacilli are somewhat swollen about their centres and show a clear space of variable size—"the shuttle form" of Artaud. This unstained portion does not represent spore formation any more than do the terminal swellings which are sometimes seen in cultures of the bacillus and which are merely degeneration forms.

As a rule, the typhoid bacillus is very motile and moves rapidly across the field of the microscope like fish in water, but some strains of the bacillus are only slightly motile. The motility is due to the presence of flagella (*vide infra*).

FIG. 216.—Film preparation of the typhoid bacillus from a gelatin culture. Carbol-fuchsin. (Oc. 2, obj. ¹⁄₁₂th, Zeiss.)

If a trace of growth from a solid medium be placed in a drop of water the bacilli separate one from another and the water is immediately rendered turbid (Chantemesse).

Staining reactions.—The typhoid bacillus stains readily with the basic aniline dyes and is gram-negative.

Staining of flagella.—The flagella may be easily stained (p. 149). Van Ermengem's or Nicolle's method is recommended as giving the best results.

In stained films the number and arrangement of the flagella can be readily made out. As a rule, each bacillus has eight to a dozen flagella, but it is not at all uncommon for individual bacilli to have as many as eighteen to twenty-four. Flagella which have been inadvertently torn away from their bacilli during the necessary manipulations will be found in every preparation.

The flagella are normally implanted regularly around the body of the organism [peritrichous] though now and again they are found arranged in tufts probably from the dragging of the surrounding liquid on these highly delicate structures. The bacilli are often agglutinated into clumps by a matrix which stains in the same manner as the flagella and it is upon this matrix that the flagella appear to be implanted.

The flagella vary in length, the average being 6–8μ (Rémy and Sugg) ; but much longer forms are to be seen. They are wavy in form and present three to eight undulations.

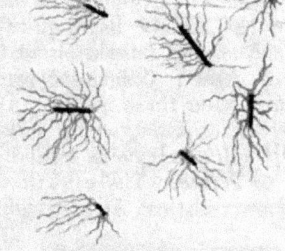

2. Cultural characteristics.

A. Conditions of growth.—The typhoid bacillus is a facultative aërobe. It grows on all the

FIG. 217.—Typhoid bacillus stained to demonstrate flagella. × 1000.

ordinary media within a wide range of temperature (4°–46° C.) the optimum being 30°–37° C. Cultures of the typhoid bacillus have no smell.

B. Characters of growth on various media. 1. Broth.—After 8–12 hours' incubation at 37° C. the medium shows a slight cloudiness, which as the growth progresses becomes more marked, and gives to the culture when examined by transmitted light a characteristic watered-silk appearance : this may be made more distinct by gently shaking the tube : later the growth becomes flocculent, falls to the bottom of the tube, and forms a very abundant sediment. Ultimately the liquid becomes clear and develops a brownish colour.

2. Gelatin.—The typhoid bacillus does not liquefy gelatin.

Stab culture.—At 20° C. growth along the line of the stab commences as early as the second day in the form of small, round, yellowish-white confluent colonies, while on the surface a thin, transparent, rather spread-out disc with iridescent margins appears ; occasionally the surface growth is represented by a thick opaque spot of very limited extent. Growth is always scanty.

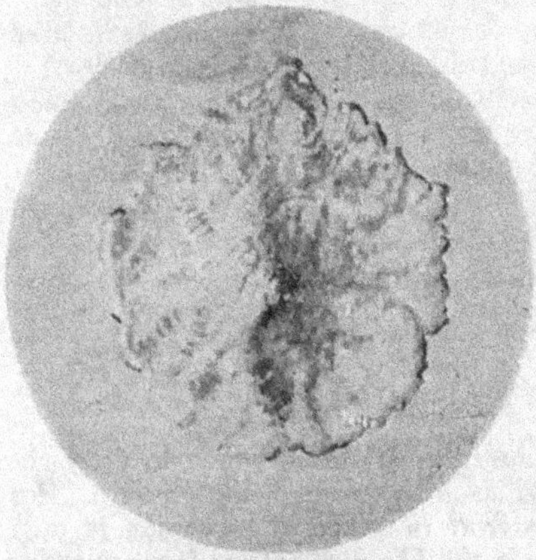

Stroke culture.—On the surface along the line of sowing the growth forms a thin transparent film with irregular margins and shot with iridescent colours ; it always remains scanty and ceases to increase after the first week. Such is the usual appearance, but sometimes a narrow, thick, opaque, yellowish-white band develops along the stroke.

In the substance of the gelatin long arborescent crystals are sometimes seen. These are due to the precipitation of phosphates.

Single colonies.—Isolated colonies on gelatin usually but not invariably

FIG. 218.—Typhoid bacillus. Photograph of a colony growing in plate culture (6 days). × 60.

present a characteristic appearance. After incubating at 20° C. for 48 hours, small circular colonies appear and soon reach the size of a pin's head and later that of a lentil, but always remain thin, bluish in colour, pearly and transparent : the edges of each colony become indented and sinuous, and at the same time ridges extend from them into the centre, which becomes thicker than the margins. These details may be made out clearly with a lens. The general appearance has been compared by German writers to an iceberg.

Colonies developing in the depth of the gelatin and sometimes even those on the surface have quite a different appearance. They are round and opaque, and remain about the size of a pin's head.

3. Agar : Coagulated serum.—There is nothing characteristic about the growth on these media. After incubating for 24 hours at 37° C. a whitish streak appears, which subsequently becomes thicker and cream-coloured. Glycerin-agar yields a more copious growth.

4. Potato.—The growth of the typhoid bacillus on potato is as a rule characteristic. At first sight there appears to be no growth at all : but on

illuminating the surface of the potato by day-light a delicate, moist, shiny deposit like the icing on cakes is seen along the line of sowing. Sometimes the culture assumes a bistre tint later.

In some cases however the growth on potato is plainly visible, being yellowish in colour and occasionally even definitely brownish. Buchner states that this appearance can be obtained at will by making the potato alkaline with a solution of carbonate of soda.

FIG. 219.—Typhoid bacillus. Culture on potato.

5. Rémy and Sugg's medium.—To avoid complications induced by variations in the chemical composition of potato, an artificial medium has been prepared by Rémy and Sugg which contains the constituent ingredients of potato. According to the authors the typhoid bacillus on this medium invariably gives a characteristic growth ; " a limited, absolutely colourless, scalloped film."

The medium is prepared as follows.
(*a*) Make a solution containing :—

Water,	1000 c.c.
Glucose,	20 grams.
Peptone,	5 ,,
Asparagin,	5 ,,
Citric acid,	0·75 gram.
Neutral potassium phosphate,	5 grams.
Magnesium sulphate,	2·5 ,,
Potassium sulphate,	2·5 ,,
Sodium chloride,	1·25 ,,
Carbonate of sodium *q.s.* to render the whole slightly alkaline.	

(*b*) To 100 c.c. of this solution add :—

Gelatin (extra quality),	10 grams.
Calcined magnesia,	2 ,,

Distribute in tubes, sterilize, slope. Sow in stroke culture.

6. Bile.—Sterilized ox-bile is a very useful medium on which to grow the typhoid bacillus (Conradi). It is used as an " enrichment medium " for obtaining [primary] cultures from material in which the bacillus is only present in small numbers, as for example the blood of enteric fever patients (p. 391. 2 (*b*)).

7. Milk.—The bacillus grows abundantly in milk without coagulating the medium.

SECTION III.—BIOLOGICAL PROPERTIES.

The difficulty of distinguishing the typhoid from the colon bacillus has rendered necessary a close study of the biological properties of the two organisms : the morphological characteristics alone are insufficient to allow of their differentiation.

1. Biochemical reactions.[1]

Action on carbohydrates.—The typhoid bacillus has a distinct action [acid without apparent gas] upon glucose, [maltose, sorbite and mannite] and also acts feebly upon lævulose and galactose, but ferments neither saccharose, lactose, [dulcite, raffinose, arabinose, erythrite, salicin, amygdalin nor inulin].

These properties furnish valuable data for the recognition of the organism, and the methods of demonstrating them will now be considered.

(*a*) Sow the bacillus in a tube of lactose-broth to which a little carbonate of lime has been added (p. 35). No gas is formed however long the culture be incubated.

(*b*) Sow on litmus-lactose-gelatin (p. 57) : the typhoid bacillus does not attack either mannite or lactose so that no acid is formed and the medium retains its blue colour (cf. *Bacillus coli*).

(*c*) Sow in Grimbert and Legros' medium. This medium has the following composition :—

Lactose (chemically pure), · · · · · · ·	20 grams.
Peptone, · · · · · · · · · ·	5 ,,
Distilled water, · · · · · · ·	1000 c.c.

Dissolve by boiling : add a little pure carbonate of lime : shake : leave for 5 minutes : filter : test the reaction, which should be neutral. Sterilize by filtering through a Chamberland bougie. Distribute into tubes and add sufficient sterilized litmus solution (p. 56).

After sowing with the typhoid bacillus and incubating, the medium retains its blue colour.

(*d*) Sow in milk.—The milk is not coagulated and if a little litmus solution be added its colour remains unchanged.

[A definite acidity is produced in the first 24 hours but this is subsequently neutralized and the medium ultimately becomes distinctly alkaline, though the time occupied in the production of an alkaline reaction varies considerably with different strains—in some cases a month may elapse before the medium is definitely alkaline.]

For these tests the milk should always be sterilized at the same known temperature : some contaminating organisms which easily coagulate milk which has been sterilized at 100° C. coagulate it more slowly and with more difficulty if it has been exposed to higher temperatures, and mistaken diagnosis may result if this fact has not been recognized (see also p. 57).

These reactions are sufficient to enable the typhoid bacillus to be distinguished from the colon bacillus (p. 393). When it is necessary to make a differential diagnosis between the typhoid and colon bacilli glucose should never be used as the fermentable agent since the typhoid bacillus has a distinct action on it.

Non-production of indol.—The typhoid bacillus never produces indol in cultures.

---

[1] Here the nature of the reactions will be briefly stated; their application to the differentiation of the typhoid and colon bacilli will form the subject of a special chapter (xxiii.).

Tests for indol.—To determine whether an organism produces indol or not, a solution of peptone must be used and not ordinary broth. The following is a medium often used for this test :—

Water, - - - - - - - - - -	100 c.c.
Witte's, Chapoteaut's or Byla's peptone, - - - -	2 grams.
Sodium chloride, - - - - - - - -	0·5 to 1 gram.

Tube in quantities of about 15 c.c. and autoclave.

After sowing, incubate for 2–8 days and apply one or other of the following tests :—

(*a*) **Salkowski's reaction.**—To the culture in peptone water add 1 c.c. of a 0·2 per cent. solution of potassium nitrite, then, slowly, 1 c.c. of a 25 per cent. solution of chemically pure sulphuric acid in water. If indol be present a rose tint appears.

Nonotte and Demanche find that the reaction is more delicate in the warm. To a peptone-water culture add 1 c.c. of a 1 in 1000 solution of nitrite of potassium and 8 drops of pure concentrated sulphuric acid and boil the upper part of the liquid. If indol be present a very distinct pink colour appears even when the amount of indol does not exceed 1 part in 4 millions : in the cold, the reaction only takes place if the amount of indol exceeds 1 part in 75,000.

(*b*) **Weyl-Legal's reaction.**—To the culture add 5 to 10 drops of a 5 per cent. solution of sodium nitro-prusside then a few drops of a 30 per cent. solution of washing soda. The solution turns brown. After a few minutes add 10 to 15 drops of glacial acetic acid ; if indol be present a characteristic blue colour appears but often only after some delay.

(*c*) **Nencki's reaction.**—To the culture add first a few drops of glacial acetic acid then 2–3 c.c. of alcohol-ether : shake and allow to stand until the ether rises : decant the layer of ether and evaporate it in a porcelain dish. To the residue add 1 to 2 drops of a 0·2 per cent. solution of potassium nitrite and a few drops of pure sulphuric acid. This method is very delicate and the least trace of indol is shown by the appearance of a rose pink colour.

(*d*) **Fleig's reaction.**—To 10 c.c. of culture add 10 c.c. of a 1 in 50 alcoholic solution of furfurol, then pure hydrochloric acid drop by drop. If indol be present the solution turns yellow. This method is very delicate.

[(*e*) **The para-dimethyl-amido-benzaldehyde test. Recommended.**—Prepare two solutions :

Solution I.—

Para-dimethyl-amido-benzaldehyde, - - - - -	4 parts.
Absolute alcohol, - - - - - - - -	380 ,,
Concentrated hydrochloric acid, - - - - -	80 ,,

Solution II.—

Saturated aqueous solution of potassium persulphate.

[To about 10 c.c. of the broth or peptone-water culture of the organism add 5 c.c. of Solution I. and then 5 c.c. of Solution II., shake the mixture and the presence of indol is indicated by the appearance, in a very short time, of a red colour, which gradually becomes darker on standing. The reaction may be accelerated by heating the mixture.]

Some peptones contain a trace of indol and to avoid all possibility of mistake Sicre recommends using a 1 per cent. solution of Byla's peptone and when testing for indol to test at the same time a tube of sterilized peptone-water as a control.

Growth on Synthetic media.—A number of synthetic media have been prepared on which the typhoid bacillus grows slowly and feebly while closely related organisms with which it may be confused grow freely.

Too much importance should not be attached to the differentiating function of these media, but, generally speaking, if growth be absent or delayed on any one of them, this is an indication sufficiently reliable to justify a suspicion of the presence of the typhoid bacillus.

For choice, the following medium, composed by Rémy and Sugg, may be used :—

Distilled water,	1000 c.c.
Glucose,	20 grams.
Nitrate of soda,	10 ,,
Magnesium sulphate,	2 ,,
Neutral phosphate of potassium,	1 gram.
Calcium chloride,	1 ,,

Inability to grow on " vaccinated " media.—Chantemesse and Widal have demonstrated the following curious property of the typhoid bacillus. If a tube of agar or gelatin be sown with the bacillus and after incubation the growth be scraped off and the medium resown with the organism the second sowing remains unfertile on incubation, the medium having been, as it were, " vaccinated " by the first growth. Unfortunately this phenomenon sometimes fails and taken alone is not a reliable test. The colon bacillus also frequently fails to grow on a medium which has been used for the growth of the typhoid bacillus.

Growth on coloured media.—D'Abundo; Nœggerath; and also Gasser have drawn attention to the property possessed by the typhoid bacillus of decolourizing during growth media stained with certain dyes.

Nœggerath's medium (p. 57) was recommended by its discoverer and later by Deschamps and Grancher as a diagnostic agent for the typhoid bacillus. When sown on the surface of gelatin plates coloured with Nœggerath's fluid, the typhoid bacillus gives rise to colonies of a purple colour while the surrounding medium becomes decolourized.

Gasser, recognizing that Nœggerath's medium gives inconstant results, substituted fuchsin-agar (p. 57). The typhoid bacillus sown on this medium and incubated at 37°–39° C. for 2 days gives red colonies, the surrounding medium being decolourized.

These reactions are unfortunately not constant and cannot be relied upon for the purpose of recognizing the typhoid bacillus.

Growth on arsenical broth.—Thionot and Brouardel found that the typhoid bacillus does not grow in broth containing arsenious acid to the extent of 0·02 gram per litre, while the colon bacillus grows not only in this medium but also when the broth contains as much as 1–2 grams of arsenious acid per litre.

Growth on artichoke.—According to Roget, the typhoid bacillus produces no apparent growth on artichoke and does not change the colour of the medium, while the colon bacillus gives a thick yellowish growth and the artichoke at the same times assumes an intense green colour.

Technique.—Remove the leaves of the artichoke but leave the choke adhering to the heart: cut into small cubes with a silver-bladed knife, place the cubes with the choke uppermost into potato tubes containing a few drops of water in the lower bulb, plug with wool and sterilize at 115° C. Sow at the junction of the choke and heart.

Growth on caffeine media.—Roth has shown that the addition of 0·5 per cent. of caffeine to media prevents the growth of the colon bacillus but has no action on the growth of the typhoid bacillus. This characteristic is not absolutely constant since some strains of the typhoid bacillus will not grow in the presence of caffeine (Courmont).

Growth on malachite-green.—According to Lœffler the addition of a small amount (about 1 in 4000) of malachite-green to culture media favours the growth of the typhoid and paratyphoid bacilli while impeding the growth of

the colon bacillus. Kiralyfi has shown that this is not a constant phenomenon : according to this observer though malachite-green inhibits the
growth of many micro-organisms—*e.g.* streptococci, staphylococci, vibrio
choleræ,—it generally has no action on the colon bacillus.

Krystal-violet has the same action as malachite-green (Drigalski and Conradi) : both typhoid and colon bacilli grow on media containing this dye while
the growth of many other organisms is inhibited.

2. Variability of flagella.

Sunlight, antiseptics in dilute solution and temperatures unsuitable to
growth have practically no influence on the number and shape of the flagella
(Rémy and Sugg). When the typhoid bacillus has been grown in culture
with the colon bacillus for some weeks the flagella are sometimes difficult to
stain (Rémy). The rarity of variation in the morphology of the flagella
is of importance in diagnosing the typhoid bacillus.

3. Viability and virulence.

Viability.—Exposure to a temperature of 60° C. kills the typhoid bacillus
in 10–20 minutes but very low temperatures have no effect on its vitality ;
thus, Prudden found the organism still alive in a block of ice which had been
kept for 3 months between − 1° and − 11° C. On the other hand, alternate
freezing and thawing rapidly kills the bacillus.

In water, the typhoid bacillus retains its vitality for some time (Strauss
and Dubarry, Chantemesse and Widal). In sterile water it has been found
to be alive after 3 months. If the water contains saprophytic micro-organisms
the typhoid bacillus disappears more quickly, but it can still be isolated after
more than 1 month.

In soil, the bacillus can survive five months and a half (Grancher and
Deschamps) : drying kills it only after 1 or 2 months (Uffelmann). Levy
and Kaiser isolated the organism from stools which after being in a cesspool
for 5 months had been spread on the surface of the ground for 15 days in
winter.

Light rapidly kills the typhoid bacillus. Cultures exposed to sunlight in
the month of May were sterilized in 4–8 hours. Vincent has found that the
blue, violet and ultra-violet rays are more efficient as bactericidal agents
than the red and ultra-red rays. Cultures spread and dried on pieces of
cloth, and then exposed to direct sunlight, were found to be sterilized in
9–26 hours (Vincent).

The typhoid bacillus is very sensitive to the action of antiseptics : the
solutions of perchloride of mercury, carbolic acid, etc. in general use will
kill the bacillus in a few minutes.

Virulence.—The great variations observed in the virulence of the typhoid
bacillus and the methods by which the virulence can be raised have already
been studied under the head of experimental inoculation.

4. The toxin of the typhoid bacillus.

The experiments conducted by Brieger and Frænkel with a view to isolating
a toxin from the typhoid bacillus gave very little result. Attempts are now
no longer made to extract a definite chemical substance from cultures. The
crude toxin found in sterilized cultures has been studied by Sanarelli, Chantemesse and others. Other observers (Macfadyen and Rowland, Besredka and
others) have prepared extracts containing an endotoxin from the bodies of
the organisms.

1. **Toxin of Sanarelli.** (α) **Method of preparation.**—Sanarelli uses a virus the virulence of which has been raised by passing it through the peritoneal cavities of guinea-pigs (p. 368). The bacillus is sown in 2 per cent. glycerin-broth and after incubating at 37° C. for a month is sterilized by heat and allowed to remain at room temperature for 8 months. The flask containing the culture is then sealed in the flame and heated to 60° C. for a few days. During this long period of maceration the intra-cellular toxin diffuses into the culture fluid and this, carefully decanted, constitutes the toxin of Sanarelli.

Gauthier and Balthazard justifiably point out that Sanarelli's toxin is a complex mixture containing substances foreign to the typhoid bacillus, and derived partly from the albuminoid substances present in the medium, partly from the dead bodies of the bacilli, etc. which have slowly undergone disintegration. It is nevertheless true that animals inoculated with the different toxins prepared by Sanarelli, Chantemesse, and Balthazard exhibit identical symptoms.

(β) **Action on laboratory animals. On rabbits.**—The toxin given subcutaneously in doses of 10 c.c. per kg. of body weight kills rabbits weighing 700–1000 grams.

Soon after inoculation the animal is seen to breathe more rapidly and to become unsteady on its legs; a general paralysis gradually comes on and about 10 hours after inoculation convulsions occur ending in death. The temperature is at first a little raised (about $\frac{1}{10}$° C.) but soon falls below normal and death takes place while the temperature is still sub-normal. The effects of the toxin vary in different animals : death is not infrequently delayed for some days and in that case is preceded by a period of cachexia of which the characteristic signs are wasting, diarrhœa, etc. *Post mortem*, the abdominal organs are found to be anæmic, and it is noticeable that there is neither congestion of the intestinal mucous membrane nor swelling of the Peyer's patches.

On mice.—1 c.c. of toxin inoculated sub-cutaneously or 0·2 c.c. intraperitoneally is generally fatal, death taking place in a few hours. *Post mortem* the spleen is enlarged and there is a small amount of a sterile effusion in the peritoneal cavity.

On guinea-pigs.—Guinea-pig inoculation is an excellent means of testing typhoid toxin—sub-cutaneously the minimal fatal dose is 1·5 c.c. per 100 grams of body weight. Intra-peritoneally the results are less constant. Sub-cutaneous inoculation of 4 or 5 c.c. of toxin per 100 grams of body weight leads to death in 15–20 hours.

From the moment of inoculation the temperature falls and continues to do so until death. About an hour after inoculation there is marked abdominal distension accompanied by extreme tenderness, the animal does not move, but sits huddled up and cries if touched : after 4 or 5 hours it is extremely dejected, its eyes are half-closed and it is seen to be in an almost uninterrupted state of tremor ; a profuse sometimes hæmorrhagic diarrhœa is often present, and finally paralysis appears, the meteorism vanishes and death takes place. *Post mortem* a variable quantity of exudate rich in leucocytes and sometimes turbid is found in the peritoneal cavity ; the spleen is enlarged, congested and friable : the walls of the small intestine are distended and completely infiltrated with blood, the mucous membrane is red and the lymphatic patches are infiltrated and congested ; the stomach and suprarenal capsules are intensely congested and ecchymosed. The intestine is full of liquid matter and contains a very virulent colon bacillus in pure culture.

On monkeys.—Monkeys are very susceptible to the toxin of the typhoid bacillus : the course of the disease and the lesions are the same as in the guinea-pig.

2. **Toxin of Chantemesse.** (α) **Method of preparation.**—Chantemesse at first recommended growing an organism of increased virulence in a maceration of spleen and bone marrow. He now, however, prefers to use a bacillus

whose virulence has been increased by growing it in collodion sacs in the peritoneal cavities of guinea-pigs, and to sow it in a solution of spleen peptone.[1]

Incubate at 37° C. : a week later the toxicity will be at a maximum. Then either filter through porcelain, or preferably, heat to 55° C., centrifuge, and decant the supernatant liquid which contains the toxin.

(β) **Properties.**—This toxin is more powerful than Sanarelli's and kills a guinea-pig weighing 500 grams in 12–24 hours when inoculated in quantities of 6 c.c. intra-peritoneally (that is about 1 c.c. per 80 grams).

It is a very unstable product, being quickly affected by air and light, and its toxic properties are diminished if it be heated to 100° C. for a few minutes. It must be stored in accurately filled tubes and kept in the dark.

3. **Toxin of Bandi.**—The bacillus after the virulence has been raised by passing it through the peritoneal cavities of a long series of guinea-pigs is sown in Lœffler's broth, incubated for 48 hours and filtered. The filtrate inoculated sub-cutaneously in quantities of 4 c.c. is sufficient to kill a guinea-pig weighing 400–500 grams.

4. **Toxin of Lépine and Lyonnet.**—A virulent culture in broth, 4–8 days old, is sterilized at 55°–60° C. for an hour. The product is toxic for dogs and horses.

5. **Toxin of Rodet, Lagriffoul and Wahly.**—Cultures of the typhoid bacillus are incubated on well aërated media for 3 days and filtered. The filtrate kills guinea-pigs when inoculated intra-peritoneally in doses of 4 c.c. per 100 grams of body weight, and rabbits when inoculated intra-venously in doses of 1 c.c. per 100 grams.

M. and Mme. Werner after growing the organism for 3 days in an aërated medium sealed the flasks and left them for 2 days at 25° C. The filtered liquid killed guinea-pigs when inoculated intra-peritoneally in quantities of ½ c.c. per 100 grams of body weight, and rabbits when inoculated intra-venously in quantities of 0·1 c.c. per 100 grams.

6. **Toxin of Moreschi.**—Moreschi grew the bacillus for 5 days in a special broth and then filtered the culture through porcelain. The filtrate when injected intra-peritoneally in doses of 0·2 c.c. killed a guinea-pig weighing 250 grams.

The special broth is prepared as follows :—

Mince 1000 grams of horse meat and 1000 grams of ox's spleen, macerate for 24 hours at room temperature in a litre of water, boil, filter, make up to 1 litre and add

Witte's peptone,	- - - - - - -	20 grams.
Plasmon,	- - - - - - -	10 ,,
Sodium chloride,	- - - - - - -	5 ,,
Ox blood,	- - - - - - -	80 ,,

Heat the mixture to 120° C. in the autoclave for 20 minutes, neutralize, and add 0·15 per cent. of caustic soda. Heat again to 120° C. Filter, tube and sterilize.

After being sub-cultured several times on this medium the bacillus grows as a very thick film on the surface while the broth remains clear. When the growth assumes these characteristics the culture has reached its maximum of toxicity.

7. **Toxin of Conradi.**—The bacillus is grown on agar for 20 hours, scraped off, mixed with a little normal saline solution and kept in the incubator at 37° C. for 24 hours. The emulsion is then diluted with more normal saline solution and filtered through a Berkefeld bougie ; the filtrate is evaporated

---

[1] The medium used by Chantemesse is prepared by macerating spleen and bone marrow in cold distilled water, filtering through porcelain and adding a little defibrinated human blood.

The solution of spleen peptone is obtained by digesting a pig's spleen and stomach in acidulated water (*vide* Martin's peptone) making slightly alkaline and sterilizing. Cultures are grown in a shallow layer of the medium contained in large wide-bottomed flasks.

in vacuo to $\frac{1}{10}$th or $\frac{1}{30}$th its original volume. The product when injected intra-peritoneally in doses of 0·2 c.c. kills a guinea-pig weighing 300 grams.

8. Toxin of Macfadyen and Rowland.—The growth on agar is scraped off and cooled to −90° C. by means of liquid air, then triturated at a very low temperature in a special apparatus. The product is diluted in normal saline solution and centrifuged. The supernatant liquid is very toxic and is fatal to guinea-pigs when inoculated intra-peritoneally in doses of 0·1 c.c.

Bassenge and Mayer obtained a similar but less toxic product by freezing the bacilli with liquid air and grinding them up in a hand mortar.

9. Toxin of Balthazard. (α) **Mode of preparation.**—The principle is the same as that underlying Macfadyen and Rowland's method.

A bacillus whose virulence has been increased by growing it in collodion sacs in the peritoneal cavities of guinea-pigs, is sown on large surfaces of agar contained in flat flasks (the agar is prepared with a 3 per cent. solution cf Defresne's peptone and contains no meat). After incubating for 24–48 hours the growth is scraped off, mixed with a little normal saline solution and rapidly centrifuged. The deposit is again shaken up with normal saline solution and centrifuged a second time. In this manner all foreign matter is removed.

The bacilli are then mixed with a 2 per cent. solution of urea or 1 per cent. solution of ammonium chloride (the effect of these solutions is to swell and break the cells and so facilitate the diffusion of the intra-cellular products). The emulsion thus obtained is distributed into tubes which are completely filled and sealed in the flame.

To facilitate the diffusion of the intra-cellular products, the tubes are now alternately frozen and thawed. They are kept at 58° C. for 8 days and daily submitted to a temperature of −21° C. for a couple of hours in a refrigerating machine (the evaporation of methyl chloride being adopted as the cooling agent). At the end of 8 days the tubes are centrifuged for 24 hours. The bacilli collect at the lower end of the centrifuge tube and the supernatant liquid carefully decanted constitutes the toxin.

(β) **Properties.**—Balthazard's method though lengthy and expensive yields a very powerful toxin, containing products of the typhoid bacillus unmixed with foreign substances. Inoculated sub-cutaneously in doses of 3 c.c. it kills rabbits weighing 2 kg. and in doses of 2 c.c. guinea-pigs weighing 200 grams. It is however less toxic than Conradi's toxin, which has the further advantage of being more easily prepared.

The action of Balthazard's toxin on animals generally is similar to that of the toxins prepared by Sanarelli and by Chantemesse ; but its action on rabbits appears to be more constant.

10. Endotoxin of Besredka.—Dried typhoid bacilli killed by heating for 1 hour at 60° C. are ground up with sodium chloride until an impalpable powder is obtained. This powder is diluted with water added drop by drop and the mixture left over-night. Next morning it is warmed in a water bath to 60°–62° C. for 2 hours, and then allowed to settle. The supernatant liquid contains the endotoxin. The average lethal dose for white mice is 0·05 c.c. intra-peritoneally. The endotoxin is destroyed only by temperatures above 127° C.

11. Typho-lysin.—As early as the second day filtered cultures of the typhoid bacillus show distinct powers of hæmolysis. This property increases with the age of the culture up to the fifteenth day when it is at its maximum (E. and P. Levy). Macfadyen and Rowland demonstrated the presence of typho-lysin in an eight-day culture grown on macerated spleen. This hæmolysin of the typhoid bacillus is not destroyed at 55° C.

The red cells of the dog are very sensitive to typho-lysin, and dogs which have been repeatedly treated with heated cultures yield an antitypholytic serum.

5. Vaccination.

A. Immunization of the lower animals.

(i) Beumer and Peipper immunized white mice by inoculating them daily for several days with increasing doses of living cultures. Guinea-pigs, rabbits, and especially goats and dogs may be vaccinated in a similar manner (Pfeiffer, Lœffler and Abel). Vincent immunized dogs and rabbits by inoculating them first with cultures heated to 60° C., then with living cultures 16 hours old, and finally with more toxic cultures 15–20 days old. The serum of animals so treated has both immunizing and agglutinating properties : immunized animals are however not immune to an intra-cerebral inoculation of typhoid toxin.

(ii) Brieger, Wassermann and Kitasato used for their immunizing experiments organisms attenuated by being grown in thymus broth (p. 34). Inoculations of a culture of a virulent bacillus grown in thymus broth and heated to 60° C. produced immunity in guinea-pigs and mice.

(iii) Sanarelli, Chantemesse and Widal, Beumer and Peipper immunized animals by inoculating them with cultures sterilized by heat.

(a) Sanarelli incubated a culture of a bacillus of increased virulence in peptone broth for a week at 37° C. and sterilized the growth at 120° C. The sterilized product possessed vaccinating properties.

Generally, it may be said that to immunize guinea-pigs weighing 400 grams it is only necessary to inoculate them several times over a period of 5 days with 16–18 c.c. of sterilized cultures. The animals are immune 4 days after the last inoculation and will then resist the inoculation of a virus of exalted virulence. During the process of immunization the animals lose a certain amount of weight but quickly recover.

It is very difficult to immunize rabbits for they are far more susceptible than guinea-pigs and death often takes place during the immunizing process ; but animals which survive the treatment are immune in a high degree.

(b) Chantemesse and Widal used broth cultures incubated at 37° C. for 15 days and then sterilized at 100° C.

Twenty c.c. are necessary to immunize a guinea-pig : the toxin should be inoculated in four doses allowing a few days to elapse between each inoculation. Immunization takes a fortnight ; after the lapse of another 8 days the test inoculation may be performed (2 c.c. of a virulent culture into the peritoneum). Not infrequently the animals die either during the immunizing process or as the result of the test inoculation.

Rabbits may be immunized in a similar manner but in these animals the process is even more difficult.

(c) Beumer and Pfeiffer immunized sheep in a like manner with cultures heated to 60° C. for an hour.

For immunizing horses Funck prefers to use cultures sterilized with carbolic acid.

(iv) Chantemesse immunized horses by injecting them with gradually increasing doses of his toxin (vide supra).

The immunization of horses is very difficult ; the inoculations whether made sub-cutaneously or intra-venously have frequently to be interrupted on account of the violence of the reaction, and it takes several years to produce a lasting immunity.

B. Human vaccination.

For many years the problem of the vaccination of the human subject against enteric fever has been under investigation. In 1896, Pfeiffer and Kolle showed that as a result of inoculating man with a small quantity of a sterilized culture [1] the serum acquired bactericidal and agglutinating properties for the typhoid bacillus.

Since then numerous methods of antityphoid inoculation have been devised, some based on the inoculation of " whole " cultures (Pfeiffer and Kolle, Wright, etc.) others on the use of extracts made from the bodies of the bacilli. Finally, Besredka has conceived a method of vaccinating with bacilli sensitized with antityphoid serum.

1. Methods based on the use of "whole" cultures.

(a) **The Wright-Leishman method.**—Wright's original method has been modified in view of the experiments of Leishman and Harrison. Wright now uses a bacillus of low virulence which he grows at 37° C. for 24–48 hours in a shallow layer of peptone broth to facilitate aëration, sterilizes by heating to 53° C. for an hour (not at 60° C. as in his original method), and then adds 0·25 per cent. of lysol to ensure its sterility.

Two inoculations into the outer surface of the arm or over the pectoral muscle are given : the first of 500 million bacilli (0·5 c.c. of vaccine), the second 10 days later of a 1000 million bacilli (1 c.c. of vaccine).

The doses prescribed by Wright and Leishman should be scrupulously observed : too small a dose will fail to produce immunity and too large a dose will be followed by a sharp reaction and may fail to vaccinate (Wright, Paladino-Blandini). It is therefore necessary to enumerate the bacillary content of the vaccine in order to standardize it.

Standardization of the vaccine.—Mix a measured volume of vaccine with an equal volume of a known dilution of blood, make a film, stain and count the number of bacilli and red cells in several fields of the microscope. The number of red cells per cubic centimetre being known, the number of bacilli is easily calculated.

(β) **Pfeiffer and Kolle's method.**—Cultures on agar 24 hours old are scraped with a platinum needle and the growth mixed with saline solution (45 c.c. for ten tubes). The emulsion is filtered through gauze, the filtrate is heated to 60° C. for 2 hours, then distributed in tubes and a little carbolic acid added. The quantity to be used for the first dose is 0·5 c.c. (corresponding to 1 loopful or 2 milligrams of fresh culture or $\frac{1}{10}$th of an agar culture) : 8 or 12 days later a second dose of 1 c.c. is given. A third dose may with advantage be given ; if this be purposed it is well in order to obviate any violent reaction to use smaller doses viz. : 0·3 c.c., 0·8 c.c., and 1 c.c.

(γ) **Bassenge and Rimpau's method.**—These authors adopt a technique similar to that of Pfeiffer and Kolle, but to avoid too violent a reaction they give four inoculations of very small quantities with an interval of 10 days between each : for the first inoculation a dose equal to $\frac{1}{30}$th of a loopful is given and then successive doses of $\frac{1}{5}$th, $\frac{1}{6}$th, and $\frac{1}{3}$th of a loopful.

(δ) **Friedberger and Moreschi's method.**—A minimal quantity ($\frac{1}{100}$th or $\frac{1}{1000}$th of a loopful) of an eighteen-hour culture on agar dried and heated to 120° C. for 2 hours is inoculated intra-venously. A single inoculation is sufficient but the intra-venous does not seem as harmless as the sub-cutaneous method and is followed by a violent reaction.

2. Methods based upon the use of bacillary extracts.

The active principle of the typhoid bacillus can be extracted by the different methods which have been studied under the head of typhoid toxin : maceration, trituration, freezing, etc. Methods of antityphoid vaccination based

---

[1] At first it was the custom to use very virulent bacilli. Wassermann has shown that there is no direct and constant relation between toxigenic and immunizing power : he suggests the use of a polyvalent vaccine prepared with a mixture of many strains of typhoid bacilli : such a vaccine is said however to have no advantage over a monovalent vaccine (Bassenge and Mayer).

on the use of bacillary extracts are complicated and do not seem to offer any particular advantages over those just considered.

(α) **Wassermann's method.**—An emulsion of cultures on agar is made in distilled water, heated to 60° C. for 24 hours, macerated for 5 days at 37° C., filtered through porcelain and dried *in vacuo* at 35° C. A single inoculation is given consisting of 0·0017 gram of the powder.

(β) **Neisser and Shiga's method.**—An emulsion of cultures on agar is made, sterilized at 60° C., macerated at 37° C. for 3 days, and then filtered. The filtrate without further preparation is used as a vaccine.

(γ) **Bassenge and Mayer's method.**—A filtrate of living cultures is used. Make an emulsion in distilled water of the growth of a very virulent bacillus on agar and, after shaking continuously for 3 days, filter. A single inoculation is given equal to the filtrate obtained from one tube of culture.

Effects of vaccination.

The results obtained in man with Wright's and with Pfeiffer and Kolle's vaccines will be chiefly quoted, as these are the best known methods and appear to give the most satisfactory results.

Two or three hours after inoculation tenderness develops about the site of inoculation, reaches its maximum in about 12 hours, and vanishes as a rule about 40 hours after inoculation.

At the same time there is some rise of temperature accompanied by stiffness of the back and limbs, headache, loss of appetite and nausea lasting twenty-four hours or so.

About the end of the first week the serum has acquired bactericidal, agglutinating, bacteriolytic and immunizing properties, and the opsonic index is raised. These newly-acquired properties rapidly increase and reach their maximum on the third day after the second inoculation.

The bactericidal and agglutinating properties persist for a long time, having been demonstrated 18 months later by Bassenge and as long as 4 years afterwards by Harrison and others. In a person previously immunized and whose serum no longer exhibits any appreciable bactericidal properties, the inoculation of a very small dose of vaccine will re-create these properties in a very high degree (Wassermann): it would therefore appear desirable to repeat the vaccinating inoculations at intervals in order to maintain and re-enforce the immunity.

Wright has drawn attention to a fact which is very important from the point of view of prophylaxis. During the first few days—less than a week—after inoculation there is a negative phase during which the resistance-capacity of the patient to the typhoid bacillus is lowered. During this period therefore vaccinated persons should not be exposed to infection, and it follows that antityphoid vaccination as now practised is not permissible in times of epidemic nor in endemic centres of the disease.

Antityphoid vaccination in the human subject has been largely practised in the English and German armies. The results are quite conclusive in favour of vaccination. Among the vaccinated the proportion of cases is markedly lower than among the unvaccinated; moreover, the cases of enteric fever which have been observed among the vaccinated have, as a rule, been less severe than among the unvaccinated, and the mortality rate is lower by one-half. The effects of antityphoid vaccination last for several years.

3. Besredka's method.

The immunity conferred by the use of antityphoid serum being very transitory, and the progress of vaccination with attenuated cultures very slow and irregular, it has been suggested by several observers (Leclainche, Calmette, Salimbeni) that mixtures of specific serum and micro-organisms might prove more effective. The results so far obtained have been only moderately encouraging, probably because there has been too much serum in the mixtures used for inoculation.

Bearing in mind the property possessed by organisms of fixing the immune body present in their specific serums, Besredka sensitizes bacilli with antityphoid serum and uses the sensitized organisms for vaccinating purposes. An emulsion of bacilli from a forty-eight-hour culture on agar is made in normal saline solution, mixed with antityphoid serum and left at 37° C. for 24 hours. The agglutinated bacilli are then centrifuged and washed several times with normal saline solution until all traces of serum have disappeared ; the emulsion is then heated in a water bath at 58° C. for half an hour.

Guinea-pigs can be rendered highly immune in about 20 hours by inoculating them sub-cutaneously with the vaccine. The immunity lasts for several months (Besredka, Paladino-Blandini) and the serum of the animals is bactericidal and prophylactic.

In man, inoculation with Besredka's vaccine produces only a very slight tenderness locally, and there is ground for hoping that this method, which confers immunity within 24 hours, will in future play an important part in the prophylaxis of enteric fever.

6. Serum therapy.

Brieger, and Wassermann and Kitasato, whose experiments have been confirmed by Sanarelli, Chantemesse and Widal and others, have shown that laboratory animals can be immunized against experimental infection by inoculating them with the serum of a vaccinated animal, and that such a serum possesses curative as well as prophylactic properties.

If a fatal dose of a culture of the typhoid bacillus be mixed with 0·5 c.c. of the serum and inoculated into the peritoneal cavity or beneath the skin of a guinea-pig the animal remains unaffected.

Guinea-pigs can be immunized in a few hours by inoculating them with 2 c.c. of the serum of a vaccinated animal. Subsequent inoculation of a dose of a virus of exalted virulence sufficient to kill a control animal is without effect on the treated animal.

Similarly, animals inoculated with an ordinarily fatal dose of culture recover if, within 3 hours of the inoculation, 1–2 c.c. of antityphoid serum be administered to them.

Chantemesse and Widal have shown that the serums of patients who have recovered from an attack of enteric fever exhibit both prophylactic and curative properties. These properties are not very well marked, and to immunize a guinea-pig about 10 c.c. of serum are necessary. Attempts to use the serum in the treatment of human enteric fever have not given conclusive results.

Artificial animal serums have been used by a great many observers in the treatment of enteric fever, but with little result (Beumer and Peipper, Shaw, Tavel, Aronson, and others).

Chantemesse and Besredka however have prepared serums which undoubtedly possess therapeutic powers.

A. Chantemesse's serum. 1. Preparation.—Chantemesse immunizes horses by repeated sub-cutaneous inoculations of his soluble toxin (p. 377) and intravenous inoculations of virulent typhoid bacilli. The process of immunization is very lengthy ; small doses should be used to begin with, and the animals must be carefully handled.

2. Properties.—Chantemesse's antityphoid serum may be repeatedly

heated to 54°–56° C. without losing any of its properties. It shows marked agglutinating power (1 in 100,000, p. 413). *In vitro* it has no bactericidal action, but *in vivo* it stimulates the leucocytes to take up and dissolve the bacilli. It protects guinea-pigs and rabbits against the inoculation of lethal doses of an exalted virus. It is markedly antitoxic (Balthazard and Chantemesse), and neutralizes toxin *in vitro*. If given as a prophylactic, 2–24 hours before the inoculation of the toxin, it will protect rabbits against the effects of four times the lethal dose of toxin. When the serum is inoculated at the same time as the toxin but at an independent site its action is less pronounced, and animals which have received more than twice the lethal dose succumb (Balthazard). When injected after the toxin the serum has still less prophylactic and curative powers, and its efficacy varies inversely as the time which has elapsed between the inoculation of the toxin and the inoculation of the serum. The prophylactic properties of the serum are short-lived and the immunity conferred lasts no longer than 10 or 12 days.

   3. **Therapeutic application.**—Chantemesse's serum which in the laboratory shows only feeble curative power has a marked influence on the phenomena of opsonization, and it is probably to this that its undoubted therapeutic properties in the treatment of enteric fever are due : according to statistics published by Chantemesse the mortality in cases treated with the serum is only 4 per cent. It is all important that the serum should be used in the early stages of the disease. Originally Chantemesse inoculated repeated doses of 5–15 c.c. sub-cutaneously, but the serum as now prepared is more active and a single inoculation of a few drops is sufficient.

   B. Besredka's serum.—An anti-endotoxic serum has been prepared by Besredka, by inoculating killed cultures followed by living cultures of the bacillus into the veins of animals.

   The serum neutralizes ten to twenty fatal doses of Besredka's endotoxin and acts as a prophylactic to the inoculation of the endotoxin.

   Montefusco, who has used Besredka's serum in the treatment of enteric fever at Naples, has obtained very satisfactory results, and thinks it will be of great value in the treatment of the disease.

7. Agglutination. Serum-diagnosis of enteric fever.

   Durham and Gruber were the first to show that [an antiserum] agglutinates [its homologous organism]. This agglutinating property, which is a reaction of infection (p. 225), is also manifested in the blood of persons suffering from or who have recovered from an attack of enteric fever. Agglutination can also be obtained with the blister fluid, milk, naturally shed tears, and occasionally even with pus, urine, bile, etc. from these persons.

   [A. S. Grünbaum first and] Widal [afterwards] utilized the agglutinating properties of the blood of enteric fever patients as a rapid and conclusive method of diagnosis—the serum diagnosis of enteric fever.

   The power of agglutinating the typhoid bacillus is developed in the blood of the patient as a rule in the early days of the illness, and while it may not infrequently be delayed, it is only very exceptionally that it is absent throughout the whole course of the disease ; (Widal and Sicard failed to get agglutination once only in 163 cases : Besson twice in 98 cases). The power of agglutination may disappear during the early weeks of convalescence, and is generally absent 6–8 months after recovery ; but occasionally it has been present as long as 3 and even 7 years after the attack.

   A positive result obtained under the conditions to be immediately described may be taken as a certain indication of enteric fever.[1] On the other hand a negative result establishes merely a probability that the disease is not enteric fever. A negative result in the early days of a suspected attack of the disease is of less value than one obtained later, for if the disease be enteric fever failure to react is then improbable. But in any case if the result be negative opportunity should always be taken to test the blood again later.

The reaction may be performed in several ways which are described as *slow* or *rapid* according to the time required : but whatever the method adopted the following rules must be observed.

General rules.—1. For the slow methods the blood must be taken under aseptic precautions from a vein at the bend of the elbow (p. 193), and may be conveniently collected in small sterile glass tubes. For the rapid methods sufficient blood can be collected in capillary tubes by pricking the finger (p. 192).

2. When the blood has to be sent some distance to a laboratory the tube in which it has been collected should be plugged with a plug of wool passed through the flame. Serum kept in the liquid condition retains its power of agglutination for a very long time. Since, however, drying has no effect on the agglutinating power of the blood, a few drops of the latter may be collected on a piece of paper or on a glass slide and allowed to dry before being sent to the laboratory, and in some cases this may be the more convenient course to adopt. For purposes of the agglutination reaction the dried blood is dissolved in a drop or two of sterile water.

In the author's experience good results have always been obtained when the blood was dried on glass but when dried on paper it seemed to lose some of its agglutinating power. Dried blood is only available for use by the rapid method.

3. The culture should always be examined microscopically to test its purity immediately before being used for the reaction. The mistakes which might arise from the use of an impure culture can be readily appreciated.

4. The serum must be added to the culture and not *vice versa* (p. 226).

A. Slow method.—The blood (collected from a vein at the bend of the elbow to ensure that a sufficient quantity is obtained) must be absolutely pure and uncontaminated, and after collection should be set aside in a sterile tube until the clot has separated ; the serum is then drawn up into a Pasteur pipette.

1. To a tube containing 6–10 c.c. of sterile broth, add ten drops of the serum and sow with a trace of a culture of the typhoid bacillus. A control consisting of a tube of broth containing no serum but sown with a trace of a typhoid culture must, of course, be put up. Incubate the tubes at 37° C. Growth in the tube to which the serum has been added will be somewhat delayed, but small clumps appear after 8 hours or so, and after about 18 hours' incubation the appearance is characteristic : the bacilli are collected together at the bottom of the tube in little whitish flocculi, which cannot be

---

[1] In suspected cases, the possibility of the patient having had previously an attack of enteric fever should always be borne in mind, because if so the blood might still retain some agglutinating power. A few rare cases are on record in which it was found that the blood of persons suffering from diseases other than enteric fever has agglutinated the typhoid bacillus in dilutions of 1 in 100 and 1 in 250. Thus in an undoubted case of pneumonia in a young man in which there was no reason to suspect a previous attack of enteric fever Besson found that the blood of the patient agglutinated the typhoid bacillus. The serum reaction in the absence of enteric infection has also been observed in a case of tuberculous meningitis (E. Mackey) and in a case of abscess of the liver (Megele). In pathological conditions in which the bile enters the blood stream, as for instance in jaundice or occlusion of the bile-duct, the blood may agglutinate the typhoid bacillus (Grünbaum, Zupnik, Kohler and others).

broken up by shaking the tube, while the broth remains perfectly clear. In the control tube, on the other hand, the broth is cloudy and shows the scintillating ripples characteristic of a growth of the typhoid bacillus.

The reaction is not always so distinct as this. Sometimes the broth instead of remaining clear shows an irregular turbidity, from which however the characteristic watered-silk appearance is absent; this turbidity is due to the precipitation of a very fine powder each grain of which when examined under the microscope is seen to be an agglomeration of bacilli. At other times the reaction may be quite characteristic at first but after incubating for 18 or 24 hours the broth is turbid above the precipitate. Naked eye appearances ought always to be supplemented by a microscopical examination by means of which the small masses of bacilli may be recognized if present and their structure defined.

2. Add the serum to a twenty-four-hour broth culture of the typhoid bacillus and incubate at 37° C. If the agglutinating power of the serum is well marked characteristic changes will take place within a few hours; the culture is at first granular, but becomes gradually clear as the bacilli fall to the bottom. When the serum is less powerfully agglutinating, clumps are formed but the broth never becomes quite clear. Naked eye appearances must, as in the previous case, always be controlled by microscopical examination.

B. Rapid method. Recommended.—This method is quicker and more sensitive than the foregoing and has the additional advantage of requiring only a few drops of blood, an amount which can be easily obtained by pricking the finger. It is therefore the method to be used in the majority of cases.

A broth or peptone culture of the typhoid bacillus is required for the reaction, and the greatest care should be exercised in the choice of a culture. In the first place it is of course essential that it be pure : secondly the growth must not be more than 24 hours old, because in old cultures clumps often form spontaneously and these will falsify the results. Spontaneously formed clumps may indeed be present even in twenty-four-hour cultures, so it is always necessary to determine by microscopical examination immediately before use that the chosen culture is satisfactory in this respect. A quantity of culture sufficient for the investigation should be drawn up into a Pasteur pipette and a little placed on a slide and examined under the microscope, the remainder, if the sample is satisfactory, being used for the serum reaction. To obviate the spontaneous formation of clumps in cultures it is better to grow the organism in a 1 or 2 per cent. solution of peptone containing no meat rather than in broth.

The method is as follows.

1. Reaction with the serum.—[(*a*) *Technique recommended.*—*1*. Take a sterile Pasteur pipette, plugged with wool and fitted with an india-rubber teat as shown in fig. 162, p. 241. Make a mark on the stem of the pipette.

[*2*. Take up 9 volumes of the peptone water culture run them up into the bulb and then take up 1 volume of the serum. Mix the culture and serum thoroughly by repeatedly expelling on to a slide and aspirating. (Dilution 1.)

[*3*. Take up 4 volumes of dilution 1 and 1 volume of serum. Mix. (Dilution 2.—1 in 50.)

[*4*. Take up equal volumes of dilution 2 and culture. Mix. (Dilution 3. —1 in 100.)

[And so on, preparing the dilutions required.

[Place a drop of each dilution on a clean cover-glass and invert the latter over the cavity in a hollow-ground slide. Lute the edge with vaseline. Place the preparations in the incubator at 37° C. Examine with a high power dry lens at the end of half an hour and again at the end of an hour. If the serum be a typhoid-agglutinating serum the bacilli will be found agglutinated together in more or less large masses.]

(β) *Another method.*—Into a small conical glass vessel introduce 10–100 drops of the culture and 1 drop of the serum. Place a drop of the mixture on a slide and cover with a cover-glass. Examine with a high power dry lens. If the serum has the power of agglutinating the typhoid bacillus, masses of agglutinated bacilli will be seen and among them a greater or smaller number of non-agglutinated bacilli. The reaction is still more distinct if the preparation be examined after 15 or 20 minutes, for compact islets of agglutinated bacilli will then be visible under the microscope. When the agglutinating property of the serum is small the reaction may only appear after the lapse of 40 minutes or an hour. The appearance is quite characteristic and renders mistakes impossible. Agglutination is assisted by a slight drying at the edge of the drop between the slide and the cover-glass.

Whole blood.—The whole blood may be used for the reaction, but before examining the preparation under the microscope time must be given to allow most of the red cells to settle, since the presence of a large number of cells detracts from the sharpness of the reaction. The method is therefore no quicker than the serum method.

Method of staining.—The preparation may be stained—so as to render the masses more distinct—and preserved for future use: for this purpose the following technique, described by Guillemin, gives good results.

Mix 1 drop of the whole blood with 9 drops of sterile broth, and add 1 drop of this to 2–5 drops of a culture of the typhoid bacillus. Spread a large drop of the mixture on a slide : place in a moist chamber for an hour or two: dry slowly, fix in alcohol-ether, treat with 10 per cent. acetic acid to dissolve the red cells, wash, stain with dilute carbol-fuchsin, wash and dry.

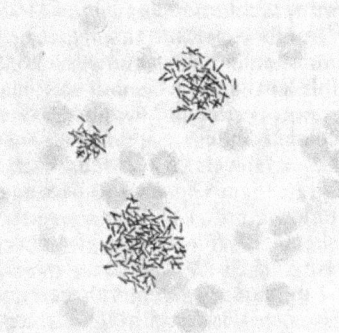

FIG. 220.—Agglutination of the typhoid bacillus by a specific serum. Jenner's stain. (Oc. 2, obj. D. Zeiss.)

2. Reaction with dried blood.—The blood collected and dried as already described is dissolved immediately before use in a drop or two of water. The solution is added to 10–50 drops of a broth culture of the typhoid bacillus contained in a conical glass vessel. It is left for a moment to allow the red cells to settle and then examined as before.

3. Reaction with dead bacilli.—The phenomenon of agglutination is not dependent upon any vital reaction of the bacilli, since it can be demonstrated with dead organisms. This fact may in certain circumstances be of practical value, because a recent culture of the typhoid bacillus is not always immediately available with which to perform the reaction, and to obtain one may involve a delay of 12 hours or so. In such a case a dead culture may be used, since experience has shown that such a culture retains its sensitiveness towards an agglutinating serum for several weeks. The following technique may be adopted in the preparation of a dead culture for this purpose (Widal and Sicard).

A sixteen- or twenty-four-hour culture of the typhoid bacillus is examined microscopically to test its purity. Formalin in the proportion of 2 drops to 15 c.c. of culture is added to kill the bacilli, which become as it were embalmed. Care must be taken to cover the cotton-wool plug of the vessel containing the culture with an india-rubber cap. Cultures killed in this way may be stored exactly as chemical reagents are stored in the laboratory. Immediately before use the tube is lightly shaken so that the bacilli shall be uniformly distributed through the medium. To a few drops of this dead culture a few drops of serum are added in the same manner as has been described above.

There are now a number of preparations of dead bacilli on the market such as Ficker's " *Typhus diagnosticum*," Stassano's emulsion, etc. These preparations allow the practitioner to perform a serum diagnosis rapidly and easily. The " *Typhus diagnosticum* " of Ficker in particular has given good results in the hands of many bacteriologists—though its reliability is questioned by de Rossi.

De Rossi advises the use of broth cultures which have been heated to 58°–60° C. for an hour. The resulting emulsion agglutinates more readily than unheated cultures and preserves its property for at least 3 months.

Tribondeau uses broth cultures killed by the addition of formalin (1–150) and stored in sealed ampoules. Under these conditions the bacilli retain their capacity for agglutination for 4 years and more.

Measurement of the agglutinating titre.—The serums of different patients vary in their agglutinating powers; sometimes this power is very feeble while in other cases it is so well marked that clumps are formed when the serum is diluted as much as 1–5000 and 1–15,000 (Jurgens).

In investigating the agglutinating property of a given serum the examination should be begun with a dilution of 1–10. Agglutination in a lower dilution than this is in no way characteristic, and indeed, since normal human serum occasionally agglutinates when diluted 1 in 10 or even 1 in 20 (*vide* colon bacillus), a reliable diagnosis can according to Rémy only be given when agglutination is found with a dilution of 1 in 50. [Moreover with some serums there appears to be an agglutination-inhibiting action when examined in low dilutions.] The degree to which the agglutinating power is developed should therefore be measured more exactly by investigating dilutions of 1 in 20, 1 in 30 [and so on to at least a dilution of 1–100].

In practice when but a small quantity of blood is available two tests suffice, one made with a dilution of 1 in 10 the other with a dilution of 1 in 50. [By the method described above (**B. 1.** *a*) if a dilution of 1 in 10 can be obtained a dilution of 1 in 100 and 1 in 500 can also be made. In our experience a dilution of 1 in 100 is the smallest dilution upon which a reliable opinion can be based in suspected cases of enteric fever.]

Widal and Sicard draw the following distinctions.

Agglutinating power very feeble if exhibited only in dilutions below 1 in 100.
Agglutinating power feeble if exhibited only in dilutions between 1 in 100 and 1 in 200.
Agglutinating power average if exhibited in dilutions 1 in 200 and 1 in 500
   „   „   marked   ,   „   1 in 500 and 1 in 2000.
   „   „   very marked „   „   above 1 in 2000.

Note.—In these measurements it is important that the drops of culture and of the serum be equal in size. A sufficient degree of accuracy is attained by the following method: take a piece of glass tubing about 20 cm. long and plug it at both ends with wool. Draw it out in the flame as though making a Pasteur pipette. Sterilize the tube without cutting it into two and then, when about to use it, file it in the middle of the capillary portion. In this way, two pipettes are obtained, which for all practical purposes will give drops of equal size: one will serve for the culture, the other for the serum.

From the point of view of prognosis, it appears that the degree to which the agglutinating power is developed [has no consistent relation to, and therefore] furnishes no reliable information as to the severity of the disease.

[Co-agglutinins in the serums of enteric patients.]

[In addition to the specific, homologous or primary agglutinins for the typhoid bacillus, the serum of enteric fever patients often contains group or heterologous or secondary agglutinins for bacilli of the paratyphoid and salmonella groups.

[In a consecutive series of 86 serums Boycott found that 59 per cent. contained secondary agglutinins and of this series 55 per cent. reacted with *B. Gaertner* and *Paratyphosus A* (Brion and Kayser), 41 per cent. with *Paratyphosus B* (Schottmüller), 33 per cent. with *Aertrycke* and 12 per cent. with *Paratyphosus B* (Schottmüller). Generally speaking the more typhoid agglutinin there is present, the more secondary agglutination is likely to be found.]

The application of the serum test to the identification of the typhoid bacillus.

The application of the agglutination reaction to the identification of the typhoid bacillus may be of considerable service but is not a test sufficiently delicate and specific to determine the identity of the bacillus with certainty.

For the purpose of testing whether a given organism be the typhoid bacillus or no it is better to use the serum of a person suffering from the disease which agglutinates quite distinctly in a dilution of 1 in 100 than an artificially prepared anti-typhoid serum.[1] Typical typhoid bacilli are agglutinated by this serum in dilutions varying from 1 in 50 to 1 in 100. Strains of the colon bacillus on the other hand are never agglutinated, or at most only in dilutions of 1 in 5 to 1 in 10. A postulate such as the following would render the diagnosis very simple : any bacillus agglutinated in a dilution of 1 in 50 may be legitimately described as a typhoid bacillus.

Unfortunately, it is now established that there are some undoubted typhoid bacilli which are not agglutinated by the serum of a person suffering from enteric fever. Rémy has shown that a typhoid bacillus which agglutinated well at first readily lost this property when grown symbiotically with the colon bacillus for a few weeks. Occasionally, strains of the typhoid bacillus isolated from the living body or from water can be agglutinated only with difficulty, and it is not until they have been sub-cultivated a certain number of times on artificial culture media that agglutination capacity is acquired (Courmont, Chantemesse, Rémy, Sacquépée and others). When a suspected typhoid bacillus has failed to give the serum reaction the following experiment may be carried out with the object of identifying the organism. Inoculate a guinea-pig every other day for a fortnight with 2 c.c. of a forty-eight-hour broth culture of the bacillus under investigation. If the blood of the guinea-pig now agglutinates an undoubted typhoid bacillus in a minimum dilution of 1 in 40 the organism which served for the inoculation of the animal may be regarded as a true typhoid bacillus. Some strains of undoubted typhoid bacilli however escape even this method of recognition (Rémy).

8. Absorption of agglutinins.

Castellani's absorption or saturation method can also be applied to the differentiation of the typhoid bacillus from closely allied organisms and by

---

[1] In the case of animals highly immunized against the typhoid bacillus, Rodet has shown that the blood not only agglutinates the typhoid bacillus in very high dilutions, but that it also has marked agglutinating properties for some strains of the colon bacillus. Pfaundler, Bruns, Kayser, have shown that very highly immunized serums agglutinate not only the organism against which the animals were immunized but also closely related species.

its means it is possible to determine whether agglutinins which have been detected in a suspected typhoid serum are specific agglutinins or co-agglutinins. (For technique see p. 436.)

9. Complement fixation.

The method of complement fixation (Bordet-Gengou reaction) is applicable to the diagnosis of enteric fever and to the identification of the typhoid bacillus. The method is described at p. 233. The results are more exact and more reliable than agglutination (Widal and Le Sourd) and the reaction gives positive results with the serum of " carriers " even when no bacilli can be detected (Schöne).

[H. R. Dean finds that the complement-fixation method affords an extremely delicate and specific means of differentiating between various members of the typhoid and paratyphoid group.   (For Dean's technique see p. 428).]

SECTION IV.—DETECTION, ISOLATION AND IDENTIFICATION OF THE TYPHOID BACILLUS.

The detection of the typhoid bacillus may be rendered difficult by the presence of other organisms in the fluid or tissue under examination.   Thus, in patients suffering from enteric fever or in patients or animals who have died from the infection the bacillus occurs in pure culture and can be readily isolated, but when it is necessary to isolate it from water, dust, stools, etc. the presence of the colon bacillus often renders the investigation by no means easy.

The methods of isolating the typhoid bacillus from water and other sources is dealt with in a separate chapter (Chap. XXIII., p. 401) and here the more simple investigations only will be considered in which the organism is assumed to be in pure culture in a fluid or tissue of the body.

1. Microscopical examination.

For purposes of microscopical examination films and sections of the spleen and other organs as well as—in the case of experimentally infected animals—

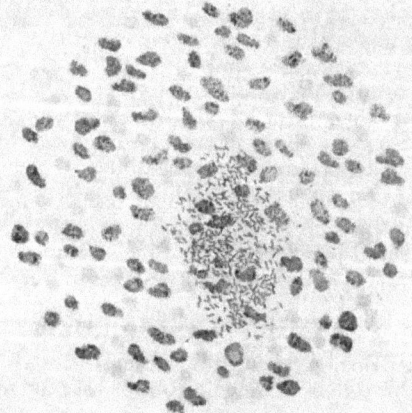

FIG. 221.—Typhoid bacillus. Section of a spleen. Carbol-thionin. (Oc. 2, obj. $\frac{1}{12}$th, Zeiss.)

films from the pus of typhoid abscesses and the peritoneal exudate, should be made.

In no case can an absolute diagnosis be made on microscopical evidence alone.

(*a*) **Films.**—Films should be stained first with methylene blue or carbol-thionin and then by Gram's method.

Films from the spleen often contain only a few organisms, but if the tissue be first incubated films made from it will be found to be very rich in bacilli. Wash the surface of the organ in a 1 in 1,000 solution of perchloride of mercury, wrap it up in a cloth wrung out of the same solution and incubate at 37° C. for 24 hours (Cornil) ; or if preferred a little of the pulp may be drawn up into a number of Pasteur pipettes and incubated (Gasser).

Gasser prefers to stain the films obtained in this way by Gram's method, using dilute carbol-fuchsin as a counterstain : the typhoid bacilli and the groundwork are then stained red while if any gram-positive organisms are present they are, of course, stained violet.

(*b*) **Sections.**—Fix the tissues to be cut in alcohol or acid perchloride solution (p. 189) and embed in paraffin. Stain the section by any of the methods applicable to the staining of gram-negative organisms—preferably with thionin or by Nicolle's tannin method (p. 217).

2. Cultures.

Culture media—broth, agar (and, for isolating the bacillus when contaminations are present, gelatin plates)—should be sown with scrapings from the spleen, with fluid exudates, products from puncture of the tonsil, urine collected under aseptic precautions, etc.

Attention has already been drawn (p. 198) to the dangers attending puncture of the spleen in the living subject. This practice should never be resorted to as a matter of routine and since the serum reaction is now available for the purposes of diagnosis (see p. 384) and the bacilli can be isolated from the blood there is no justification for running the risk attending the operation.

Examination of the blood. (*a*) **Courmont's method.**—Collect the blood aseptically by puncture of a vein at the bend of the elbow (p. 193) and sow 2–4 c.c. immediately in a large volume (200–300 c.c.) of ordinary broth. Incubate at 37° C. If no turbidity appears after 24 hours shake the flask to promote the growth of the organism and then incubate again, and examine the culture daily. In some cases when the typhoid blood has powerful agglutinating properties growth may be delayed, being invisible until the third or fourth day, and may then occur solely as clumps in the deposit which forms at the bottom of the liquid.

(β) **Busquet's method.**—To minimize the inconvenience caused by the agglutinating and bactericidal properties of the blood, Busquet sows a number of flasks each containing 250 c.c. of peptone broth with a few drops only of blood.

Sacquépée and Perquis adopt the additional precaution of defibrinating the blood as it leaves the vein.

(γ) **Lafforgue's method.**—Lafforgue eliminates the serum, which contains the bactericidal substances. The blood is rendered non-coagulable by the addition of sodium citrate (2 drops of a 1 in 5 solution of sodium citrate to 2 c.c. of blood) and then centrifuged. The deposit only is sown in broth in the proportion of 20 c.c. of broth to the deposit from 2 c.c. of blood. Under these conditions the typhoid bacillus grows rapidly.

(δ) **Conradi's method.**—Conradi has shown that bile is an excellent medium, since it renders the blood non-coagulable and inhibits its bactericidal action. Cultures in bile-containing media are recommended for the detection of the typhoid bacillus in blood.

The principle may be applied in different ways. The simplest methods are those of Zeidler and Kayser.

1. Methods of Zeidler and Kayser.—Zeidler adds 1 c.c. of blood (30 drops

to 5 c.c. of ox bile previously sterilized in the autoclave. The mixture is incubated at 37° C. for 12–24 hours and then plated out on malachite green agar (p. 409) or better, litmus-lactose-agar.

Kayser adopts a similar technique and sows 2·5 c.c. of fresh blood in 5 c.c. of sterile ox bile and then plates on Conradi-Drigalski's medium (p. 407).

2. Conradi's technique.—Conradi uses ox bile containing 10 per cent. of peptone and 10 per cent. of glycerine. The mixture is sterilized at 100° C. for 2 hours. A small clot of the suspected blood is added to 5 c.c. of the sterilized medium and incubated at 37° C. for about 15 hours. Plates of litmus-lactose-agar are then sown with the cultures.

3. Roosen-Runge's technique.—Glycocholate of sodium is used instead of bile. The medium is an ordinary agar medium to which 10 grams per litre of glycocholate of sodium have been added.

4. Dünschmann's technique.—Dünschmann is of opinion that the valuable constituent in bile is taurocholate and not glycocholate of sodium. He recommends the following medium :

Gelatin, - - - - - - - - - -	5 grams.
Agar, - - - - - - - - - -	30 ,,
Lactose, - - - - - - - - - -	40 ,,
Peptone, - - - - - - - - -	10 ,,
Taurocholate of sodium, - - - - - - - -	20 ,,
Water, - - - - - - - - - -	1000 c.c.

When only a small quantity of blood is available it is used for sowing surface plates on bile-salt-agar.

Detection in sputum.—To detect the typhoid bacillus in sputum employ one of the methods described in Chap. XXIII.

CHAPTER XXII.

BACILLUS COLI.

In man and the lower animals the colon bacillus, which was originally described by Escherich, is a normal inhabitant of the alimentary canal where it makes its appearance a few hours after birth.

In the intestines of healthy human subjects the colon bacillus is associated with numerous other micro-organisms (at least fifty different species are present, many of them being anaërobes) ; it is also frequently found in the mouth—twenty-five out of sixty-five cases (Grimbert and Choquet).

Though often only slightly virulent when isolated from the healthy intestine, the colon bacillus in certain circumstances and in a diseased environment may acquire a high degree of virulence, as for instance in all febrile conditions, in enteric fever, and in the majority of diseases of the intestine ; and may then act as the causal agent of a number of diseases affecting man.

It is, for instance, the cause of secondary infections in enteric fever, dysentery and cholera.

In some cases of septicæmia the colon bacillus is the organism present in the blood, and while as a rule bacillæmic conditions due to it are not severe, they may on occasions present all the clinical features of enteric fever. The bacillus is also the cause of some attacks of enteritis, of some cases of choleraic and infantile diarrhœa, etc. Peritonitis may sometimes be due to the colon bacillus (as, for instance, peritonitis resulting from perforation of the gut or following strangulated hernia, and peritonitis unaccompanied by perforation). Invading the biliary passages this organism determines suppurative cholangitis and possibly infective jaundice, and is responsible for some cases of sore throat, broncho-pneumonia, endocarditis, pericarditis, and meningitis. It is the causal agent in a number of infections of the urinary passages and must be identified with the *urinary bacillus of Clado.* In women it plays an important part in determining pathological conditions of the true pelvis (such as salpingitis and metritis). Finally it is responsible for most *post mortem* and agonic *ante mortem* infections.

The colon bacillus is found in the soil, in water contaminated with animal excreta, and in dust.

SECTION I.—EXPERIMENTAL INOCULATION.

The colon bacillus is [usually] pathogenic to guinea-pigs, rabbits, mice, and other animals. Though often avirulent when isolated from the stools of healthy persons, its virulence can be rapidly increased by passing it through the peritoneal cavities of a series of guinea-pigs. Guinea-pigs are the most suitable animals for the study of the experimental disease.

A virulent strain may easily be obtained by suturing the anus of a guinea-pig. The animal dies of intestinal obstruction, and a pure culture of a very virulent colon bacillus can be isolated from the cloudy peritoneal exudate. Should the exudate as is sometimes the case contain a few other organisms mixed with the colon bacillus, a pure culture of the latter can be readily obtained by plating on gelatin.

Sub-cutaneous inoculation of a colon bacillus of low virulence into guinea-pigs, rabbits or mice leads, as a rule, to the formation of an abscess which resolves spontaneously. Intra-peritoneal inoculation produces a more severe, but not usually fatal infection.

The inoculation of a virulent strain, on the other hand, usually gives rise to an acute disease in these animals : the effects will be described in detail.

1. Guinea-pigs. (α) **Intra-peritoneal inoculation.**—The inoculation of a few drops of a broth culture into the peritoneal cavity causes death in about 20 hours with symptoms of sub-acute peritonitis and a sub-normal temperature. *Post mortem*, there is a generalized peritonitis with a copious, turbid exudate : the coils of the intestine are covered with a purulent fibrinous exudate : the lumen of the gut is filled with diarrhœal matter, the walls are swollen and congested and occasionally show some mucous ecchymoses, the Peyer's patches are swollen and the spleen enlarged ; in females the organs of generation are congested and it is not uncommon to find the uterus filled with an hæmorrhagic exudate. The organism can be isolated from the blood and internal organs.

(β) **Intra-pleural inoculation.**—Death supervenes in 24 hours. *Post mortem* there is an excess of fluid, sometimes blood-stained, in the pleura, with fibrinous deposit on the lungs, pericardial effusion, congestion of the lungs and intestine and swelling of the spleen. The bacillus is present in the blood and internal organs.

(γ) **Sub-cutaneous inoculation.**—This leads to a less severe infection than the preceding and much larger doses of culture (1-2 c.c.) are required to produce a fatal result. A swelling forms at the site of inoculation, the bacillus becomes disseminated and death takes place in 48 hours. *Post mortem* the Peyer's patches and the spleen are swollen and the intestine congested and ecchymosed.

2. Mice.—Mice, though less susceptible, succumb to the inoculation of cultures of the bacillus. The lesions are similar to those in guinea-pigs.

3. Rabbits.—Rabbits also are less susceptible than guinea-pigs, and much larger doses must be used to produce death ; *post mortem* the lesions are similar in the two cases.

When a small dose is inoculated sub-cutaneously the animal does not die for several days and *post mortem* suppurative foci will be found in the liver, spleen and mesenteric glands.

Intra-venous inoculation usually leads to a rapidly fatal infection ; the rabbit suffers from a colon bacillæmia resulting in the production of the usual lesions in the walls of the intestine and spleen.

Sometimes the animal may survive the intra-venous inoculation of a few drops of a broth culture for several months. In such cases an atrophic paralysis appears as the result of an *anterior poliomyelitis* (Gilbert and Lion); this affection of the cord is not necessarily fatal, and the rabbits sometimes recover even though the symptoms may have been very marked.

SECTION II.—MORPHOLOGY.

1. Microscopical appearance.

The colon bacillus, like the typhoid bacillus, is a small rod-shaped organism with rounded ends. Morphologically, the two organisms are identical and subject to the same variations; spindle-shaped forms and pseudo-sporing forms are met with equally in the two cases.

Staining methods.—Like the typhoid bacillus, the colon bacillus is gram-negative and stains with the ordinary dyes.

Motility.—As a rule, the colon bacillus is less motile than the typhoid bacillus.

The motility varies greatly in strains from different sources; in some cases indeed the bacilli are non-motile, in others the movements are slow and limited, while in others again the organisms are almost as motile as the typhoid bacillus.

Flagella.—The flagella of the colon bacillus offer many points of contrast with those of the typhoid bacillus. They can be stained by the same methods as the latter but successful preparations are more difficult to obtain.

The number of flagella is always smaller than in the case of the typhoid bacillus: the colon bacillus has usually about four to six flagella and it is quite the exception to find as many as twelve.

The flagella may be arranged all round the surface [peritrichous]: but more commonly they are seen arranged in one or two bunches attached to points on the surface, generally towards one end [lophotrichous]. The flagella rarely exceed $3-5\mu$ in length being only an half to a third as long as those of the typhoid bacillus: they are not so wavy and undulating and are never seen in the tangled bunches so characteristic of the typhoid bacillus.

2. Cultural characteristics.

A. Conditions of growth.—The conditions under which growth takes place are the same for the typhoid and colon bacilli: both are able to grow at 45° C., but given equal opportunities the colon bacillus grows rather more quickly. Its growth is accompanied by an unpleasant fæcal odour which is characteristic of the organism.

B. Characteristics of growth on various media. 1. Broth.—In cultures incubated at 37° C. growth is visible in 6–8 hours and has in general the same characteristics as the growth of the typhoid bacillus; a greyish pellicle however often forms on the surface of the medium which is only exceptionally seen in cultures of the typhoid bacillus.

2. Gelatin.—The colon bacillus does not liquefy gelatin.

(α) *Stab cultures.*—In cultures incubated at 20° C. growth is visible in 24 hours. The small colonies which form along the line of the stab become opaque and soon unite to form a continuous line of growth. On the surface, a thick whitish pellicle of creamy consistence forms and may extend to the side of the tube. In short, the growth of the colon bacillus is, as a rule, both more copious and more rapid than that of the typhoid bacillus but the differences may not be very marked and cannot be relied upon for purposes of differentiation.

(β) *Stroke culture.*—After incubating for 30 hours a thin, bluish layer with pinked edges appears which subsequently becomes whitish and opaque.

In typical cases, the growth is more abundant and more opaque than that of the typhoid bacillus.

(γ) *Isolated colonies.*—As a rule, isolated colonies are small and lenticular and their margins indented ; at first they are bluish and transparent but later become white and opaque and are larger than those of the typhoid bacillus. Frequently, however, the colonies remain transparent and preserve the " iceberg " appearance already noted as characteristic of the typhoid bacillus.

Colonies which develop in the depth of the gelatin have the appearance of small whitish opaque grains.

3. Agar and coagulated serum.—On these media the colon bacillus forms a whitish layer with no characteristic feature. Gas-bubbles sometimes form in the depth of the medium and increasing in size lift up the medium.

4. Potato.—As a rule, the growth is at first yellowish and then later becomes brown, thick, raised and moist ; but some strains of the colon bacillus give a thin colourless pellicle indistinguishable from the growth of the typhoid bacillus. The quality and variety of the potato used have much to do with the appearance of the growth.

On Rémy and Sugg's solid medium the colon bacillus invariably gives rise to an abundant, thick growth which may be glairy or dry and which is always of a dirty yellow or brown colour.

5. Milk.—Milk is coagulated in 24–30 hours when incubated at 37° C.

SECTION III.—BIOLOGICAL PROPERTIES.

1. Biochemical reactions.

1. Action on carbohydrates.—In both aërobic and anaërobic culture the colon bacillus decomposes lævulose, lactose, saccharose, maltose, glucose, erythrite, and mannite, with the formation of acid (formic, acetic, butyric, lactic), gas (hydrogen, carbon-dioxide), and ethyl alcohol. These reactions are invaluable for the purpose of identifying the organism. The technique has been described in connexion with the action of the typhoid bacillus on sugars (p. 373).

Attention must be drawn to the fact that the colon bacillus under certain conditions, particularly when it is grown in symbiosis with the typhoid bacillus (Rémy), may lose its power of splitting up sugars with the formation of acid and gas. Grimbert and Legros have found however, that in some cases where dysgonic influences have affected the fermentation properties of the colon bacillus these properties though markedly diminished, are not altogether lost ; they have been able to show, for instance, that milk will be coagulated if in a shallow layer, and that lactose if present in sufficient quantity is feebly, but nevertheless definitely attacked.

(*a*) Action on lactose-broth containing calcium carbonate.—When sown on this medium and incubated at 37° C. for 12–20 hours the colon bacillus decomposes the lactose with the formation of acids, which in turn attack the calcium carbonate and give rise to numerous bubbles of carbon-dioxide.

(*b*) Action on litmus in presence of a carbohydrate fermented by the organism.—Litmus-lactose-gelatin and litmus-mannite-gelatin. The blue colour of the litmus is first changed to red and later assumes a peculiar colour somewhat resembling that of the skin of an onion.

(*c*) Action on Grimbert and Legros' medium (p. 373).—The colour of the medium is rapidly changed to red.

(*d*) Milk.—Milk is rapidly coagulated (*vide* p. 373).

[(*e*) Litmus milk.—The litmus is first turned pink and subsequently bleached.]

Numerous more or less ingenious methods have been devised to illustrate the fermentation properties of the colon bacillus. Thus for instance, to an agar or other medium containing lactose, a substance (*e.g.* fluorescin) is added which is altered or intensified in colour by the acids formed out of the lactose : in other cases a reagent is selected for addition to the medium which is coloured in alkaline solutions but colourless in acid solutions (*e.g.* phenol-phthaleïn). Ramond's method may be described as an example.

(*f*) **Ramond's method.**—Take a tube of gelatin containing 4 per cent. of lactose and after melting it—being careful not to apply too much heat—add sufficient aqueous solution of acid fuchsin (Rubin S.) to impart a red-cerise colour to the gelatin, then just decolourize with a saturated aqueous solution of sodium carbonate—2–3 drops are sufficient—filter, sterilize at 105° C. for 5 minutes and pour the now colourless medium into a sterile Petri dish. The typhoid bacillus produces no change of colour when sown on this medium, while on the other hand the colon bacillus, in virtue of the acids formed from the lactose which neutralize the sodium carbonate, regenerates the red tint so that a characteristic rose - coloured area develops

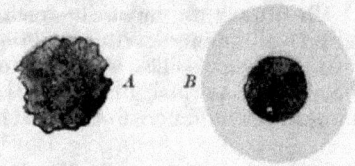

FIG. 222.—Appearances presented by the typhoid bacillus (*A*) and the colon bacillus (*B*) when grown on Ramond's agar. (After Gauthié.)

around colonies of this organism. This method is not so delicate as that with litmus-tinted media.

2. Action on neutral red.—Neutral red in culture media is reduced and decolourized by the colon bacillus. The typhoid bacillus has no action on the dye.

Liquefy a tube containing 10 c.c. of ordinary agar (or glucose-agar), add 3 or 4 drops of a sterile saturated aqueous solution of neutral red, and when the medium has cooled and set sow it with the colon bacillus in stab culture and incubate at 37° C. for 24 hours. The medium will now no longer be red but will exhibit a greenish fluorescence, and on further incubation this will soon change to a canary yellow colour. This reaction has been adapted by Savage to the detection of the colon bacillus in water (p. 411).

3. Indol formation.—An important and very constant characteristic of the colon bacillus is the formation of indol in culture media.

The value of the indol reaction in the diagnosis of the colon bacillus has been called in question by some authors on the ground that they not infrequently fail to find any indol in cultures of this organism : and Rémy has shown that when the colon bacillus is grown with the typhoid bacillus the former may lose its capacity to produce indol.

Recent work demonstrates that the negative results obtained by the earlier observers were due to the imperfections of their technique. " The property of producing indol is far less variable than is generally believed," and the indol reaction furnishes one of the best tests there is for identifying the organism, provided that the following precautions be observed, viz. :—

1. That peptone water and not ordinary broth be used as the culture medium.

2. That the culture be examined between the third and the eighth day but never later.

3. That the test be performed exactly as described at p. 374.

4. Cultures on synthetic media.—The colon bacillus as a rule grows luxuriantly in the different liquid media of Nægeli, Maasse, Fränkel, Rémy and Sugg (p. 375).

5. Growth on vaccinated media.—If the colon bacillus be sown on a tube of agar or gelatin on which the typhoid bacillus has already been grown and scraped off as described above, some amount of growth generally takes place which though distinct is less abundant than on tubes of new media.

6. Growth on coloured media.—The colon bacillus decolourizes both

Nœggerath's medium and fuchsin-agar. The typhoid bacillus gives similar results.

7. Growth on arseniated broth.—A typical colon bacillus grows in broth containing as much as 2 grams of arsenious acid per litre (Thoinot and Brouardel).

8. Growth on artichoke.—A typical colon bacillus grows luxuriantly on artichoke, and turns the medium green (p. 375).

9. Growth on media containing caffeine.—The colon bacillus does not grow on media containing 0·5 per cent. of caffeine (p. 408).

10. Growth on malachite-green media.—According to Lœffler the addition of a small quantity of malachite-green to culture media prevents the growth of the colon bacillus, but does not interfere with the growth of the typhoid bacillus. As a matter of fact, the colon bacillus grows on media containing either malachite green or crystal violet (pp. 409 and 407).

2. Variability of flagella.

The variability of the flagella is very limited, their characteristics being little influenced by antiseptics, temperatures unfavourable to growth, etc. (Rémy and Sugg).

Examination of the flagella should never be neglected when it is desired to identify the colon bacillus.

3. Vitality and Virulence.

Vitality.—All that has been said with regard to the vitality of the typhoid bacillus is equally applicable to the colon bacillus.

Virulence.—The virulence of the colon bacillus is subject to great variation (*vide* experimental inoculation, p. 394).

4. Toxin.

Malvos has shown that porcelain-filtered broth cultures are toxic. Broth cultures also yield a toxic precipitate when heated with sulphate of ammonia. As a rule, the toxin is not very harmful and large doses of filtered cultures must be inoculated to produce a fatal result in experimental animals.

The inoculation of a large dose of toxin into the ear-vein of a rabbit produces the following symptoms : At first there is muscular weakness, sub-normal temperature, drowsiness and coma : later, convulsions set in and finally a generalized tetanic condition which continues till the animal dies (Gilbert). A smaller dose produces a chronic intoxication with diarrhœa, drowsiness and wasting, the animal often dying of cachexia.

In guinea-pigs, the inoculation of large quantities of toxin into the peritoneal cavity is followed by a sub-normal temperature and leads to collapse and death (Boix). The blood may contain organisms (especially the colon bacillus) which have found their way from the intestinal canal (Achard and Renault).

Colilysin.—In suitable media the colon bacillus forms an hæmolytic substance (Kayser).

Colilysin is only produced in any quantity if the broth has a markedly acid reaction (80 c.c. of decinormal oxalic acid per litre).

The hæmolysin is present after incubating for 2 days at 37° C. but continues to increase in amount until the fourth day and remains at its maximum until the end of the second week.

Colilysin is a powerful solvent of dog red-cells ; it has less action on horse, ox, and rabbit cells, and very little and in some cases no action at all on the red cells of other animals (man, guinea-pigs, birds etc.). Colilysin can be kept for months at the ordinary temperature of the laboratory and is not destroyed by heating to 120° C. for half an hour.

Some normal serums (those of man, the horse etc.) neutralize the hæmolytic property of colilysin: and an anti-colilysin can be readily produced by inoculating various animals sub-cutaneously with four-day old broth cultures of the colon bacillus.

5. Vaccination and serum therapy.

Guinea-pigs and rabbits can be immunized by repeatedly inoculating them either with small doses of living and virulent organisms or with filtered cultures of similar strains. Albarran and Mosny produced a very high degree of immunity in dogs and rabbits by repeatedly inoculating them with small doses of filtered cultures and with the filtrates derived from macerating the internal organs of animals dead of a colon bacillus infection. Rodet immunized horses and sheep by inoculating them repeatedly with increasing doses of living or dead cultures.

The serum of vaccinated animals has marked immunizing properties and also, to some extent, therapeutic properties. These properties are manifested against the strains used for immunization but may be wanting against strains from other sources.

Antityphoid serum is neither prophylactic nor curative for the colon bacillus.

According to the experiments of Sanarelli and some other observers animals vaccinated against the colon bacillus should be immune to both the colon and typhoid bacilli, and the serum of the animals should immunize against the typhoid bacillus. These results have however not been confirmed.

6. Agglutination.

(a) The serum of animals infected with the colon bacillus or immunized against that organism, as well as the serum of persons suffering from infections due to the colon bacillus, have the property of agglutinating the bacillus. The agglutination reaction is always obtained with the strain producing the infection, but the results are often negative if other than the infecting organism be employed for the reaction, though the latter may be an authentic colon bacillus. This method of diagnosis cannot therefore be relied upon. The capacity of the colon bacillus to agglutinate is increased to a very marked extent by sub-culturing it on artificial media (Rodet).

(b) The colon bacillus is not agglutinated by the serum of animals vaccinated against the typhoid bacillus nor by the serum of persons suffering from enteric fever. But for this reaction to be of any value it is important that certain precautions be observed (vide footnote on p. 389).

All human serums whether taken from enteric fever patients or not exert a slight agglutinating action on the colon bacillus when diluted five or ten times. Unless this fact be borne in mind it may lead to error. All mistakes may be avoided by adopting the following methods.

Determine carefully first of all the agglutinating power of the typhoid serum which is to be used in the reaction: then mix a drop of the highest dilution of the serum which will definitely agglutinate the typhoid bacillus with a culture of the colon bacillus. Thus, for example, if the highest dilution in which a given typhoid serum will agglutinate the typhoid bacillus be 1–100 this dilution of the serum should be used in testing the suspected colon bacillus. Under these conditions the agglutination of the colon bacillus is never observed, and the serum reaction can be employed as an excellent means for differentiating the two organisms provided that it be always remembered that a strain of the typhoid bacillus which is not agglutinated by a typhoid serum may very occasionally be encountered.

SECTION IV.—DETECTION, ISOLATION AND IDENTIFICATION.

The methods of detecting the colon bacillus in the tissues and fluids of the body are similar in principle to those employed for the detection of the typhoid bacillus. These methods as well as the differentiating tests, etc. are fully dealt with in Chap. XXIII.

It must be remembered that the colon bacillus often multiplies in the body immediately after death, and even during the last few hours of life : the finding of the colon bacillus in the tissues or fluids under these conditions is therefore of no diagnostic value whatever.

The bacillus of Green Diarrhœa.

According to Lesage and Thiercelin the bacillus of green diarrhœa is merely a chromogenic variety of the colon bacillus. The organism is found in practically pure culture in the stools of children suffering from the disease.

Experimental inoculation.—The organism is only slightly pathogenic for laboratory animals. Rabbits, when inoculated intra-venously or fed with cultures of the bacillus, suffer from an attack of green diarrhœa from which they recover in a few days.

Microscopical appearance.—Morphologically the bacillus is a short rod-shaped organism with rounded ends in every way similar to the colon bacillus.

Cultures.—The bacillus of green diarrhœa is a facultative aërobe. It grows on all the ordinary media and gives rise to a disagreeable odour. The green colouring matter is only produced in aërobic culture.

A pure culture is very easily obtained by plating a trace of the stool of an infected child on gelatin.

Broth.—At first the medium is uniformly cloudy but later a greenish sediment is deposited.

Gelatin is not liquefied. In stab culture, the bacillus gives rise to a scanty whitish growth in the substance of this medium and on the surface to a small greenish lenticular pellicle. On sloped gelatin, the growth is poor, greenish in colour and has a tendency to spread away from the line of sowing : after a few days the gelatin is tinted uniformly green. Isolated colonies form small greenish granular points.

On agar.—The growth is poor, greenish in colour and spreading. The agar acquires a green tint.

On potato.—The growth is luxuriant, covers the whole surface of the medium and is of a dirty green mucous appearance.

Milk is rapidly coagulated.

Carbohydrate media are strongly fermented.

CHAPTER XXIII

THE ISOLATION OF THE-TYPHOID AND COLON BACILLI FROM WATER, STOOLS, ETC. AND THE METHODS OF IDENTIFYING THE TWO ORGANISMS.

Introduction.

THE isolation of the typhoid bacillus from water, etc. in which it is mixed with other species of organisms, and especially when the colon bacillus is also present, presents certain difficulties which may be summed up under four headings.

1. On gelatin media, at the ordinary temperature of the atmosphere, colonies of the typhoid bacillus develop slowly (requiring about 48 hours) while saprophytic organisms which liquefy the medium grow more quickly and so put an end to the investigation.

2. The colon bacillus very often retards the growth of the typhoid bacillus when the two organisms are sown together on artificial culture media, with the result that the presence of the latter may pass unnoticed. There is, in fact, a true antagonism between the colon bacillus and the typhoid bacillus (Grimbert). A similar antagonism also exists between certain other micro-organic species and the typhoid bacillus when sown together on artificial media (Besson).

3. Rémy, though he does not admit that the typhoid bacillus is crowded out by the colon bacillus, nevertheless lays stress on the difficulty of isolating the former when the latter organism is also present. He shows that by growing the two organisms together their properties may be profoundly modified : thus the typhoid bacillus occasionally loses its property of being agglutinated by a specific serum, and the colon bacillus may under like conditions lose its indol-producing and fermentation properties.

4. The ordinary method of gelatin-plating only permits of a very small quantity of a suspected water being sown and it is therefore possible that if the typhoid bacillus be present only in small numbers as compared with other organisms, it may escape notice.

It is not a matter for surprise therefore to find that much experimental work has been done with a view to perfecting a method or methods of detecting with certainty the presence of the typhoid bacillus in material in which it may be suspected to occur.

SECTION I.—THE ISOLATION OF THE TYPHOID AND COLON BACILLI.

1. Original methods.

Under this heading will be briefly considered various methods which though in use until recently do not give dependable results, being practically useless for detecting the typhoid bacillus when the latter is mixed with the colon bacillus. These methods are now almost entirely discarded.

(a) **Rodet's method.**—Rodet showed that the typhoid and colon bacilli would grow at 45° C. while most other organisms failed to do so, and on this fact based the following method of analysis. To a flask containing sterilized broth he added 20–100 c.c. of the suspected water and incubated at 45° C. for 20–24 hours. If on taking the flask out of the incubator the broth was cloudy a strong presumption was raised that the typhoid or colon bacillus or both were present in the water. Microscopical examination of the culture and, if need be, isolation on gelatin plates removed all doubt.

(b) **Method of Chantemesse and Widal.**—Chantemesse and Widal found that both typhoid and colon bacilli would grow in artificial media containing 2·5 grams of carbolic acid per litre, and utilized the fact in order to detect these organisms in water.

To tubes containing 20 c.c. of liquefied gelatin add 1 c.c. of a 5 per cent. solution of carbolic acid and a few drops of the water to be examined and pour plates. Unfortunately a certain number of organisms develop in the plates which, as they grow, liquefy the medium and consequently soon put an end to the experiment. A large number of plates must be sown with each of the suspected samples because only a very small amount of water can be used for each plate.

(c) **Vincent's method.**—Vincent devised a method, which for a long time was in general use, based upon a combination of the two preceding observations. He used broth containing 0·1 per cent. of carbolic acid as the culture medium and incubated the cultures at 41·5°–42° C.

To each of half-a-dozen tubes containing 10 c.c. of broth add, immediately before use, 5 drops of a 5 per cent. solution of carbolic acid. Sow with 0·5–1 c.c. of the suspected water, cover with india-rubber caps to prevent evaporation of the carbolic acid, and incubate at 41·5° or 42° C. If the medium in any of the tubes becomes cloudy after incubating for 12 or 20 hours, transfer a little of the culture to a fresh tube of carbolic-broth and incubate it similarly at 41·5° C. As a rule, when the suspected water contains the colon bacillus the first sub-culture yields a pure growth of the latter organism. It must, however, be borne in mind that some saprophytes (*Bacillus subtilis*, *Bacillus mesentericus*, *B. luteus*, the white streptococcus of water, *Proteus vulgaris*, etc.) will also grow under these conditions. These latter organisms cannot be excluded by further sub-cultivation in carbolic-broth because once they become accustomed to carbolic media they grow in them just as well as the colon bacillus. A watered silk appearance in the tubes is a fairly reliable indication of the presence of the colon or typhoid bacillus, but the investigation must always be carried further by microscopical examination and isolation on gelatin. It is well to remember that in carbolic-broth the colon and typhoid bacilli often occur as very short rods (cocco-bacilli) arranged in pairs and devoid of motility.

(d) **Method of Péré.**—This is merely Vincent's method modified in such a way as to allow large quantities of the suspected water to be examined.

Prepare a concentrated broth (meat, 1000 grams, water 1000 grams, and peptone 50 grams), distribute in quantities of 50 c.c. in a series of flasks, and autoclave.

To each flask add 3 c.c. of a 5 per cent. solution of carbolic acid and 100 c.c. of the suspected water. Sow five or six flasks and incubate them at 41° C. As soon as the medium becomes cloudy (15–20 hours) sow a series of broth tubes each containing 0·1 per cent. carbolic acid with a trace of the growth from any of the flasks that may be cloudy. Incubate at 41° C. and continue the experiment as in Vincent's method.

(e) **Method of Pouchet and Bonjean.**—This also is a modification of Vincent's method. To each of a series of flasks containing 100 c.c. of sterile broth add 150 c.c. of the water to be examined and 5 c.c. of a 5 per cent. solution of carbolic acid. Incubate at 42° C.

If the medium becomes cloudy in any of the flasks sow sub-cultures for three generations in 0·1 per cent. carbolic acid broth and incubate at 42° C. Finally, sow a tube of ordinary broth from the last carbolic broth culture, incubate at 36° C. for 8 days and then inoculate a guinea-pig with 0·3 c.c. of culture per 100 grams of animal. If the animal die sow cultures with fragments of the internal organs and heart blood.

The five methods just described are available for the isolation of the typhoid bacillus provided that the colon bacillus is not also present but if, as is most often the case, the two organisms are present together the isolation of the former is impossible by these means.

2. Elsner's method and its modifications.

A. Elsner's method.

The method is available according to Elsner for the isolation of the typhoid bacillus from sources such as water or stools in which the colon bacillus is also present.

The technique is based upon the fact that the typhoid and the colon bacilli grow, to the exclusion of most other organisms, on a potato-jelly containing iodide of potassium. Disappointing results are however frequently obtained ; sometimes the plates are rapidly liquefied and the experiment brought to an end ; at other times the typhoid bacillus cannot be found even though it has been purposely introduced into a sample of water as a control. Several attempts have been made to improve the method, and these will be considered subsequently.

Technique. A. Isolation from water.—1. Prepare and sterilize :—(i) a number of tubes each containing 10 c.c. of potato gelatin (p. 41).
(ii) The following solution :—

Distilled water,	50 grams.
Potassium iodide,	10 „

2. Immediately before use, melt the potato-gelatin tubes and add 1 c.c. (20 drops) of the iodide solution.
The gelatin will then contain 1 per cent. of iodide.
3. Sow ten to fifteen tubes each with 0·5 or 1 c.c. of the suspected water and plate.
4. According to Elsner, the colon bacillus appears on these plates as early as the second day (at 22° C.) as circular, opaque, slightly brown colonies while the typhoid bacillus does not develop until the plates have been incubated for 4 days and then as smaller, transparent, barely visible colonies. Other organisms fail to grow.

As a matter of fact, various organisms other than the typhoid and colon bacilli, and some of which liquefy the gelatin, do grow on the medium ; and then again the colonies of the typhoid bacillus are not so easily differentiated as Elsner makes out. It must be distinctly realized that Elsner's medium possesses no specific property which ensures the development of the typhoid and colon bacilli to the exclusion of other organisms. Its only advantage is that it allows the typhoid bacillus an equal opportunity with the colon bacillus to grow. It is necessary, therefore, to examine carefully every colony on the plates which does not liquefy the medium and which does not form pigment. This is easily done by transferring them each to a separate tube of broth and then incubating at 37° C. After incubating for 24 hours the morphology of the organisms is determined by examining the cultures microscopically and only those tubes which show short, gram-negative bacilli with rounded ends need be reserved for the further tests to be described later.

If any of the broth cultures prove to be impure they must be plated out again on Elsner's jelly. Sow a loopful of the broth in a fresh tube of the jelly, a drop of this on a second tube, and three drops of the second into a third tube (p. 77).

B. Isolation from stools.—The technique to be adopted in this case is similar to that just described. Dilute a loopful of the stool in a tube of sterile water and use a drop of the dilution to sow a tube of Elsner's gelatin : mix thoroughly and transfer a drop to a second tube and from the second tube two or three drops to a third tube. Pour plates and incubate. All the non-liquefying colonies which develop must be picked off for further investigation in the manner described above.

B. Grimbert's method.

Grimbert attributes the failure of Elsner's method partly to the want of uniformity of the medium due to variations in the chemical composition of potatoes, and partly to the fact that Elsner did not test the reaction of his medium. According to Grimbert the addition of iodide of potassium is not essential: ordinary gelatin can be used if the reaction be such that 10 c.c. are neutralized by 5 c.c. of lime water, though it is better to have a medium of constant chemical composition. Grimbert's medium is used in the same way as Elsner's, but the colonies are more slow in developing and the earliest do not appear before the third day. The method, as a matter of fact, has hardly any advantage over Elsner's original method.

Technique.—To 1,000 c.c. of water add :—

Maltose, -	1 gram.
Soluble starch,	2 grams.
Asparagin,	2 ,,
Neutral phosphate of potassium.	2 ,,
Potassium sulphate, -	2 ,,
Magnesium sulphate,	2 ,,
Ammonium bimalate,	2 ,,
Magnesium carbonate,	1 gram.

Dissolve 15 per cent. of gelatin in the mixture, clear with white of egg, heat to 115°, filter, and test the reaction thus: dilute 10 c.c. of the gelatin with 50 c.c. of warm distilled water, add a few drops of an alcoholic solution of phenol-phthalein, then run in lime water until a permanent rose pink colour is obtained. If more than 3 c.c. of lime water are required to neutralize the gelatin reduce the acidity by the addition of a small quantity of normal soda solution until 10 c.c. of the gelatin are neutralized with 5 c.c. of lime water.

Immediately before use 1 per cent. of iodide or bromide of potassium may be added.

C. Rémy's method.

Rémy suggests the use of a medium which is more nutritive and less acid than Grimbert's. By means of his " differential gelatin " he has been able to isolate the typhoid bacillus from stools in all the cases of enteric fever which he has investigated.

This " differential gelatin " has no greater selective property than Elsner's medium and the majority of micro-organisms grow in it. Still, liquefying species are to some extent checked and the inhibiting influence of the colon bacillus on the typhoid bacillus is not apparent on this medium.

Technique.—Preparation of the " differential gelatin." To a litre of water in a flask add :

Asparagin,	6 grams.
Oxalic acid, -	0·5 gram.
Lactic acid,	0·15 ,,
Citric acid, -	0·15 ,,
Di-sodium phosphate,	5 grams.
Potassium sulphate, -	1·25 ,,
Sodium chloride,	2 ,,
Witte's peptone,	30 ,,

Heat to 110° C. for 15 minutes, and on taking the flask out of the autoclave pour the boiling liquid into another flask containing 120–150 grams of best quality gelatin. Shake the flask until the gelatin is dissolved, add soda solution until the mixture is slightly alkaline, heat in the autoclave again to 110° C. for 15 minutes, then add sufficient half-normal sulphuric acid [1] to render the medium acid to such an extent that 10 c.c. require the addition of 0·2 c.c. of half-normal solution of soda to neutralize; [2] mix by shaking well, then heat in the steamer at 100° C. for 10 minutes and filter.

[1] A normal solution contains 98 grams H_2SO_4 per litre.
[2] This acidity is equivalent to 0·5 gram of H_2SO_4 per litre.

After filtration, test the reaction again thus : mix 10 c.c. of gelatin with 100 c.c. of distilled water, add a few drops of phenol-phthaleïn solution, and from a 1 c.c. pipette graduated in tenths of a cubic centimetre run in a half-normal solution of soda ; the red colour should appear when 0·2 c.c. of the solution have been added.

The desired degree of acidity being obtained, dissolve 2·5 grams of magnesium sulphate for every litre of gelatin. Tube in quantities of 10 c.c. and sterilize on three successive occasions at 100° C.

Immediately before use add to each tube 1 c.c. of a sterile 35 per cent. solution of lactose and 0·1 c.c. of a 2·5 per cent. solution of carbolic acid.

Method of sowing.—Rémy's gelatin is used in the same way as Elsner's. It is advisable, first of all, to sow the suspected water in a broth containing 0·5 per cent. of sulphuric and carbolic acids, and after incubating at 30° C. for 24 hours to use this culture for sowing the gelatin plates by the dilution method. Colonies of the colon and typhoid bacilli appear in the plates after incubating for 2 days.

Cultural characteristics. *The colon bacillus.*—Colonies in the depth of the medium are rounded, ovoid or fusiform and of a yellowish-brown colour. Minute bubbles of gas are occasionally formed. Colonies on the surface which are sometimes transparent and bluish at first, rapidly become opaque : some of them are hemispherical and of a yellowish-brown colour while others have irregular margins and tend to spread.

The typhoid bacillus.—In the depth of the medium the colonies are bluish-white, smaller than the colonies of the colon bacillus and form no gas. Surface colonies are not well seen until the third day : at first " they are rather like moulds in appearance " but later spread out, become more bluish in colour and may attain the size of a threepenny-piece.

The differences between the colonies of the typhoid and colon bacilli are frequently very slight and many sub-cultures may have to be made before the nature of the organism can be definitely determined.

D. Besson's method.

Elsner's gelatin method has two great disadvantages. In the first place, it is only available for the analysis of small quantities of water even though the number of plates used be large—which is in itself a disadvantage—and, secondly, the medium does not prevent the growth of saprophytic organisms, which sometimes liquefy the plates as early as the second day and so put an end to the experiment. With the object of simplifying and at the same time rendering the method more efficient as a means of water analysis, Besson, in 1896, introduced certain modifications which he claims improve it in that they rapidly eliminate saprophytes and permit the use of large volumes of water.

1. Weigh out 30 grams of peptone (Chapoteaut) and 5 grams common salt, add a litre of water and dissolve in the steamer ; then, without neutralizing, heat in the autoclave to 115° C., filter, tube in quantities of 10 c.c. and sterilize at 115° C.

2. When an experiment is to be done, take ten tubes of the peptone water and to each add 20–30 drops of freshly prepared Gram's iodine solution (p. 143) and 10 c.c. of the water to be examined.

The amount of iodine solution to be added varies a little with the composition of the peptone. The first few drops will be rapidly decolourized but when about 20–25 drops have been added the medium assumes a pale brownish-pink colour which disappears in 5–6 minutes. When this occurs sufficient iodine has been added.

3. Incubate the tubes at 37°–38° C. Under these conditions the colon bacillus produces a visible growth in 8–12 hours and the typhoid bacillus in about 15–20 hours while other organisms do not appear until later. The tubes should be examined at frequent intervals.

4. After incubating for 18 hours pick out the tubes which are cloudy and sow sub-cultures in iodine-peptone-water.

5. Incubate the latter for 15 or 20 hours then plate a few drops from each tube on litmus-lactose-agar. At the same time sow sub-cultures in ordinary broth for inoculation later.

6. The plates are to be examined and the colonies tested as described above.

With this method Besson has succeeded in isolating the colon bacillus, the typhoid bacillus, and Friedländer's bacillus from water.

3. Methods based on precipitation.

When a chemical precipitate is produced in a liquid containing micro-organisms, a large proportion of the latter are carried down mechanically with the precipitate, so that if the latter be collected the organisms originally present in the liquid are concentrated in a small volume. The principle here involved is the basis of several methods for the detection of the colon and typhoid bacilli in water. Their only advantage is in the concentration of the micro-organisms, since the nature of the organisms and the presence of the typhoid bacillus can only be definitely established by carrying out a series of experiments on ordinary lines using the precipitate as the original material.

A. Vallet's method.—Pour 20 c.c. of the water under examination into a sterile tube, add 4 drops of a saturated aqueous solution of hyposulphite of sodium and 4 drops of a saturated solution of lead nitrate. A precipitate forms which carries down with it the majority of the organisms present in the water (the chemicals used have no bactericidal action on the typhoid bacillus). Centrifuge the mixture and suspend the deposit in a few drops of the hyposulphite solution. Sow the liquid—which now contains all the organisms originally present in the 20 c.c. of water—on Elsner's gelatin (*vide ante*).

B. Schueder's method.—In Schueder's method the fluid is not centrifuged. Pour 2 litres of the water under examination into a tall vessel and add 20 c.c. of a 7·75 per cent. solution of hyposulphite of soda and 20 c.c. of a 10 per cent. solution of lead nitrate. Mix intimately and allow to stand for 20 hours. Decant the supernatant liquid and suspend the precipitate in 14 c.c. of a saturated solution of hyposulphite. Sow the emulsion in quantities of 0·5 c.c. on a number of litmus-lactose-agar plates.

C. Ficker's method.—Ficker precipitates the organisms with sulphate of iron. To 2 litres of the suspected water add 8 c.c. of a 10 per cent. solution of soda and 7 c.c. of a 10 per cent. solution of sulphate of iron. The precipitation takes 2 or 3 hours to complete. Centrifuge the precipitate and dissolve the deposit in one-half its volume of a 25 per cent. solution of neutral tartrate of potassium. Sow the solution on Conradi and Drigalski's medium by the dilution method (p. 407).

D. Müller's method.—The precipitation is effected by means of oxychloride of iron which acts more quickly than the sulphate and does not require that the water shall be made alkaline.

To 3 litres of the water under examination add 5 c.c. of the oxychloride solution. Precipitation is complete in about half an hour. Collect the precipitate on a filter and sow, without re-dissolving, either on Conradi-Drigalski plates or, better (Nicter), on malachite-green-agar plates (p. 409).

4. Method based on the motility of the typhoid bacillus.

Cambier, in examining water for the presence of typhoid bacillus, relies upon the property possessed by the organism of rapidly passing through porous membranes (p. 155).

Place a porous porcelain bougie in a large test-tube. Half fill both the bougie and the tube with ordinary broth, and sterilize in the autoclave. Sow the suspected water in the lumen of the bougie and incubate at 37° C. As soon as the broth in the test-tube becomes cloudy sow a little of it on any of the ordinary media used for differentiating the typhoid bacillus.

This method is not very reliable since some strains of the colon bacillus also pass very rapidly through the walls of a porous bougie, and the various modifications of the method which have been introduced seem to have little to recommend them.

5. Chantemesse's methods.

Chantemesse has introduced two methods of isolating the typhoid bacillus, both of which depend upon obtaining surface colonies on carbolic agar. For isolating the organism from stools and water the second is not only more rapid but is simpler.

First method.—Filter 5 or 6 litres of the suspected water through a Chamberland bougie. Wash the surface of the filter in 200 c.c. of a 3 per cent. peptone solution. Incubate the latter at 37° C. and arrange the culture so that air can be bubbled through it while incubating. Add more peptone solution at the end of 12 hours and again at the end of 24 hours. Then centrifuge the culture. The typhoid bacilli being motile and isolated (*i.e.* not grouped in clumps) remain in suspension while non-motile organisms and those massed together in zooglea masses go to the bottom of the vessel. With the supernatant liquid sow a number of Esmarch's roll tubes by the dilution method using carbolic-agar as the medium.

Carbolic-agar.—Dissolve 30 grams of peptone and 20 grams of agar in a litre of water, and make feebly alkaline (p. 31), tube in quantities of 10 c.c. and sterilize. Immediately before use melt the agar and add four drops of a 5 per cent. solution of carbolic acid to each tube (=0·1 per cent. of carbolic acid).

Incubate the cultures at 37° C. Growth appears in about 16–20 hours. Sow all colonies at all resembling the typhoid bacillus on the various media used for differentiating the organism.

Second method (recommended).—Sow the suspected material directly on litmus-lactose-carbolic-agar.

Litmus-lactose-carbolic-agar.—Prepare agar as above and add 2 per cent. lactose. Tube in quantities of 10 c.c. and sterilize. Just before use melt a number of tubes of lactose-agar and to each add 1 c.c. of sterile neutral litmus solution and 4 drops of a 5 per cent. solution of carbolic acid. Mix thoroughly, pour into Petri dishes in thin layers (1–2 mm. deep) and allow to set.

Dip a fine sterile badger-hair brush in a tube of sterile water to which a trace of the suspected stool has been added, and without recharging it sow in succession six surface plates of litmus-lactose-carbolic-agar.

If water is to be examined, filter it through a Chamberland bougie and sow litmus-lactose-carbolic-agar plates with the deposit left on the bougie. For spreading the plates use a glass rod bent at a right angle (Drigalski's spatula).

Incubate the plates at 37° C. and after 12–15 hours numerous colonies will be found on the plates some red (the colon bacillus) and some blue (the typhoid bacillus). Test the blue colonies by the agglutination reaction.

6. Conradi-Drigalski's method.

This method, in principle the same as that of Chantemesse, is in very general use in Germany.

The suspected material is sown on the surface of agar containing lactose and litmus. Crystal-violet is used in place of the carbolic acid in Chantemesse's medium, and is found to be just as effective in restraining the majority of organisms while allowing the growth of the colon and typhoid bacilli.

Conradi and Drigalski's medium.

(*a*) **Preparation.**—Macerate 1500 grams of minced beef in 2 litres of water for 24 hours: boil the mixture for an hour: filter: make up to 2 litres with water: add—

Peptone (Witte),	20	grams.
Nutrose,	20	,,
Salt,	10	,,

boil, add 60 grams of agar; heat until the agar is dissolved: make feebly alkaline to litmus paper : autoclave for an hour at 120° C., filter in the steamer and sterilize for a quarter of an hour at 115° C.   Prepare

Litmus solution (Kahlbaum). -   -   -   -   -   -   300 c.c.
Lactose, -   -   -   -   -   -   -   -   -   30 grams.

Sterilize at 100° C. for 15 minutes.

Mix the agar and litmus solution while they are both hot. If the colour of the litmus indicate that the medium is acid add sufficient 10 per cent. soda solution to render it faintly alkaline and then a further 4 c.c. of warm 10 per cent. solution of sodium hydroxide.   Lastly, add to the mixture, 20 c.c. of a hot sterile (0·1 per cent.) solution of crystal-violet B, Höchst.

(β) **Mode of use.**—Pour the Conradi-Drigalski agar carefully, without contaminating it, into large Petri dishes (15–20 cm. in diameter).   Sow the suspected material on the surface of the agar (*vide ante* Chantemesse's method).   Incubate the plates at 37° C.   The typhoid bacillus gives blue transparent colonies and the colon bacillus red opaque colonies.

Hagemann's medium.

This is a modification of the preceding.

Liebig's extract (Lemco). -   -   -   -   -   -   10 grams.
Peptone (Witte),   -   -   -   -   -   -   -   10   ,,
Salt,   -   -   -   -   -   -   -   -   -   3   ,,
Water, -   -   -   -   -   -   -   -   -   600 c.c.

Boil.   Add 500 c.c. of milk.   Boil and dissolve 20 grams of agar in the hot liquid. Heat to 120° C. in the autoclave for half an hour.   Filter in the steamer and distribute in Erlenmeyer flasks.   Sterilize.   When required for use, liquefy the agar, make slightly alkaline with soda solution, add a few cubic centimetres of litmus and finally three drops of a 1 per cent. alcoholic solution of crystal violet.

7. Method of Endo.

The principle of the method depends upon the fact that if sulphite of sodium be added to agar containing fuchsin the medium is decolourized, and if the decolourized medium be sown with the colon bacillus the acids produced by the organism restore the colour of the fuchsin and the colonies of the organism acquire a red colour, while under similar conditions the colonies of the typhoid and paratyphoid bacilli are colourless.

The agar medium of Endo is used in exactly the same way as Chantemesse's and Conradi-Drigalski's agar.   It is very easy to prepare and gives good results.

Prepare a litre of peptone broth in the ordinary way, add 30 grams of agar and dissolve in the steamer.   Filter.   Make absolutely neutral to litmus paper then add 10 c.c. of a 10 per cent. solution of sodium bicarbonate.

Add 10 grams of chemically pure lactose, and 5 c.c. of a filtered saturated alcoholic solution of fuchsin which imparts a red colour to the medium.   Now add 25 c.c. of a freshly prepared 10 per cent. solution of sodium sulphite.   Decolourization commences at once and is complete after sterilization.   Distribute in quantities of 15 c.c. in tubes, sterilize at 115° C. and store in the dark.   When required for use melt the agar and pour into Petri dishes.

After incubating for 15 hours at 37° C. colonies of the colon bacillus on this medium have red centres and after 24 hours are entirely red with a greenish iridescence.

8. Methods based on the use of caffeine.

As already mentioned (p. 375) Roth has shown that the addition of 0·5 per cent. of caffeine to culture media checks the growth of the colon bacillus but does not interfere with that of the typhoid bacillus.   This fact has been successfully applied by Roth, Hoffmann and others to the isolation of the typhoid bacillus from water and stools.   According to Courmont and Lacomme however, the method is uncertain since some strains of the typhoid bacillus do not grow on caffeine-containing media.   The results should always be controlled by some other method.

Roth's technique.—Prepare broth in the ordinary way and add sufficient soda solution to give a permanent pink colour with phenol-phthaleïn. Add 80–100 c.c. of a 1 per cent. solution of caffeine to every 100 c.c. of broth.

Sow the fluid with the material to be examined and incubate at 37° C. for 24 hours, then plate traces of the culture on gelatin.

Ficker's technique.—To 100 c.c. of a 3 per cent. peptone-meat-broth add 0·6 gram of pure caffeine and 0·00007 gram of crystal-violet (0·7 c.c. of a 0·01 per cent. solution). Sow the fluid with the suspected material, incubate at 37° C. for 12 or 13 hours and sow Conradi-Drigalski plates with the culture obtained.

Lubenau's technique.—Lubenau sows in 100 c.c. of Ficker's broth containing 0·3 per cent. of caffeine, incubates for 13 hours, adds 100 c.c. of broth containing 0·6 per cent. of caffeine, incubates for a second period of 13 hours and adds 100 c.c. of broth containing 0·9 per cent. of caffeine. Before and after the second addition of broth he sows surface plate cultures on litmus-lactose-caffeine-agar for purposes of isolation.

Lubenau's caffeine-agar.—Prepare a litre of 6 per cent. peptone-beef-broth, dissolve 40–60 grams of agar in the broth, make neutral to litmus, heat to 120° C., filter and sterilize. After sterilization and while the agar is still hot add 60 c.c. of litmus solution, 5 grams of lactose and finally 110 c.c. of a 6 per cent. solution of pure caffeine. Distribute in Petri dishes.

Gathgens' technique.—To a litre of Endo's medium add 33 c.c. of a 10 per cent. solution of pure caffeine. Distribute in tubes in quantities of 15 c.c. which can afterwards be used to prepare plates.

9. Malachite green media.

Lœffler has stated that the addition of a certain quantity of malachite green to culture media impedes the development of the colon bacillus while having no effect on the growth of the typhoid and paratyphoid bacilli.

Unfortunately malachite green media are not so selective as Lœffler believed, for though the growth of a large number of organisms (*streptococci, staphylococci,* cholera vibrios, etc.) is inhibited the colon bacillus will often grow (Kiralyfi).

These media are difficult to prepare; if too much green be added the typhoid bacillus is inhibited, if too little, the colon bacillus grows as rapidly as the typhoid bacillus. It is essential to use a chemically pure compound and the amount to be added to the agar varies very much with the different commercial preparations. A series of experiments should be done to determine the quantity (1 in 4,000 to 1 in 6,000) to be added (Lentz and Tietz, Schindler). These different complications render the method of little practical value, and to make it more efficient Lœffler has recently advised the addition of ox bile to his malachite green media (*vide infra*). As the result of his own experience Fürth concludes that methods based upon the use of malachite green are inferior to Conradi-Drigalski's method.

Growth on malachite green media diminishes the agglutinability of the typhoid bacillus.

Of the different malachite green methods Leuch's seems to be the best.

Leuch's technique.—Prepare an agar medium with :—

Beef,	500 grams.
Water,	1 litre.
Common salt,	5 grams.
Dextrin.	10 ,,
Agar,	30–40 ,,

Neutralize, using litmus as the indicator. Add 5 c.c. of a normal solution of sodium carbonate and 100 c.c. of a 10 per cent. solution of nutrose. After filtering and sterilizing add 16–18 c.c. of a 1 per cent. solution of malachite green.

Surface cultures are sown in Petri dishes (p. 407). Colonies of the typhoid bacillus destroy the colour of the medium and a characteristic yellowish zone forms around them.

To identify an organism isolated on this medium Lœffler advises the use of a so-called *typhoid solution*.

Lœffler's typhoid solution.—This fluid is coagulated by the typhoid bacillus in 16–24 hours and floating on the coagulum is a clear green liquid. The colon bacillus produces not an homogeneous coagulum but a greenish mass adhering to the sides of the tube.

The solution consists of :

Distilled water,	100 c.c.
Nutrose,	1 gram.
Glucose,	1 ,,
Peptone,	2 grams.
Lactose, -	5 ,,
2 per cent. solution of chemically pure malachite green,	1 c.c.
Normal soda solution,	1·5 ,,

Peabody and Pratt's technique.—Sow the suspected fluid in broth containing 0·1 per cent. of malachite green, incubate at 37° C. for 18 hours and isolate on Conradi-Drigalski plates.

10. Method based upon the use of China green.

[Werbitzki recommends the addition of China green to agar (1·4–1·5 c.c. of a 0·2 per cent. solution per 100 c.c. of agar) for the purpose of restraining the growth of the colon bacillus when attempting the isolation of the typhoid and paratyphoid bacilli from such material as stools.]

11. Methods based on the use of bile.

The adjuvant properties of bile for the typhoid bacillus have been applied to the detection of the bacillus in water and stools. Bile may be used alone (either as such or in the form of bile salts) or mixed with malachite green.

Dünschmann's technique.—Prepare the following medium :

Distilled water,	100 c.c.
Agar,	3 grams.
Peptone,	3 ,,
Lactose, -	3 ,
Gelatin, -	1 gram.
Taurocholate of sodium, -	1 ,,

Heat to 120° C. Filter and distribute the medium in tubes in quantities of 10 c.c. After sterilization, add 1 c.c. of a sterile solution of litmus to each tube.

To use the medium, pour the contents of four or five tubes into a similar number of Petri dishes and with a Drigalski's spatula (p. 407) charged with the suspected material—and without recharging—sow surface cultures on each dish.

Jackson and Melia's technique.—The suspected material is enriched by growing in ox bile and the culture used to sow plates of Hesse's agar.

Hesse's agar.—In a litre of boiling water dissolve :

Liebig's extract,	5 grams.
Peptone,	10 ,,
Sodium chloride,	8·5 ,,
Agar, -	30 ,,

Heat to 120° C., filter, distribute in tubes (10 c.c. in each) and sterilize in the autoclave.

Method.—Sow the suspected material in 5 c.c. of bile and incubate at 37° C. for 24 hours. Take eight test-tubes each containing 9 c.c. of sterile water : to the first add 1 c.c. of the bile culture, to the second 1 c.c. of the first, to the third 1 c.c. of the second and so on. Take eight tubes of Hesse's agar, liquefy the medium and to one add 1 c.c. of the first dilution to the second 1 c.c. of the second dilution and so on, and pour plates.

Lœffler's technique.—A mixture of bile and malachite green is used. Sow the material on plates of Leuchs' nutrose-agar (p. 409) containing 3 per cent. of ox bile and 1·9 per cent. of a 0·2 per cent. solution of malachite green.

Padlewsky's technique.—This method also depends upon the use of a mixture of bile and malachite green. Sow on plates prepared with the following medium:

Distilled water,	100 c.c.
Agar,	3 grams.
Peptone,	3 ,,
Ox bile,	5 ,,
Lactose,	1 gram.

The medium should be slightly alkaline to litmus. After sterilization cool to 65° C. and add, firstly a mixture of

1 per cent. aqueous solution of malachite green,	0·5 c.c.
Ox bile,	0·5 ,,

then,

10 per cent. aqueous solution of sulphite of sodium,	1 c.c.

12. Method based upon the use of brilliant green.

Conradi substitutes for malachite green a mixture of *Brillantgrün-Kristall* and picric acid. An agar containing these dyes favours the growth of the typhoid and paratyphoid bacilli while inhibiting that of most other organisms. The colon bacillus either does not grow at all or only in very small numbers.

Prepare a slightly modified Hesse's agar:

Water,	1 litre.
Peptone,	10 grams.
Agar,	30 ,,
Liebig's extract,	20 ,,

Make alkaline, heat and filter. For every 1·5 litres of agar add

0·1 per cent. aqueous solution of *Brillantgrün-Kristall* (extra pure, Höchst),	10 c.c.
1 per cent. aqueous solution of picric acid,	10 ,,

On this medium, the colonies of the typhoid bacillus are bright green and transparent and thicker in the centre than at the margins: colonies of the paratyphoid bacilli are larger, and yellowish-green in colour.

13. Method based upon the use of neutral red.

Savage has applied the property possessed by the colon bacillus of reducing neutral red to the detection of that organism in water (p. 397).

The method is only applicable to the detection of the colon bacillus and does not indicate the presence of the typhoid bacillus.

The reduction of neutral red, however, is not, as was formerly thought to be the case, a specific property of the colon and paratyphoid bacilli: *B. pyocyaneus*, *B. fluorescens*, *B. enteritidis* and some of the harmless saprophytic organisms found in water give fluorescence in Savage's neutral-red broth; while, on the other hand, some strains of the colon bacillus exert hardly any decolourizing action on neutral red (Sicre, Vincent). Savage's method is therefore unreliable.

Savage's technique.—Prepare broth thus:—

Water,	1 litre.
Beef,	250 grams.

Boil, make up to 1 litre and add

Peptone (Defresne),	20 grams.
Common salt,	20 ,,
Glucose,	5 ,,

Boil, cool, decant and add 10 c.c. of a 5 per cent. solution of neutral-red. Distribute in tubes and sterilize. The medium should be ruby-red in colour.

Method of analysis.—Sow a number of tubes of the medium with different quantities (1 c.c. to 10 c.c.) of the suspected water. Incubate at 37° C. for 24 hours. The presence of the colon bacillus is indicated by a beautiful green fluorescence or a canary yellow tint according as to whether the water contains few or many colon bacilli.

14. Methods based on the agglutination of the typhoid bacillus.

Chantemesse, Windelbandt, Schepilewsky have made use of the agglutinating properties of antityphoid serum for isolating the typhoid bacillus. This method is very delicate and permits of the isolation of the bacillus from mixtures in which it is present in great dilution.

Windelbandt's technique.—To 10 c.c. of sterile broth add 1 c.c. of the water under examination. Incubate the mixture for 3–5 days. By this time the growth is very abundant, the broth is cloudy and the surface covered with a pellicle. Remove the surface growth and add to the remainder a few drops of a powerfully agglutinating anti-typhoid serum. The agglutinated typhoid bacilli fall to the bottom of the tube. Centrifuge the broth culture, collect the deposit and dilute it with a little normal saline solution. Sow litmus-lactose-agar plates with the diluted deposit.

Chantemesse's technique.—A simple method devised by Chantemesse consists in adding 30 grams of peptone to 1 litre of the suspected water, neutralizing and incubating for 20 hours. If little clumps form filter through paper and add anti-typhoid serum to the filtrate. After standing for 2 hours decant the liquid, filter the deposit through paper, and sow the clumps retained on the filter on Chantemesse's agar (p. 407).

Altschüller's technique.—Altschüller adds peptone and salt to the suspected water and incubates for 24 hours. Ten c.c. of the culture are now transferred to a test-tube the lower end of which is drawn out and opened and attached to a piece of india-rubber tubing closed by a clip. A few drops of a typhoid immune serum are added and a precipitate is soon formed which collects in the narrow drawn-out part of the tube. By releasing the clip the deposit can be run into a tube containing sterile peptone water. The mixture is shaken and then incubated at 37° C. The typhoid bacillus grows rapidly and is unaccompanied by other organisms.

[15. MacConkey's media.]

[The basis of MacConkey's media consists of a stock solution composed of :

Sodium taurocholate (commercial),	0·5 gram.
Peptone (Witte),	2·0 grams.
Distilled water,	100 ,,

For liquid media there is added to this stock solution 0·5 per cent. of a 1 per cent. solution of neutral red and 0·5 per cent. of glucose or 1 per cent. of the other carbohydrates or alcohols, as the case may be, and the medium is distributed into Durham's fermentation tubes and sterilized in the steamer for 10 minutes on each of two days, great care being taken not to overheat the medium. If it be thought advisable white of egg may be used to clear the medium.

[Bile-salt-agar is made by dissolving 1·5–2 per cent. agar in the stock solution. This is best done in the autoclave. The medium is cleared with white of egg and filtered. After filtration the same amount of neutral red is added as in the case of the liquid media (MacConkey).

[A consideration of the fermentation reactions of the various organisms shows that by the use of certain carbohydrates or alcohols either alone or in combination organisms can be separated by means of colour reactions. MacConkey's medium forms a most useful nutritive medium to which to add these substances.]

SECTION II.—THE IDENTIFICATION OF THE TYPHOID AND COLON BACILLI.

(i) An organism may be suspected to belong to the typhoid-colon group if it have the following characteristics :

1. A bacillus with rounded ends, generally motile, decolourized by Gram's method, having no capsule [and not forming spores].

2. Cloudiness with a watered-silk appearance in broth culture.

3. No liquefaction of gelatin.

(ii) If conforming to these requirements it remains to determine if the organism (which must of course be investigated in pure culture) be a typhoid or colon bacillus.

It is now that difficulties arise, though if what has been said in the foregoing chapters be recalled it seems impossible to confuse typical specimens of the two organisms. The motility, the characters of the flagella, the appearance of the growth on potato, the indol reaction, a study of the fermentation properties and the agglutination test should furnish a sure means of diagnosis.

Unfortunately some strains of the colon bacillus readily lose—when grown, for instance, symbiotically with the typhoid bacillus (Rémy) or a Pasteurella (Lesage), etc.—their capacity to produce indol and some of their fermentation properties ; other strains are very motile, while others again yield a very scanty growth on potato and on gelatin grow like the typhoid bacillus. Similarly, some strains of the typhoid bacillus are only slightly motile, and their flagella can only be stained with difficulty : others give a slightly pigmented growth on potato resembling cultures of the colon bacillus : finally, when grown in the presence of the colon bacillus or after passing through human tissues some strains lose their characteristic property of being agglutinated by antityphoid serum. Hence a certain amount of confusion arises which is further increased by the existence of a whole group of bacilli very closely related to the typhoid bacillus and having properties intermediate between it and the colon bacilli (paratyphoid bacilli, Chap. XXV.).

It will therefore be clear that for accurate diagnosis it is necessary to study several of the characteristics of the organism. The table below gives a list of the tests on which the diagnosis should be based. The investigation of the fermentation reactions, the production of indol, the characters of the flagella and the agglutinating properties will in the majority of cases afford sufficient information upon which to determine whether the organism is a typhoid or a colon bacillus. When an organism has all the characteristics of the typhoid bacillus except that it is not agglutinated by an antityphoid serum it must be tested as indicated under 14 in the table. If the serum of a guinea-pig which has been inoculated every other day for a fortnight with 2 c.c. of a forty-eight-hour old broth culture of the organism under investigation agglutinate an authentic typhoid bacillus in a dilution of at least 1 in 40 (Rémy) the organism must be regarded as a strain of the typhoid bacillus. Finally the two last tests (15 and 16) in the table will be found very valuable and should permit of the identification of the organism in even the most difficult cases.

Method of Diagnosis.	Colon Bacillus.	Typhoid Bacillus.
1. Culture in carbonated lactose-broth.	Abundant gas-formation (12–36 hrs.).	No gas formation.
2. Stroke culture on litmus-lactose-gelatin.	The colour of the litmus is first changed to red then to a pale brown along the stroke.	The colour of the litmus is unchanged.
3. Single colonies on litmus-lactose-agar.	Red colonies.	Blue or violet colonies.

METHOD OF DIAGNOSIS.	COLON PACILLUS.	TYPHOID BACILLUS.
4. Growth in milk.	Coagulation in 24 – 36 hours.[1]	No coagulation.
5. Growth on potato.	Thick brownish growth (inconstant).	Thin colourless glazed growth (inconstant).
6. Action on neutral-red.	Reduced.	Not reduced.
7. Growth on artichoke.	Abundant growth, the medium becoming green (inconstant).	No apparent growth. No change in colour of medium.
8. Growth on the synthetic media of Nœgeli, Rémy and Sugg, Fränkel and others.	Copious and rapid growth (inconstant).	A poor growth appearing slowly (inconstant).
9. Growth in peptone water.	Indol.	No indol.
10. Single colonies on de-colourized fuchsin-agar (Endo).	Red colonies.	Colourless colonies.
11. Growth on (a) Caffeine media. (b) Malachite green.	No growth (possible exceptions).	Growth (possible exceptions).
12. Flagella.	Flagella short and few in number (3 to 4 on each bacillus).	Numerous (8–18), long, wavy, undulating flagella.
13. Action of anti-typhoid serum (using the serum in its highest agglutinating dilution).	No agglutination.	Distinct agglutination (possible exceptions).[2]
14. Serum of a guinea-pig immunized with the organism.	The serum does not agglutinate a true typhoid bacillus in a dilution of 1 in 40.	The serum agglutinates a true typhoid bacillus in a dilution of 1 in 40 (some possible exceptions).
15. Simultaneous inoculation of anti-typhoid serum.	If the bacillus is virulent the simultaneous inoculation of antityphoid serum does not protect the animal.	If the bacillus is virulent, the simultaneous inoculation of antityphoid serum protects the animal.
16. Complement fixation.	No deviation of complement with a heated antityphoid serum.	Complement is deviated with a heated antityphoid serum.

[1] Should no coagulation occur sow the bacillus in a shallow layer of milk. Some strains of the colon bacillus only coagulate milk under these conditions.

[2] Typhoid bacilli recently isolated from the body occasionally fail to agglutinate until they have been sub-cultured several times in broth.

CHAPTER XXIV.

THE PNEUMOBACILLUS OF FRIEDLÄNDER.

THOUGH the pneumobacillus is not—as Friedländer believed—the infecting agent in acute lobar pneumonia, it nevertheless occupies an important place in human pathology and may be the cause of any of the following diseases, viz. : broncho-pneumonia, pericarditis, pleurisy, peritonitis meningitis, otitis, parotiditis, dacryocystitis, stomatitis, orchitis, and epididymitis ; and is further responsible for many suppurative conditions. Ch. Nicolle and Hébert have drawn attention to the fact that some cases of pseudo-membranous sore throats are due to the pneumobacillus ; it is also associated at times with the diphtheria bacillus ; and finally, it may occasionally cause an hæmorrhagic type of septicæmia (Weichselbaum, Netter).

It is present in the saliva of many persons (4·5 per cent. according to Netter). In the circumambient media the bacillus appears to be widely distributed ; Uffelmann found it in the air, Emmerich in dust, Grimbert in water, Besson in samples of water from many and various sources.

No valid distinction can now be drawn between the pneumobacillus and the bacillus described by Escherich as the *Bacillus lactis aërogenes*: the proof of their identity was sketched by Denys and Martin and extended by Grimbert and Legros.[1] These researches were confirmed by Bertarelli ; he considered the *Bacillus lactis aërogenes* to be merely a variety of the pneumobacillus.

The *Bacillus lactis aërogenes* has been found in stools, in soil, water, and air. It is one of the causes of the fermentation of milk and seems to be responsible for some cases of enteritis in breast-fed children. It plays an important rôle in urinary infections (Morelle, Worsburg, Heyse, etc.).

---

[1] Without enlarging upon the facts which have led to the conclusion that the two organisms are identical, the following characters which according to Grimbert and Legros they possess in common may just be mentioned. They are both non-motile encapsulated bacilli, do not liquefy gelatin, do not produce indol, ferment the same sugars and have the same action upon animals.

SECTION I.—EXPERIMENTAL INOCULATION.

Mice and guinea-pigs are very susceptible to the inoculation of virulent strains of the pneumobacillus (*vide infra*). Rabbits are distinctly more immune.

Mice.—If a few drops of a culture be inoculated sub-cutaneously into a mouse they lead to the formation of an abscess containing creamy, ropy pus : the bacillus then becomes generalized and the animal dies in 1–3 days. *Post mortem*, the spleen is enlarged and the bacillus can be isolated from the blood and internal organs. Intra-pulmonary inoculation results in the formation of foci of broncho-pneumonia and terminates in death.

Guinea-pigs.—Sub-cutaneous inoculation of a small dose of culture leads to the formation of an abscess at the site of inoculation. Doses of 1 c.c. of a broth culture prove fatal : an abscess forms at the site of inoculation and death supervenes more or less rapidly with lesions of broncho-pneumonia and generalization of the bacillus.

Rabbits.—A dose of several c.c. of a broth culture injected into the marginal vein of the ear of a rabbit leads to the death of the animal in a few days. The bacillus may be found in the blood and internal organs : lesions of hæmorrhagic septicæmia are sometimes present. Sub-cutaneous inoculation is followed by a less severe disease.

Ch. Nicolle and Hébert by abrading the mucous membrane of the vulva of a rabbit and infecting the abraded area produced a swelling of the labia majora which was accompanied by a white discharge rich in pneumobacilli.

Pigeons.—Pigeons are only slightly susceptible. The inoculation of very virulent strains into the peritoneum is however fatal.

SECTION II.—MORPHOLOGY.

1. Microscopical appearance.

The pneumobacillus is a rather broad, rod-shaped organism of which the length does not exceed on an average 1–2μ. Sometimes however in cultures,

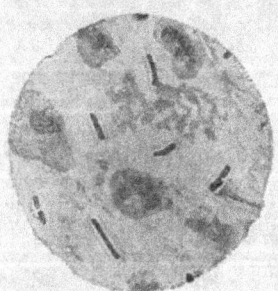

besides the cocco-bacillary forms, other long and even filamentous bacilli may be seen. The bacilli are often arranged in pairs : they are non-motile and never form spores.

In pus, sputum and blood, the pneumobacillus has a distinct capsule. The capsule is less distinct, but can nevertheless be demonstrated, in artificial cultures on solid media (Grimbert, Nicolle and Hébert).

Staining reactions.—The pneumobacillus is easily stained by the basic aniline dyes. It is gram-negative. The capsules may be stained by the method described in the chapter dealing with the pneumococcus.

FIG. 223.—Pneumobacillus of Friedländer. Sputum. (Carbol-thionin. ×1000.)

2. Cultural characteristics.

Conditions of growth.—The pneumobacillus is a facultative anaërobe and grows on all the ordinary media, which should be slightly acid for preference. Cultures can be obtained above 15° C. ; the optimum temperature is about 37° C.

Characters of growth on the ordinary media. Broth.—After incubating for 24 hours at 37° C. the medium is slightly cloudy, and on the surface a viscous pellicle is formed which makes a ring round the tube just above the surface of the liquid. On further incubation the pellicle falls to the bottom of the tube leaving the broth cloudy and viscous.

Gelatin. *Stab culture.*—Incubated at 20° C., a small raised white growth is formed on the surface of the gelatin after 48 hours : the growth later extends along the line of the stab and assumes a typical nail-line appearance. The medium is not liquefied. Bubbles of gas often form around the growth.

Isolated colonies.—Small round granular whitish colonies, which become somewhat raised, appear towards the third day.

Agar. Coagulated serum.—Growth on these media takes the form of a thick white viscous layer.

Potato.—A thick yellowish and viscous streak is formed and gas is also produced.

Milk.—The medium is coagulated sometimes rapidly and at other times more slowly. In the first sub-culture some strains of the bacillus do not coagulate milk but on further sub-cultivation they quickly acquire this property (Denys and Martin).

FIG. 224.—Pneumo-bacillus. Stab culture in gelatin. 7th day.

SECTION III.—BIOLOGICAL PROPERTIES.

1. Vitality and virulence.—Cultures of the pneumo-bacillus are rapidly killed at 60°–80° C., but in dry albuminous matter the bacillus is much more resistant : it seems to retain both its virulence and its vitality for a long time in water and soil. The virulence of different strains of the pneumobacillus is subject to considerable variations ; it is possible that there are different varieties of the organism (the *Bacillus lactis aërogenes* would be one of these varieties, see p. 415).

2. Toxin.—Filtered cultures contain a toxin which is fatal to rabbits and produces symptoms of paralysis. *Post mortem* the intestines are congested and show small hæmorrhages.

3. Bio-chemical reactions. Indol. Nitrites.—In a neutral 3 per cent. solution of peptone the pneumobacillus does not produce indol. It forms nitrites out of nitrates.

Fermentation reactions.—The pneumobacillus ferments glycerin, and certain of the carbohydrates, viz. : glucose, galactose, arabinose, mannite, dulcite, saccharose, lactose, maltose, raffinose and dextrin, but is without action on erythrite. Frankland has described a strain which does not ferment glycerin.

Grimbert recommends the following medium for the study of the fermentation reactions :—

Test substance, - - - - - - - - -	3 grams.
Dry peptone, - - - - - - - - -	2 ,,
Water, - - - - - - - - - -	100 c.c
Calcium carbonate, - - - - - - - -	*Quantum sufficit.*

The formation of gas is naked-eye evidence of fermentation. If the calcium carbonate be omitted and litmus solution added the blue colour of the latter is changed to red during the fermentation. Glycerin is more slowly broken up.

2 D

SECTION IV.—DETECTION, ISOLATION AND IDENTIFICATION OF THE PNEUMOBACILLUS.

I. In sputum.—(*a*) Prepare films and stain with carbol-thionin or carbol-gentian-violet. Gram's stain must also of course be used.

(*b*) Inoculate a mouse with a trace of the sputum.

II. In blood, pus, etc.—Microscopical examination and cultures, supplemented by the inoculation of a mouse, will render the identification of the organism easy.

III. In pseudo-membranous sore throats.—(*a*) Scrapings from the false membrane should be stained with a single stain and by Gram's method and examined microscopically.

(*b*) Cultures should be sown on coagulated serum as in the case of diphtheria. In 15–20 hours fairly large, round, greyish, viscous colonies appear which can be easily recognized with the naked eye and under the microscope.

IV. In water.—Adopt the method of cultivation in dilute carbolic acid or on peptone salt agar (Chap. LXV.). After two or three passages, pour gelatin plates on which the round raised dull white colonies of the pneumobacillus will easily be recognized. Sow one of these colonies in broth, and after 48 hours' incubation test the virulence of the culture on mice.

Differential diagnosis from the pneumococcus.—The pneumobacillus is easily differentiated from the pneumococcus by its cultural characteristics and by the fact that it is gram-negative.

Differential diagnosis from the colon bacillus.—In water examination the pneumobacillus is likely to be confused with the colon bacillus, but the mistake may easily be avoided by bearing in mind the following points.

PNEUMOBACILLUS.	COLON BACILLUS.
Absence of motility in broth culture.	Motile.
Encapsulated. The capsule is very well marked in fluids and tissues, but less visible in cultures.	Non-encapsulated.
No indol formation in peptone water.	Forms indol.
Ferments glycerin.	Does not ferment glycerin.

The bacillus of rhinoscleroma.

The bacillus of rhinoscleroma was discovered by V. Frisch. It is found in the nasal and pharyngeal lesions of rhinoscleroma, and may also multiply in the deeper tissues of the nasal fossæ. Rona found the organism in pure culture in the enlarged sub-maxillary glands in a case of the disease.

The bacillus of rhinoscleroma is very similar to the pneumobacillus : Netter and Gunther regard them as varieties of the same species. Their biological characteristics however justify their being regarded as different organisms (Paltauf and Bertarelli).

Microscopical appearance.—In sections of rhinoscleroma nodules, encapsulated cocco-bacilli resembling the pneumobacillus in shape and size are seen in the interior of certain very large cells (cells of Mickulicz) which have an excentrically placed crescent-shaped nucleus. The fluid of the nodules does not appear on microscopical examination to contain the organism, but by sowing cultures its presence can be demonstrated.

Cultures.—The bacillus of rhinoscleroma grows on all the ordinary laboratory media. Unlike the pneumobacillus it does not grow on slightly acid media, and does not ferment carbohydrates ; further, its cultures are much more scanty than those of the pneumobacillus (Paltauf).

In cultures, the bacillus of rhinoscleroma always forms capsules which may be easily demonstrated by the following method. Dilute a little of the growth in a 1 per cent. solution of acetic acid, spread on a slide, dry, and stain with aniline-violet : examine in water.

On *broth*, *agar* and *serum*—The growth on these media is very similar to the growth of the pneumobacillus but more scanty.

On *gelatin* the growth is thread-like and very limited. The tylotate appearance so characteristic of the pneumobacillus is never produced.

On *milk*. The medium is not coagulated.

Experimental inoculation.—Laboratory animals are not susceptible to inoculation with the bacillus of rhinoscleroma.

Löwenberg's bacillus.

(*The bacillus of ozæna.*)

The bacillus found in the mucous exudates in ozæna by Löwenberg and Abel is no longer regarded as the cause of ozæna. It resembles the pneumobacillus so closely that it seems necessary to regard the two organisms as identical (Viollet, de Simoni, and others).

The microscopical appearance, the cultural characteristics and the results of inoculation are the same in both cases. The only differences between them seem to be that the bacillus of ozæna does not ferment all the carbohydrates which are fermented by the pneumobacillus and does not coagulate milk.

CHAPTER XXV.

THE PARATYPHOID BACILLI.[1]
(THE PARACOLON BACILLI.)

UNDER the heading of paratyphoid bacilli are described certain organisms
which in many respects resemble the typhoid and the colon bacilli : they
are all gram-negative motile bacilli which do not form spores and do not
liquefy gelatin. From the clinical standpoint also, though the symptoms
are markedly different, there is a certain resemblance in that the diseases
produced by the typhoid and paratyphoid bacilli and probably also by
the colon bacillus are all primarily of a septicæmic nature.

The name **paratyphoid** was introduced by Achard and Bensaude in 1896
to describe an organism (paratyphoid B)[2] resembling the typhoid bacillus,
which they had isolated from a case of osteomyelitis following an attack of
a disease clinically indistinguishable from enteric fever, and from the urine
of another case of a similar disease.

In 1897 Besson isolated a similar organism from a case of pericarditis following a disease which had been diagnosed as enteric fever.

In 1897 Widal and Nobécourt also found a similar organism (paratyphoid
B)[2] in pus from a thyroid abscess in which there were no symptoms of a general
infection. To this organism they gave the name "**para-colon**" bacillus.

Gwyn in 1898 was the first to isolate a "para-colon" bacillus from the
blood of a person suffering from a disease clinically indistinguishable from
enteric fever.

In 1900-1 Schottmüller undertook an investigation into the nature of the
organisms present in the blood of cases which had the clinical symptoms of
enteric fever. In addition to the typhoid bacillus he found two other species
of organisms, closely related to the typhoid bacillus and to each other, to
which Brion and Kayser gave the names **paratyphoid A** and **paratyphoid B** ;
the latter being the more frequently found. To these organisms then the
term "paratyphoid " is properly applied : organisms, that is, which have many

[1] This part of the subject has been entirely rewritten.
[2] See Boycott, *Journal of Hygiene*, vi. 33 *et seq.*

of the bacteriological characteristics of the typhoid and colon bacilli and which give rise to a clinical disease having all the symptoms of enteric fever.

Further study however soon revealed the fact that these paratyphoid bacilli were, at all events in the laboratory, very closely related to if not identical with organisms which had been isolated from certain septicæmic diseases in animals accompanied by hæmorrhages—the **hæmorrhagic septicæmia group.** The first of this group to be described was that isolated by Salmon and Theobald Smith in 1885 from swine suffering from hog-cholera and known as the bacillus of hog-cholera.[1]

The paratyphoid bacilli especially the B variety, had also many characteristics in common with an organism isolated by Gaertner in 1888 at Frankenhausen from an epidemic of **food-poisoning**, and known as the *bacillus enteritidis Gaertner.*

Closely related also to the paratyphoid bacilli is an organism known as the *bacillus enteritidis Aertrycke,*[2] isolated in 1898 by de Nobele from an epidemic of food-poisoning at Aertrycke in Belgium and by Durham at Hatton in England. By its cultural characteristics this organism cannot be distinguished from the *bacillus enteritidis Gaertner,*[3] but as Durham showed by an application of the agglutination reaction, then recently introduced, the two could be sharply differentiated.

Hence in the first quinquennium of the century a number of organisms were known which from the laboratory point of view were all very like each other, but which—and this seemed remarkable—gave rise to different diseases. The paratyphoid bacilli A and B caused a septicæmic disease clinically almost if not quite identical with enteric fever ; the hæmorrhagic septicæmia group caused a septicæmia and diarrhœa in animals ; while the group consisting of Gaertner's and Durham's and de Nobele's bacilli were—and still are—regarded as the cause of epidemics of food-poisoning.

It was easy to distinguish the paratyphoid A bacillus from the other bacilli mentioned both by its cultural characteristics and by its agglutination reactions. The gaertner bacillus also could by its agglutination reactions be distinguished from the aertrycke bacillus, the bacillus of hog-cholera and the bacillus paratyphoid B.

Then difficulties arose as to the nature of the three last-named organisms. The bacillus of hog-cholera was soon shown to be identical with the aertrycke bacillus ; and the relationship of the latter bacillus to the paratyphoid B bacillus therefore alone remained to be determined. The position in 1906 was summarized by Boycott : " On the whole, the distinction between hog-cholera [aertrycke] and paratyphoid B, though slender, seems to be real. The morbific relations to man are different, for while the former gives rise to a sudden acute illness (food-poisoning), paratyphoid B causes a disease with no clear clinical distinctions from enteric fever." [4]

With a view to studying Castellani's absorption reaction Bainbridge took the paratyphoid bacilli as a suitable group upon which to work. By the aid of this reaction he has now made it clear that the bacillus paratyphoid B

[1] At the time, these authorities believed the organism to be the cause of hog-cholera and their opinion was accepted by other observers subsequently. Hence the name by which it is still very commonly known, the bacillus of hog-cholera, *bacillus suipestifer,* or *bacillus choleræ suis.* In 1903 however the researches of de Schweintz, Dorset and others showed that hog-cholera is not to be ascribed to the Salmon-Smith bacillus but to a filter-passing organism (Chap. LXIV.), the hog-cholera bacillus being merely a secondary infection.

[2] This bacillus will in future be described as " the aertrycke bacillus."

[3] In future referred to as " the gaertner bacillus."

[4] See however Paratyphoid B as a cause of food-poisoning p. 432.

is separate and distinct from the aertrycke bacillus, and this observation has been confirmed by Dean from a study of complement fixation reactions.

Bainbridge has further shown that unless the absorption tests be applied the paratyphoid B bacillus cannot be differentiated from the aertrycke bacillus, and that in practically all cases the so-called paratyphoid B bacillus isolated from cases of food-poisoning is in reality the aertrycke bacillus. The paratyphoid B bacillus can however give rise to acute gastro-enteritis though this would at present seem to be a very uncommon association (p. 432).

Bainbridge's investigations have also demonstrated that a number of the bacteria causing diseases in the lower animals (*vide post*) are not separate species, but are either identical with the paratyphoid B bacillus, the aertrycke bacillus or the gaertner bacillus, or are impure cultures of two or more of these organisms. The various rat and mice viruses are therefore shown by laboratory procedures, as well as by practical experience, not to be so harmless to man and the domestic animals as they are claimed to be.

Lignières proposed to designate all those organisms which had the morphological and cultural attributes of the bacillus of hog-cholera [*bacillus aertrycke*] by the name **Salmonella** after Salmon to whom the discovery of that organism is due. This term has met with some acceptance on the Continent, and is a convenient appellation under which to include a number of organisms very closely related bacteriologically, though clinically the diseases to which they give rise generally differ. It forms an appropriate classification for purposes of practical bacteriology and will therefore be adopted here. The Salmonella group, used in its original sense as defined above, includes the following organisms: the *bacillus paratyphosus B, bacillus enteritidis Aertrycke* (syn. *bacillus suipestifer*), and *bacillus enteritidis Gaertner*: as well as a number of organisms which have received specific names but which have now been shown to be identical with one or other of the preceding: these are *bacillus danysz, bacillus typhi murium, bacillus psittacosis* and *bacillus icteroides*.

Other names also have been proposed for the group of organisms discussed in this chapter: Theobald Smith suggested "the hog-cholera group"; Durham, the "intermediate group," and Trautmann, the "paratyphoid group."

The paratyphoid A bacillus is obviously excluded on cultural grounds from the Salmonella group. The "paratyphoid bacilli" will therefore be dealt with under two headings (1) The bacillus paratyphosus A (2) The Salmonella group.

The paratyphoid bacilli may then be grouped thus—

GROUP.	SPECIFIC ORGANISMS.	SYNONYMS.
I.	Bacillus paratyphosus A.	
II.	The Salmonella group :—	
	(i) Bacillus paratyphosus B.	
	(ii) Bacillus enteritidis Aertrycke.	Bacillus of hog cholera. B. suipestifer. B. choleræ suis.
	(iii) Bacillus enteritidis Gaertner.	

The organisms known as Bacillus danysz, B. typhi murium, B. psittacosis and B. icteroides are either identical with one or other of the members of the Salmonella group or are mixed cultures of two or more of these organisms.

CHAPTER XXVI.[1]

BACILLUS PARATYPHOSUS A.

THE paratyphoid A bacillus was first described by Schottmüller who isolated it from the blood of patients suffering from a disease clinically indistinguishable from enteric fever.

Paratyphoid A fever is a septicæmia characterized by " a mild pyrexia simulating enteric fever, marked by no acute gastric or intestinal symptoms and rarely fatal " (Firth). The lymphatic system is less affected than in enteric fever though one case is recorded where a single perforation was found (Grattan and Wood).

The bacillus has never been isolated in England (Bainbridge)[2] and comparatively few cases of paratyphoid A infection have been recorded on the Continent of Europe. In America its distribution is uncertain : but it is worth noting that in one year in the Allegheny General Hospital the relation of paratyphoid A fever (48 cases) to enteric fever was 8 to 11 (Proescher and Roddy).

In India, on the other hand, paratyphoid A fever is very prevalent. Grattan and Wood estimate that one-third of the cases of " simple continued fever " in India are cases of paratyphoid A fever, and so constantly is the A variety of the bacillus found that paratyphoid fever in that country connotes an infection with the paratyphoid A bacillus (Firth).

The bacillus apparently remains in the system for a time after an attack of paratyphoid A fever and " carriers " would appear to be the chief agent in the dissemination of the disease (Firth). Convalescents are usually infective for a comparatively short period, and " chronic carriers " (persons in whom the bacillus remains more than 3 months), would seem to be uncommon.

---

[1] This chapter has been rewritten.

[2] Bainbridge, F. A., The Milroy Lectures, Royal College of Physicians, *Lancet*, 1912, i.

The paratyphoid A bacillus has been recovered from the gall bladder after death and during operations for gall stones or cholecystitis; it has also been isolated once from an abdominal abscess and once from an apparently healthy man. Bainbridge states that it has been isolated from a case of acute enteritis. The organism has never yet been recovered from other than human tissues.

SECTION I.—EXPERIMENTAL INOCULATION.

All strains of the paratyphoid A bacillus are virulent for laboratory animals. The inoculation of 4 c.c. of a broth culture sub-cutaneously is fatal to guinea-pigs (Brion and Kayser): in mice, inoculation produces a fatal disease accompanied with symptoms of acute enteritis.

Guinea-pigs and mice are easily infected with a fatal disease by feeding them on cultures of the bacillus.

SECTION II.—MORPHOLOGY.

1. Microscopical appearance.

The paratyphoid A bacillus is a short stout rod-shaped organism with rounded ends often having the appearance of a cocco-bacillus: in old cultures long filamentous forms are occasionally seen. It is very motile and is provided with from four to ten delicate flagella. Morphologically it is indistinguishable from the other bacilli of the typhoid-colon group.

Staining reactions.—The paratyphoid A bacillus stains readily with the ordinary basic aniline dyes and occasionally exhibits polar staining. The bacillus is decolourized by Gram's method.

2. Cultural characteristics.

The paratyphoid A bacillus is a facultative aërobe and grows readily on the ordinary media in a manner very like the typhoid bacillus.

Broth.—The medium is rendered cloudy and has a watered silk appearance.

Gelatin.—The colonies are iceberg-like, translucent and bluish: in stroke culture the growth is thin and streaked with blue. The medium is not liquefied.

Potato.—On potato the bacillus gives a barely visible glaze.

Artichoke.—Generally colourless: a green colour may be produced after some time.

Milk.—Milk is not coagulated.

Litmus milk.—In litmus milk, acid is formed, the colour of the litmus being changed to pink (p. 373): the acidity is permanent. No clot is formed. *The permanent acidity without clot is peculiar* to this member of the typhoid-colon group.

Litmus whey.—A slight but permanent acidity indicated by the change in colour of the litmus from amethyst to pink.

SECTION III.—BIOLOGICAL PROPERTIES.

1. Biochemical reactions.

(a) **Fermentation of carbohydrates.**—The paratyphoid A bacillus produces acid and gas in glucose but has no action (producing neither acid nor gas) on lactose. The bacillus is thus easily differentiated from the typhoid bacillus on the one hand and from the colon bacillus on the other.

It forms acid and gas also in lævulose, maltose,[1] galactose, mannite, dulcite, sorbite and glycerin : but neither acid nor gas in raffinose, saccharose and lactose.

The paratyphoid A bacillus does not ferment carbohydrates so powerfully as the paratyphoid B bacillus.

(β) **Neutral-red media.**—The paratyphoid A bacillus like the colon bacillus reduces neutral-red, and in media containing the dye may give rise to a greenish fluorescence, but the reaction is less marked than with the colon bacillus and the Salmonella group. Fluorescence in neutral-red media is however a very inconstant change ; the best medium for the reaction is agar containing 1 per cent. of glucose and 1 c.c. per litre of a saturated aqueous solution of neutral-red.

(γ) **Endo's medium.**—On fuchsin-agar decolourized with sodium sulphite the bacillus, like the typhoid bacillus, gives colourless colonies.

(δ) **Caffeine media.**—According to Ducamp the paratyphoid A bacillus will not grow in broth containing 0·5 per cent. of caffeine (*vide B. paratyphosus B*).

(ε) **Malachite green media.**—Malachite green is decolourized by the paratyphoid A bacillus (1 week) but more slowly than by the bacilli of the Salmonella group (48 hours).

(ζ) **Vaccinated media.**—The paratyphoid A bacillus does not grow on media which have already served for the growth of the typhoid, colon, paratyphoid A or paratyphoid B bacilli.

(η) **Indol.**—The paratyphoid A bacillus forms no indol in culture.

2. Virulence.

Sacquépée and Chevrel were able to increase the virulence of the bacillus by passing it through a series of animals by sub-cutaneous inoculation. The inoculation of 0·5 c.c. of a twenty-four-hour culture of the exalted virus in broth was sufficient to kill guinea-pigs.

The virulence is lost somewhat readily in culture.

3. Toxin.

The paratyphoid A bacillus produces a soluble toxin in culture media. Cultures sterilized at 60° C. are pyogenic when inoculated beneath the skin of guinea-pigs. Cultures of an exalted virus sterilized by heat or filtered through porcelain kill guinea-pigs when inoculated sub-cutaneously in doses of 3–10 c.c.

4. Vaccination.

Guinea-pigs and white rats can be easily vaccinated against the paratyphoid A bacillus by inoculating them with attenuated or sterilized cultures. From the experiments of Cushing and of Sacquépée and Chevrel it would appear that animals immunized against the paratyphoid A bacillus are also immunized but to a lesser degree against the typhoid bacillus (intervaccination or group immunization). The serum of immunized animals is distinctly immunizing and bacteriolytic for paratyphoid A.

Human vaccination.—In man, vaccination with Wright's typhoid vaccine affords no immunity against an infection with the paratyphoid A bacillus.

Prophylactic vaccination of the human subject though suggested (Bainbridge, Leishman) has not yet been attempted. Certain preliminary laboratory experiments have however been quite recently recorded by Cummins and Cumming.

[1] The amount of gas formed out of maltose is always small whatever the organism.

5. Agglutination.

The serum of vaccinated animals and of persons suffering from paratyphoid A fever will agglutinate the paratyphoid A bacillus. The serum-diagnosis of paratyphoid A fever however requires considerable skill and care on the part of the observer.

Agglutination with the serum of immunized animals.—The serum of animals highly immunized against the paratyphoid A bacillus contains both specific agglutinins and group agglutinins. Not only does such a serum agglutinate the paratyphoid A bacillus but it agglutinates also the typhoid bacillus and other related bacilli; but if the limits of agglutination be determined it will be found that the serum agglutinates the paratyphoid A bacillus in a much higher dilution than it agglutinates the typhoid or any other related bacillus.

Thus an anti-paratyphoid A serum agglutinates all strains of the paratyphoid A bacillus in dilutions of 1–1,000, 1–5,000 and even 1–40,000. On the other hand it has very little agglutinating action on strains of the paratyphoid B bacillus, on the gaertner bacillus, or on the aertrycke bacillus and only agglutinates the typhoid bacillus in dilutions of 1–200, 1–100 or 1–20.

The following table [1] illustrates this :—

Agglutination limits after incubation for 2 hours at 42° C. Macroscopic method. In all cases control tubes showed no agglutination.

SERUM.	EMULSION OF BACILLI.			
	Paratyphoid A.	Paratyphoid B.	Aertrycke.	Gaertner.
Paratyphoid A.	50,000	< 100	< 100	< 100

Conversely, experimental typhoid serums have little action on strains of the paratyphoid A bacillus and only agglutinate them in low dilution.

A similar statement is justifiable for paratyphoid B, aertrycke and gaertner serums.

To sum up : *By means of the agglutination reaction with an experimental serum the paratyphoid A bacillus can with certainty be identified, being clearly differentiated from the typhoid, paratyphoid B, gaertner and aertrycke bacilli.*

Agglutination reaction with human serum.—The conditions are somewhat different when working with an human serum.

The experience in cases of paratyphoid fever in India is as follows :—The agglutination titre is usually low (1–20 to 1–40) and the reaction commonly transient: it may be quite as high for the typhoid bacillus as during an ordinary attack of enteric fever and the co-agglutinin for this bacillus may remain after the specific agglutinin has vanished. Rarely, the specific agglutinin alone is present ; and sometimes both it and the co-agglutinin for the typhoid bacillus are present in so small an amount and for so short a time as to be easily overlooked. Finally it is possible in undoubted cases of paratyphoid A fever for the serum to contain co-agglutinins for the typhoid and paratyphoid B bacilli but no specific agglutinin (Firth).[2]

The group agglutinin for the typhoid bacillus quickly disappears in para-

[1] Bainbridge, F. A., *Journal of Pathology and Bacteriology*, xiii. p. 341.
[2] *Journal of the Royal Army Medical Corps.*

typhoid A fever, in contrast to the persistence of the specific agglutinin which follows enteric fever (Firth).

Moreover, Grattan and Wood[1] record that "In some cases the limits of co-agglutination for the typhoid bacillus exceeded the limits of specific agglutination : and that in other cases again, at one period of the disease the limits of co-agglutination for the typhoid bacillus exceeded the limits of specific agglutination, and at another period the limits of specific agglutination exceeded the limits of co-agglutination."

These observers find that "antityphoid inoculation seldom if ever produces co-agglutinins for the paratyphoid A bacillus. Hence in a typhoid-vaccinated individual a serum reaction against the paratyphoid A bacillus in a dilution of 1–20 is strong evidence of paratyphoid A fever. "And in inoculated persons an attack of paratyphoid A fever raises the titre of agglutination for the typhoid bacillus about the 8th day, while the agglutinins for the paratyphoid A bacillus do not appear much before the 12th day."

6. Absorption tests.

To explain the phenomena just described it is necessary to assume the presence of group agglutinins, or, in those cases where co-agglutination is very marked, the existence of a double infection.

Castellani's method of absorption of agglutinins may be used for diagnostic purposes when the results of the agglutination tests are doubtful.

Let us take the case of a serum which has but little agglutinating action on the typhoid bacillus but agglutinates the paratyphoid A bacillus in higher dilution (1–40 to 1–100). To such a serum add paratyphoid A bacilli in sufficient quantity to remove the whole of the paratyphoid A agglutinins and centrifuge. Then test the agglutinating action of the clear supernatant fluid on both the typhoid and paratyphoid A bacilli (to ascertain that the agglutinins for the paratyphoid A bacillus have been removed). If the typhoid bacillus be agglutinated it may be assumed that agglutinins were present in the serum for both the typhoid and paratyphoid A bacilli ; but if on the other hand the typhoid bacillus is not agglutinated, then the original action of the serum on the typhoid bacillus was due to the presence in it of a co-agglutinin. But in cases of paratyphoid A fever, when the serum agglutinates both the typhoid and paratyphoid A bacilli, absorption with the paratyphoid A bacillus removes all agglutinins—both specific and group ; while absorption with the typhoid bacillus removes merely the co-agglutinin for the typhoid bacillus, leaving the specific agglutinin for the paratyphoid A bacillus intact even though it may in the first instance have only been demonstrable in a dilution of 1–20 (Harvey and Wood).

If a person inoculated against enteric fever and whose serum agglutinates the typhoid bacillus become infected with a paratyphoid A infection, absorption of the serum with the paratyphoid A bacillus only reduces the titre and does not entirely remove the agglutinins specific for the typhoid bacillus (Harvey and Wood).

If in a case of paratyphoid A fever the serum agglutinate the typhoid bacillus but not the paratyphoid A bacillus, absorption with the latter will remove the whole of the agglutinin (co-agglutinin) for the former, whereas the agglutinins for the typhoid bacillus in cases of enteric fever are not removed by absorption with the paratyphoid A bacillus (Harvey and Wood).

Another difference between the specific and group agglutinins for the paratyphoid A bacillus is that when the serum-emulsion mixture is put up in sedimentation tubes the specific agglutination appears almost immediately while the co-agglutination does not appear for some hours (Harvey and Wood).

To sum up : *The serum of persons suffering from paratyphoid A fever usually*

[1] *Journal of the Royal Army Medical Corps.*

agglutinates the paratyphoid A bacillus only in low dilution (1–20 to 1–40). Co-agglutinins for related organisms and especially for the typhoid bacillus are as a rule also present in the serum: the amount of the co-agglutinin is very variable often exceeding the titre of the specific agglutinin, and moreover may be present to the exclusion of the latter. Absorption tests will allow of a correct diagnosis.

7. Complement fixation.

The serum of animals immunized with the paratyphoid A bacillus contains an immune body which is fixed by both the paratyphoid A and typhoid bacilli, but the converse does not hold good ; the immune body in the serum of animals immunized with the typhoid bacillus is not fixed by the paratyphoid A bacillus (Rieux and Sacquépée).

In healthy men inoculated against enteric fever, specific amboceptors for the paratyphoid A bacillus were present in amounts apparently equal to the specific amboceptors for the typhoid bacillus, though in no case had the individuals tested suffered from paratyphoid A fever (Grattan and Wood).

If the dosage of antigen and antibody be carefully determined, the paratyphoid A bacillus can be clearly and absolutely differentiated from the typhoid and paratyphoid B bacilli by the complement fixation method (H. R. Dean).

Dean's method. Preparation of extract.—Agar cultures were sown in Roux bottles, incubated at 37° C. for 48 hours and then emulsified in 20 c.c. of distilled water. The emulsion was tubed in quantities of 5 c.c., and in some cases heated to 60° C. (this heating had no effect on the properties of the extract); then after being thoroughly shaken it was frozen hard, and subsequently thawed slowly at room temperature. After freezing and thawing twice the emulsion was shaken all night in a shaking machine, again frozen and thawed twice, shaken again all night and then centrifuged until the supernatant liquid was clear or only slightly opalescent. The extract was then stored in the cold room.

Antiserum.—Rabbits were inoculated intra-venously four or five times at intervals of 4 or 5 days with saline emulsions of 24-hour agar cultures. The animals were tested on the eighth day after the last inoculation and if the serum was satisfactory were bled on the ninth or tenth day. The serum was heated at 56° C. for half an hour and stored in quantities of 2 c.c.

Complement.—The guinea-pig serum was prepared on the day of use.

Hæmolytic system.—Sheep red cells and rabbit-sheep serum.

Experimental data.—The bulk of fluid in each tube was 2·5 c.c.—0·5 c.c. of diluted bacterial extract (antigen), 0·5 c.c. of diluted serum (antibody) and 0·5 c.c. of a 1 in 10 dilution of fresh guinea-pig serum (complement). After incubation there were added 0·5 c.c. of the dilution of hæmolytic serum (determined on the day of the experiment) and 0·5 c.c. of a 1 in 20 suspension of washed sheep cells.

The dilution of antiserum necessary for a differentiation experiment can be ascertained by the titration of the antiserum with the homologous extract. As a rule satisfactory differentiation can be obtained with the greatest dilution of antiserum which is found to give a thoroughly satisfactory reaction with the homologous extract.

With low dilutions of antiserum a marked group reaction is obtained and differentiation is impossible. The group antibodies can however be removed by absorption (*vide supra*).

SECTION IV.—THE DIAGNOSIS OF PARATYPHOID A. INFECTIONS.

The isolation and identification of the bacillus.

The diagnosis of a paratyphoid A infection should be based upon the demonstration of the specific organism in the tissues or excreta.

Any of the methods described for the isolation of the typhoid bacillus are quite applicable to the isolation of the paratyphoid A bacillus.

Agglutination reactions with the patients' serum supplemented by absorption tests should form a part of the diagnosis in every case.

1. The most important step is to demonstrate the presence of the bacillus in the blood stream. (The isolation of the paratyphoid A bacillus is comparatively easy during the first 4 or 5 days of the pyrexia: the chances of a successful blood culture are greatly diminished by the 8th day, even when the usual 5 c.c. of blood is withdrawn—cf. enteric fever (Grattan and Wood).)

Grattan and Wood sow the blood (5 c.c.) on sterilized ox bile and after incubating for 24 hours at 37° C. plate the growth on Conradi-Drigalski's original medium, which they consider better than the more recent selective media [1]; they incubate again and then pick off colonies which resemble those of the typhoid bacillus, which is indistinguishable on the Conradi-Drigalski medium from the paratyphoid A bacillus.

The organism isolated must be fully identified. Firstly, it must be shown to belong to the paratyphoid group by a careful study of its morphological, staining, cultural and fermentation reactions. (The characteristics of the paratyphoid A bacillus may be conveniently summarized here: it is a short stumpy gram-negative bacillus which does not liquefy gelatin, forms no indol, does not clot milk but turns the medium permanently acid, ferments glucose, mannite and dulcite with production of acid and gas, but has no action on lactose and cane sugar.) Then the bacillus must be examined, as regards its agglutination reactions, with known typhoid, paratyphoid A and paratyphoid B serums.

In the case of an organism isolated from the blood these tests are sufficient for identification (Grattan and Wood).

2. To isolate the organism from the urine or from the excreta some of the material should be sown in dulcite broth or dulcite peptone water. Dulcite is the enrichment medium *par excellence* for the paratyphoid group (Boycott). After incubation—2 or 3 days sometimes elapse before the paratyphoid A bacillus produces gas in dulcite media, but ultimately the amount formed is considerable—some of the culture may be plated on Conradi-Drigalski's or M'Conkey's medium and the organisms isolated tested as above. But according to Grattan and Wood when a bacillus resembling the paratyphoid A bacillus is isolated from the stools it is necessary to carry the identification a step further (by means of absorption tests) than in the case of a similar organism isolated from the blood.

A 24-hour growth of the suspected organism on agar is emulsified in about 0·2 c.c. of paratyphoid A serum—which is diluted according to its titre, the object being to have an excess of bacilli for the amount of agglutinin present: *e.g.* a serum having a titre of 1–300 may be diluted ten times.

Incubate the emulsion for 2 hours at 37° C., centrifuge and put up the clear supernatant liquid in a series of dilutions in sedimentation tubes and test its agglutinating action with a known paratyphoid A bacillus.

If the organism used for absorption be the paratyphoid A bacillus then it will have completely removed the specific paratyphoid A agglutinins from

---

[1] None of these media will differentiate with any degree of certainty the typhoid bacillus from the paratyphoid bacilli. That however is immaterial since it is unlikely that they will be present together. The main use of these media is to differentiate at sight the typhoid and paratyphoid bacilli from the colon bacillus which is invariably present and usually in large numbers.

the serum. Grattan and Wood say :—" before accepting a suspected organism we require that it shall completely remove the agglutinin specific for the paratyphoid A bacillus. As controls we have frequently tested heterologous organisms such as the typhoid, paratyphoid B and colon bacilli against our paratyphoid A serum but have never removed the agglutinin specific for the paratyphoid A bacillus."

3. The reactions of the serum of patients suffering from paratyphoid A fever have been described above.

The pseudo-paratyphoid A bacillus.

This organism appears to be a common inhabitant of the intestines of pigs (Morgan) and has once been isolated from the human subject during some investigations on the cause of summer diarrhœa (Morgan).

The pseudo-paratyphoid A bacillus is culturally identical with the paratyphoid A bacillus but differs from the latter in that it is not agglutinated by a specific paratyphoid A serum.

www.ingramcontent.com/pod-product-compliance
Lightning Source LLC
Chambersburg PA
CBHW081254170526
45165CB00011B/3307